NATURAL HISTORY
UNIVERSAL LIBRARY

西方博物学大系

主编：江晓原

BRITISH ZOOLOGY

不列颠动物志

[英] 托马斯・彭南特 著

华东师范大学出版社

图书在版编目(CIP)数据

不列颠动物志 = British Zoology : 英文 / (英) 托马斯 · 彭南特 (Thomas Pennant)著. — 上海 : 华东师范大学出版社, 2018
(寰宇文献)
ISBN 978-7-5675-7989-7

Ⅰ. ①不… Ⅱ. ①托… Ⅲ. ①动物志-英国-英文 Ⅳ. ①Q958.556.1

中国版本图书馆CIP数据核字(2018)第154595号

不列颠动物志
British Zoology
(英) 托马斯 · 彭南特 (Thomas Pennant)

特约策划　黄曙辉　徐　辰
责任编辑　庞　坚
特约编辑　许　倩
装帧设计　刘怡霖

出版发行　华东师范大学出版社
社　　址　上海市中山北路3663号　邮编 200062
网　　址　www.ecnupress.com.cn
电　　话　021-60821666　行政传真　021-62572105
客服电话　021-62865537
门市(邮购)电话　021-62869887
地　　址　上海市中山北路3663号华东师范大学校内先锋路口
网　　店　http://hdsdcbs.tmall.com/

印 刷 者　虎彩印艺股份有限公司
开　　本　787 × 1092　16开
印　　张　143.5
版　　次　2018年8月第1版
印　　次　2018年8月第1次
书　　号　ISBN 978-7-5675-7989-7
定　　价　2200.00元 (精装全二册)

出 版 人　王　焰

总　目

《西方博物学大系》总序

江晓原

《西方博物学大系》收录博物学著作超过一百种，时间跨度为15世纪至1919年，作者分布于16个国家，写作语种有英语、法语、拉丁语、德语、弗莱芒语等，涉及对象包括植物、昆虫、软体动物、两栖动物、爬行动物、哺乳动物、鸟类和人类等，西方博物学史上的经典著作大备于此编。

中西方“博物”传统及观念之异同

今天中文里的“博物学“一词，学者们认为对应的英语词汇是Natural History，考其本义，在中国传统文化中并无现成对应词汇。在中国传统文化中原有“博物”一词，与“自然史”当然并不精确相同，甚至还有着相当大的区别，但是在“搜集自然界的物品”这种最原始的意义上，两者确实也大有相通之处，故以“博物学”对译Natural History一词，大体仍属可取，而且已被广泛接受。

已故科学史前辈刘祖慰教授尝言：古代中国人处理知识，如开中药铺，有数十上百小抽屉，将百药分门别类放入其中，即心安矣。刘教授言此，其辞若有憾焉——认为中国人不致力于寻求世界“所以然之理”，故不如西方之分析传统优越。然而古代中国人这种处理知识的风格，正与西方的博物学相通。

与此相对，西方的分析传统致力于探求各种现象和物体之间的相互关系，试图以此解释宇宙运行的原因。自古希腊开始，西方哲人即孜孜不倦建构各种几何模型，欲用以说明宇宙如何运行，其中最典型的代表，即为托勒密（Ptolemy）的宇宙体系。

比较两者，差别即在于：古代中国人主要关心外部世界“如何”运行，而以希腊为源头的西方知识传统（西方并非没有别的知识传统，只是未能光大而已）更关心世界“为何”如此运行。在线性发展无限进步的科学主义观念体系中，我们习惯于认为“为何”是在解决了“如何”之后的更高境界，故西方的分析传统比中国的传统更高明。

然而考之古代实际情形，如此简单的优劣结论未必能够成立。例如以天文学言之，古代东西方世界天文学的终极问题是共同的：给定任意地点和时刻，计算出太阳、月亮和五大行星（七政）的位置。古代中国人虽不致力于建立几何模型去解释七政“为何”如此运行，但他们用抽象的周期叠加（古代巴比伦也使用类似方法），同样能在足够高的精度上计算并预报任意给定地点和时刻的七政位置。而通过持续观察天象变化以统计、收集各种天象周期，同样可视之为富有博物学色彩的活动。

还有一点需要注意：虽然我们已经接受了用“博物学”来对译 Natural History，但中国的博物传统，确实和西方的博物学有一个重大差别——即中国的博物传统是可以容纳怪力乱神的，而西方的博物学基本上没有怪力乱神的位置。

古代中国人的博物传统不限于“多识于鸟兽草木之名”。体现此种传统的典型著作，首推晋代张华《博物志》一书。书名“博物”，其义尽显。此书从内容到分类，无不充分体现它作为中国博物传统的代表资格。

《博物志》中内容，大致可分为五类：一、山川地理知识；二、奇禽异兽描述；三、古代神话材料；四、历史人物传说；五、神仙方伎故事。这五大类，完全符合中国文化中的博物传统，深合中国古代博物传统之旨。第一类，其中涉及宇宙学说，甚至还有“地动”思想，故为科学史家所重视。第二类，其中甚至出现了中国古代长期流传的“守宫砂”传说的早期文献：相传守宫砂点在处女胳膊上，永不褪色，只有性交之后才会自动消失。第三类，古代神话传说，其中甚至包括可猜想为现代“连体人”的记载。第四类，各种著名历史人物，比如三位著名刺客的传说，此三名刺客及所刺对象，历史上皆实有其人。第五类，包括各种古代方术传说，比如中国古代房中养生学说，房中术史上的传说人物之一“青牛道士封君达”等等。前两类与西方的博物学较为接近，但每一类都会带怪力乱神色彩。

“所有的科学不是物理学就是集邮”

在许多人心目中，画画花草图案，做做昆虫标本，拍拍植物照片，这类博物学活动，和精密的数理科学，比如天文学、物理学等等，那是无法同日而语的。博物学显得那么的初级、简单，甚至幼稚。这种观念，实际上是将“数理程度”作为唯一的标尺，用来衡量一切知识。但凡能够使用数学工具来描述的，或能够进行物理实验的，那就是“硬”科学。使用的数学工具越高深越复杂，似乎就越“硬”；物理实验设备越庞大，花费的金钱越多，似乎就越“高端”、越“先进”……

这样的观念，当然带着浓厚的“物理学沙文主义”色彩，在很多情况下是不正确的。而实际上，即使我们暂且同意上述“物理学沙文主义”的观念，博物学的“科学地位”也仍然可以保住。作为一个学天体物理专业出身，因而经常徜徉在“物理学沙文主义”幻影之下的人，我很乐意指出这样一个事实：现代天文学家们的研究工作中，仍然有绘制星图，编制星表，以及为此进行的巡天观测等等活动，这些活动和博物学家“寻花问柳”，绘制植物或昆虫图谱，本质上是完全一致的。

这里我们不妨重温物理学家卢瑟福（Ernest Rutherford）的金句：“所有的科学不是物理学就是集邮（All science is either physics or stamp collecting）。”卢瑟福的这个金句堪称“物理学沙文主义”的极致，连天文学也没被他放在眼里。不过，按照中国传统的“博物”理念，集邮毫无疑问应该是博物学的一部分——尽管古代并没有邮票。卢瑟福的金句也可以从另一个角度来解读：既然在卢瑟福眼里天文学和博物学都只是“集邮”，那岂不就可以将博物学和天文学相提并论了？

如果我们摆脱了科学主义的语境，则西方模式的优越性将进一步被消解。例如，按照霍金（Stephen Hawking）在《大设计》（*The Grand Design*）中的意见，他所认同的是一种“依赖模型的实在论（model-dependent realism）”，即“不存在与图像或理论无关的实在性概念（There is no picture- or theory-independent concept of reality）”。在这样的认识中，我们以前所坚信的外部世界的客观性，已经不复存在。既然几何模型只不过是对外部世界图像的人为建构，则古代中国人干脆放弃这种建构直奔应用（毕竟在实际应用中我们只需要知道七政"如何"运行），又有何不可？

传说中的"神农尝百草"故事，也可以在类似意义下得到新的解读："尝百草"当然是富有博物学色彩的活动，神农通过这一活动，得知哪些草能够治病，哪些不能，然而在这个传说中，神农显然没有致力于解释"为何"某些草能够治病而另一些则不能，更不会去建立"模型"以说明之。

"帝国科学"的原罪

今日学者有倡言"博物学复兴"者，用意可有多种，诸如缓解压力、亲近自然、保护环境、绿色生活、可持续发展、科学主义解毒剂等等，皆属美善。编印《西方博物学大系》也是意欲为"博物学复兴"添一助力。

然而，对于这些博物学著作，有一点似乎从未见学者指出过，而鄙意以为，当我们披阅把玩欣赏这些著作时，意识到这一点是必须的。

这百余种著作的时间跨度为15世纪至1919年，注意这个时间跨度，正是西方列强"帝国科学"大行其道的时代。遥想当年，帝国的科学家们乘上帝国的军舰——达尔文在皇家海军"小猎犬号"上就是这样的场景之一，前往那些已经成为帝国的殖民地或还未成为殖民地的"未开化"的遥远地方，通常都是踌躇满志、充满优越感的。

作为一个典型的例子，英国学者法拉在（Patricia Fara）《性、植物学与帝国：林奈与班克斯》（*Sex, Botany and Empire, The Story of Carl Linnaeus and Joseph Banks*）一书中讲述了英国植物学家班克斯（Joseph Banks）的故事。1768年8月15日，班克斯告别未婚妻，登上了澳大利亚军舰"奋进号"。此次"奋进号"的远航是受英国海军部和皇家学会资助，目的是前往南太平洋的塔希提岛（Tahiti，法属海外自治领，另一个常见的译名是"大溪地"）观测一次比较罕见的金星凌日。舰长库克（James Cook）是西方殖民史上最著名的舰长之一，多次远航探险，开拓海外殖民地。他还被认为是澳大利亚和夏威夷群岛的"发现"者，如今以他命名的群岛、海峡、山峰等不胜枚举。

当"奋进号"停靠塔希提岛时，班克斯一下就被当地美丽的土著女性迷昏了，他在她们的温柔乡里纵情狂欢，连库克舰长都看不下去了，“道德愤怒情绪偷偷溜进了他的日志当中，他发现自己根本不可能不去批评所见到的滥交行为”，而班克斯纵欲到了“连嫖妓都毫无激情”的地步——这是别人讽刺班克斯的说法，因为对于那时常年航行于茫茫大海上的男性来说，上岸嫖妓通常是一项能够唤起“激情”的活动。

而在“帝国科学”的宏大叙事中，科学家的私德是无关紧要的，人们关注的是科学家做出的科学发现。所以，尽管一面是班克斯在塔希提岛纵欲滥交，一面是他留在故乡的未婚妻正泪眼婆娑地“为远去的心上人绣织背心”，这样典型的“渣男”行径要是放在今天，非被互联网上的口水淹死不可，但是“班克斯很快从他们的分离之苦中走了出来，在外近三年，他活得倒十分滋润”。

法拉不无讽刺地指出了“帝国科学”的实质：“班克斯接管了当地的女性和植物，而库克则保护了大英帝国在太平洋上的殖民地。”甚至对班克斯的植物学本身也调侃了一番：“即使是植物学方面的科学术语也充满了性指涉。……这个体系主要依靠花朵之中雌雄生殖器官的数量来进行分类。”据说“要保护年轻妇女不受植物学教育的浸染，他们严令禁止各种各样的植物采集探险活动。”这简直就是将植物学看成一种“涉黄”的淫秽色情活动了。

在意识形态强烈影响着我们学术话语的时代，上面的故事通常是这样被描述的：库克舰长的“奋进号”军舰对殖民地和尚未成为殖民地的那些地方的所谓“访问”，其实是殖民者耀武扬威的侵略，搭载着达尔文的“小猎犬号”军舰也是同样行径；班克斯和当地女性的纵欲狂欢，当然是殖民者对土著妇女令人发指的蹂躏；即使是他采集当地植物标本的“科学考察”，也可以视为殖民者“窃取当地经济情报”的罪恶行为。

后来改革开放，上面那种意识形态话语被抛弃了，但似乎又走向了另一个极端，完全忘记或有意回避殖民者和帝国主义这个层面，只歌颂这些军舰上的科学家的伟大发现和成就，例如达尔文随着“小猎犬号”的航行，早已成为一曲祥和优美的科学颂歌。

其实达尔文也未能免俗，他在远航中也乐意与土著女性打打交道，当然他没有像班克斯那样滥情纵欲。在达尔文为“小猎犬号”远航写的《环球游记》中，我们读到：“回程途中我们遇到一群黑人姑娘在聚会，……我们笑着看了很久，还给了她们一些钱，这着实令她们欣喜一番，拿着钱尖声大笑起来，很远还能听到那愉悦的笑声。”

有趣的是，在班克斯在塔希提岛纵欲六十多年后，达尔文随着“小猎犬号”也来到了塔希提岛，岛上的土著女性同样引起了达尔文的注意，在《环球游记》中他写道：“我对这里妇女的外貌感到有些失望，然而她们却很爱美，把一朵白花或者红花戴在脑后的髪髻上……”接着他以居高临下的笔调描述了当地女性的几种发饰。

用今天的眼光来看，这些在别的民族土地上采集植物动物标本、测量地质水文数据等等的“科学考察”行为，有没有合法性问题？有没有侵犯主权的问题？这些行为得到当地人的同意了吗？当地人知道这些行为的性质和意义吗？他们有知情权吗？……这些问题，在今天的国际交往中，确实都是存在的。

也许有人会为这些帝国科学家辩解说：那时当地土著尚在未开化或半开化状态中，他们哪有“国家主权”的意识啊？他们也没有制止帝国科学家的考察活动啊？但是，这样的辩解是无法成立的。

姑不论当地土著当时究竟有没有试图制止帝国科学家的“科学考察”行为，现在早已不得而知，只要殖民者没有记录下来，我们通常就无法知道。况且殖民者有军舰有枪炮，土著就是想制止也无能为力。正如法拉所描述的：“在几个塔希提人被杀之后，一套行之有效的易货贸易体制建立了起来。”

即使土著因为无知而没有制止帝国科学家的“科学考察”行为，这事也很像一个成年人闯进别人的家，难道因为那家只有不懂事的小孩子，闯入者就可以随便打探那家的隐私、拿走那家的东西、甚至将那家的房屋土地据为己有吗？事实上，很多情况下殖民者就是这样干的。所以，所谓的“帝国科学”，其实是有着原罪的。

如果沿用上述比喻，现在的局面是，家家户户都不会只有不懂事的孩子了，所以任何外来者要想进行“科学探索”，他也得和这家主人达成共识，得到这家主人的允许才能够进行。即使这种共识的达成依赖于利益的交换，至少也不能单方面强加于人。

博物学在今日中国

博物学在今日中国之复兴，北京大学刘华杰教授提倡之功殊不可没。自刘教授大力提倡之后，各界人士纷纷跟进，仿佛昔日蔡锷在云南起兵反袁之“滇黔首义，薄海同钦，一檄遥传，景从恐后”光景，这当然是和博物学本身特点密切相关的。

无论在西方还是在中国，无论在过去还是在当下，为何博物学在它繁荣时尚的阶段，就会应者云集？深究起来，恐怕和博物学本身的特点有关。博物学没有复杂的理论结构，它的专业训练也相对容易，至少没有天文学、物理学那样的数理“门槛”，所以和一些数理学科相比，博物学可以有更多的自学成才者。这次编印的《西方博物学大系》，卷帙浩繁，蔚为大观，同样说明了这一点。

最后，还有一点明显的差别必须在此处强调指出：用刘华杰教授喜欢的术语来说，《西方博物学大系》所收入的百余种著作，绝大部分属于“一阶”性质的工作，即直接对博物学作出了贡献的著作。事实上，这也是它们被收入《西方博物学大系》的主要理由之一。而在中国国内目前已经相当热的博物学时尚潮流中，绝大部分已经出版的书籍，不是属于“二阶”性质（比如介绍西方的博物学成就），就是文学性的吟风咏月野草闲花。

要寻找中国当代学者在博物学方面的“一阶”著作，如果有之，以笔者之孤陋寡闻，唯有刘华杰教授的《檀岛花事——夏威夷植物日记》三卷，可以当之。这是刘教授在夏威夷群岛实地考察当地植物的成果，不仅属于直接对博物学作出贡献之作，而且至少在形式上将昔日“帝国科学”的逻辑反其道而用之，岂不快哉！

2018 年 6 月 5 日
于上海交通大学
科学史与科学文化研究院

《不列颠动物志》是托马斯·彭南特（Thomas Pennant，1726 –1798）的一部博物学著作。彭南特是英国著名博物学家、旅行家、作家兼古物专家，生于一个优渥的威尔士绅士家庭，自幼受到良好的博雅教育，终其一生居住在弗林特郡惠特福德近旁的道宁赫尔。他少年时就喜爱英国博物学家弗朗西斯·威洛比的《鸟类志》，立志探索自然的奥秘。十八岁时，彭南特进入牛津大学，先后就读于王后学院和奥里尔学院。因衣食无忧，他求学只为满足求知欲，因而毕业时没拿任何学位，直到四十五岁时才因其动物学领域的成就获得了一个荣誉学位。1747 年，他造访康沃尔时爱上了古物研究和化石，进而成为古物学会的会员。三十四岁成婚后，他更为富裕，开始资助一些科学研究，并成为有名的标本和古物收藏家。林奈赏识他的论文，举荐他加入瑞典皇家科学学会，他从此成为林奈的终生笔友。1761 年，他开始撰写《不列颠动物志》；1765 年，这部大作已见雏形之际，其妻去世，对他造成沉重打击。为抚平丧妻之痛，他出发游历欧洲大陆，与各地博物学大家交游，稿件则由他的出版商本杰明·怀特负责整理。1766 年至 1767 年，《不列颠动物志》分“四足兽”“鸟类”“爬行类”“甲壳动物”四卷以四开本刊行，书中配有 130 余幅精致插画，用散文式的优美笔触记录了产自英国的诸多常见生物。本书令彭南特声誉鹊起，也是他一系列博物学名著的发轫之作，出版后，他被选为皇家学会会员。此后，他又陆续出版了《四足动物志》《北极动物志》和《印度动物志》等书，广受好评。

今据原版影印。

BRITISH ZOOLOGY.

CLASS I.

QUADRUPEDS.

qui fint, qui in urbe fua hofpites, in patria fua peregrini, et cognitione femper
ieri effe velint, fibi per me placeant, fibi dormiant, non ego illis hæc confcripfi,
n illis vigilavi. ———————— Camdeni Brit. Præfat.

LONDON.
Printed for Benj. White,
MDCCLXXVI.

PREFACE.

AT a time, when the ſtudy of natural hiſtory ſeems to revive in *Europe*; and the pens of ſeveral illuſtrious foreigners have been employed in enumerating the productions of their reſpective countries, we are unwilling that our own iſland ſhould remain inſenſible to its particular advantages; we are deſirous of diverting the aſtoniſhment of our countrymen at the gifts of nature beſtowed on other kingdoms, to a contemplation of thoſe with which (at leſt with equal bounty) ſhe has enriched our own.

A judicious Foreigner has well remarked, that an *Engliſhman* is excuſable ſhould he be ignorant of the papal hiſtory, where it does not relate to *Great Britain*; but inexcuſable ſhould he neglect inquiries into the origin

a 2 of

of parlements, the limitation of the royal prerogative, and the gradual deviation from the feodal to the present syftem of government.

The obfervation is certainly juft, and the application appears too obvious to be pointed out; yet the generality of mankind can reft contented with ignorance of their native foil, while a paffion for novelty attracts them to a fuperficial examination of the wonders of *Mexico*, or *Japan*; but thefe fhould be told, that fuch a paffion is a fure criterion of a weak judgement: utility, truth and certainty, fhould alone be the point at which fcience fhould aim; and what knowlege can be more ufeful than of thofe objects with which we are moft intimately connected? and where can we reafon with greater certainty on fuch points, than in our own country, where a conftant recourfe may be had to the fpecimen of what we have under confideration? But thefe, and many other arguments for examining into the productions of our own ifland, may here be waved, as the admirable LINNÆUS has difplayed them at large in an oration *, which for mafterly reafoning,

* *Amœn. Acad. Tom.* II. p. 409. *Stillingfleet's Swedifh* Tracts, Tr. 1.

and

and happy ingenuity, may vie with the beſt compoſitions.

Yet, as that great naturaliſt has, in the ſame tract, publiſhed an eulogium on *Sweden*; and as an incitement to his countrymen to apply themſelves to the ſtudy of nature, enumerated the natural productions of that kingdom; we ſhall here attempt a parallel, and point out to the *Britiſh* reader, his native riches; many of which were probably unknown to him, or perhaps ſlightly regarded.

Do the heights of *Torſburg*, or *Swucku*, afford more inſtruction to the naturaliſt than the mountains of *Cumberland*, or *Caernarvonſhire*? whoſe ſides are covered with a rich variety of uncommon vegetables, while their bowels are replete with the moſt uſeful minerals. The *Derbyſhire* hills, abounding in all the magnificence of caves and cliffs; the mountains of *Kerry*, and that ſurprizing harbour the *Bullers* of *Buchan* *, may well be oppoſed to the rocks of *Blackulla*, or the caverns of *Skiula*. *Sweden* can no where produce a parallel to that happy combination of grandeur and beauty in *Keſwick* † vale, or

* Between *Aberdeen* and *Peterhead*.

† In *Cumberland*.

Killarny * lake; nor can *Europe* ſhew a natural wonder equal to the *Giant's Cauſeway* in the north of *Ireland*.

The excellence and number of our ſprings (whether medicinal or incruſting) are well known to common inquirers.

Our minerals are as great in quantity, as rich in quality: of gold, indeed, we cannot produce many ſpecimens, yet ſufficient to ſhew that it is found in this iſland †; but ſilver is found in great abundance in our lead ores, and veins of native ſilver in the copper ore of *Muckrus*, on the lake of *Killarny*. The hæmatites iron ores of *Cumberland*, and the beautiful columnar iron ores of the foreſt of *Dean*, are ſufficient to diſplay our riches in that uſeful commodity. No country produces ſo great a quantity of tin as *Cornwall*; and that county, and ſeveral others in the north have been long noted for their inexhauſtible veins of copper; nor leſs famous

* In the county of *Kerry*.

† That our country produces gold, appears in Dr. *Borlaſe's* Hiſtory of *Cornwall*, p. 214. So late as the year 1753, ſeveral pieces were found in what the miners call *ſtream tin*; one ſpecimen was as thick as a gooſe quill; others weighed to the value of ſeventeen ſhillings, twenty-ſeven ſhillings, and another even to the value of three guineas.

are

are the lead mines of *Derbyshire*, *Cardiganshire* and *Flintshire*, which have been worked for ages, yet shew no sign of the decline of their stores.

In all these, nature sports with great luxuriancy; the crystallized lead ore of *Tralee* *, the fibrous lead ore of *Tipperary*; the laminated lead ore of *Lord Hoptoun*'s mines; the crystalized tins, and the figured ores of *Zink*, are equally noted for their elegance, scarcity, and richness.

The ore of *Zink*, or *Lapis Calaminaris*, is found in vast quantities in the counties of *Somerset* and *Flint*; while *black lead* or *wadd*, a substance scarce known in other kingdoms, abounds in the mountains of *Cumberland*.

To the *Swedish Petroleum*, we may oppose the Well at *Pitchford*, and that of *St. Catherine*'s near *Edinburgh*. Our amber and our jet, together with our inexhaustible strata of coal found in so many parts of this kingdom, will, in the article of bitumens, give us the superiority over these so much boasted productions of *Sweden*.

* In the county of *Kerry*.

To avoid a tedious enumeration, we ſhall only mention our wonderful mines of rock ſalt; our allum and our vitriol works; our various marbles, alabaſters, and ſtones; our moſt excellent clays and earths *; all which articles, and many more unnoted here, might have furniſhed us with an ample field for panegyric.

Our botanical productions are not leſs abundant; but the works of *Ray*, which have lately been much enlarged and methodized, according to the *Linnæan* ſyſtem, by the ingenious Mr. *Hudſon*, in his *Flora Anglica*, are a ſufficient diſplay of our vegetable riches.

Our Zoology would be a copious ſubject to enlarge on, but the work in hand reſtrains us from anticipating our reader's curioſity. We might expatiate on the clouds of *Soland* geeſe which breed on the *Baſs iſland*, or *Puffins* on that of *Prieſtholme:* on our fiſh, and other marine animals; on our inſects, and the various other ſenſitive productions of this kingdom; but we forbear a

* If the inquiſitive reader is deſirous of a farther account of the number and excellence of our ſubterraneous productions, we refer him to the learned Dr. *Woodward*'s Catalogue of the *Engliſh Foſſils*, *London* 1729, particularly to p. 5.

parade

parade of uſeleſs declamation, and ſhall only add, that as few countries receive more advantages from their natural breed of quadrupeds, unmixed with any beaſt that preys on man, ſo, few can boaſt a greater variety of birds, whether local, or migratory.

This is a general view of the natural hiſtory of our own country; why then ſhould we neglect inquiring into the various benefits that reſult from theſe inſtances of the wiſdom of our Creator, which his divine munificence has ſo liberally, and ſo immediately placed before us? Such a neglect is certainly highly to be blamed, for (to expreſs ourſelves in the words of an eminent writer) "the Creator did not beſtow ſo much "curioſity, and workmanſhip on his crea-"tures, to be looked on with a careleſs in-"curious eye, eſpecially to have them ſlight-"ed or contemned; but to be admired by "the rational part of the world, to magnify "his own power to all the world, and the "ages thereof; and ſince the works of the "creation are all of them ſo many demon-"ſtrations of the infinite wiſdom and power "of God, they may ſerve to us, as ſo ma-"ny arguments exciting us to a conſtant fear "of

"of the Deity, and a ſteady and hearty obe-
"dience to all his laws." *

Much might be added to this ſubject, if conſidered in a theological light; but ſince the writings of *Boyle*, *Ray*, and *Derham*, fully prove that the ſtudy of natural hiſtory enforces the theory of religion and practice of morality, we had better refer to their works in general, than mangle them by imperfect quotations.

To exalt our veneration towards the Almighty, is the principal end of this ſublime ſcience; and next to that, the various benefits reſulting from it to human ſociety deſerve our ſerious conſideration.

To give an obvious inſtance: what wonderful changes have been made in human affairs by the diſcovery of an obſcure mineral. The antients, ignorant of the application of the magnet, timidly attempted a mere coaſting navigation; while we, better informed of the uſes of it, traverſe the wideſt oceans, and by the diſcovery of the new world, have layed open to ſcience, an inexhauſtible fund of matter.

The riſe and progreſs of medicine, kept

* *Derham*'s Phyſ. Theol. Book XI. c. 24.

pace

pace with the advancement of this moſt important diſcovery; and though neceſſity was the parent of the mechanic arts, yet they alſo throve, and grew to maturity, under the ſame influence.

Many more inſtances might be added to this brief view of the utility of natural knowlege; but we ſhall only give ſome of its uſes in the polite arts, which have hitherto been too little connected with it.

To inſtance particularly in painting, its uſes are very extenſive: the permanency of colors depends on the goodneſs of the pigments; but the various animal, vegetable, and foſſil ſubſtances (out of which they are made) can only be known by repeated trials; yet the greateſt artiſts have failed in this reſpect: the ſhadows of the divine *Raphael* have acquired an uniform blackneſs, which obſcures the fineſt productions of his pencil, while the paintings of *Holbein*, *Durer*, and the *Venetian-ſchool*, (who were admirably ſkilled in the knowlege of pigments) ſtill exiſt in their primitive freſhneſs.

But theſe advantages are ſmall, compared to thoſe derived from the knowlege of nature in the repreſentation of objects: painting is an an imitation of nature; now, who can imitate without consulting the original? But to come to what is more particularly the object of our inquiries; animal and vegetable life are the essence of landscape, and often are secondary objects in historical paintings; even the sculptor in his limited province would do well to acquire a correctness of design with a perfect knowlege of the muscles of animals. But the painter should have all this and more; he should be acquainted with all their various tints, their manner of living, their peculiar motions or attitudes, and their places of abode *, or he will fall into manifest errors.

> Plurimus inde labor tabulas imitando juvabit
> Egregias, operumque typos, sed plura docebit
> *Natura ante oculos præsens*, nam firmat et auget
> Vim genii, ex illâque artem experientia complet †.

* That great artist, Mr. *Ridinger*, of *Ausburg*, exceeds all others in the three last particulars; nothing can equal his prints of animals for propriety of attitudes, for a just idea of their way of life, and for the beautiful and natural scenery that accompanies them. His finest works are, his *Wilde Thiere*, *Kleine Thiere*, and *Jagdbare Thiere*; but there are scarce any of his performances that can fail giving pleasure to all admirers of nature represented as herself.

† *Fresnoy de arte graph. lin.* 537.

Descrip-

Defcriptive poetry is ftill more indebted to natural knowlege, than either painting or fculpture: the poet has the whole creation for his range; nor can his art exift without borrowing metaphors, allufions, or defcriptions from the face of nature, which is the only fund of great ideas. The depths of the feas, the internal caverns of the earth, and the planetary fyftem are out of the painter's reach; but can fupply the poet with the fublimeft conceptions: nor is the knowlege of animals and vegetables lefs requifite, while his creative pen adds life and motion to every object.

From hence it may be eafily inferred, that an acquaintance with the works of nature is equally neceffary to form a genuine and correct tafte for any of the above mentioned arts. Tafte is no more than a quick fenfibility of imagination refined by judgement, and corrected by experience; but experience is another term for knowlege *, and to judge of natural images, we muft acquire the fame knowlege, and by the fame means as the painter, the poet, or the fculptor.

* See the Effay on the origin of our ideas of the fublime and beautiful.

Thus

Thus far natural hiſtory in general ſeems connected with the polite arts; but were we to deſcend into all its particular uſes in common life, we ſhould exceed the bounds of a preface: it will be therefore neceſſary to confine our inquiries to the inveſtigation of a ſingle part of the material world, which few are ſo ignorant as not to know is divided into the animal, vegetable, and foſſil kingdoms.

Vaſt would be the extent of the inquiries into each of theſe; but though ambition may tempt us to pervade the whole field of ſcience, yet a little experience will open to our views the immenſe tracts of natural knowlege, and we ſhall find it an arduous taſk only to inveſtigate a ſingle province, ſo as to ſpeak with preciſion and certainty; without which there can be no real improvements in natural hiſtory.

For theſe reaſons, a partial examination of this ſcience is all that a conſiderate mind will aim at, which may perhaps be moſt naturally guided to give the preference to the moſt exalted ſubject of it.

Zoology is the nobleſt part of natural hiſtory, as it comprehends all ſenſitive beings, from

from reaſoning man, through every ſpecies of animal life, till it deſcends to that point where ſenſe is wholly extinct, and vegetation commences: and certainly none will deny, that life, and voluntary motion are ſuperior to a mere vegetating principle, or the more inactive ſtate of the foſſil kingdom.

Should we follow the train of reflections which naturally ariſe from the contemplation of animals, they would ſwell this preface into a volume: and ſhould we only mention the various uſes of *Britiſh* animals in common life, yet even theſe would greatly exceed the bounds to which we have thought it right to limit ourſelves. The knowlege of *Diætetics* is a neceſſary branch of medicine, as by a proper attention to that article, an obſtinate diſtemper may be eradicated, when common remedies have failed; but this can never be attained, without the ſtudy of Zoology, which aſſiſts us greatly in learning the different qualities of animal food; and how far a difference of nutriment may contribute to cure the diſeaſe.

Cloathing is eſſential, not only to our comfort, but ſubſiſtence; and the number of our manufactures, relative to this ſingle article,

ticle, demand our care for their extenſion and improvement; eſpecially as the maintenance of thouſands depends on theſe important branches of commerce; yet theſe may be enlarged, by diſcovering new properties in animals, or by the farther cultivation of thoſe already diſcovered. The ſcience of Zoology is requiſite for each of theſe; and if we reflect but a little on the unwearied diligence of our rivals the *French*, we ſhould attend to every ſiſter ſcience that may any ways preſerve our ſuperiority in manufactures and commerce.

Domeſtic œconomy is an object of equal conſequence; and the author* of the *Calendar* of *Flora* has eſtabliſhed the uſes of Zoology in this particular, with undeniable evidence. This excellent writer has united a happy invention, with the moſt ſolid judgment, and certainly merits the higheſt commendations, as a friend of human kind. Our ingenious countryman, and worthy friend, the late Mr. *Stillingfleet*, in the ſame year purſued almoſt the ſame plan as far as his time would permit, with equal ſucceſs,

* *Alex. Mal. Berger.*

and

and manifeſtly proved the utility of the project, in a learned diſcourſe prefixed to his work *.

If then Zoology can ſuggeſt ſo many hints towards enlarging and improving our manufactures and agriculture; we ſhall not think our time miſapplied, in offering to the publick, the NATURAL HISTORY of the *Quadrupeds* and *Birds* of GREAT BRITAIN. ' This compilation had its peculiar difficulties; but the labor of travelling through a dry arrangement of the ſubject, was very frequently alleviated by the beautiful ſpecimens we met with in our progreſs: beſides, we own with pleaſure that we have been greatly aided by the lovers of natural hiſtory, who ſince the appearance of the firſt edition have contributed to enrich the preſent with ſeveral valuable obſervations; by collecting and digeſting theſe materials, we have not only rendered the work more complete, but are alſo encouraged to trace the *Britiſh Zoology* through ſome of the remaining claſſes.

Let therefore every merit that may appear in the preſent edition, and every error that

* *Swediſh* Tracts, tranſlated from the *Amæn. Acad.* ſecond edition.

may have been ſuppreſſed from the former, be attributed to the kind informations we have received from our learned and ingenious friends; among whom we are ambitious of naming the Honorable *Daines Barrington*; the Reverend Sir *John Cullum*, Baronet; the Reverend Mr. *George Aſhby*, and the Reverend Mr. *Green* of *Cambridge*; *William Conſtable*, Eſquire; *Joſeph Banks*, Eſquire; the late *Benjamin Stillingfleet*, Eſquire; *Thomas Falconer*, Eſquire, of *Cheſter*; Doctor *John Reinold Forſter*; the Reverend Doctor *Buckworth*; the Reverend Mr. *Hugh Davies*, of *Beaumaris*; Mr. *Travis*, Surgeon, of *Scarborough*; Mr. *Latham*, Surgeon, of *Dartford*; *Thomas Tofield*, of *Yorkſhire*, Eſquire; Mr. *Plymly*, of *Longnor*, *Shropſhire*; *Owen Holland*, Eſquire, of *Conway*; *Henry Seymer*, Eſquire, of *Hanford*, *Wilts*; Doctor *Lyſons*, of *Gloceſter*; Doctor *Solander*; the late Mr. *Peter Collinſon*; the Reverend Mr. *White*, of *Selborn*, *Hants*; and that Father of *Britiſh* Ornithologiſts, the late Mr. *George Edwards*, of the College of Phyſicians.

In the proſecution of our plan, we ſhall, to avoid the perplexity ariſing from forming

a new

a new ſyſtem, adopt (as far as relates to the *Quadrupeds* and *Birds)* that of the ineſtimable *Ray*, who advanced the ſtudy of nature far beyond all that went before him; and whoſe abilities, integrity, and mildneſs, were no leſs an ornament to the human ſpecies in general, than to his own country in particular. Yet, as this excellent man was in a manner the founder of ſyſtematic Zoology, ſo later diſcoveries have made a few improvements on his labors: wherever then, he is miſtaken in the arrangement, we ſhall attempt a reform, aſſiſted by the more modern ſyſtems, all of which owe their riſe to the plan chalked out by our illuſtrious countryman. It is unneceſſary to detain the reader in this place with the reaſons for our deviation from the order we obſerved in our laſt edition, for they are given at large in the Prefaces to our *Synopſis of Quadrupeds* and *Genera of Birds* *.

We have, in our deſcriptions, wholly omitted the anatomy of animals; as that part, unleſs executed with the greateſt ſkill, would be no ſmall blemiſh to the reſt of this

* Printed at *Edinburgh*, 1773.

performance; but the reader may judge of the extent of our plan, by the following heads: the character of the genus ſhall firſt be mentioned: then the ſpecific name: the ſynonyms from different authors; and the genera in which thoſe authors have placed the animal. The names ſhall be given in ſeveral *European* languages *; and we ſhall conclude with a brief, but ſufficient deſcription, adding at the ſame time, the various uſes, and natural hiſtory of each individual.

If this plan ſucceeds, in promoting the knowlege of nature in this kingdom, we ſhall think ourſelves amply rewarded. Could our exhortations avail, we ſhould recommend this ſtudy moſt earneſtly to every country gentleman. To thoſe of an active turn, we might ſay, that ſo pleaſing and uſeful an employment would relieve the *tædium* ariſing from

* In the ornithology the *European* names are prefixed to the author referred to in the ſynonyms,

Italian	to	Aldrovand, Olina, or Zinanni.
French		Briſſon, or de Buffon.
German		Geſner, or Kramer.
Swediſh		the Fauna Suecica.
Daniſh and *Norwegian*		Brunnich.
Carniolan		Scopoli.

a ſame-a ſameneſs of diverſions; every object would produce ſome new obſervation, and while they might ſeem only to gratify themſelves with a preſent indulgence, they would be laying up a fund of uſeful knowlege; they would find their ideas ſenſibly enlarged, till they comprehended the whole of domeſtic œconomy, and the wiſe order of Providence.

To thoſe of a ſedentary diſpoſition, this ſtudy would not only prove agreeable, but ſalutary: men of that turn of mind are with difficulty drawn from their books, to partake of the neceſſary enjoyments of air and exerciſe; and even when thus compelled, they profit leſs by it than men of an illiberal education. But this inconvenience would be remedied, could we induce them to obſerve and reliſh the wonders of nature; aided by philoſophy, they would find in the woods and fields a ſeries of objects, that would give to exerciſe charms unknown before; and enraptured with the ſcene, they will be ready to exclame with the poet:

> On every thorn, delightful wiſdom grows;
> In every rill, a ſweet inſtruction flows. YOUNG.

Thus would the contemplative naturaliſt learn from all he ſaw, to love his Creator for

b 3 his

his goodneſs; to repoſe an implicit confidence in his wiſdom; and to revere his awful omnipotence. We ſhall dwell no longer on this ſubject, than to draw this important concluſion; that health of body, and a chearful contentment of mind, are the general effects of theſe amuſements. The latter is produced by a ſerious and pleaſing inveſtigation of the bounties of an all-wiſe and beneficent Providence; as conſtant and regular exerciſe is the beſt preſervative of the former.

Downing,
March 1. 1776.

THOMAS PENNANT.

EXPLANATION OF REFERENCES.

Ælian. an. var. — CLAUDII *Æliani* Opera quæ extant omnia, Cura & Opera *Conradi Geſneri* Tigurini, fol. *Tiguri*, 1556.

Alb. — Nat. Hiſt. of Birds, by *Eleazer Albin*, 3 vol. 4to. *London*, 1738.

Aldr. av. — *Ulyſſis Aldrovandi* Ornithologia, fol. *Francofurti*, 1610, 1613.

Amæn. acad. — *Caroli Linnæi* Amænitates Academicæ, 6 tom. 8vo. *Lugd. Bat.* & *Holmiæ*, 1749, &c.

Ariſt. hiſt. — *Ariſtotelis* Hiſtoria de Animalibus, *Julio Cæſare Scaligero* interprete, fol. *Toloſæ*, 1619.

Ariſtoph. — *Ariſtophanis* Comœdiæ undecim, *Gr.* & *Lat.* cum Scholiis antiquis, fol. *Amſtelodami*, 1710.

Barbot. — Deſcription of the Coaſts of South and North *Guinea*, and *Angola*, by *John Barbot*, in *Churchill*'s Coll. of Voyages, Vol. V.

Bell's Travels, — into *Perſia*, *China*, &c. 2 vol. 8vo. 1764.

Belon av. — L'Hiſtoire de la Nature des Oiſeaux, avec leurs Deſcriptions & naifs Portraits, par *Pierre Belon*, fol. *Paris*, 1555.

Belon obſ. — Les Obſervations de pluſieurs Singularites & Choſes memorables trouvees en *Grece*, *Aſie* & *Judie*, par *Pierre Belon*, fol. *Paris*, 1555.

b 4 *Belon.*

Belon.	La Nature & Diverſité des Poiſſons, &c. 8vo. tranſverſ. par *Pierre Belon. Paris*, 1555.
Borlaſe's Corn.	Nat. Hiſt. of *Cornwall*, by *William Borlaſe*, A. M. fol. *London*, 1758.
Briſſon quad.	Regnum animale in Claſſes IX. diſtributum, a *D. Briſſon*, 8vo. *Lugd. Bat.* 1762.
Briſſon av.	Ornithologie, ou Methode contenant la Diviſion des Oiſeaux, &c. Ouvrage enrichi des Figures, par M. *Briſſon*, 6 tom. 4to. *Paris*, 1760.
Br. Zool.	*Britiſh* Zoology. Claſs I. Quadrupeds. II. Birds. Illuſtrated with 132 Plates, imperial Paper. *London*, 1766.
Br. or *Brunnich.*	M. Th. *Brunnichii* Ornithologia Borealis, 8vo. *Copenhagen*, 1764.
Br. Monog.	A Hiſtory of the Eider-Duck, in *Daniſh*, by Mr. *Brunnich*, 12mo. *Copenhagen*, 1763.
De Buffon,	Hiſt. Nat. generale & particuliere, avec la Deſcription du Cabinet du Roy, par M. *De Buffon*, 15 tom. 4to. *a Paris*, 1749, &c.
Caii opuſc.	*Joannis Caii* Britanni Opuſcula, a *S. Jebb* edita, 8vo. *Londini*, 1729.
Camden.	*Camden*'s Britannia, publiſhed by Biſhop *Gibſon*, 2 vol. fol. 3d edition. *London*, 1753.
Cat. Carol.	Nat. Hiſt. of *Carolina* and the *Bahama* Iſlands, by *Mark Cateſby*, 2 vol. fol. *London*, 1731.
Charlton ex.	*Gualteri Charletoni* Exercitationes de Differentiis, &c. Animalium, fol. *Londini*, 1677.

Cluſ.

Cluſ. ex. — *Caroli Cluſii* Exoticorum Libri X. fol. *Antverpiæ*, 1605.

Crantz's Greenl. — Hiſtory of *Greenland*, &c. by *David Crantz*. Tranſlated from the *High Dutch*. 2 vol. 8vo. *London*, 1767.

Dale's hiſt. — of *Harwich* and *Dover-court*, by *Sam. Dale*, 4to. *London*, 1730.

Egede's Greenl. — Deſcription of *Greenland*, by *Hans Egede*, Miſſionary in that Country for twenty Years. Tranſlated from the *Daniſh*, 8vo. *London*, 1745.

Edw. — Nat. Hiſt. of Birds and other rare and undeſcribed Animals, by *George Edwards*, 7 vol. 4to. *London*, 1743, &c.

Faun. Suec. — *Caroli Linnæi* Fauna Suecica, ſiſtens Animalia *Sueciæ* Regni, 8vo. *Holmiæ*, 1761.

Friſch. — A Hiſtory of the Birds of *Germany*, with colored Plates, and Deſcriptions in the *German* Language, 2 vol. fol. by *John Leonard Friſch*. Printed at *Berlin*, 1734, &c.

GESNER QUAD. — *Conrad. Geſneri* Hiſtoria Quadrupedum, fol. *Frankfort*, 1603.

GESNER AV. — *Geſner* de Avium Natura, fol. *Francofurti*, 1585.

Geſner icon. — Icones Animalium Quadr. vivip. & ovip. quæ in Hiſt. Animalium *Conradi Geſneri* Libri I. & II. deſcribuntur, fol. *Tiguri*, 1560.

Girald. Cam. — Itinerarium *Cambriæ*, Auctore *Sil. Giraldo Cambrenſe*, cum Annot. *Poveli*, 12mo. *Londini*, 1585.

Grew's muſ. — Catalogue of the Rarities belonging to the Royal Society, by Dr. *N. Grew*, fol. *London*, 1685.

Gunner.

Gunner. Det *Trondhiemfte* Gelfkabs Skrifter. *Kiobenthavn*, 1761.

Haffelquift's itin. *Fred. Haffelquiftii* Iter Palæftinum, 8vo. *Holmiæ*, 1757.

Hift. d'Oif. Hiftoire Naturelle des Oifeaux.

This is a continuation of the Natural Hiftory by M. *de Buffon*; and is to be included in five *quarto* volumes. Three only at this time are publifhed. The two firft volumes are the joint performances of M. *de Buffon* and M. *Monbeillard*; the remaining three will be written by the laft.

Hor. Ice. Nat. Hift. of *Iceland*, by *N. Horrebow*. Tranflated from the *Danifh*, fol. *London*, 1758.

Jonfton's Nat. Hift. *Johannis Jonftoni*, M. D. Hiftoria Naturalis, 2 tom. fol. *Amftelodami*, 1657.

Klein quad. *Jac. Theod. Klein* Quadrupedum Difpofitio, brevifque Hift. Nat. 4to. *Lipfiæ*, 1751.

Klein av. *J. Theod. Klein* Hiftoriæ Avium Prodromus, 4to. *Lubecæ*, 1750.

Klein ftem. *J. Theod. Klein* Stemmata avium, 40 Tabulis Æneis ornata, 4to. *Lipfiæ*, 1759.

Kramer. *Gulielmi Henrici Kramer* Elenchus Vegetabilium & Animalium per *Auftriam Inferiorem* obfervatorum, 8vo. *Viennæ*, *Pragæ* & *Tergefti*, 1756.

LIN. SYST. *Caroli Linnæi* Syftema Naturæ, edit. 12, reformata, 8vo. *Holmiæ*, 1766.

Marten's Spitzberg. Voyage into *Spitzbergen* and *Greenland*, by *Fred. Marten. London*, 1694.

Martin's Weft. Ifles. Defcription of the Weftern Iflands of *Scotland*, by *M. Martin*, 2d edit. 8vo. *London*, 1716.

Martin's

Martin's St. Kilda. Voyage to *St. Kilda*, by *M. Martin*, 4th edit. 8vo. *London*, 1753.

Merret pinax. Pinax Rerum Naturalium Britannicarum, Authore *Christoph. Merret*, 12mo. *Londini*, 1667.

Meyer's an. A Work wrote in *German*, containing 200 colored Plates of various Animals, with the Skeleton of each, by *John-Daniel Meyer*, Miniature Painter, at *Nuremberg*, 2 vol. fol. 1748.

Morton's Northampt. Hist. Nat. of *Northamptonshire*, by *John Morton*, A. M. fol. *London*, 1712.

Nov. Com. Petrop. Novi Commentarii Academiæ Scientiarium imperialis *Petropolitanæ*, 7 tom. 4to. *Petropoli*, 1750, &c.

Olina. Uccelliera overo Discorso della Natura e Proprieta di diversi Uccelli e in particolare di que'che Cantano. Opera di *Gio. Petro Olina*, fol. in *Roma*, 1684.

Plin. Nat. Hist. *Plinii* Historia Naturalis, cum Notis *Harduini*, 2 tom. fol. *Paris*, 1723.

Pl. Enl. Colored Figures of Birds, Reptiles and Insects, publishing at *Paris*, under the Title of *Planches Enluminées*.

Pontoppidan. Nat. Hist. of *Norway*, by the Right Reverend *Eric Pontoppidan*, Bishop of *Bergen*. Translated from the *Danish*, fol. *London*, 1755.

Prosp. Alpin. *Prosperi Alpini* Historiæ *Ægypti* Pars prima & secunda, 2 tom. 4to. *Lugd. Bat.* 1735.

RAII SYN. QUAD. *Raii* Synopsis methodica Anim. Quadrupedum & Serpentini Generis, 8vo. *Londini*, 1693.

RAII SYN. AV. *Raii* Synopsis methodica Avium & Piscium, 8vo. *London*, 1713.

Russel's

Ruſſel's Alep. The Natural Hiſtory of *Aleppo* and the Parts adjacent, by *Alexander Ruſſel*, M. D. 4to. *London*, 1756.

SCOPOLI. Annus. I. Hiſtorico-Naturalis, *Johannis Antonii Scopoli. Lipſiæ*, 1769.

Sib. Muſ. *Alberti Sebæ* Rerum Naturalium Theſaurus, 4 tom. fol. *Amſterdam*, 1734, &c.

Sib. Scot. Prodromus Hiſtoriæ Naturalis Scotiæ, Auctore *Roberto Sibbaldo*, M. D. Eq. Aur. fol. *Edinburgi*, 1684.

Sib. Hiſt. Fife. Hiſtory of the Sheriffdoms of *Fife* and *Kinroſs*, by Sir *Robert Sibbald. Edinburgh*, fol. 1710.

Smith's Kerry. Natural and Civil Hiſtory of the County of *Kerry*, 8vo. *Dublin*, 1756.

Syn. Quad. Synopſis of Quadrupeds, containing Deſcriptions of 292 Animals, with 31 Plates, 8vo. 1771. by *Thomas Pennant*, Eſquire.

Turner. Avium præcipuarum quarum apud *Plinium* & *Ariſtotelem* Mentio eſt, brevis & ſuccincta Hiſtoria, per Dm. *Gulielmum Turnerum*, Artium & Medicinæ Doctorem, 12mo. *Coloniæ*, 1544. *N. B.* This Book is not paged.

WIL. ORN. The Ornithology of Mr. *Francis Willughby*; publiſhed by Mr. *Ray*, fol. *London*, 1678.

Worm. Muſ. Muſeum Wormianum, fol. *Amſtelodami*, 1655.

Zinanni. Delle uova e dei Nidi degli Uccelli, Libro primo del Conte *Giuſeppe Zinanni*, in *Venezia*, 1737.

CLASS

CLASS I.

QUADRUPEDS.

CLASS I.

QUADRUPEDS.

Div. I. HOOFED.
II. DIGITATED.
III. PINNATED.
IV. WINGED.

Div. I. Sect. I. WHOLE HOOFED.

Genus
I. HORSE.

Sect. II. CLOVEN HOOFED.

II. OX.
III. SHEEP.
IV. GOAT.
V. DEER.
VI. HOG.

Div.

Div. II. DIGITATED.

Sect. I. With large canine teeth, ſeparated from the cutting teeth.

Six cutting teeth in each jaw.

Rapacious, carnivorous.

Genus

VII. DOG.
VIII. CAT.
IX. BADGER.
X. WEESEL.
XI. OTTER.

Sect. II. With only two cutting teeth in each jaw.

Uſually herbivorous, frugivorous.

XII. HARE.
XIII. SQUIRREL.
XIV. DORMOUSE.
XV. RAT.
XVI. SHREW.
XVII. MOLE.
XVIII. URCHIN.

Div.

DIV. III. PINNATED.

GENUS

XIX. SEAL.

DIV. IV. WINGED.

XX. BAT.

c

E R R A T A.

Page 12, line 3, *for* infect *read* infest. P. 15, note, *for* maritima *read* lanceolata. P. 18, margin, *for* DOMESTIC *read* 3. DOMESTIC. P. 25, l. 9, *for* co racles *read* coracles. *Ibid.* note, *for Stanley read Stavely*. P. 46, l. 13, *for* were *read* are. *Ibid.* l. 15, *for* agreed *read* agree. P. 78, l. 22, *for* out ricks *read* oat ricks. P. 79, l. 4, *for* our *read* other. P. 89*, 90*, 91*, 92*, 93*, 94*, 95*, 96*. P. 101, l. 9, *for* second satire fourth book *read* fourth satire second book. P. 102, l. 8, *for Boadicia read Boadicea*. P. 115, *running title*, *for* NORWAY RAT *read* BROWN RAT. P. 128, l. 9, *for* Europæus *read* Europæa. P. 137, l. 12, *for* tectis *read* rectis. P. 170, *running title*, *for* ERNE *read* CINEREOUS. P. 175, *margin*, *for* NEOT *read* NEST. P. 181, l. 5, *for* fuscis *read* fasciis. *Ibid.* l. 21, *for* tips, all *read* tips of all. P. 186, l. 25, *after* twenty-seven *add* inches. P. 193, l. 12, *dele (the male)*. P. 199, l. 10, *for* fine *read* five. P. 202, *after the character of the genus add* EARED OWLS. P. 203, *dele* EARED OWLS. P. 210, l. 18, *for* diffre *read* differ P. 220, l. 14, *for* illice *read* ilice. P. 222, note, *for Melolantha read Melolontha*. *Ibid. for Rosel read Ræsel.* P. 241, l. 5, *for* clifts *read* clefts. P. 250, l. 17, *for* disturb *read* disturbed. P. 262, l. 9, *for* Cocque *read* Coq. P. 263, 264, 265, *running title*, WOOD GROUS. P. 269, 270, *running title*, RED GROUS. P. 275, l. 25, *for* Sâr *read* Jâr. P. 286, l. 1, *for* quarts *read* pints. *Ibid.* l. 10, *for* canne patiere *read* canne petiere. P. 294, l. 1, *for* is *read* was. P. 326, l. 18, *for* Sparrow *read* Bunting. P. 328, l. 2, *for* breast *read* belly. P. 351, l. 13, *for* atri capilla *read* atricapilla. *Ibid.* l. 24, *for* with white bar *read* with a white bar. P. 384, l. *penult. dele* ? P. 401, last l. *for* breed *read* breeder. P. 404, l. 27, *after Indian read* air. P. 406, l. 1, *for* monographics *read* monographies. *Ibid.* l. 8, *for* tribes *read* tribe. P. 446, l. 22, *for* pair *read* pairs. P. 461, last l. *for* such as employed *read* such as are, &c. P. 485, last l. *for* table *read* tables. *Ibid. dele* preceding this class. P. 515, l. 8, *for* house *read* holes. P. 528, l. 6, *for* above knee *read* above the knee. P. 546, l. 18, *for* Larus Minuta *read* Sterna Minuta. P. 550, l. 26, *for* unctious *read* unctuous. P. 563, note, *for* Knat *read* Gnat. P. 619, l. 23, *dele*) *and place it in the preceding line after* suspects. P. 630, l. 13, *for* one *read* the. P. 644, l. 14, *for* fleetnese *read* fleetness. *Ibid.* l. 21, *for* at time *read* at the time. P. 645, l. 8, *for* cartamea *read* carta mea. P. 652, note †, *for* let flutter *read* let it flutter. *For* Hist. d' Oys. *read* Hist. d' Ois. *passim.*

THE Book-binder is requested to place the Plates according to the numbers affixed to the figures which refer to the descriptions.

PLATES

TO

BRITISH ZOOLOGY.

VOL. I. OCTAVO.

PLATES.

XXXVII.

PLATES.

LV.

PLATES.

Pl. 1.

No 1

CLASS I.

QUADRUPEDS.

Div. I. HOOFED.

Sect. I. WHOLE HOOFED.
II. CLOVEN HOOFED.

SECT. I.

Hoof consisting of one piece. Six cutting teeth in each jaw.

I. HORSE.

Raii syn. quad. 62.
Merret pinax. 166.
Gesn. quad. 404.
Klein quad. 4.
De Buffon iv. 174.

Equus auriculis brevibus erectis, juba longa. *Brisson quad.* 69.
Eq. Caballus. *Lin. syst.* 100.
Eq. cauda undique setosa. *Faun. Suec.* 47.
Br. Zool. 1. *Syn. quad.* No. 1.

1. Generous.

	Horse.	Mare.	Gelding.
Brit.	March, Ceffyl	Caseg	Dispaiddfarch
Fren.	Le Cheval	La Cavale, Jument	Cheval ongre
Ital.	Cavallo	Cavalla	
Span.	Cavallo	Yegua	
Port.	Cavallo	Egoa	
Germ.	Pferd	Stut, Motsch	
Dut.	Paerd, Hengst	Merrie	
Swed.	Hæst	Stood, Horss	
Dan.	Hæst, Oeg, Hingst	Stod-Hæst, Hoppe	

THE breed of horses in *Great Britain* is as mixed as that of its inhabitants: The frequent introduction of foreign horses has given us a variety, that no single country can boast of: moſt other kingdoms produce only one kind, while ours, by a judicious mixture of the ſeveral ſpecies, by the happy difference of our ſoils, and by our ſuperior ſkill in management, may triumph over the reſt of *Europe*, in having brought each quality of this noble animal to the higheſt perfection.

SWIFTNESS.

In the annals of *Newmarket*, may be found inſtances of horſes that have literally out-ſtripped the wind, as the celebrated *M. Condamine* has lately ſhewn in his remarks * on thoſe of *Great Britain*. *Childers* † is an amazing inſtance of rapidity, his ſpeed having been more than once exerted equal to $82\frac{1}{2}$ feet in a ſecond, or near a mile in a minute: The ſame horſe has alſo run the round courſe at *Newmarket*, (which is about 400 yards leſs than 4 miles) in ſix minutes and forty ſeconds; in which caſe his fleetneſs is to that of the ſwifteſt *Barb*, as four to three; the former, according to Doctor *Maty*'s computation, covering at every bound a ſpace of ground equal in length to twenty-three feet royal, the latter only that of eighteen feet and a half royal.

Horſes of this kind, derive their origin from

* In his tour to Italy, 190.

† M. *Condamine* illuſtrates his remarks with the horſe, *Starling*; but the report of his ſpeed being doubtful, we chuſe to inſtance the ſpeed of *Childers*, as indiſputable and univerſally known.

Arabia;

Arabia; the ſeat of the pureſt, and moſt generous breed. *

The ſpecies uſed in hunting, is a happy combination of the former with others ſuperior in ſtrength, but inferior in point of ſpeed and lineage: an union of both is neceſſary; for the fatigues of the chace muſt be ſupported by the ſpirit of the one, as well as by the vigor of the other.

No country can bring a parallel to the ſtrength and ſize of our horſes deſtined for the draught; or to the activity and ſtrength united of thoſe that form our cavalry.

STRENGTH.

In our capital there are inſtances of ſingle horſes that are able to draw on a plain, for a ſmall ſpace, the weight of three tuns; but could with eaſe, and for a continuance draw half that weight †. The pack-horſes of *Yorkſhire*, employed in conveying the manufactures of that county to the moſt remote parts of the kingdom, uſually carry a burden of 420 pounds; and that indifferently over the higheſt hills of the north, as well as the moſt level roads; but the moſt remarkable proof of the ſtrength of our *Britiſh* horſes, is to be drawn from that of our mill-horſes: ſome of theſe will carry at

* For a particular account of the *Arabian* horſes, the reader is referred to No. I. in the Appendix to this volume.

† *Hollingſhed* makes it a matter of boaſt, that in his time, five horſes could draw with eaſe for a long journey 3000lb. weight.

one load thirteen meaſures, which at a moderate computation of 70 pounds each, will amount to 910; a weight ſuperior to that which the leſſer ſort of camels will bear: this will appear leſs ſurpriſing, as theſe horſes are by degrees accuſtomed to the weight; and the diſtance they travel no greater than to and from the adjacent hamlets.

British Cavalry.

Our cavalry in the late campaigns, (when they had opportunity) ſhewed over thoſe of our allies, as well as of the *French*, a great ſuperiority both of ſtrength and activity: the enemy was broken through by the impetuous charge of our ſquadrons; while the *German* horſes, from their great weight, and inactive make, were unable to ſecond our efforts; though thoſe troops were actuated by the nobleſt ardor.

Antient.

The preſent cavalry of this iſland only ſupports its antient glory; it was eminent in the earlieſt times: our ſcythed* chariots, and the activity† and good diſcipline of our horſes, even ſtruck terror into *Cæſar*'s legions: and the *Britains*, as ſoon as they became civilized enough to coin, took care to repreſent on their money the animal for which they were ſo celebrated. It is now impoſſible to trace out this ſpecies; for thoſe which exiſt among the *indigenæ* of *Great Britain*, ſuch as the little horſes

* *Covinos* vocant, quorum falcatis axibus utuntur. *Pomp. Mela*, lib. iii. c. 6.

† *Cæſar. Com.* lib. iv. *Strabo.* lib. iv.

horſes of *Wales* and *Cornwal*, the hobbies of *Ireland*, and the ſhelties of *Scotland*, though admirably well adapted to the uſes of thoſe countries, could never have been equal to the work of war; but probably we had even then a larger and ſtronger breed in the more fertile and luxuriant parts of the iſland. Thoſe we employ for that purpoſe, or for the draught, are an offſpring of the *German* or *Flemiſh* breed, meliorated by our ſoil, and a judicious culture.

The *Engliſh* were ever attentive to an exact culture of theſe animals; and in very early times ſet a high value on their breed. The eſteem that our horſes were held in by foreigners ſo long ago as the reign of *Athelſtan*, may be collected from a law of that monarch prohibiting their exportation, except they were deſigned as preſents. Theſe muſt have been the native kind, or the prohibition would have been needleſs, for our commerce was at that time too limited to receive improvement from any but the *German* kind, to which country their own breed could be of no value.

But when our intercourſe with the other parts of *Europe* was enlarged, we ſoon layed hold of the advantages this gave of improving our breed. *Roger de Beleſme*, Earl of *Shrewſbury* *, is the firſt that is on record: he introduced the *Spaniſh* ſtallions into his eſtate in *Powiſland*, from which that

* Created by *William* the *Conqueror*.

B 3 part

part of *Wales* was for many ages celebrated for a ſwift and generous race of horſes. *Giraldus Cambrenſis*, who lived in the reign of *Henry* II. takes notice of it *; and *Michael Drayton*, cotemporary with *Shakeſpear*, ſings their excellence in the ſixth part of his *Polyolbion*. This kind was probably deſtined to mount our gallant nobility, or courteous knights for feats of *Chivalry*, in the generous conteſts of the tilt-yard. From theſe ſprung, to ſpeak the language of the times, the *Flower* of *Courſers*, whoſe elegant form added charms to the rider; and whoſe activity and managed dexterity gained him the palm in that field of gallantry and romantic honor.

RACES.

Notwithſtanding my former ſuppoſition, races were known in *England* in very early times. *Fitz-Stephen*, who wrote in the days of *Henry* II. mentions the great delight that the citizens of *London* took in the diverſion. But by his words, it appears not to have been deſigned for the purpoſes of gaming, but merely to have ſprung from a generous emulation of ſhewing a ſuperior ſkill in horſemanſhip.

Races appear to have been in vogue in the reign of Queen *Elizabeth*, and to have been carried

* In hæc tertia *Walliæ* portione quæ *Powiſia* dicitur ſunt equitia peroptima, et equi emiſſaria laudatiſſima, de *Hiſpanienſium* equorum generoſitate, quos olim Comes *Slopeſburiæ Robertus de Beleſme* in fines iſtos adduci curaverat, originaliter propagati. *Itin. Camb.* 222.

to

to ſuch exceſs as to injure the fortunes of the nobility. The famous *George* Earl of *Cumberland* is recorded to have waſted more of his eſtate than any of his anceſtors; and chiefly by his extreme love to horſe-races, tiltings, and other expenſive diverſions. It is probable that the parſimonious Queen did not approve of it; for races are not among the diverſions exhibited at *Kennelworth* by her favorite *Leiceſter*. In the following reign, were places allotted for the ſport: *Croydon* in the South, and *Garterly* in *Yorkſhire*, were celebrated courſes. *Cambden* alſo ſays, that in 1607 there were races near *York*, and the prize was a little golden bell.

Not that we deny this diverſion to be known in theſe kingdoms in earlier times; we only aſſert a different mode of it, gentlemen being then their own jockies, and riding their own horſes. Lord *Herbert* of *Cherbury* enumerates it among the ſports that gallant philoſopher thought unworthy of a man of honor. "The exerciſe, (ſays he) I do "not approve of, is running of horſes, there being "much cheating in that kind; neither do I ſee "why a brave man ſhould delight in a creature "whoſe chief uſe is to help him to run away*."

The increaſe of our inhabitants, and the extent

* The Life of *Edward* Lord *Herbert* of *Cherbury*, publiſhed by Mr. *Walpole*, p. 51.

Jarvis Markham, who wrote on the management of horſes 1599, mentions running horſes; but thoſe were only deſigned for matches between gentleman and gentleman.

B 4 of

of our manufactures, together with the former neglect of internal navigation to convey those manufactures, multiplied the number of our horses: an excess of wealth, before unknown in these islands, increased the luxury of carriages, and added to the necessity of an extraordinary culture of these animals: their high reputation abroad, has also made them a branch of commerce, and proved another cause of their vast increase.

As no kingdom can boast of parallel circumstances, so none can vie with us in the number of these noble quadrupeds; it would be extremely difficult to guess at the exact amount of them, or to form a periodical account of their increase: the number seems very fluctuating: *William Fitz-Stephen* relates, that in the reign of King *Stephen*, *London* alone poured out 20,000 horsemen in the wars of those times: yet we find that in the beginning of Queen *Elizabeth*'s reign *, the whole kingdom could not supply 2000 horses to form our cavalry: and even in the year 1588, when the nation was in the most imminent danger from the *Spanish* invasion, all the cavalry which the nation could then furnish amounted only to 3000: to account for this difference we must imagine, that the number of horses which took the field in *Stephen*'s reign was no more than an undisciplined

* Vide Sir *Edward Harwood*'s memorial. *Harleian Misc.* iv. 255. The number mentioned by *Fitz-Stephens* is probably erroneous, and ought to be read 2000.

rabble;

rabble; the few that appeared under the banners of *Elizabeth*, a corps well formed, and ſuch as might be oppoſed to ſo formidable an enemy as was then expected: but ſuch is their preſent increaſe, that in the late war, the number employed was 13,575; and ſuch is our improvement in the breed of horſes, that moſt of thoſe which are uſed in our waggons and carriages * of different kinds, might be applied to the ſame purpoſe: of thoſe, our capital alone employs near 22,000.

The learned M. *de Buffon* has almoſt exhauſted the ſubject of the natural hiſtory of the horſe, and the other domeſtic animals; and left very little for after writers to add. We may obſerve, that this moſt noble and uſeful quadruped is endowed with every quality that can make it ſubſervient to the uſes of mankind; and thoſe qualities appear in a more exalted, or in a leſs degree, in proportion to our various neceſſities.

Undaunted courage, added to a docility half reaſoning, is given to ſome, which fits them for military ſervices. The ſpirit and emulation ſo apparent in others, furniſh us with that ſpecies, which is admirably adapted for the courſe; or, the more noble and generous pleaſure of the chace.

Patience and perſeverance appear ſtrongly in that moſt uſeful kind deſtined to bear the burdens

* It may be alſo obſerved, that the uſe of coaches was not introduced into *England* till the year 1564.

we

we impoſe on them; or that employed in the ſlavery of the draught.

Though endowed with vaſt ſtrength, and great powers, they very rarely exert either to their maſter's prejudice; but on the contrary, will endure fatigues, even to death, for our benefit. Providence has implanted in them a benevolent diſpoſition, and a fear of the human race, together with a certain conſciouſneſs of the ſervices we can render them. Moſt of the hoofed quadrupeds are domeſtic, becauſe neceſſity compels them to ſeek our protection: wild beaſts are provided with feet and claws, adapted to the forming dens and retreats from the inclemency of the weather; but the former, deſtitute of theſe advantages, are obliged to run to us for artificial ſhelter, and harveſted proviſions; as nature, in theſe climates, does not throughout the year ſupply them with neceſſary food.

But ſtill, many of our tame animals muſt by accident endure the rigor of the ſeaſon: to prevent which inconvenience, their feet (for the extremities ſuffer firſt by cold) are protected by ſtrong hoofs of a horny ſubſtance.

The tail too is guarded with long buſhy hair that protects it in both extremes of weather; during the ſummer it ſerves by its pliancy and agility, to bruſh off the ſwarms of inſects, which are perpetually attempting either to ſting them, or to depoſit their eggs in the *rectum*; the ſame length of hair contributes to guard them from the cold in winter.

winter. But we, by the abſurd and cruel cuſtom of docking, a practice peculiar to our country, deprive theſe animals of both advantages: in the laſt war our cavalry ſuffered ſo much on that account, that we now ſeem ſenſible of the error, and if we may judge from ſome recent orders in reſpect to that branch of the ſervice*, it will for the future be corrected.

Thus is the horſe provided againſt the two greateſt evils he is ſubject to from the ſeaſons: his natural diſeaſes are few; but our ill uſage, or neglect, or, which is very frequent, our over care of him, bring on a numerous train, which are often fatal.

* The following remark of a noble writer on this ſubject is too ſenſible to be omitted.

'I muſt own I am not poſſeſſed with the *Engliſh* rage of cut-'ting off all extremities from horſes. I venture to declare I 'ſhould be well pleaſed if their tails, at leaſt a ſwitch or a 'nag tail, (but better if the whole) was left on. It is hardly 'credible what a difference, eſpecially at a certain ſeaſon of 'the year, this ſingle alteration would make in our cavalry, 'which though naturally ſuperior to all other I have ever 'ſeen, are however, long before the end of the campaign, 'for want of that natural defence againſt the flies, inferior to 'all: conſtantly ſweating and fretting at the picquet, tor-'mented and ſtung off their meat and ſtomachs, miſerable 'and helpleſs; while the foreign cavalry bruſh off the ver-'min, are cool and at eaſe, and mend daily, inſtead of pe-'riſhing as ours do almoſt viſibly in the eye of the be-'holder.'

Method of breaking Horſes, &c. by *Henry* Earl of *Pembroke*, p. 68.

Among

Among the diftempers he is naturally fubject to, are the worms, the bots, and the ftone: the fpecies of worms that infect him are the *lumbrici*, and *afcarides*; both thefe refemble thofe found in human bodies, only larger: the bots are the *erucæ*, or caterpillars of the *oeftrus*, or gadfly: thefe are found both in the *rectum*, and in the ftomach, and when in the latter bring on convulfions, that often terminate in death.

The ftone is a difeafe the horfe is not frequently fubject to; yet we have feen two examples of it; the one in a horfe near *Highwycombe*, that voided fixteen *calculi*, each of an inch and a half diameter; the other was of a ftone taken out of the bladder of a horfe, and depofited in the cabinet of the late Dr. *Mead*; weighing eleven ounces *. Thefe ftones are formed of feveral crufts, each very fmooth and gloffy; their form triangular; but their edges rounded, as if by collifion againft each other.

The all-wife Creator hath finely limited the feveral fervices of domeftic animals towards the human race; and ordered that the parts of fuch, which in their lives have been the moft ufeful, fhould after death contribute the left to our benefit. The chief ufe that the *exuviæ* of the horfe can be applied to, is for collars, traces, and other parts of the harnefs; and thus, even after death, he preferves fome analogy with his former employ. The

* *Mufeum Meadianum*, p. 261.

hair

hair of the mane is of uſe in making wigs; of the tail in making the bottoms of chairs, floor-cloths, and cords; and to the angler in making lines.

2. ASS.

Aſinus, *Raii ſyn. quad.* 63.
Geſn. quad. 5.
Klein. quad. 6.
De Buffon iv. 377.
Equus auriculis longis flaccidis, juba brevi. *Briſſon quad.* 70.
Equus aſinus. *Lin. ſyſt.* 100.
Eq. caudæ extremitate ſetoſa cruce nigra ſuper humeros. *Faun. Suec.* 35 *.
Br. Zool. 5. *Syn. quad.* No. 3.

Brit. Aſyn, *fœm.* Aſen
Fren. L'Ane, *f.* L'Aneſſe
Ital. Aſino, Miccio. *f.* Miccia
Span. Aſno, Borrico. *f.* Borrica
Port. Aſno, Burro. *f.* Aſna, Burra
Germ. Eſel
Dut. Eezel
Swed. Aſna
Dan. Aſen, Eſel.

THIS animal, tho' now ſo common in all parts of theſe iſlands, was entirely loſt among us during the reign of queen *Elizabeth*; *Hollingſhed* † informing us that in his time, " *our lande did yeelde no aſſes.*" But we are not to ſuppoſe ſo uſeful an animal was unknown in theſe kingdoms before that period; for mention is made of them ſo early as the time of king ‡ *Ethelred*, above four hundred

* *Habitat in magnatum prædiis rarius.* Faun. Suec. 35. *edit.* 1746. We imagine that ſince that time the ſpecies is there extinct, for *Linnæus* has quite omitted it in the laſt edition of the *Fauna Suecica.*

† 109.

‡ When the price of a mule or young aſs was 12 s. *Chron. precioſum*, 51.

years

years preceding; and again in the reign of * *Henry* III. ſo that it muſt have been owing to ſome accident, that the race was extinct during the days of *Elizabeth*. We are not certain of the time it was again introduced; probably in the ſucceeding reign, when our intercourſe with *Spain* was renewed; in which country this animal was greatly uſed, and where the ſpecies is in great perfection.

The aſs is originally a native of *Arabia*, and other parts of the *Eaſt*: a warm climate produces the largeſt and the beſt, their ſize and ſpirit declining in proportion as they advance into colder regions. "With difficulty," ſays Mr. *Adanſon*, ſpeaking of the aſſes of *Senegal*, "did I know this "animal, ſo different did it appear from thoſe of "*Europe*: the hair was fine, and of a bright mouſe "color, and the black liſt that croſſes the back "and ſhoulders had a good effect. Theſe were the "aſſes brought by the *Moors* from the interior "parts of the country †." The migration of theſe beaſts has been very ſlow; we ſee how recent their return is in *Great Britain*: in *Sweden* they are even at preſent a ſort of rarity, nor does it appear by the laſt hiſtory of *Norway* ‡, that they had yet reached that country. They are at preſent naturalized in

* In 1217, when the *Camerarius* of St. *Alban*'s loſt two aſſes, &c. *Chr. pr.* 60.

† *Voy. Senegal.* 212.

‡ *Pontoppidan*'s Nat. Hiſtory of *Norway*.

this

this kingdom; our climate and ſoil ſeems to agree with them; the breed is ſpread thro' all parts; and their utility is more and more experienced.

They are now introduced into many ſervices that were before allotted to horſes; which will prove of the utmoſt uſe in ſaving thoſe noble animals for worthier purpoſes. Many of our richeſt mines are in ſituations almoſt inacceſſible to horſes; but where theſe ſurefooted creatures may be employed to advantage, in conveying our mineral treaſures to their reſpective marts: we may add too, that ſince our horſes are become a conſiderable article of commerce, and bring annually great ſums into theſe kingdoms, the cultivation of an animal that will in many caſes ſupply the place of the former, and enable us to enlarge our exports, certainly merits our attention.

The qualities of this animal are ſo well known, that we need not expatiate on them; its patience and perſeverance under labor, and its indifference in reſpect to food, need not be mentioned; any weed or thiſtle contents it: if it gives the preference to any vegetable, it is to the *Plantane*; for which we have often ſeen it neglect every other herb in the paſture. The narrow-leaved *Plantane** is greedily eat by horſes and cows: of late years it has been greatly cultivated and ſowed with clover in North *Wales*, particularly in *Angleſea*, where

* *Plantago maritima. Fl. Angl.* 52.

the

the ſeed is harveſted, and thence diſperſed thro' other parts of the principality.

MULE.

Mulus, *Raii ſyn. quad.* 64.
Geſn. quad. 702.
Aſinus biformis, *Klein. quad.* 6.
Charlton ex. 4.
Equus auriculis longis erectis, juba brevi. *Briſſon quad.* 71.
Equus mulus, *Lin. ſyſt.* 101.
Faun. Suec. 35. edit. 1.
Br. Zool. 6.

Brit. Mul, *fæm.* Mules
Fren. Le Mulet
Ital. Mula
Span. Mulo
Port. Mula
Germ. Maulthier, Mauleſel
Dut. Muyl-Eeſel
Swed. Mulaſna
Dan. Muule, *v.* Muul-Eſel.

THIS uſeful and hardy animal is the off-ſpring of the horſe and aſs, or aſs and mare; thoſe produced between the two laſt are eſteemed the beſt, as the mule is obſerved to partake leſs of the male than the female parent; not but they almoſt always inherit in ſome degree the obſtinacy of the parent aſs, tho' it muſt be confeſſed that this vice is heightened by their being injudiciouſly broke: inſtead of mild uſage, which gently corrects the worſt qualities, the mule is treated with cruelty from the firſt; and is ſo habituated to blows, that it is never mounted or loaded without expectation of ill treatment; ſo that the unhappy animal either prepares to retaliate, or in the terror of bad uſage, becomes invincibly retrograde. Could we prevale on our countrymen to conſider this animal in the light its uſeful qualities merit, and pay due atten- tion

tion to its breaking, they might with ſucceſs form it for the ſaddle, the draught, or the burden. The ſize and ſtrength of our breed is at preſent ſo improved by the importation of the *Spaniſh* male aſſes, that we ſhall ſoon have numbers that may be adapted to each of thoſe uſes. Perſons of the firſt quality in *Spain* are drawn by them; for one of which (as Mr. *Clarke* informs us*) fifty or ſixty guineas is no uncommon price; nor is it ſurprizing, if we conſider how far they excel the horſe in draught, in a mountanous country; the mule being able to tread ſecurely where the former can hardly ſtand.

This brief account may be cloſed with the general obſervation, that neither mules nor the ſpurious offſpring of any other animal generate any farther: all theſe productions may be looked on as monſters; therefore nature, to preſerve the original ſpecies of animals entire and pure, wiſely ſtops, in inſtance of deviation, the powers of propagation.

* Letters on the *Spaniſh* nation.

DIV. I. SECT. II. CLOVEN HOOFED.

* With horns.

** Without horns.

II. *OX. Horns bending out laterally.
Eight cutting teeth in the lower jaw, none in the upper.
Skin along the lower ſide of the neck pendulous.

DOMESTIC. *Raii ſyn. quad.* 70.
Merret pinax. 166.
Geſn. quad. 25, 26, 92.
Taurus domeſticus. *Klein. quad.* 10.
Charlton ex. 8.

Bos cornibus levibus teretibus, ſurſum reflexis. *Briſſon quad.* 52.
Bos taurus. *Lin. ſyſt.* 98.
Bos cornibus teretibus flexis. *Faun. Suec.* 46.
Br. Zool. 7. *Syn. quad.* No. 4.

	BULL.	COW.	OX.	CALF.
Brit.	Tarw	Buwch	Ych, Eidion	Llo
Fren.	Le Taureau	La Vache	Le Bœuf	Veau
Ital.	Toro	Vacca	Bue	Vitello
Span.	Toro	Vaca	Buey	Ternera
Port.	Touro	Vaca	Boy	Vitela
Germ.	Stier	Kuh	Ochs	Kalb
Dut.	Stier, Bul	Koe	Os	Kalff
Swed.	Tiur	Ko	Noot	Kalff
Dan.	Tyr	Koe	Oxe, Stud	Kalv

THE climate of *Great-Britain* is above all others productive of the greateſt variety and abundance of wholeſome vegetables, which, to crown our happineſs, are almoſt equally diffuſed thro'

HIGHLAND BULL.

LANCASHIRE COW.

Moses Griffiths Del.

thro' all its parts: this general fertility is owing to thoſe clouded ſkies, which foreigners miſtakenly urge as a reproach on our country; but let us chearfully endure a temporary gloom, which cloaths not only our meadows but our hills with the richeſt verdure. To this we owe the number, variety, and excellence of our cattle, the richneſs of our dairies, and innumerable other advantages. *Cæſar* (the earlieſt writer who deſcribes this iſland of *Great-Britain*) ſpeaks of the numbers of our cattle, and adds that we neglected tillage, but lived on milk and fleſh*. *Strabo* takes notice of our plenty of milk, but ſays we were ignorant of the art of making cheeſe†. *Mela* informs us, that the wealth of the *Britains* conſiſted in cattle: and in his account of *Ireland* reports that ſuch was the richneſs of the paſtures in that kingdom, that the cattle would even burſt if they were ſuffered to feed in them long at a time‡.

This preference of paſturage to tillage was delivered down from our *Britiſh* anceſtors to much later times; and continued equally prevalent during the whole period of our feodal government:

* *Lib.* 5. † *Lib.* 4.

‡ Adeo luxurioſa herbis non lætis modo ſed etiam dulcibus, ut ſe exigua parte diei pecora impleant, ut niſi pabulo prohibeantur, diutius paſta diſſiliant. Lib. iii. c. 6.

Hollinſhed ſays, (but we know not on what authority,) that the *Romans* preferred the *Britiſh* cattle to thoſe of *Liguria. Deſc. Br.* 109.

C 2 the

the chieftain, whofe power and fafety depended on the promptnefs of his vaffals to execute his commands, found it his intereft to encourage thofe employments that favoured that difpofition; that vaffal, who made it his glory to fly at the firft call to the ftandard of his chieftain, was fure to prefer that employ, which might be tranfacted by his family with equal fuccefs during his abfence. Tillage would require an attendance incompatible with the fervices he owed the baron, while the former occupation not only gave leifure for thofe duties, but furnifhed the hofpitable board of his lord with ample provifion, of which the vaffal was equal partaker. The reliques of the larder of the elder *Spencer* are evident proofs of the plenty of cattle in his days; for after his winter provifions may have been fuppofed to have been moftly confumed, there were found, fo late as the month of *May*, in falt, the carcafes of not fewer than 80 beeves, 600 bacons, and 600 muttons *. The accounts of the feveral great feafts in after times, afford amazing inftances of the quantity of cattle that were confumed in them. This was owing partly to the continued attachment of the people to grazing †; partly to the preference that the *Englifh* at all times gave to animal food. The quan-

* *Hume*'s hiftory of *England* ii. 153.

† *Polyd. Virgil Hift. Angl.* vol. i. 5. who wrote in the time of *Henry* the VIII. fays *Angli plures pecuarii quam aratores*.

-tity

tity of cattle that appear from the lateſt calculation to have been conſumed in our metropolis, is a ſufficient argument of the vaſt plenty of theſe times; particularly when we conſider the great advancement of tillage, and the numberleſs variety of proviſions, unknown to paſt ages, that are now introduced into theſe kingdoms from all parts of the world *.

Our breed of horned cattle has in general been ſo much improved by a foreign mixture, that it is difficult to point out the original kind of theſe iſlands. Thoſe which may be ſuppoſed to have been purely *Britiſh* are far inferior in ſize to thoſe on the northern part of the *European* continent: the cattle of the highlands of *Scotland* are exceeding ſmall, and many of them, males as well as females, are hornleſs: the *Welſh* runts are much larger: the black cattle of *Cornwall* are of the ſame ſize with the laſt. The large ſpecies that is now cultivated through moſt parts of *Great-Britain* are either entirely of foreign extraction, or our own improved by a croſs with the foreign kind. The *Lincolnſhire* kind derive their ſize from the *Holſtein*

* That inquiſitive and accurate hiſtorian *Maitland* furniſhes us with this table of the quantity of cattle that were conſumed in *London* above 30 years ago, when that city was far leſs populous than it is at preſent.

Beeves	98,244.	Pigs	52,000.
Calves	194,760.	Sheep and Lambs	711,123.
Hogs	186,932.		

breed; and the large hornlefs cattle that are bred in fome parts of *England* come originally from *Poland.*

About two hundred and fifty years ago there was found in *Scotland* a wild race of cattle, which were of a pure white color, and had (if we may credit *Boethius*) manes like lions. I cannot but give credit to the relation; having feen in the woods of *Drumlanrig* in *N. Britain,* and in the park belonging to *Chillingham* caftle in *Northumberland,* herds of cattle probably derived from the favage breed. They have loft their manes; but retain their color and fiercenefs: they were of a middle fize; long leg'd; and had black muzzles, and ears: their horns fine, and with a bold and elegant bend. The keeper of thofe at *Chillingham* faid, that the weight of the ox was 38 ftones: of the cow 28: that their hides were more efteemed by the tanners than thofe of the tame; and they would give fix-pence per ftone more for them. Thefe cattle were wild as any deer: on being approached would inftantly take to flight and galop away at full fpeed: never mix with the tame fpecies; nor come near the houfe unlefs conftrained by hunger in very fevere weather. When it is neceffary to kill any they are always fhot: if the keeper only wounds the beaft, he muft take care to keep behind fome tree, or his life would be in danger from the furious attacks of the animal; which will never defift till a period is put to its life.

Frequent

Frequent mention is made of our ſavage cattle by hiſtorians. One relates that *Robert Bruce* was (in chacing theſe animals) preſerved from the rage of a wild Bull by the intrepidity of one of his courtiers, from which he and his lineage acquired the name of *Turn-Bull. Fitz-Stephen* * names theſe animals *(Uri-Sylveſtres)* among thoſe that harbored in the great foreſt that in his time lay adjacent to *London.* Another enumerates among the proviſions at the great feaſt of *Nevil* † archbiſhop of *York*, ſix wild Bulls; and *Sibbald* aſſures us that in his days a wild and white ſpecies was found in the mountains of *Scotland*, but agreeing in form with the common ſort. I believe theſe to have been the *Biſontes jubati* of *Pliny* found then in *Germany*, and might have been common to the continent and our iſland: the loſs of their ſavage vigor by confinement might occaſion ſome change in the external appearance, as is frequent with wild animals deprived of liberty; and to that we may aſcribe their loſs of mane. The *Urus* of the *Hercynian* foreſt deſcribed by *Cæſar*, book VI. was of this kind, the ſame which is called by the modern *Germans*, *Aurochs*, i. e. Bos ſylveſtris ‡.

The ox is the only horned animal in theſe iſlands

* A Monk who lived in the reign of *Henry* II. and wrote a Hiſtory of *London*, preſerved in *Leland's itin.* VIII.

† *Leland's Collectanea.* vi.

‡ *Geſner Quad.* 144. In *Fitz-Stephen*, *Urus* is printed *Urſus.*

that will apply his ſtrength to the ſervice of mankind. It is now generally allowed, that in many caſes oxen are more profitable in the draught than horſes; their food, harneſs, and ſhoes being cheaper, and ſhould they be lamed or grow old, an old working beaſt will be as good meat, and fatten as well as a young one.

There is ſcarce any part of this animal without its uſe. The blood, fat, marrow, hide, hair, horns, hoofs, milk, creme, butter, cheeſe, whey, urine, liver, gall, ſpleen, bones, and dung, have each their particular uſe in manufactures, commerce and medicine.

The ſkin has been of great uſe in all ages. The antient *Britains*, before they knew a better method, built their boats with oſiers, and covered them with the hides of bulls, which ſerved for ſhort * coaſting voyages.

> Primum cana ſalix madefacto vimine parvam
> Texitur in Puppim, cæſoque induta juvenco,
> Vectoris patiens, tumidum ſuper emicat amnem:
> Sic *Venetus* ſtagnante *Pado*, fuſoque *Britannus*
> Navigat oceano. *Lucan.* lib. iv. 131.

* That theſe *vitilia navigia*, as *Pliny* calls them, were not made for long voyages, is evident not only from their ſtructure, but from the account given by *Solinus*, that the crew never eat during the time they were at ſea. *Vide C. Junii Solini polyhiſtor.* 56.

The

The bending willow into barks they twine;
Then line the work with ſpoils of ſlaughter'd kine.
Such are the floats *Venetian* fiſhers know,
Where in dull marſhes ſtands the ſettling *Po*;
On ſuch to neighboring *Gaul*, allured by gain,
The bolder *Britons* croſs the ſwelling main. *Rowe.*

Veſſels of this kind are ſtill in uſe on the *Iriſh* lakes; and on the *Dee* and *Severn*: in *Ireland* they are called *Curach*, in *England Co racles*, from the *Britiſh Cwrwgl*, a word ſignifying a boat of that ſtructure.

At preſent, the hide, when tanned and curried, ſerves for boots, ſhoes, and numberleſs other conveniences of life.

Vellum is made of calves ſkin, and goldbeaters ſkin is made of a thin vellum, or a finer part of the ox's guts. The hair mixed with lime is a neceſſary article in building. Of the horns are made combs, boxes, handles for knives, and drinking veſſels; and when ſoftened by water, obeying the manufacturer's hand, they are formed into pellucid laminæ for the ſides of lanthorns. Theſe laſt conveniences we owe to our great king *Alfred*, who firſt invented them to preſerve his candle time meaſurers, from the wind *; or (as other writers will have it) the tapers that were ſet up before the reliques in the miſerable tattered churches of that time †.

* *Anderſon's hiſt. commerce*, I. 45.
† *Stanley's hiſt. of churches*, 103.

In

In medicine, the horns were employed as alexipharmics or antidotes againſt poiſon, the plague, or the ſmall-pox; they have been dignified with the title of *Engliſh bezoar*; and are ſaid to have been found to anſwer the end of the oriental kind: the chips of the hoofs, and paring of the raw hides, ſerve to make carpenters glue.

The bones are uſed by mechanics, where ivory is too expenſive; by which the common people are ſerved with many neat conveniencies at an eaſy rate. From the *tibia* and *carpus* bones is procured an oil much uſed by coach-makers and others in dreſſing and cleaning harneſs, and all trappings belonging to a coach; and the bones calcined, afford a fit matter for teſts for the uſe of the refiner in the ſmelting trade.

The blood is uſed as an excellent manure for fruit trees *; and is the baſis of that fine color, the *Pruſſian* blue.

The fat, tallow, and ſuet, furniſh us with light; and are alſo uſed to precipitate the ſalt that is drawn from briny ſprings. The gall, liver, ſpleen and urine, have alſo their place in the *materia medica*.

The uſes of butter, cheeſe, creme and milk, in domeſtic œconomy; and the excellence of the latter, in furniſhing a palatable nutriment for moſt people, whoſe organs of digeſtion are weakened, are too obvious to be inſiſted on.

* *Evelyn's* phil. diſc. of earth, p. 319..

Horns

Horns twisted spirally, and pointing outwards. III. SHEEP.
Eight cutting teeth in the lower jaw, none in the upper.

4. FLEECY.

Ovis, *Raii syn. quad.* 73.
Gesn. quad. 71.
Ovis aries, ovis anglica mutica cauda scrotoque ad genua pendulis. *Lin. syst.* 97.
Ovis cornibus compressis lunatis. *Faun. Suec.* 45.
Aries, &c. *Klein. quad.* 13.
Aries laniger cauda rotunda brevi *Brisson quad.* 48.
De Buffon. v. 1. *tab.* 1, 2.
Br. Zool. 10. *Syn. quad.* No. 8.

	MALE.	FEMALE.	LAMB.
Brit.	Hwrd. Maharen	Dafad	Oen
Fren.	Le Belier	La brebis	L'Agneau
Ital.	Montone	Pecora	Agnello
Span.	Carnero	Oveja	Cordero
Port.	Caneiro	Ovelha	Cordeiro
Germ.	Widder	Schaaf	Lamm
Dut.	Ram	Schaep	Lam
Swed.	Wadur	Faar	Lamb
Dan.	Vædder, Være	Faar	Lam, *agna* Gimmer Lam.

IT does not appear from any of the early writers, that the breed of this animal was cultivated for the sake of the wool among the *Britains*; the inhabitants of the inland parts of this island either went entirely naked, or were only clothed with skins. Those who lived on the sea coasts, and were the most civilized, affected the manners of the *Gauls*, and wore like them a sort of garments made of coarse wool, called *Brachæ*. These they probably had from *Gaul*, there not being the lest traces of manufactures

manufactures among the *Britains*, in the histories of those times.

On the coins or money of the *Britains* are seen impressed the figures of the horse, the bull and the hog, the marks of the tributes exacted from them by the conquerors*. The Reverend Mr. *Pegge* was so kind as to inform me that he has seen on the coins of *Cunobelin* that of a sheep. Since that is the case, it is probable that our ancestors were possessed of the animal, but made no farther use of it than to strip off the skin, and wrap themselves in it, and with the wool inmost, obtain a comfortable protection against the cold of the winter season.

This neglect of manufacture, may be easily accounted for, in an uncivilized nation whose wants were few, and those easily satisfied; but what is more surprising, when after a long period we had cultivated a breed of sheep, whose fleeces were superior to those of other countries; we still neglected to promote a woollen manufacture at home. That valuable branch of business lay for a considerable time in foreign hands; and we were obliged to import the cloth manufactured from our own materials. There seems indeed to have been many unavailing efforts made by our monarchs to preserve both the wool and the manufacture of it among ourselves: *Henry* the second, by a patent

* *Cambden*. I. Preface. cxiii.

granted

granted to the weavers in *London*, directed that if any cloth was found made of a mixture of *Spaniſh* wool, it ſhould be burnt by the mayor *: yet ſo little did the weaving buſineſs advance, that *Edward* the third was obliged to permit the importation of foreign cloth in the beginning of his reign; but ſoon after, by encouraging foreign artificers to ſettle in *England*, and inſtruct the natives in their trade, the manufacture increaſed ſo greatly as to enable him to prohibit the wear of foreign cloth. Yet, to ſhew the uncommercial genius of the people, the effects of this prohibition were checked by another law, as prejudicial to trade as the former was ſalutary; this was an act of the ſame reign, againſt exporting woollen goods manufactured at home, under heavy penalties; while the exportation of wool was not only allowed but encouraged. This overſight was not ſoon rectified, for it appears that, on the alliance that *Edward* the fourth made with the king of *Arragon*, he preſented the latter with ſome ewes and rams of the *Coteſwold* kind; which is a proof of their excellency, ſince they were thought acceptable to a monarch, whoſe dominions were ſo noted for the fineneſs of their fleeces †.

In the firſt year of *Richard* the third, and in the two ſucceeding reigns, our woollen manufactures

* *Stow* 419.

† *Rapin* i. 605. in the note. *Stow's Annales*, 696.

received

received ſome improvements*; but the grand riſe of all its proſperity is to be dated from the reign of queen *Elizabeth*, when the tyranny of the duke of *Alva* in the *Netherlands* drove numbers of artificers for refuge into this country, who were the founders of that immenſe manufacture we carry on at preſent. We have ſtrong inducements to be more particular on the modern ſtate of our woollen manufactures; but we deſiſt, from a fear of digreſſing too far; our enquiries muſt be limited to points that have a more immediate reference to the ſtudy of *Zoology*.

No country is better ſupplied with materials, and thoſe adapted to every ſpecies of the clothing buſineſs, than *Great-Britain*; and though the ſheep of theſe iſlands afford fleeces of different degrees of goodneſs, yet there are not any but what may be uſed in ſome branch of it. *Herefordſhire*, *Devonſhire*, and *Coteſwold downs* are noted for producing ſheep with remarkably fine fleeces; the *Lincolnſhire* and *Warwickſhire* kind, which are very large, exceed any for the quantity and goodneſs of their wool. The former county yields the largeſt ſheep in theſe iſlands, where it is no uncommon thing to give fifty guineas for a ram, and a guinea for the admiſſion of a ewe to one of the valuable males;

* In that of *Richard*, two-yard cloths were firſt made. In that of *Henry* the VIII. an *Italian* taught us the uſe of the diſtaff. Kerſies were alſo firſt made in *England* about that time.

or

or twenty guineas for the uſe of it for a certain number of ewes during one ſeaſon. *Suffolk* alſo breeds a very valuable kind. The fleeces of the northern parts of this kingdom are inferior in fineneſs to thoſe of the ſouth; but ſtill are of great value in different branches of our manufactures. The *Yorkſhire* hills furniſh the looms of that county with large quantities of wool; and that which is taken from the neck and ſhoulders, is uſed (mixed with *Spaniſh* wool) in ſome of their fineſt cloths.

Wales yields but a coarſe wool; yet it is of more extenſive uſe than the fineſt *Segovian* fleeces; for rich and poor, age and youth, health and infirmities, all confeſs the univerſal benefit of the flannel manufacture.

The ſheep of *Ireland* vary like thoſe of *Great-Britain*. Thoſe of the ſouth and eaſt being large, and their fleſh rank. Thoſe of the north, and the mountainous parts ſmall, and their fleſh ſweet. The fleeces in the ſame manner differ in degrees of value.

Scotland breeds a ſmall kind, and their fleeces are coarſe. *Sibbald* (after *Boethius*) ſpeaks of a breed in the iſle of *Rona*, covered with blue wool; of another kind in the iſle of *Hirta*, larger than the biggeſt he goat, with tails hanging almoſt to the ground, and horns as thick, and longer than thoſe of an ox*. He mentions another kind, which is clothed

* *Gmelin* deſcribes an animal he found in *Siberia*, that in many particulars agrees with this; he calls it *Rupicapra cornubus*

clothed with a mixture of wool and hair; and a fourth ſpecies, whoſe fleſh and fleeces are yellow, and their teeth of the colour of gold; but the truth of theſe relations ought to be enquired into, as no other writer has mentioned them, except the credulous *Boethius*. Yet the laſt particular is not to be rejected: for notwithſtanding I cannot inſtance the teeth of ſheep, yet I ſaw in the ſummer of 1772, at *Athol* houſe, the jaws of an ox, with teeth thickly incruſted with a gold colored *pyrites*; and the ſame might have happened to thoſe of ſheep had they fed in the ſame grounds, which were in the valley beneath the houſe.

Beſides the fleece, there is ſcarce any part of this animal but what is uſeful to mankind. The fleſh is a delicate and wholeſome food. The ſkin dreſſed, forms different parts of our apparel; and is uſed for covers of books. The entrails, properly prepared and twiſted, ſerve for ſtrings for various mu-

bus arietinis; *Linnæus* ſtyles it *Capra ammon. Syſt.* 97. and *Geſner*, p. 934. imagines it to be the *Muſimon* of the antients; the horns of the *Siberian* animal are two yards long, their weight above thirty pounds. As we have ſo good authority for the exiſtence of ſuch a quadruped, we might venture to give credit to *Boethius*'s account, that the ſame kind was once found in *Hirta*: but having thrice within theſe few years had opportunity of examining the *Muſimon*, we found that both in the form of the horns, and the ſhortneſs of the tail, it had the greateſt agreement with the goat, in which *genus* we have placed it No. 11. of our *Synopſis*, with the trivial name of *Siberian*.

ſical

ſical inſtruments. The bones calcined (like other bones in general) form materials for teſts for the refiner. The milk is thicker than that of cows; and conſequently yields a greater quantity of butter and cheeſe; and in ſome places is ſo rich, that it will not produce the cheeſe without a mixture of water to make it part from the whey. The dung is a remarkably rich manure; inſomuch that the folding of ſheep is become too uſeful a branch of huſbandry for the farmer to neglect. To conclude, whether we conſider the advantages that reſult from this animal to individuals in particular, or to theſe kingdoms in general, we may with *Columella* conſider this in one ſenſe, as the firſt of the domeſtic animals. *Poſt majores quadrupedes ovilli pecoris ſecunda ratio eſt; quæ prima ſit ſi ad utilitatis magnitudinem referas. Nam id præcipue contra frigoris violentiam protegit, corporibuſque noſtris liberaliora præbet velamina; et etiam elegantium menſas jucundis et numeroſis dapibus exornat* *.

The ſheep as to its nature, is a moſt innocent mild and ſimple animal; and conſcious of its own defenceleſs ſtate, remarkably timid: if attacked when attended by its lamb, it will make ſome ſhew of defence, by ſtamping with its feet, and puſhing with its head: it is a gregarious animal, is fond of any jingling noiſe, for which reaſon the

* *De re ruſtica, lib.* vii. *c.* 2.

leader of the flock has in many places a bell hung round its neck, which the others will conſtantly follow: it is ſubject to many diſeaſes: ſome ariſe from inſects which depoſite their eggs in different parts of the animal; others are cauſed by their being kept in wet paſtures; for as the ſheep requires but little drink, it is naturally fond of a dry ſoil. The dropſy, vertigo (the *pendro* of the Welſh) the pthiſick, jaundice, and worms in the liver * annually make great havoke among our flocks: for the firſt diſeaſe, the ſhepherd finds a remedy by turning the infected into fields of broom; which plant has been alſo found to be very efficacious in the ſame diſorder among the human ſpecies.

The ſheep is alſo infeſted by different ſorts of inſects: like the horſe it has its peculiar *Oeſtrus* or *Gadfly*, which depoſits its eggs above the noſe in the frontal ſinuſes; when thoſe turn into maggots they become exceſſive painful, and cauſe thoſe violent agitations that we ſo often ſee the animal in. The *French* ſhepherds make a common practice of eaſing the ſheep, by trepanning and taking out the maggot; this practice is ſometimes uſed by the *Engliſh* ſhepherds, but not always with the ſame ſucceſs: beſides theſe inſects, the ſheep is troubled with a kind of tick and louſe, which magpies and ſtarlings contribute to eaſe it of, by lighting on its back, and picking the inſects off.

* Faſciola hepatica, *Lin. ſyſt.* 648.

Horns

Pl. III
Nº
GOAT.
M Griffiths del

Horns bending backwards and almoſt cloſe at their baſe. IV. GOAT.

Eight cutting teeth in the upper jaw, none in the lower.

Male generally bearded.

5. DOMESTIC.

Raii ſyn. quad. 77.
Meyer's an. i. *Tab.* 68.
Charlton ex. 9.
Klein quad. 15.
Geſn. quad. 266. 268.
De Buffon. v. 59. *Tab.* 8. 9.

Hircus cornibus interius cultratis, exterius rotundatis, infra carinatis, arcuatis. *Briſſon quad.* 38.
Capra Hircus, *Lin ſyſt.* 94.
Capra cornibus carinatis arcuatis, *Faun. Suec.* 44.
Br. Zool. 13. *Syn. quad* p. 14.

	MALE.	FEMALE.	KID.
Brit.	Bwch	Gafr	Mynn
Fren.	Le Bouc	La Chevre	Chevreau
Ital.	Becco	Capra	Capretto
Span.	Cabron	Cabra	Cabrito
Port.	Cabram	Cabra	Cabrito
Germ.	Bock	Geiſz	Bocklein
Dut.	Bok	Giyt	
Swed.	Bock	Geet	Kiidh
Dan.	Buk, Geedebuk	Geed	Kid

THE goat is the moſt local of any of our domeſtic animals, confining itſelf to the mountanous parts of theſe iſlands: his moſt beloved food is the tops of the boughs, or the tender bark of young trees; on which account he is ſo prejudicial to plantations, that it would be imprudent to draw him from his native rocks, except ſome method could be thought on to obviate this evil.

evil. We have been informed, that there is a freeholder in the pariſh of *Trawſvynnyd*, in *Merionethſhire*, who hath, for ſeveral years paſt, broke the teeth of his goats ſhort off with a pair of pincers, to preſerve his trees. This practice has certainly efficacy ſufficient to prevent the miſchief, and may be recommended to thoſe who keep them for their ſingularity; but ought by no means to be encouraged, when thoſe animals are preſerved for the ſake of their milk, as the great ſalubrity of that medicine ariſes from their promiſcuous feeding.

This quadruped contributes in various inſtances to the neceſſities of human life; as food, as phyſick, and as cloathing: the whiteſt wigs are made of its hair; for which purpoſe that of the he-goat is moſt in requeſt; the whiteſt and cleareſt is ſelected from that which grows on the haunches, where it is longeſt and thickeſt; a good ſkin well haired is ſold for a guinea; though a ſkin of bad hue, and ſo yellow as to baffle the barber's ſkill to bleach, will not fetch above eighteen-pence, or two ſhillings.

The *Welch* goats are far ſuperior in ſize, and in length and fineneſs of hair, to thoſe of other mountanous countries. Their uſual color is white: thoſe of *France* and the *Alps* are ſhort-haired, reddiſh, and their horns ſmall. We have ſeen the horns of a *Cambrian* he-goat three feet two inches long, and three feet from tip to tip.

The

The ſuet of the goat is in great eſteem, as well as the hair. Many of the inhabitants of *Caernarvonſhire* ſuffer theſe animals to run wild on the rocks during winter as well as ſummer; and kill them in *October*, for the ſake of their fat, either by ſhooting them with bullets, or running them down with dogs like deer. The goats killed for this purpoſe, are about four or five years old. Their ſuet will make candles, far ſuperior in whiteneſs and goodneſs to thoſe made from that of the ſheep or the ox, and accordingly brings a much greater price in the market: nor are the horns without their uſe, the country people making of them excellent handles for tucks and penknives. The ſkin is peculiarly well adapted for the glove manufactory, eſpecially that of the kid: abroad it is dreſſed and made into ſtockings, bed-ticks, bolſters *, bed-hangings, ſheets, and even ſhirts. In the army it covers the horſeman's arms, and carries the foot-ſoldier's proviſions. As it takes a dye better than any other ſkin, it was formerly much uſed for hangings in the houſes of people of fortune, being ſuſceptible of the richeſt colors; and when flowered and ornamented with gold and ſilver, became an elegant and ſuperb furniture.

* Bolſters made of the hair of a goat were in uſe in the days of *Saul*; as appears from I. *Samuel*, c. 19. v. 13. The ſpecies very probably was the *Angora* goat, which is only found in the *Eaſt*, and whoſe ſoft and ſilky hair ſupplied a moſt luxurious couch. *Vide Syn. quad.* p. 15.

The flesh is of great use to the inhabitants of the country where it resides; and affords them a cheap and plentiful provision in the winter months, when the kids are brought to market. The haunches of the goat are frequently salted and dried, and supply all the uses of bacon: this by the natives is called *Coch yr wden*, or hung venison.

The meat of a splayed goat of six or seven years old, (which is called *Hyfr)* is reckoned the best; being generally very sweet and fat. This makes an excellent pasty; goes under the name of rock venison, and is little inferior to that of the deer. Thus nature provides even on the tops of high and craggy mountains, not only necessaries, but delicacies for the inhabitants.

The milk of the goat is sweet, nourishing and medicinal: it is an excellent succedaneum for ass's milk; and has (with a tea-spoon ful of hartshorn drank warm in bed in the morning, and at four o'clock in the afternoon, and repeated for some time) been a cure for pthisical people, before they were gone too far. In some of the mountanous parts of *Scotland* and *Ireland*, the milk is made into whey; which has done wonders in this and other cases, where coolers and restoratives are necessary: and to many of those places, there is a great resort of patients of all ranks, as there is in *England* to the *Spaws* or *Baths*. It is not surprizing that the milk of this animal is so salutary, as it brouzes only on the tops, tendrils andflowers of the mountain

tain ſhrubs, and medicinal herbs; rejecting the groſſer parts. The blood of the he-goat dried, is a great recipe in ſome families for the pleuriſy and inflammatory diſorders *.

Cheeſe made of goats milk, is much valued in ſome of our mountanous countries, when kept to proper age; but has a peculiar taſte and flavor.

The rutting ſeaſon of theſe animals, is from the beginning of *September* to *November*; at that time the males drive whole flocks of the females continually from place to place, and fill the whole atmoſphere around them with their ſtrong and ungrateful odor; which though as diſagreeable as aſſa fœtida itſelf, yet may be conducive to prevent many diſtempers, and to cure nervous and hyſterical ones. Horſes are imagined to be much refreſhed with it; on which account many perſons keep a he-goat in their ſtuds or ſtables.

Goats go with young four months and a half, and bring forth from the latter end of *February* to the latter end of *April*: Having only two teats, they bear generally but two young, and ſometimes three; and in good warm paſtures there have been inſtances, though rare, of their bringing four at a time: both young and old are affected by the weather: a rainy ſeaſon makes them thin; a dry ſunny one makes them fat and blythe: their ex-

* This remedy is taken notice of even by Dr. *Mead* in his *monita medica*, p. 35. under the article *pleuritis*. The *Germans* uſe that of the *Stein-boc*, or *Ibex*.

D 4 ceſſive

ceſſive venery prevents longævity, for they ſeldom live in our climate above eleven or twelve years.

Theſe animals with amazing ſwiftneſs and ſafety, climb up the moſt rugged rocks, and aſcend the moſt dangerous places: they can ſtand unmoved on the higheſt precipices, and ſo balance their centre of gravity, as to fix themſelves in ſuch ſituations with ſecurity and firmneſs; ſo that we ſeldom hear of their falling, or breaking their necks. When two are yoked together, as is frequently practiſed, they will, as if by conſent, take large and hazardous leaps; yet ſo well time their mutual efforts, as rarely to miſcarry in the attempt.

The origin of the domeſtic goat is the *Steinboc*, *Ibex* or wild goat, *Syn. quad.* No. 9. a ſpecies now found only in the *Alps*, and in *Crete*.

Horns

Horns upright, ſolid, branched, annually deciduous. Eight cutting teeth in the lower jaw, none in the upper.

V. DEER.

6. STAG.

Red Deer, Stag or Hart.
Cervus *Raii ſyn. quad.* 84.
Charlt. ex. 11.
Meyer's an. Tab. 22.
Geſner quad. 326.
Grew's Muſeum, 21.
De Buffon, Tom. vi. 63. Tab. 9, 10.

Cervus cornibus teretibus ad latera incurvis. *Briſſon quad.* 58.
Cervus Elaphus. *Lin. ſyſt.* 93.
C. cornibus ramoſis teretibus recurvatis. *Faun. Suec.* 40.
C. nobilis. *Klein. quad.* 23.
Br. Zool. 15. *Syn. quad.* No. 38.

	STAG.	HIND.	YOUNG, or CALF.
Brit.	Carw	Ewig	Elain
Fren.	Le Cerf	La Biche	Faon
Ital.	Cervio	Cervia	
Span.	Ciervo	Cierva	
Port.	Cervo	Cerva	
Germ.	Hirtz, Hirſch	Hind	Hinde kalb
Dutch,	Hart	Hinde	
Swed.	Hiort, Kronhiort	Hind	
Dan.	Kronhiort	Hind	Kid, or Hind kalv

7. FALLOW.

Platycerata. *Plinii,* lib. xi. c. 37.
Eurycerata. *Oppian Cyneg.* lib. 11. lin. 293.
Fallow deer, or buck; cervus platyceros. *Raii ſyn. quad.* 85.
Dama vulgaris. *Geſner quad.* 307.
Meyer's an. Tom. i. Tab. 71.
De Buffon. Tom. vi. 161. Tab. 27, 28.

Cervus cornuum unica et altiore ſummitate palmata. *Briſſon quad.* 62.
Cervus dama. Cervus cornibus ramoſis recurvatis compreſſis: ſummitatibus palmatis. *Lin. ſyſt.* 93.
Faun. Suec. 42. *Br. Zool.* 15. *Syn. quad.* No. 37.
Cervus palmatus. *Klein. quad.* 25.

Brit.

	BUCK.	DOE.	FAWN.
Brit.	Hydd	Hyddes	Elain
Fren.	Le Dain	La Daine	Faon
Ital.	Daino		Cerbiatto
Span.	Gamo, Corza		Venadito
Port.	Corza		Veado
Germ.	Damhirſch		
Swed.	Dof, Dof hiort		
Dan.	Daae Dijr		

AT firſt, the beaſts of chace had this whole iſland for their range; they knew no other limits than that of the ocean; nor confeſſed any particular maſter. When the *Saxons* had eſtabliſhed themſelves in the *Heptarchy*, they were reſerved by each ſovereign for his own particular diverſion: hunting and war in thoſe uncivilized ages were the only employ of the great; their active, but uncultivated minds, being ſuſceptible of no pleaſures but thoſe of a violent kind, ſuch as gave exerciſe to their bodies, and prevented the pain of thinking.

But as the *Saxon* kings only appropriated thoſe lands to the uſe of foreſts which were unoccupied; ſo no individuals received any injury: but when the conqueſt had ſettled the *Norman* line on the throne, this paſſion for the chace was carried to an exceſs, which involved every civil right in a general ruin: it ſuperſeded the conſideration of religion even in a ſuperſtitious age: the village communities, nay, even the moſt ſacred edifices were turned into one vaſt waſte, to make room for animals, the objects of a lawleſs tyrant's pleaſure. The new

new foreſt in *Hampſhire* is too trite an inſtance to be dwelt on : ſanguinary laws were enacted to preſerve the game ; and in the reigns of *William Rufus*, and *Henry* the firſt, it was leſs criminal to deſtroy one of the human ſpecies than a beaſt of chaſe*. Thus it continued while the *Norman* line filled the throne ; but when the *Saxon* line was reſtored under *Henry* the ſecond, the rigor of the foreſt laws was immediately ſoftened.

When our barons began to form a power, they clamed a vaſt, but more limited tract for a diverſion that the *Engliſh* were always fond of. They were very jealous of any encroachments on their reſpective bounds, which were often the cauſe of deadly feuds : ſuch a one gave cauſe to the fatal day of *Chevy-chace*, a fact, which though recorded only in a ballad, may, from what we know of the manners of the times, be founded on truth ; not that it was attended with all the circumſtances the author of that natural, but heroic compoſition hath given it, for on that day neither a *Percy* nor a *Douglas* fell : here the poet ſeems to have clamed his privilege, and mixed with this fray ſome of the events of the battle of *Otterbourne*.

When property became happily more divided by the relaxation of the feodal tenures, theſe extenſive

* An antient hiſtorian ſpeaks thus of the penalties incurred ; *Si cervum aut aprum oculos eis evellebat ; amavit enim feras tanquam erat pater earum. M. Paris.* 11.

hunting

hunting-grounds became more limited; and as tillage and husbandry increased, the beasts of chace were obliged to give way to others more useful to the community. The vast tracts of land before dedicated to hunting, were then contracted; and in proportion as the useful arts gained ground, either lost their original destination, or gave rise to the invention of *Parks*. Liberty and the arts seem coeval, for when once the latter got footing, the former protected the labors of the industrious from being ruined by the licentiousness of the sportsman, or being devoured by the objects of his diversion: for this reason, the subjects of a despotic government still experience the inconveniences of vast wastes, and forests, the terrors of the neighbouring husbandmen *; while in our well-regulated monarchy, very few chaces remain: we still indulge ourselves in the generous pleasure of hunting, but confine the deer-kind to parks, of which *England* boasts of more than any other kingdom in *Europe*. Our equal laws allow every man his pleasure; but confine them in such bounds, as prevents them from being injurious to the meanest of the community. Before the reformation, our prelates seem to have guarded sufficiently against the want of this amusement, the see of *Norwich* in particular, being pos-

* In *Germany* the peasants are often obliged to watch their grounds the whole night, to preserve the fences and corn from being destroyed by the deer.

sessed

ſeſſed about that time of thirteen parks*. They ſeem to have forgot good king *Edgar*'s advice, *Docemus etiam ut ſacerdos non ſit venator neque accipitrarius neque potator, ſed incumbat ſuis libris ſicut ordinem ipſius decet* †.

It was cuſtomary to ſalt the veniſon for preſervation, like other meat. *Rymer* preſerves a warrant of *Edward* III. ordering ſixty deer to be killed for that purpoſe.

The ſtag and buck agree in their nature; only the latter being more tender is eaſier tamed, and made familiar. The firſt is become leſs common than it was formerly; its exceſſive vitiouſneſs during the rutting ſeaſon, and the badneſs of its fleſh, induce moſt people to part with the ſpecies. Stags are ſtill found wild in the highlands of *Scotland*, in herds of four or five hundred together, ranging at full liberty over the vaſt hills of the north. Some grow to a great ſize: when I was at *Invercauld* Mr. *Farquharſon* aſſured me that he knew an inſtance of one that weighed eighteen ſtone *Scots*, or three hundred and fourteen pounds, excluſive of the entrails, head and ſkin. Formerly the great highland chieftains uſed to hunt with the magnificence of an eaſtern monarch; aſſembling four or five thouſand of their clan, who drove the deer into the toils, or to the ſtation their lairds had pla-

* *Peacham's Compleat Gentleman*, 261. † *Leges Saxon.* 87.

ced

ced themſelves in : but as this pretence was frequently uſed to collect their vaſſals for rebellious purpoſes, an act was paſſed prohibiting any aſſemblies of this nature. Stags are likewiſe met with on the moors that border on *Cornwal* and *Devonſhire*, and in *Ireland* on the mountains of *Kerry*, where they add greatly to the magnificence of the romantic ſcenery of the lake of *Killarny*.

The ſtags of *Ireland* during its uncultivated ſtate, and while it remained an almoſt boundleſs tract of foreſt, had an exact agreement in habit, with thoſe that range at preſent through the wilds of *America*. They were leſs in body, but very fat; and their horns of a ſize far ſuperior to thoſe of *Europe*, but in form agreed in all points. Old *Giraldus* ſpeaks with much preciſion of thoſe of *Ireland*, *Cervos præ nimia pinguedine minus fugere prævalentes, quanto minores ſunt corporis quantitate, tanto præcellentius efferuntur, capitis et cornuum dignitate* *.

We have in *England* two varieties of fallow-deer which are ſaid to be of foreign origin: The beautiful ſpotted kind, and the very deep brown ſort, that are now ſo common in ſeveral parts of this kingdom. Theſe were introduced here by king

* *Topogr. Hiberniæ*. c. 19. *Lawſon* in his hiſtory of *Carolina* p. 123, mentions the fatneſs of the *American* ſtags, and their inferiority of ſize to the *European*. I have often ſeen their horns, which vaſtly exceed thoſe of our country in ſize, and number of antlers.

James

James the *firſt* out of *Norway* *, where he paſſed ſome time when he viſited his intended bride *Mary* of *Denmark* †. He obſerved their hardineſs; and that they could endure, even in that ſevere climate, the winter without fodder. He firſt brought ſome into *Scotland*, and from thence tranſported them into his chaces of *Enfield* and *Epping*, to be near his palace of *Theobalds*; for it is well known, that monarch was in one part of his character the *Nimrod* of his days, fond to exceſs of hunting, that image of war, although he deteſted the reality. No country produces the fallow-deer in quantities equal to *England*. In *France* they are ſcarcely known, but are ſometimes found in the north ‡ of *Europe*. In *Spain* they are extremely large. They are met with in *Greece*, the *Holy Land* ||, and in *China* §; but in every country except our own are in a ſtate of nature, unconfined by man.

They are not natives of *America*; for the deer known in our colonies by that name are a diſtinct ſpecies, a ſort of ſtag, as we have remarked p. 51. of our *Synopſis* of quadrupeds.

The uſes of theſe animals are almoſt ſimilar; the ſkin of the buck and doe is ſufficiently known to

* This we relate on the authority of Mr. *Peter Collinſon*.

† One of the *Welch* names of this animal (*Gievr-danas*, or *Daniſh* goat) implies that it was brought from ſome of the *Daniſh* dominions. *Ed. Llwyd. Ph. tr.* No. 334.

‡ *Pontop. Norway.* II. 9. *Faun. Suec.* ſp. 42.

|| *Haſſelquiſt. itin.* 290. § *Du Halde* hiſt. *China.* I. 315.

every

every one; and the horns of the ſtag are of great uſe in mechanics; they, as well as the horns of the reſt of the deer kind, being exceſſively compact, ſolid, hard and weighty; and make excellent handles for couteaus, knives, and ſeveral other utenſils. They abound in that ſalt, which is the baſis of the ſpirit of *Hartſhorn*; and the remains (after the ſalts are extracted) being calcined, become a valuable aſtringent in fluxes, which is known by the name of burnt *Hartſhorn*. Beſides theſe uſes in mechanics and medicine, there is an inſtance in *Giraldus Cambrenſis*, of a counteſs of *Cheſter*, who kept milch hindes, and made cheeſe of their milk, ſome of which ſhe preſented to archbiſhop *Baldwin*, in his itinerary through *Wales*, in the year 1188 *.

* *Girald. Camb. Itin.* p. 216.

Δορκας

Pl. IV.

M. Griffith pinx.

P. Maze

8. ROE.

Δορκας, *Aristotelis* de Part. lib. iii. c. 2.
Iorcas, Dorcas, *Oppian* Cyneg. lib. ii. lin. 296. 315.
Caprea, *Plinii*, lib. xi. c. 37.
Capreolus Vulgo. *Raii syn. quad.* 89.
Camd. Brit. ii. 771.
Meyer's anim. ii. Tab. 73.
Capreolus, *Sib. Scot. pars* 3. 9.
Caprea, capreolus, Dorcas. *Gesner' quad.* 296.
Merret pinax. 166.

Cervus cornibus teretibus, erectis. *Brisson quad.* 61.
De Buffon, Tom. vi. 289. *Tab.* 32, 33.
Cervus minimus, *Klein quad.* 24.
Cervus capreolus, *Lin. syst.* 94.
C. Cornibus ramosis teretibus erectis, summitate bifida, *Faun. Suec.* 43. *Br. Zool.* 18. *Syn. quad.* No. 43. *Tour in Scotland.* 288 *Tab.* xiv.

Brit. Iwrch, *fæm.* Iyrchell
Fren. Le Chevreuil
Ital. Capriolo
Span. Zorlito, Cabronzillo montes
Port. Cabra montes
Ger. Rehbock, *fæm.* Rehgeefs
Swed. Radiur, Rabock
Dan. Raaedijr Raaebuk

THE roebuck prefers a mountanous woody country to a plain one; was formerly very common in *Wales*, in the north of *England*, and in *Scotland*; but at prefent the fpecies no longer exifts in any part of *Great-Britain*, except in the *Scottifh* highlands. In *France* they are more frequent; they are alfo found in *Italy*, *Sweden*, and *Norway*; and in *Afia* they are met with in *Siberia**. The firft that are met with in *Great-Britain* are in the woods on the fouth fide of Loch *Rannoch*, in *Perthfhire*: the laft in thofe of *Langwal*, on the

* *Bell*'s Travels.

ſouthern borders of *Cathneſs*: but they are moſt numerous in the beautifull foreſts of *Invercauld*, in the midſt of the *Grampian* hills. They are unknown in *Ireland*.

This is the leſt of the deer kind, being only three feet nine inches long, and two feet three inches high before, and two feet ſeven behind. The weight from 50 to 60 lb. The horns are from eight to nine inches long, upright, round, and divided into only three branches; their lower part is ſulcated lengthways, and extremely rugged; of this part is made handles for couteaus, knives, *&c.* The horns of a young buck in its ſecond year are quite plain: in its third year a branch appears; but in the fourth its head is complete. The body is covered during winter with very long hair, well adapted to the rigor of the highland air; the lower part of each hair is aſh-color; near the ends is a narrow bar of black, and the points are yellow: The hairs on the face are black, tipped with aſh-color; the ears are long, their inſides of a pale yellow, and covered with long hair; the ſpaces bordering on the eyes and mouth are black. During ſummer its coat has a very different appearance, being very ſhort and ſmooth, and of a bright reddiſh color.

The cheſt, belly, and legs, and the inſide of the thighs, are of a yellowiſh white; the rump is of a pure white: the tail is very ſhort. On the outſide of of the hind leg, below the joint, is a tuft of long hair.

The make of the roebuck is very elegant, and formed for agility. Thefe animals do not keep in herds like other deer, but only in families; they bring two fawns at a time, which the female is obliged to conceal from the buck while they are very young. The flefh of this creature is reckoned a delicate food.

It is a tender animal, incapable of bearing great cold. *M. de Buffon* tells us that in the hard winter of 1709, the fpecies in *Burgundy* were almoft deftroyed, and many years paft before it was reftored again. I was informed in *Scotland*, that it is very difficult to rear the fawns; it being computed that eight out of ten of thofe that are taken from their parents die.

Wild roes during fummer feed on grafs, and are very fond of the *rubus faxatilis*, called in the highlands the *roebuck* berry; but in winter time, when the ground is covered with fnow, they brouze on the tender branches of fir and birch.

In the old *Welfh* laws, a roebuck was valued at the fame price as a fhe-goat; a ftag at the price of an ox; and a fallow deer was efteemed equal to that of a cow; or, as fome fay, a he-goat *.

FOSSIL HORNS.

It will not be foreign to the prefent fubject, to mention the vaft horns frequently found in *Ireland*,

* *Leges Wallicæ*, 258.

E 2 and

and others ſometimes met with in our own kingdom. The latter are evidently of the ſtag kind, but much ſtronger, thicker, heavier, and furniſhed with fewer antlers than thoſe of the preſent race; of thoſe ſome have been found on the ſea-coaſt of *Lancaſhire* *, and a ſingle horn was dug a few years ago out of the ſands near *Cheſter*. Thoſe found in † *Ireland* muſt be referred to the elk kind, but of a ſpecies different from the *European*, being provided with brow antlers which that wants: neither are they of the *Mooſe* deer or *American*, which entirely agrees with the elk of *Europe*, as I have found by compariſon. Entire ſkeletons of this animal are ſometimes met with, lodged in a white marle. Some of theſe horns are near twelve feet between tip and tip ‡. Not the fainteſt account (traditional or hiſtoric) is left of the exiſtence of theſe animals in our kingdom; ſo that they may poſſibly be ranked among thoſe remains which foſſiliſts diſtinguiſh by the title of diluvian.

Mr. *Graham*, factor to the *Hudſon*'s *Bay company*, once gave me hopes of diſcovering the living animal. He informed me that he had received

* *Ph. Tr.* No. 422.

† No. 227. *Boate's Nat. Hiſt. Ireland*, 137.

‡ A pair of this ſize is preſerved at Sir *Patrick Bellew's*, Bart. in the county of *Louth*. The great difference between the Mooſe horns and the Foſſil is ſhewen in Plates VII. and IX. of my *Synopſis of Quadrupeds*.

accounts

accounts from the *Indians* who reſort to the facto-ries, that there is found a deer, about ſeven or eight hundred miles weſt of *York fort*, which they call *Waſkeſſeu*, and ſay is vaſtly ſuperior in ſize to the common *Mooſe*. But as yet nothing has tranſpi-red relating to ſo magnificent an animal. The dif-ference of ſize between the modern *Mooſe*, and the owners of the foſſil horns may be eſtimated by the following account. The largeſt horns of the *Ame-rican Mooſe* ever brought over, are only thirty-two inches long, and thirty-four between tip and tip. The length of one of the foſſil horns is ſix feet four inches. The ſpace between tip and tip near twelve feet. The largeſt *Mooſe* deſcribed by any authentic voyager does not exceed the ſize of a great horſe; that which I ſaw (a female) was fif-teen hands high. But we muſt ſearch for much larger animals to ſupport the weight of our foſſil horns. If *Joſſelyn*'s or *Dudly*'s *Mooſe* of twelve feet in height ever exiſted *, we may ſuppoſe that to have been a ſpecies, which as population advanc-ed, retired into diſtant parts, into depths of woods unknown but to diſtant *Indians*.

* Voy. to *New England*, 88. *New England* Rarities, 19. See alſo Mr. *Dudly*'s account in Ph. Tranſ. abridg. VII. 447.

E 3 ** Without

** Without horns.

VI. HOG. Divided hoofs.
Cutting teeth in both jaws.

9. COMMON. Sus, seu Porcus domesticus. *Raii syn. quad.* 92.
Gesner quad. 872.
Charlton ex. 14.
Sus caudatus auriculis oblongis acutis, cauda pilosa. *Brisson quad.* 74.
De Buffon, Tom. V. 99. Tab. 6, 7.
Klein quad. 25.
Sus scrofa. *Lin. syst.* 102.
Sus dorso antice setoso, cauda pilosa.
Faun. Suec. 21.
Br. Zool. 19. *Syn. quad.* No. 54.

	BOAR.	SOW.	HOG.
Brit.	Baedd	Hwch	Mochyn
Fren.	Le Verrat	La Truye	Porc
Ital.	Verro	Porca	Porco
Span.	Berraco	Puerca	Puerco
Port.		Porca	Porco
Germ.	Eber	Sau	Barg
Dut.	Beer	Soch	Varken
Swed.		Swiin	
Dan.	Orne	Soë	

ACCORDING to common appearances, the hog is certainly the most impure and filthy of all quadrupeds: we should however reflect that filthiness is an idea merely relative to ourselves; but we form a partial judgment from our own sensations, and overlook that wise maxim of Providence, that every part of the creation should have its respective inhabitants. By this œconomy of nature, the earth is never overstocked, nor any part of of the creation ufelefs. This obfervation may be exemplified in the animal before us; the hog alone devouring what is the refufe of all the reft, and contributing not only to remove what would be a nuifance to the human race, but alfo converting the moft naufeous offals into the richeft nutriment: for this reafon its ftomach is capacious, and its gluttony exceffive; not that its palate is infenfible to the difference of eatables; for where it finds variety, it will reject the worft with as diftinguifhing a tafte as other quadrupeds *.

This animal has (not unaptly) been compared to a mifer, who is ufelefs and rapacious in his life, but on his death becomes of public ufe, by the very effects of his fordid manners. The hog during life renders little fervice to mankind, except in removing that filth which other animals reject: his more than common brutality, urges him to devour even his own off-fpring. All other domeftic quadrupeds fhew fome degree of refpect to mankind; and even a fort of tendernefs for us in our

* The ingenious author of the *Pan Suecus*, has proved this beyond contradiction, having with great induftry drawn up tables of the number of vegetables, which each domeftic animal chufes, or rejects: and it is found that the hog eats but 72, and refufes 171 plants,

	eats	rejects	
The Ox	276.	218.	
Goat	449.	126.	
Sheep	387.	141.	
Horfe	262.	212.	*Amæn. Acad.* ii. 203.

helplefs years; but this animal will devour infants, whenever it has opportunity.

The parts of this animal are finely adapted to its way of life. As its method of feeding is by turning up the earth with its nofe for roots of different kinds; fo nature has given it a more prone form than other animals; a ftrong brawny neck; eyes fmall, and placed high in the head; a long fnout, nofe callous and tough, and a quick fenfe of fmelling to trace out its food. Its inteftines have a ftrong refemblance to thofe of the human fpecies; a circumftance that fhould mortify our pride. The external form of its body is very unweildy; yet, by the ftrength of its tendons, the wild boar (which is only a variety of the common kind) is enabled to fly from the hunters with amazing agility: the back toe on the feet of this animal prevents its flipping while it defcends declivities, and muft be of fingular ufe when purfued: yet, notwithftanding its powers of motion, it is by nature ftupid, inactive, and drowfy; much inclined to increafe in fat, which is difpofed in a different manner from other animals, and forms a regular coat over the whole body. It is reftlefs at a change of weather, and in certain high winds is fo agitated as to run violently, fcreaming horribly at the fame time: it is fond of wallowing in the dirt, either to cool its furfeited body, or to deftroy the lice, ticks, and other infects with which it is infefted. Its difeafes generally arife from intemperance; meafles, impof-

tumes,

tumes, and ſcrophulous complaints are reckoned among them. *Linnæus* obſerves that its fleſh is wholeſome food for athletic conſtitutions, or thoſe that uſe much exerciſe; but bad for ſuch as lead a ſedentary life: it is though of moſt univerſal uſe, and furniſhes numberleſs materials for epicuriſm, among which brawn is a kind peculiar to *England**. The fleſh of the hog is an article of the firſt importance to a naval and commercial nation, for it takes ſalt better than any other kind, and conſequently is capable of being preſerved longer. The lard is of great uſe in medicine, being an ingredient in various ſorts of plaiſters, either pure, or in the form of pomatum; and the briſtles are formed into bruſhes of ſeveral kinds.

This animal has been applied to an uſe in this iſland, which ſeems peculiar to *Minorca* and the part of *Murray* which lies between the *Spey* and *Elgin*. It has been there converted into a beaſt of draught; for I have been aſſured by a miniſter of that country, eye witneſs to the fact, that he had on his firſt coming into his pariſh ſeen a cow, a ſow, and two *Trogues* (young horſes) yoked together, and drawing a plough in a light ſandy ſoil; and that the ſow was the beſt drawer of the four. In *Minorca* the aſs and the hog are common help-mates, and are yoked together in order to turn up the land.

The wild-boar was formerly a native of our coun- try,

* *Hollingſhed Deſcr. Brit.* 109.

try, as appears from the laws of *Hoel dda* *, who permitted his grand huntſman to chace that animal from the middle of *November* to thebeginning of *December*. *William* the Conqueror puniſhed with the loſs of their eyes, any that were convicted of killing the wild-boar, the ſtag, or the roebuck †; and *Fitz-Stephen* tells us, that the vaſt foreſt that in his time grew on the north ſide of *London*, was the retreat of ſtags, fallow-deer, wild-boars, and bulls. *Charles* I. turned out wild-boars in the *New Foreſt*, *Hampſhire*, but they were deſtroyed in the civil wars.

* *Leges Wallicæ*. 41. † *Leges Saxon*. 292.

DIV.

DIV. II. SECT. I.

DIGITATED QUADRUPEDS,

With large canine teeth, ſeparated from the cutting teeth.
Six cutting teeth in each jaw.
Rapacious, carnivorous.

VII. DOG.

Six cutting teeth, and two canine.
Five toes before, four behind.
Blunt claws. Long viſage.

10. FAITHFULL.

Canis, *Raii ſyn. quad.* 175.
Charlton ex. 26.
Merret pinax, 168.
Geſner quad. 160, 249, 250.
Canis domeſticus. *Briſſon quad.* 170.
De Buffon, Tom. v. p. 185.
Klein quad. 63.
Canis familiaris. *Lin. ſyſt.* 56.
Canis cauda recurva. *Faun. Suec.* 5.
Brit. Zool. 23. *Syn. quad.* No. 110.

Brit. Ci, *fæm.* Gaſt
Fren. Le Chien
Ital. Cane
Span. Perro
Port. Cam
Germ. Hund
Dut. Hond
Swed. Hund
Dan. Hund, *fæm.* Tæve

DR. *Caius*, an *Engliſh* phyſician, who flouriſhed in the reign of queen *Elizabeth*, has left among ſeveral other tracts relating to natural hiſtory, one written expreſsly on the ſpecies of *Britiſh* dogs: they were wrote for the uſe of his learned ed friend *Geſner*; with whom he kept a ſtrict correſpondence; and whoſe death he laments in a very elegant and pathetic manner.

Beſides a brief account of the variety of dogs then exiſting in this country, he has added a ſyſtematic table of them: his method is ſo judicious, that we ſhall make uſe of the ſame; explain it by a brief account of each kind; and point out thoſe that are no longer in uſe among us.

SYNOPSIS OF BRITISH DOGS.

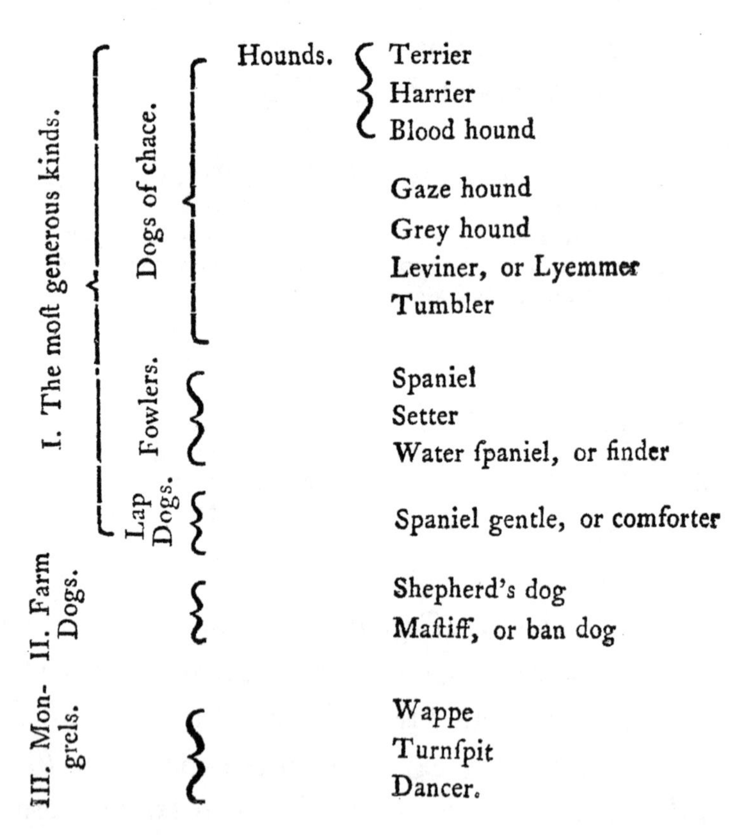

The

The firſt variety is the *Terrarius* or *Terier*, which takes its name from its ſubterraneous employ; being a ſmall kind of hound, uſed to force the fox, or other beaſts of prey, out of their holes; (and in former times) rabbets out of their burrows into nets.

The *Leverarius*, or *Harrier*, is a ſpecies well known at preſent; it derives its name from its uſe, that of hunting the hare; but under this head may be placed the fox-hound, which is only a ſtronger and fleeter variety, applied to a different chace*.

The *Sanguinarius*, or *Bloodhound*, or the *Sleuthounde*† of the *Scots*, was a dog of great uſe, and in high eſteem with our anceſtors: its employ was to recover any game that had eſcaped wounded from the hunter; or been killed and ſtole out of the foreſt. It was remarkable for the acuteneſs of its ſmell, tracing the loſt beaſt by the blood it had ſpilt; from whence the name is derived: This

* Prince *Griffith ap Conan* (who began his reign in the year 1079) divided hunting into three kinds: the firſt and nobleſt ſort was the *Helfa ddolef*, which is hunting for the melody of the cry, or notes of the pack: The ſecond ſort was the *Helfa gyfartha*, or hunting when the animal ſtood at bay: The laſt kind was the *Helfa gyffredin*, i. e. common hunting; which was no more than the right any perſon had, who happened accidentally to come in at the death of the game, to claim a ſhare. *Lewis*'s *Hiſt. of Wales*, 56.

† From the *Saxon Slot* the impreſſion that a deer leaves of its foot in the mire, and *hund* a dog. So they derive their name from following the track.

ſpecies

ſpecies could, with the utmoſt certainty, diſcover the thief by following his footſteps, let the diſtance of his flight be ever ſo great; and through the moſt ſecret and thickeſt coverts: nor would it ceaſe its purſuit, till it had taken the felon. They were likewiſe uſed by *Wallace* and *Bruce* during the civil wars. The poetical hiſtorians of the two heroes, frequently relate very curious paſſages on this ſubject; of the ſervice theſe dogs were of to their maſters, and the eſcapes they had from thoſe of the enemy. The bloodhound was in great requeſt on the confines of *England* and *Scotland*; where the borderers were continually preying on the herds and flocks of their neighbors. The true bloodhound was large, ſtrong, muſcular, broad breaſted, of a ſtern countenance, of a deep tan-color, and generally marked with a black ſpot above each eye.

The next diviſion of this ſpecies of dogs, comprehends thoſe that hunt by the eye; and whoſe ſucceſs depends either upon the quickneſs of their ſight, their ſwiftneſs, or their ſubtility.

The *Agaſæus*, or Gazehound, was the firſt: it chaced indifferently the fox, hare, or buck. It would ſelect from the herd the fatteſt and faireſt deer; purſue it by the eye; and if loſt for a time, recover it again by its ſingular diſtinguiſhing faculty; and ſhould the beaſt rejoin the herd, this dog would fix unerringly on the ſame. This ſpecies is now loſt, or at leſt unknown to us.

It muſt be obſerved that the *Agaſæus* of Dr. *Caius*, is

is a very different ſpecies from the *Agaſſeus* of *Oppian*, for which it might be miſtaken from the ſimilitude of names: this he deſcribes as a ſmall kind of dog, peculiar to *Great-Britain*; and then goes on with theſe words;

Γυρὸν, ἀσαρκότατον, λασιότριχον, ὄμμασι νωθές.

Curvum, macilentum, hiſpidum, oculis pigrum.

what he adds afterwards, ſtill marks the difference more ſtrongly;

Ῥίνεσι δ' αὖτε μάλιστα πανέξοχός ἐστιν ἀγασσεύς.

Naribus autem longè præſtantiſſimus eſt agaſſeus.

From *Oppian's* whole deſcription, it is plain he meant our Beagle *.

The next kind is the *Leporarius*, or Gre-hound. Dr. *Caius* informs us, that it takes its name *quod præcipui gradus ſit inter canes*; the firſt in rank among dogs: that it was formerly eſteemed ſo, appears from the foreſt laws of king *Canute*; who enacted, that no one under the degree of a gentleman ſhould preſume to keep a gre-hound; and ſtill more ſtrongly from an old *Welſh* ſaying; *Wrth ei Walch, ei Farch, a'i Filgi, yr adwaenir Bonheddig:* Which ſignifies, that you may know a gentleman by his hawk, his horſe and his gre-hound.

* *Opp. Cyneg. lib.* i. *lin.* 473. 476.
Nemeſianus alſo celebrates our dogs.
Diviſa *Britannia* mittit
Veloces, noſtrique orbis venantibus aptos.

Froiſ-

Froissart relates a fact not much to the credit of the fidelity of this species: when that unhappy Prince *Richard* the second was taken in *Flint* castle, his favorite gre-hound immediately deserted him, and fawned on his rival *Bolingbroke*; as if he understood, and foresaw the misfortunes of the former *. The story is so singular, that we give it in the note in the words of the historian.

* Le Roy *Richard* avoit ung levrier lequel on nommoit *Math*, tres beau levrier oultre mesure, & ne vouloit ce chien cognoistre nul homme hors le Roi, et quand le Roy vouloit chevaucher, celuy qui lavoit en garde le laissoit aller, et ce levrier venoit tantost devers le Roy le festoyer ce luy mettoient incontinent quil estoit eschappé les deux pieds sur les epaules. Et adoncques advint que le Roy et le conte *Derby* parlans ensemble en la place de la court dudit chasteau, et leur chevaulx tous sellez, car ils vouloient monter a cheval, ce levrier nomme *Math* qui estoit coustumier de faire au Roy ce que dist est, laissa le Roy et sen vint au duc de *Lenclastre*, et luy fist toutes telles contenances que paravant il avoit acoustume de faire au Roy, et lui assist les deux pieds sur le col, et le commenca moult grandement a cherir, le duc de *Lenclastre* qui point ne cognoissoit ce levrier, demanda au Roy, et que veult ce levrier faire, cousin, dist le Roy, ce vous est une grant signifiance & a moy petite. Comment dist duc lentendez vous. Je lentends dist le Roy, le levrier vous festoye et receult au jourdhuy comme Roy *d'Angleterre* que vous serez et ien seray depose, et le levrier en a cognoissance naturelle. Si le tenez deles vous, car il vous suyura et meslongera. Le duc de *Lenclastre* entendit bien ceste parolle et fist chere au levrier le quel oncques depuis ne voulut suyvre *Richard de Bourdeaulx* suyvit le duc de *Lenclastre*. *Chronicque de Froissart*, tom. iv. Fueillet 72. Edition de Paris, 1530.

The

The variety called the *Highland* gre-hound, and now become very ſcarce, is of a very great ſize, ſtrong, deep cheſted, and covered with long and rough hair. This kind was much eſteemed in former days, and uſed in great numbers by the powerfull chieftains in their magnificent hunting matches. It had as ſagacious noſtrils as the *Blood-hound*, and was as fierce. This ſeems to be the kind *Boethius* ſtyles, *genus venaticum cum celerrimum tum audaciſſimum: nec modo in feras, ſed in hoſtes etiam latroneſque; præſertim ſi dominum ductoremve injuriam affici cernat aut in eos concitetur.*

The third ſpecies is the *Levinarius*, or *Lorarius*; The Leviner or Lyemmer: the firſt name is derived from the lightneſs of the kind; the other from the old word *Lyemme*, a thong: this ſpecies being uſed to be led in a thong, and ſlipped at the game. Our author ſays, that this dog was a kind that hunted both by ſcent and ſight; and in the form of its body obſerved a medium between the hound, and the gre-hound. This probably is the kind now known to us by the name of the *Iriſh* gre-hound, a dog now extremely ſcarce in that kingdom, the late king of *Poland* having procured from them as many as poſſible. I have ſeen two or three in the whole iſland: they were of the kind called by *M. de Buffon Le grand Danois*, and probably imported there by the *Danes* who long poſſeſſed that kingdom. Their uſe ſeems originally to have been for the chaſe of wolves with which

Ireland ſwarmed till the latter end of the laſt century. As ſoon as thoſe animals were extirpated, the numbers of the dogs decreaſed; for from that period, they were kept only for ſtate.

The *Vertagus*, or Tumbler, is a fourth ſpecies; which took its prey by mere ſubtility, depending neither on the ſagacity of its noſe, nor its ſwiftneſs: if it came into a warren, it neither barked, nor ran on the rabbets; but by a ſeeming neglect of them, or attention to ſomething elſe, deceived the object till it got within reach, ſo as to take it by a ſudden ſpring. This dog was leſs than the hound; more ſcraggy, and had prickt up ears; and by Dr. *Caius*'s deſcription ſeems to anſwer to the modern lurcher.

The third diviſion of the more generous dogs, comprehends thoſe which were uſed in fowling; firſt, the *Hiſpaniolus* or ſpaniel: from the name it may be ſuppoſed, that we were indebted to *Spain* for this breed: there were two varieties of this kind, the firſt uſed in hawking, to ſpring the game, which are the ſame with our ſtarters.

The other variety was uſed only for the net, and was called *Index*, or the ſetter; a kind well known at preſent. This kingdom has long been remarkable for producing dogs of this ſort, particular care having been taken to preſerve the breed in the utmoſt purity. They are ſtill diſtinguiſhed by the name of *Engliſh* ſpaniels; ſo that notwithſtanding the

the derivation of the name, it is probable they are natives of *Great Britain.* We may ſtrengthen our ſuſpicion by ſaying that the firſt who broke a dog to the net was an *Engliſh* nobleman of a moſt diſtinguiſhed character, the great *Robert Dudly* Duke of *Northumberland**. The Pointer, which is a dog of foreign extraction, was unknown to our anceſtors.

The *Aquaticus*, or Fynder, was another ſpecies uſed in fowling; was the ſame as our water ſpaniel; and was uſed to find or recover the game that was ſhot.

The *Melitæus*, or *Fotor*; the ſpaniel gentle or comforter of Dr. *Caius* (the modern lap dog) was the laſt of this diviſion. The *Malteſe* little dogs were as much eſteemed by the fine ladies of paſt times, as thoſe of *Bologna* are among the modern. Old *Hollingſhed* is ridiculouſly ſevere on the fair of his days, for their exceſſive paſſion for theſe little animals; which is ſufficient to prove it was in his time † a novelty.

The ſecond grand diviſion of dogs comprehends the *Ruſtici*; or thoſe that were uſed in the country.

The firſt ſpecies is the *Paſtoralis*, or ſhepherd's dog; which is the ſame that is uſed at preſent, either in guarding our flocks, or in driving herds of cattle. This kind is ſo well trained for thoſe

* *Wood*'s *Ath. Ox.* II. 27.

† In the reign of Queen *Elizabeth.*

purpofes, as to attend to every part of the herd be it ever fo large; confine them to the road, and force in every ftraggler without doing it the leaft injury.

The next is the *Villaticus*, or *Catenarius*; the *maf-tiff* or *band* dog; a fpecies of great fize and ftrength, and a very loud barker. *Manwood* fays *, it derives its name from *mafe thefefe*, being fuppofed to frighten away robbers by its tremendous voice. *Caius* tells us that three of thefe were reckoned a match for a bear; and four for a lion: but from an experiment made in the Tower by *James* the firft, that noble quadruped was found an unequal match to only three. Two of the dogs were difabled in the combat, but the third forced the lion to feek for fafety by flight †. The *Englifh* bull dog feems to belong to this fpecies; and probably is the dog our author mentions under the title of *Laniarius*. *Great-Britain* was fo noted for its maftiffs, that the *Roman* Emperors appointed an officer in this ifland with the title of *Procurator Cynegii* ‡, whofe fole bufinefs was to breed, and tranfmit from hence to the *Amphitheatre*, fuch as would prove equal to the combats of the place,

> Magnaque taurorum fracturi colla *Britanni* ||.
> And *Britifh* dogs fubdue the ftouteft bulls.

* *Manwood's* *Foreft Law.*

† *Stow's* *Annals*, 1427.

‡ *Camd. Brit.* in *Hampfhire.*

|| *Claudian* de laude *Stilichonis. Lib.* iii. *Lin.* 301.

Gratius

Gratius ſpeaks in high terms of the excellency of the *Britiſh* dogs,

Atque ipſos libeat penetrare *Britannos?*
O quanta eſt merces et quantum impendia ſupra!
Si non ad ſpeciem mentituroſque decores
Protinus: hæc una eſt catulis jactura *Britannis.*
At magnum cum venit opus, promendaque virtus,
Et vocat extremo præceps diſcrimine Mavors,
Non tunc egregios tantum admirere *Moloſſos* *.

If *Britain*'s diſtant coaſt we dare explore,
How much beyond the coſt the valued ſtore;
If ſhape and beauty not alone we prize,
Which nature to the *Britiſh* hound denies:
But when the mighty toil the huntſman warms,
And all the ſoul is rouſed by fierce alarms,
When Mars calls furious to th' enſanguin'd field
Even bold *Moloſſians* then to theſe muſt yield.

Strabo tells us, that the maſtiffs of *Britain* were trained for war, and were uſed by the *Gauls* in their battles †: and it is certain a well-trained maſtiff might be of conſiderable uſe in diſtreſſing ſuch half-armed and irregular combatants as the adverſaries of the *Gauls* ſeem generally to have been before the *Romans* conquered them.

The laſt diviſion is that of the *Degeneres* or *Curs.* The firſt of theſe was the *Wappe*, a name derived

* *Gratii* Cynegeticon. *Lin.* 175.
† *Strabo. Lib.* iv.

from its note: its only uſe was to alarm the family, by barking, if any perſon approached the houſe. Of this claſs was the *Verſator*, or turnſpit; and laſtly the *Saltator*, or dancing dog; or ſuch as was taught variety of tricks, and carried about by idle people as a ſhew. Theſe *Degeneres* were of no certain ſhape, being mongrels or mixtures of all kinds of dogs.

We ſhould now, according to our plan, after enumerating the ſeveral varieties of *Britiſh* dogs, give its general natural hiſtory; but ſince *Linnæus* has already performed it to our hand, we ſhall adopt his ſenſe, tranſlating his very words (wherever we may) with literal exactneſs.

" The dog eats fleſh, and farinaceous vege-
" tables, but not greens: its ſtomach digeſts bones:
" it uſes the tops of graſs as a vomit. It voids
" its excrements on a ſtone: the *album græcum* is
" one of the greateſt encouragers of putrefaction.
" It laps up its drink with its tongue: it voids
" its urine ſideways, by lifting up one of its hind
" legs; and is moſt diuretic in the company of
" a ſtrange dog. *Odorat anum alterius:* its ſcent
" is moſt exquiſite, when its noſe is moiſt: it treads
" lightly on its toes; ſcarce ever ſweats; but when
" hot lolls out its tongue. It generally walks
" frequently round the place it intends to lye down
" on: its ſenſe of hearing is very quick when aſleep:
" it dreams. *Procis rixantibus crudelis: catulit cum*
" *variis: mordet illa illos: cohæret copula junctus:*

it

" it goes with young ſixty-three days; and common-
" ly brings from four to eight at a time: the male
" puppies reſemble the dog, the female the bitch.
" It is the moſt faithful of all animals: is very
" docible: hates ſtrange dogs: will ſnap at a ſtone
" thrown at it: will howl at certain muſical notes:
" all (except the *S. American* kind) will bark at
" ſtrangers: dogs are rejected by the *Maho-*
" *metans.*"

11. FOX.

Vulpes. *Raii ſyn. quad.* 177
Morton's Northampt. 444.
Meyer's an. i. Tab. 36.
Canis fulvus, pilis cinereis intermixtis. *Briſſon quad.* 173.
De Buffon. Tom. vii. 75. Tab. 6.
Geſner quad. 966.
Vulpes auctorum. *Haſſelquiſt Itin.* 191.
Canis vulpes. *Lin. ſyſt.* 59.
Canis Alopex. C. cauda recta apice nigro. vulpes campeſtris. *ibid.*
Canis cauda recta apice albo, *Faun. Suec.* 7.
Vulpes vulgaris. *Klein quad.* 73.
Br. Zool. 28. *ſyn. quad. N.* 112.

Brit. Llwynog, *fœm* Llwynoges.
Fren. Le Renard
Ital. Volpe
Span. Raposa
Port. Rapoza
Germ. Fuchs
Dut. Vos
Swed. Raff
Dan. Rev

THE fox is a crafty, lively, and libidinous animal: it breeds only once in a year (except ſome accident befals its firſt litter;) and brings four or five young, which, like puppies, are born blind. It is a common received opinion, that this animal will produce with the dog kind; which may be well founded; ſince it has been proved that the congenerous wolf will*. Mr. *Brook*, animal-merchant in *Holborn*, turned a wolf to a *Pomeranian* bitch then in heat: the congreſs was immediate, with the circumſtances uſual with the canine ſpecies. The bitch brought ten whelps, one of which I afterwards ſaw at the Duke of *Gordon's* in *Scotland*. It bore a great reſemblance to the male parent, and had much of its nature: being ſlipped at a weak deer, it inſtantly caught at the animal's throat and killed it. The fox ſleeps much in the day, but is in motion the whole night in ſearch of prey. It will feed on fleſh of any kind, but its favourite food is lambs, rabbets, hares, poultry, and feathered game. It will, when urged by hunger, eat carrots and inſects; and thoſe that live near the ſea-coaſts, will, for want of other food, eat crabs, ſhrimps, or

* *M. de Buffon* aſſerts the contrary, and gives the following account of the experiment he had made. *J'en fis garder trois pendant deux ans, une femelle & deux mâles: on tenta inutilement de les faires accoupler avec des chiennes; quoiqu'ils n'euſſent jamais vû de femelle de leur eſpece, et qu'ils paruſſent preſſés du beſoin de jouir, ils ne pûrent s'y determiner, ils refuſerent toutes les chiennes, mais des qu'on leur preſenta leur femelle légitime, ils la couvrirent, quoiqu'enchainés, et elle produiſit quatre petits. Hiſt. Naturelle*, vii. 81. The ſame experiments were tried with a bitch and a male fox, and with a dog and a female wolf, and as *M. de Buffon* ſays with the ſame ill ſucceſs. Vol. v. 210, 212. but the fact juſt cited, proves the poſſibility paſt conteſt.

ſhell

ſhell fiſh. In *France* and *Italy*, it does incredible damage in the vineyards, by feeding on the grapes, of which it is very fond. The fox is a great deſtroyer of rats, and field mice; and like the cat, will play with them a conſiderable time, before it puts them to death.

When the fox has acquired a larger prey than it can devour at once, it never begins to feed till it has ſecured the reſt, which it does with great addreſs. It digs holes in different places, returns to the ſpot where it had left the booty; and (ſuppoſing a whole flock of poultry to have been its prey) will bring them one by one, and thruſt them in with its noſe, and then conceal them by ramming the looſe earth on them, till the calls of hunger incite him to pay them another viſit.

Of all animals the fox has the moſt ſignificant eye, by which it expreſſes every paſſion of love, fear, hatred, &c. It is remarkably playful, but like all other ſavage creatures half reclamed, will on the leſt offence bite thoſe it is moſt familiar with.

It is a great admirer of its buſhy tail, with which it frequently amuſes and exerciſes itſelf by running in circles to catch it: and in cold weather wraps it round its noſe.

The ſmell of this animal in general is very ſtrong, but that of the urine is moſt remarkably fœtid. This ſeems ſo offenſive even to itſelf, that it will take the trouble of digging a hole in the ground, ſtretching its body at full length over it, and there, after

after depofiting its water, cover it over with the earth, as the cat does its dung. The fmell is fo offenfive, that it has often proved the means of the fox's efcape from the dogs, who have fo ftrong an averfion to the filthy *effluvia*, as to avoid encountering the animal it came from. It is faid that the fox makes ufe of its urine as an expedient to force the cleanly badger from its habitation: whether that is the means is rather doubtful; but that the fox makes ufe of the badger's hole is certain: not through want of ability to form its own retreat; but to fave itfelf fome trouble: for after the expulfion of the firft inhabitant, the fox improves, as well as enlarges it confiderably, adding feveral chambers; and providently making feveral entrances to fecure a retreat from every quarter. In warm weather it will quit its habitation for the fake of bafking in the fun, or to enjoy the frefh air; but then it rarely lies expofed, but chufes fome thick brake, and generally of gorfe, that it may reft fecure from furprize. Crows, magpies, and other birds, who confider the fox as their common enemy, will often, by their notes of anger, point out its retreat.

This animal is common in all parts of *Great Britain*, and fo well known as not to require a defcription. The fkin is furnifhed with a foft and warm fur, which in many parts of *Europe* is ufed to make muffs and line cloaths. Vaft numbers are taken in *Le Vallais*, and the *Alpine* parts of *Switzerland*. At *Laufanne* there are furriers who are

in

Pl. V.
The WOLF.
Desmoulins pinx.
Mouzell fec.

in poſſeſſion of between two and three thouſand ſkins, all taken in one winter.

There are three varieties of foxes found in the mountanous parts of theſe iſlands, which differ a little in form, but not in color, from each other. Theſe are diſtinguiſhed in *Wales*, by as many different names. The *Milgi* or *gre-hound fox*, is the largeſt, talleſt, and boldeſt; and will attack a grown ſheep or wether: the *maſtiff fox* is leſs, but more ſtrongly built: the *Corgi*, or *cur fox*, is the leſt, and lurks about hedges, out-houſes, *&c.* and is the moſt pernicious of the three to the feathered tribe. The firſt of theſe varieties has a white tag or tip to the tail: the laſt a black. The number of theſe animals in general would ſoon become intolerable, if they were not proſcribed, having a certain reward ſet on their heads.

WOLF.

In this place we ſhould introduce the wolf, a congenerous animal, if we had not fortunately a juſt right to omit it in a hiſtory of *Britiſh* quadrupeds. It was, as appears by *Hollingſhed**, very noxious to the flocks in *Scotland* in 1577; nor was it entirely extirpated till about 1680, when the laſt wolf fell by the hand of the famous *Sir Ewin Cameron.* We may therefore with confidence aſſert the non-exiſtence of thoſe animals, notwithſtanding *M. de Buffon* maintains that the *Engliſh pretend* to the contrary †.

* *Diſc. Scot.* 10.

† *Tom.* vii.

It

It has been a received opinion, that the other parts of theſe kingdoms were in early times delivered from this peſt by the care of king *Edgar*. In *England* he attempted to effect it by commuting the puniſhments for certain crimes into the acceptance of a number of wolves tongues from each criminal: in *Wales* by converting the tax of gold and ſilver into an annual tribute of 300 wolves heads. Notwithſtanding theſe his endeavours, and the aſſertions of ſome authors, his ſcheme proproved abortive. We find that ſome centuries after the reign of that *Saxon* monarch, theſe animals were again increaſed to ſuch a degree, as to become the object of royal attention; accordingly *Edward* the firſt iſſued out his mandate to *Peter Corbet* to ſuperintend and aſſiſt in the deſtruction of them in the ſeveral counties of *Glouceſter*, *Worceſter*, *Hereford*, *Salop*, and *Stafford* *: and in the adja-

* Pro *Petro Corbet, de Lupis Capiendis.*

Rex, omnibus Ballivis, &c. Sciatis quod injunximus dilecto et fideli noſtro Petro Corbet *quod in omnibus foreſtis et parcis et aliis locis intra comitatus noſtros* Glouceſter, Wygorn, Hereford, Salop, *et* Stafford, *in quibus lupi poterunt inveniri lupos cum hominibus canibus et ingeniis ſuis capiat et deſtruat modis omnibus quibus viderit expedire.*

Et ideo vobis mandamus quod eidem intendentes et auxiliantes eſtis. Teſte rege apud Weſtm. 14 Maii A. D. 1281. Rymer, vol. i. pars 2. p. 192.

By the grant of liberties from king *John*, to the inhabitants of *Devonſhire*, it appears that theſe animals were not then extirpated, even in that ſouthern country. *vide Appendix* No.

cent

cent county of *Derby*, as *Camden*, p. 902, informs us, certain perſons at *Wormhill* held their lands by the duty of hunting and taking the wolves that infeſted the country, whence they were ſtiled *Wolve hunt*. To look back into the *Saxon* times we find that in *Athelſtan*'s reign wolves abounded ſo in *Yorkſhire*, that a retreat was built at *Flixton* in that county, *to defend paſſengers from the wolves, that they ſhould not be devoured by them*: and ſuch ravages did thoſe animals make during winter, particularly in *January* when the cold was ſevereſt, that our *Saxon anceſtors* diſtinguiſhed that month by the title *wolf moneth**. They alſo called an outlaw *Wolfſhed*, as being out of the protection of the law, proſcribed, and as liable to be killed as that deſtructive beaſt. *Et tunc gerunt* caput lupinum, *ita quod ſine judiciali inquiſitione rite pereant*. Bracton lib. iii. Tr. 11. c. 11. alſo Knighton 2356.

They infeſted *Ireland* many centuries after their extinction in *England*, for there are accounts of ſome being found there as late as the year 1710; the laſt preſentment for killing of wolves being made in the county of *Cork* about that time †.

BEAR.

The Bear, another voracious beaſt, was once an inhabitant of this iſland, as appears from different authorities: to begin with the more antient, *Martial* informs us, that the *Caledonian* bears were

* *Verſtegan*'s *Antiq*. 59.

† *Smith*'s *hiſt*. *Cork*. II. 226.

uſed

uſed to heighten the torments of the unhappy ſufferers on the croſs.

> Nuda *Caledonio* ſic pectora præbuit urſo
> Non falſâ pendens in cruce *Laureolus* *.

And *Plutarch* relates, that Bears were tranſported from *Britain* to *Rome*, where they were much admired †. Mr. *Llwyd* ‡ alſo diſcovered in ſome old *Welſh* MS. relating to hunting, that this animal was reckoned among our beaſts of chace, and that its fleſh was held in the ſame eſteem with that of the hare or boar. Many places in *Wales* ſtill retain the name of *Pennarth*, or the bear's head, another evidence of their exiſtence in our country. It does not appear how long they continued in that principality; but there is proof of their infeſting *Scotland* ſo late as the year 1057 ||, when a *Gordon*, in reward for his valor for killing a fierce bear, was directed by the King to carry three *Bears' heads* on his banner. They are ſtill found in the mountanous parts of *France*, particularly about the *grande Chartreuſe* in *Dauphinè*, where they make great havoke among the out-ricks of the poor farmers. Long after their extirpation out of this kingdom, theſe animals were imported for an end, that

* *Martial. Lib. Spect. ep.* 7.

† *Plutarch*, as cited by *Camden*, p. 1227. ‡ *Raii ſyn. quad.* 214.

|| Hiſt. of the *Gordons*. I, 2.

does

does no credit to the manners of the times: bear-baiting in all its cruelty was a favorite paſtime with our anceſtors. We find it in queen *Elizabeth*'s days, exhibited (tempered with our merry diſports) as an entertainment for an ambaſſador, and again among the various amuſements prepared for her majeſty at the princely *Kenelworth*.

Our nobility alſo kept their bear-ward: twenty ſhillings was the annual reward of that officer from his lord the fifth earl of NORTHUMBERLAND, 'when 'he comyth to my lorde in *criſtmas* with his lord-'ſhippes beeſts for makynge of his lordſchip paſ-'tyme the ſaid xii days *.

It will not be foreign to the ſubject here to add, that our monarchs in very early times kept up the ſtate of a menagery of exotic animals. *Henry* I. had his lions, leopards, lynxes, and porpentines (porcupines) in his park at *Woodſtock* †. The emperor *Frederick* ſent to Henry III. a preſent of three leopards in token of his royal ſhield of arms, wherein three leopards were pictured ‡. The ſame prince had alſo an elephant which (with its keeper) was maintained at the expence of the ſheriffs of *London* for the time being ||. The other animals had their keeper, a man of faſhion, who was allowed ſix-pence a day for himſelf and ſix-pence for each beaſt.

* *Northumberland Houſhold Book.*

† *Stow's hiſt London* 1, 79. ‡ *Ibid.*

|| *Idem.* 118.

Six

VIII. CAT. Six cutting teeth and two canine in each jaw.
Five toes before; four behind.
Sharp hooked claws, lodged in a ſheath, that may be exerted at pleaſure.
Round head: ſhort viſage: rough tongue.

12. WILD. Felis pilis ex fuſco flavicante, et albido variegatis veſtita, cauda annulis alternatim nigris et ex ſordide albo flavicantibus cincta. *Briſſon quad.* 192.
De Buffon, Tom. vi. 20. Tab I.
Morton Northampt. 443.
Geſner quad. 325.
Catus ſylveſtris ferus vel feralis eques arborum, *Klein quad.* 75.
Br. Zool. 22. *Syn. quad.* No. 133.

Brit. Cath goed
Fren. Le Chat Sauvage
Span. Gato Montis
Germ. Wilde katze, Boumritter
Dan. Vild kat

THIS animal does not differ ſpecifically from the tame cat; the latter being originally of the ſame kind, but altered in color, and in ſome other trifling accidents, as are common to animals reclamed from the woods and domeſticated.

The cat in its ſavage ſtate is three or four times as large as the houſe-cat; the head larger, and the face flatter. The teeth and claws, tremendous: its muſcles very ſtrong, as being formed for rapine: the tail is of a moderate length, but very

very thick, marked with alternate bars of black and white, the end always black: the hips and hind part of the lower joints of the leg, are always black: the fur is very ſoft and fine. The general color of theſe animals is of a yellowiſh white, mixed with a deep grey: theſe colors, though they appear at firſt ſight confuſedly blended together, yet on a cloſe inſpection will be found to be diſpoſed like the ſtreaks on the ſkin of the tiger, pointing from the back downwards, riſing from a black liſt that runs from the head along the middle of the back to the tail.

This animal may be called the *Britiſh* tiger; it is the fierceſt, and moſt deſtructive beaſt we have; making dreadful havoke among our poultry, lambs, and kids. It inhabits the moſt mountanous and woody parts of theſe iſlands, living moſtly in trees, and feeding only by night. It multiplies as faſt as our common cats; and often the females of the latter will quit their domeſtic mates, and return home pregnant by the former.

They are taken either in traps, or by ſhooting: in the latter caſe it is very dangerous only to wound them, for they will attack the perſon who injured them, and have ſtrength enough to be no deſpicable enemy. Wild cats were formerly reckoned among the beaſts of chace; as appears by the charter of *Richard* the ſecond, to the abbot of *Peterborough*, giving him leave to hunt the hare, fox, and wild cat. The uſe of the fur was in lining of robes; but it was esteemed not of the most luxurious kind; for it was ordained 'that no abbess 'or nun should use more costly apparel than such 'as is made of lambs or cats skins *.' In much earlier times it was also the object of the sportsman's diversion.

Felemque minacem
Arboris in trunco longis præfigere telis.
Nemesiani Cynegeticon, L. 55.

Domestic. Felis domestica seu catus. *Raii syn. quad.* 170.
Charlton ex. 20.
Meyer's an. i. Tab. 15.
Gesner quad. 317.
Brisson quad. 191.
De Buffon, Tom. vi. 3. Tab. 2.
Felis catus, *Lin. syst.* 62.
Felis cauda elongata, auribus æqualibus. *Faun. Suec.* 9.
Br. Zool. 21. *Syn. quad.* No. 133.

Brit. Cath, *mas.* Gwr cath
Fren. Le Chat
Ital. Gatto
Span. Gato
Port. Gato
Germ. Katz
Dut. Cyperse Kat. Huyskat.
Swed. Katta
Dan. Kat.

THIS animal is so well known as to make a description of it unnecessary. It is an useful, but deceitful domestic; active, neat, sedate, intent on its prey. When pleased purres and moves its tail: when angry spits, hisses, and strikes with its foot. When walking, it draws in its claws: it drinks

* Archbp. *William Corboyl's* canons, A. D. 1127. quoted by Mr. *T. Row* in Gent. Mag. *April* 1774.

drinks little: is fond of fiſh: it waſhes its face with its fore-foot, (*Linnæus* ſays at the approach of a ſtorm:) the female is remarkably ſalacious; a piteous, ſqualling, jarring lover. Its eyes ſhine in the night: its hair when rubbed in the dark e-mits fire: it is even proverbially tenacious of life: always lights on its feet: is fond of perfumes; *Marum*, *Cat-mint*, *valerian*, &c *.

Our anceſtors ſeem to have had a high ſenſe of the utility of this animal. That excellent Prince *Hoel dda*, or *Howel* the *Good*, did not think it beneath him (among his laws relating to the prices, &c. of animals †,) to include that of the cat; and to deſcribe the qualities it ought to have. The price of a kitling before it could ſee, was to be a penny; till it caught a mouſe two-pence; when it commenced mouſer four-pence. It was required beſides, that it ſhould be perfect in its ſenſes of hearing and ſeeing, be a good mouſer, have the claws whole, and be a good nurſe: but if it fail-ed in any of theſe qualities, the ſeller was to forfeit to the buyer the third part of its value. If any one ſtole or killed the cat that guarded the Prince's granary, he was to forfeit a milch ewe, its fleece and lamb; or as much wheat as when poured on the cat ſuſpended by its tail (the head touching the floor) would form a heap high enough to cover

* Vide *Lin. ſyſt.*

† *Leges Wallicæ*, p. 247, 248.

G 2 the

the tip of the former *. This laſt quotation is not only curious, as being an evidence of the ſimplicity of ancient manners, but it almoſt proves to a demonſtration that cats are not aborigines of theſe iſlands; or known to the earlieſt inhabitants. The large prices ſet on them, (if we conſider the high value of ſpecies at that time †) and the great care taken of the improvement and breed of an animal that multiplies ſo faſt, are almoſt certain proofs of their being little known at that period.

* Sir *Ed. Coke* in his Reports, mentions the ſame kind of puniſhment anciently for killing a ſwan, by ſuſpending it by the bill, &c. Vide, *Caſe des Swannes.*

† *Howel dda* died in the year 948, after a reign of thirty-three years over *South Wales*, and eight years over all *Wales*.

Six

IX. BADGER.

Six cutting teeth, two canine, in each jaw.
Five toes before; five behind: very long ſtrait claws on the forefeet.
A tranſverſe orifice between the tail, and the anus.

13. COMMON.

Badger, Brock, Gray, Pate, Taxus ſive Meles. *Raii ſyn. quad.* 185.
Meyer's an. i. Tab. 31.
Sib. Scot. 11.
Meles pilis ex ſordidè albo et nigro variegatis veſtita, capite tæniis alternatim albis et nigris variegato. *Briſſon quad.* 183.
De Buffon, Tom. viii. Tab. 7. p. 104.
Geſner quad. 686.
Urſus meles. Urſus cauda concolore, corpore ſupra cinereo, ſubtus nigro, faſcia longitudinali per oculos aureſque nigra. *Lin. ſyſt.* 70.
Coati cauda brevi. *Klein quad.* 73.
Meles unguibus anticis longiſſimis. *Faun. Suec.* 20.
Br. Zool. 30. *Syn quad. No.* 142.

Brit. Pryf Llwyd, Pryf penfrith
Fren. Le Taiſſon, Le Blaireau
Ital. Taſſo
Span. Texon
Port. Texugo
Germ. Tachs
Dut. Varkens Das
Swed. Graf Suin
Dan. Grevlin, Brok

THOUGH the badger is a beaſt of great ſtrength, and is furniſhed with ſtrong teeth, as if formed for rapine, yet it is found to be an animal perfectly inoffenſive: roots, fruits, graſs, inſects, and frogs are its food: it is charged with deſtroying lambs and rabbets; but, on enquiry, there ſeems to be no other reaſon to think it a beaſt of prey, than from the analogy there is between its

its teeth and thoſe of carnivorous animals. Nature denied the badger the ſpeed and activity requiſite to eſcape its enemies, ſo hath ſupplied it with ſuch weapons of offence that ſcarce any creature would hazard the attacking it; few animals defend themſelves better, or bite harder: when purſued, they ſoon come to bay, and fight with great obſtinacy. It is an indolent animal, and ſleeps much, for which reaſon it is always found very fat. It burrows under ground, like the fox; and forms ſeveral different apartments, though with only one entrance, carrying in its mouth graſs in order to form a bed for its young. It confines itſelf to its hole during the whole day, feeding only at night: it is ſo cleanly an animal as never to obey the calls of nature in its apartments; but goes out for that purpoſe: it breeds only once in a year, and brings four or five at a time.

Descrip. The uſual length of the badger, is two feet ſix inches, excluſive of the tail, which is but ſix inches long: the weight fifteen pounds. The eyes are very ſmall: the ears ſhort and rounded: the neck ſhort: the whole ſhape of the body clumſy and thick; which being covered with long coarſe hairs like briſtles, makes it appear ſtill more aukward. The mouth is furniſhed with ſix cutting teeth and two canine teeth in each jaw; the lower has five grinders on each ſide, the upper five; in all thirty four.

The noſe, chin, lower ſides of the cheeks, and the

the middle of the forehead, are white: each ear and eye is inclosed in a pyramidal bed of black; the base of which incloses the former; the point extends beyond the eye to the nose: the hairs on the body are of three colors; the bottoms of a dirty yellowish white; the middle black; the ends ash-colored, or grey; from whence the proverb, As grey as a badger. The hairs which cover the tail are very long, and of the same colors with those of the body: the throat and under parts of the body are black: the legs and feet of the same color, are very short, strong and thick: each foot is divided into five toes; those on the fore feet are armed with long claws, well adapted for digging; in walking the badger treads on its heel, like the bear; which brings the belly very near the ground. Immediately below the tail, between that and the anus, is a narrow transverse orifice, which opens in a kind of pouch, from whence exudes a white substance of a very fœtid smell; this seems peculiar to the badger and the *Hyæna*.

This animal is not mentioned by *Aristotle*, not that it was unknown to the ancients, for *Pliny* takes notice of it *.

Naturalists once distinguished the badger by the name of the swine-badger, and the dog-badger; from the supposed resemblance of their heads to

* Alia solertia in metu *Melibus*, sufflatæ cutis distentu ictus hominum et morsus canum arcent. *Lib.* viii. c. 38.

those animals, and so divided them into two species: but the most accurate observers have been able to discover only one kind; that, whose head and nose resemble those of the dog.

The skin of the badger, when dressed with the hair on, is used for pistol furniture. The Highlanders make their pendent pouches of it. The hair is frequently used for making brushes to soften the shades in painting, which are called sweetening tools. These animals are also hunted in the winter nights for the sake of their flesh; for the hind quarters may be made into hams, not inferior in goodness to the best bacon. The fat is in great request for ointments and salves.

In *China* it seems to be more common food than in *Europe*: for Mr. *Bell* * says, he has seen about a dozen at one time in the markets at *Pekin*; and that the *Chinese* are very fond of them. It does not appear that this animal is found in the hotter parts of *Asia*; but is confined to the cold, or the temperate parts of the world.

* *Bell's* Travels, I. 83.

Six

Pl. VI. No. 1

FITCHET.

MARTIN. No.

X. WEESEL.

Six cutting teeth, two canine, in each jaw.
Sharp noſe, ſlender bodies.
Five toes before, five behind.

14. FITCHET.

Putorius. Polecat or Fitchet. *Raii ſyn. quad.* 199.
Meyer's an. ii. Tab. 6.
Charlton ex. 20.
Geſner quad. 767.
Muſtela pilis in exortu ex cinereo albidis, colore nigricante terminatis, oris circumferentia alba. *Briſſon quad.* 180.
De Buffon, Tom. vii. 199. Tab. 23.
Muſtela putorius, *Lin. ſyſt.* 67.
Muſtela fœtida, *Klein quad.* 63.
Muſtela flaveſcente nigricans, ore albo, collari flaveſcente. *Faun. Suec.* 16.
Br. Zool. 37. *Syn. quad.* No. 152.

Brit.	Ffwlbard	*Germ.*	Iltis, ulk, Buntſing
Fren.	Le Putois	*Dut.*	Bonſing
Ital.	Foetta, Puzolo	*Swed.*	Iller
Span.	Putoro	*Dan.*	Ilder

DESCRIP.

THE length of this animal is about ſeventeen inches, excluſive of the tail; that of the tail ſix. The ſhape of this animal in particular, as well as of the whole genus, is long and ſlender; the noſe ſharp-pointed, and the legs ſhort: in fine, admirably formed for inſinuating itſelf into the ſmalleſt holes and paſſages, in ſearch of prey: it is very nimble and active, runs very faſt, will creep up the ſides of walls with great agility, and ſpring with vaſt force. In running, the belly ſeems to touch the ground: in preparing to jump, it arches its back, which aſſiſts it greatly in that action.

The

The ears are ſhort, rounded and tipt with white: the circumference of the mouth, that is to ſay, the ends of the lower and upper mandibles are white: the head, throat, breaſt, legs and thighs, are wholly of a deep chocolate color, almoſt black. The ſides are covered with hairs of two colors; the ends of which are of a blackiſh hue, like the other parts; the middle of a full tawny color: in others cinereous.

The toes are long, and ſeparated to the very origin: the tail is covered with pretty long hair.

MANNERS.

The fitchet is very deſtructive to young game of all kinds, and to poultry: they generally reſide in woods, or thick brakes; burrowing under ground, forming a ſhallow retreat, about two yards in length; which commonly ends, for its ſecurity, among the roots of ſome large trees. It will ſometimes lodge under hay-ricks, and in barns: in the winter it frequents houſes, and makes a common practice of robbing the dairy of the milk: it alſo makes great havoke in warrens.

It will bring five or ſix young at a time. Warreners aſſert, that the fitchet will mix with the ferret; and they are ſometimes obliged to procure an intercourſe between theſe animals, to improve the breed of the latter, which by long confinement will abate its ſavage nature, and become leſs eager after rabbets, and conſequently leſs uſeful. M. *de Buffon* denies that it will admit the fitchet; yet gives the figure of a variety under the name of the *Ferret*

Polecat,

Polecat *, which has much the appearance of being a ſpurious offſpring. But to put the matter out of diſpute, the following fact need only be related: The Rev. Mr. *Lewis*, Vicar of *Llanſowel* in *Caermarthenſhire*, had a tame female ferret, which was permitted to go about the houſe: at length it abſented itſelf for ſeveral days; and on its return proved with young: it produced nine, of a deep brown color, more reſembling the fitchet than the ferret. What makes the matter more certain is, that Mr. *Lewis* had no male of this ſpecies for it to couple with; neither was there any within three miles, and thoſe cloſely confined.

The ferret agrees with the fitchet in many reſpects, particularly in its thirſt after the blood of rabbets. It may be added, that the ferret comes originally from *Africa* †; and is only cultivated in *Great Britain*.

Though the ſmell of the fitchet, when alive, is rank and diſagreeable, even to a proverb; yet the ſkin is dreſt with the hair on, and uſed as other furs for tippets, *&c.* and is alſo ſent abroad to line cloaths.

* *La Furet Putois*, Tom. vii. Tab. 25

† Κὰι γαλας αγρίας ἃς ἡ λυβύη φερει. *Strabo*, *Lib.* iii. p. 144, *Edit. Caſaubon.*

Martes,

15. MARTIN. Martes, alias Foyna. The Martin and Martlet. *Raii syn. quad.* 200.
Meyer's an. ii. Tab. 4.
Martin, or Martern. *Charlton exer.* 20.
The Mertrick. *Martin's West. Isles*, 36.
Gesner quad. 764.
Mustela pilis in exortu albidis castaneo colore terminatis vestita, gutture albo. *Brisson quad.* 178.
De Buffon, Tom. vii. 161. Tab. 18.
Mustela martes. *Lin. syst.* 67.
M. martes. *Klein. quad.* 64.
M. fulvo-nigricans gula pallida. *Faun. Suec.* 15.
Br. Zool. 38. *Syn. quad.* No. 154.

Brit.	Bela graig	*Germ.*	Hausſ marder, stein marder
Fren.	La Fouine	*Dut.*	Marter
Ital.	Foina, Fouina	*Swed.*	Mard
Span.	Marta, Gibellina	*Dan.*	Maar.

MANNERS. THIS is the most beautifull of the *British* beasts of prey: its head is small, and elegantly formed: its eyes lively: and all its motions shew great grace, as well as agility: when taken young, it is easily tamed, is extremely playful, and in constant good humour: nature will recur, if it gets loose; for it will immediately take advantage of its liberty, and retire to its proper haunts. It makes great havoke among poultry, game, *&c.* and will eat mice, rats, and moles. With us it inhabits woods, and makes its lodge in the hollows of trees; and brings from four to six young at a time.

DESCRIP. The martin is about eighteen inches long; the tail ten, or, if the measurement be taken to the end of the hair at the point, twelve inches.

The

Pl. VIII

No. 19

OTTER.

BADGER.

No. 13

The ears are broad, rounded and open: the back, ſides, and tail, are covered with a fine thick down, and with long hair intermixed: the bottom is aſh-colored: the middle of a bright cheſnut color: the tips black: the head brown, with ſome ſlight caſt of red: the legs and upper ſides of the feet are of a chocolate color: the palms, or under ſides, are covered with thick down like that on the body: the feet are broad: the claws white, large and ſharp; well adapted for climbing trees, which in this country are its conſtant reſidence. The throat and breaſt are white: the belly of the ſame color with the back, but rather paler: the hair on the tail is very long; eſpecially at the end, where it appears much thicker than near the origin of it: the hair in that part is alſo darker. But martins vary in their colors, inclining more or leſs to aſh-color, according to their ages or the ſeaſons they are taken in.

FINE SMELL.

The ſkin and excrements of this animal have a fine muſky ſcent; and are entirely free from that rankneſs which diſtinguiſhes the other ſpecies of this genus: the ſkin is a valuable fur; and much uſed for linings to the gowns of magiſtrates.

Martes

16. PINE MARTIN.

Martes abietum. *Raii ſyn. quad.* 200.
Meyer's an. ii. Tab. 5.
Martes ſylveſtris. *Geſner quad.* 765.
Muſtela pilis in exortu ex cinereo albidis caſtaneo colore terminatis veſtita, gutture flavo. *Briſſon quad.* 179.
De Buffon, Tom. vii. 186. Tab. 22.
Br. Zool. 39. *Syn. quad.* No. 155.

Brit. Bela goed
Fren. La Marte
Ital. Marta, Martura, Martora, Martorello
Span. Marta
Port.
Germ. Feld-marder, wild-marder
Dut. Marter
Swed.

THIS ſpecies is found in *Great Britain*; but is much leſs common in *England* than the former: it is ſometimes taken in the counties of *Merioneth* and *Caernarvon*, as I was informed by my late worthy friend Mr. *W. Morris*, where it is diſtinguiſhed from the other kind, by the name of *bela goed*, or wood martin, it being ſuppoſed entirely to inhabit the woods; the *bela graig* to dwell only among the rocks. Tho' this is ſo rare in theſe parts, yet in *Scotland* it is the only kind; where it inhabits the fir foreſts, building its neſt at the top of the trees *. It loves a cold climate, and is found in much greater numbers in the north of *Europe*, than in the other parts. *North*

FUR.

America abounds with theſe animals. Prodigious numbers of their ſkins are annually imported from

* Vide *Sibbald's Hiſt. Scot.* Part II. Lib. iii. p. 11.

No. 17

WEESEL.

ERMINE.

No. 1

M Griffiths del

from *Hudſon's bay* and *Canada*. In one of the company's ſales * not fewer than 12,370 good ſkins, and 2360 damaged ones were ſold; and about the ſame time, the *French* brought into the port of *Rochelle* from *Canada*, not leſs than 30,325.

The principal differences between this and the former kind, conſiſt in the ſize, this being leſs: the breaſt too is yellow; the color of the body much darker, and the fur in general greatly ſuperior in fineneſs, beauty, and value.

17. COMMON.

The Weaſel or Weeſel. Muſtela vulgaris: in *Yorkſhire*, the Fitchet or Foumart. *Raii ſyn. quad.* 195.
Girald. Cambrenſ. 149.
The Whitred. *Sib. Scot.* 11.

Muſtela ſupra rutila, infra alba. *Briſſon quad.* 173.
De Buffon, Tom. vii. 235. Tab. 29.
Geſner quad. 753.
Muſtela vulgaris. *Klein quad.* 62.
Br. Zool. 39. *Syn. quad.* No. 150.

Brit.	Bronwen	*Germ.*	Wiſel
Fren.	La Belette	*Dut.*	Weezel
Ital.	Donnola, Ballottula, Benula	*Swed.*	Veſla
Span.	Comadreia	*Dan.*	Væſel
Port.	Doninha		

DESCRIP.

THIS ſpecies is the leſt of the weeſel kind; the length of the head and body not exceeding ſix, or at moſt ſeven inches. The tail is only two inches and a half long, and ends in a point: the ears are large; and the lower parts of them are doubled in.

* In 1743. Vide *Dobbs*'s account of *Hudſon's bay*, 200.

The

COLOR. The whole upper part of the body, the head, tail, legs, and feet are of a very pale tawny brown. The whole under ſide of the body from the chin to the tail is white; but beneath the corners of the mouth on each jaw is a ſpot of brown.

PREY. This, like the reſt of the kind, is very deſtructive to young birds, poultry, and young rabbets; and beſides is a great devourer of eggs. It does not eat its prey on the place; but after killing it, by one bite near the head, carries it off to its young, or its retreat. The weeſel alſo preys upon moles, as appears by their being ſometimes caught in the mole-traps. It is a remarkably active animal, and will run up the ſides of walls with ſuch facility, that ſcarce any place is ſecure from it; and its body is ſo ſmall, that there is ſcarce any hole but what is pervious to it. This ſpecies is much more domeſtic than the others; frequenting outhouſes, barns, and granaries; where, to make as it were ſome atonement for its depredations among our tame fowl, it ſoon clears its haunts from rats and mice, being infinitely more an enemy to them than the cat itſelf. It brings five or ſix young at a time: its ſkin and excrements are moſt intolerably fœtid.

This animal is confounded by *Linnæus* with the Stoat or Ermine. He ſeems unacquainted with our weeſel in its brown color; but deſcribes it in the white ſtate under the title of *Snomus*, or *Muſtela nivalis.*

.IX.

The **MUSIMON.**

The **BEAVER.**

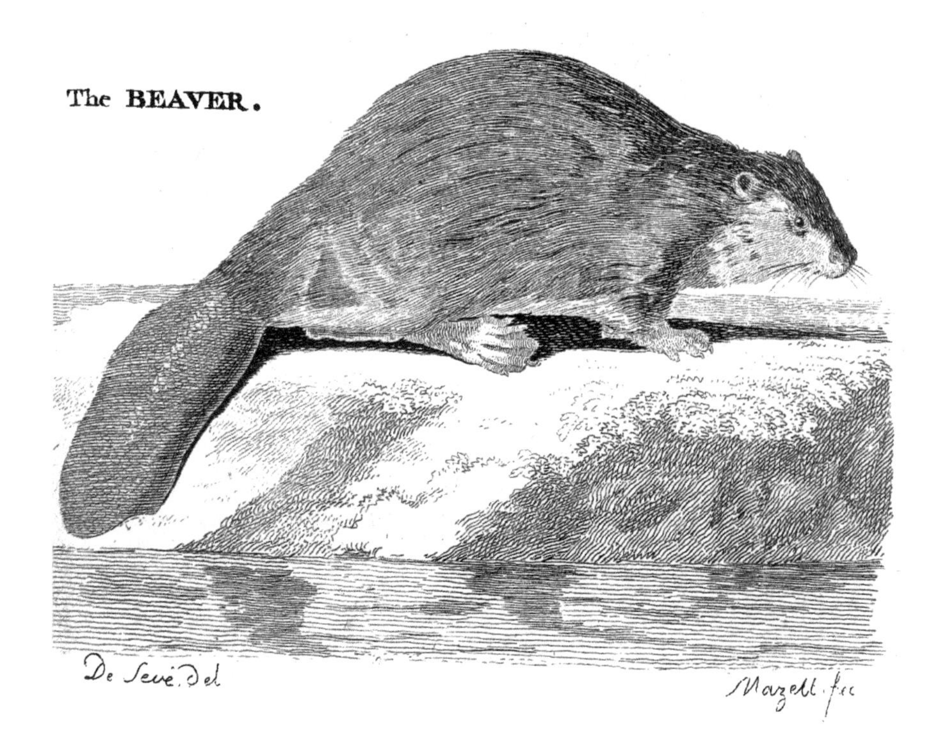

De Sevè. Del

Mazell. fec

nivalis *. I have met with it in that circumſtance, in the iſle of *Ilay.*

18. STOAT, OR ERMINE.

Muſtela candida, animal ermineum, *Raii ſyn quad.* 198
Mort. Northampt. 442.
Meyer's an. ii. Tab. 23, 24.
Muſtela hieme alba, æſtate ſupra rutila infra alba, caudæ apice nigro. *Briſſon quad.* 176.
De Buffon, vii. 240. *Tab.* 29. *Fig.* 2. *Tab.* 31. *Fig.* 1.
Geſner quad. 753.
Muſtela erminea. M. plantis fiſſis, caudæ apice atro. *Lin. ſyſt.* 68. *Faun. Suec.* 17.
Pontop. Norway. Part ii. p. 25.
Br. Zool. 40. *Syn. quad.* No. 151.

Brit.	Carlwm
Fren.	L'Hermine, Le Roſelet
Ital.	Armellino
Span.	Armino, Armelina
Germ.	Hermelin, *Klein.* 63.
Swed.	Hermelin, Lekatt
Dut.	Hermilyn
Dan.	Hermelin, Lekat

DESCRIP.

THE length of the ſtoat to the origin of the tail, is ten inches: that of the tail is five inches and a half. The colors bear ſo near a reſemblance to thoſe of the weeſel, as to cauſe them to be confounded together by the generality of common obſervers; the weeſel being uſually miſtaken for a ſmall ſtoat: but theſe animals have evident and invariable ſpecific differences, by which they may be eaſily known. Firſt, by the ſize; the weeſel being ever leſs than the ſtoat: ſecondly, the tail of the latter is always tipt with black, is longer in

* *Similima* Ermineo *ſed dimidio minor, caudæ apice pilo vix uno alterove albo. Faun. Suec.* No. 18. *Syſt. Nat.* 69.

proportion to the bulk of the animal, and more hairy; whereas the tail of the weesel is shorter, and of the same color with the body: thirdly, the edges of the ears, and the ends of the toes in this animal, are of a yellowish white. It may be added, that the stoat haunts woods, hedges and meadows; especially where there are brooks, whose sides are covered with small bushes; and sometimes (but less frequently than the weesel) inhabits barns, and other buildings.

ERMINES.

In the most northern parts of *Europe*, these animals regularly change their color in winter; and become totally white, except the end of the tail, which continues invariably black; and in that state are called *Ermines*: I am informed that the same is observed in the highlands of *Scotland*. The skins and tails are a very valuable article of commerce in *Norway*, *Lapland*, *Russia*, and other cold countries; where they are found in prodigious numbers. They are also very common in *Kamtschatka* and *Siberia**. In *Siberia* they burrow in the fields, and are taken in traps baited with flesh. In *Norway* † they are either shot with blunt arrows, or taken in traps made of two flat stones, one being propped up with a stick, to which is fastned a baited string, which when the animals nibble, the stone falls down and crushes them to death. The *Laplanders* take them in the same manner, only in-

HOW TAKEN.

* *Bell's Travels*, i. 199. † *Hist. Norway*, ii. 25.

stead

ſtead of ſtones make uſe of two logs of wood*. The ſtoat is ſometimes found white in *Great-Britain*, but not frequently: and then it is called a white weeſel. That animal is alſo found white; but may be eaſily diſtinguiſhed from the other in the ermine ſtate, by the tail, which in he weeſel is of a light tawny brown. With us the former is obſerved to begin to change its color from brown to white in *November*, and to begin to reſume the brown the beginning of *March*.

The natural hiſtory of this creature is much the ſame with that of the weeſel, its food being birds, rabbets, mice, *&c.* its agilitythe ſame, and its ſcent equally fetid: it is much more common in *England* than that animal.

* *Oeuvres de Maupertuis*, iii. 187.

XI. OTTER. Six cutting teeth, two canine, in each jaw. Five toes on each foot; each toe palmated.

19. OTTER. Le Loutre, *Belon* 26. *pl.* 27
Lutra. The otter. *Raii ſyn. quad.* 187.
Grew's Muſ. 16.
Morton's Northampt. 444.
Sib. Scot. 10.
Geſner quad. 687.
Lutra caſtanei coloris. *Briſſon quad.* 201.
De Buffon, Tom. vii. 134. Tab. 11. xiii. 322.
Muſtela lutra. *Lin. ſyſt.* 66.
Pontop. Norw. 2. 27.
Lutra digitis omnibus æqualibus. *Faun. Suec.* 12.
Br. Zool. 32. *Syn. quad.* No. 138.

Brit.	Dyfrgi	*Germ.*	Otter, Fiſch Otter
Fren.	Le Loutre	*Dut.*	Otter
Ital.	Lodra, Lodria, Lontra.	*Swed.*	Utter
Span.	Nutria	*Dan.*	Odder
Port.			

DESCRIP. THE uſual length of this animal is three feet three inches, including the tail, which is ſixteen inches long.

The head and noſe are broad and flat, the neck ſhort, and equal in thickneſs to the head: the body long: the tail broad at the baſe, tapers off to a point at the end, and is the whole way compreſſed horizontally. The eyes are very ſmall, and placed nearer the noſe than is uſual in quadrupeds: the ears extremely ſhort, and their orifice narrow: the opening of the mouth is ſmall, the lips muſcular, and capable of being brought very cloſe

close together: the nose and the corners of the mouth are furnished with very long whiskers; so that the whole appearance of the otter is something terrible: it has thirty-six teeth, six cutting and two canine above and below; of the former the middlemost are the least: it has besides five grinders on each side in both jaws. The legs are very short, but remarkably strong, broad, and muscular; the joints articulated so loosely, that the animal is capable of turning them quite back, and bringing them on a line with the body, so as to perform the office of fins. Each foot is furnished with five toes, connected by strong broad webs, like those of water fowl. Thus nature in every article has had attention to the way of life she had allotted to an animal, whose food is fish; and whose haunts must necessarily be about waters.

FUR.

The color of the otter is entirely a deep brown, except two small spots of white on each side the nose, and another under the chin. The skin of this animal is very valuable, if killed in the winter; and is greatly used in cold countries for lining cloaths: but in *England* it is only used for covers for pistol furniture. The best furs of this kind come from the northern part of *Europe*, and *America*. Those of *N. America* are larger than the *European* otters. The *Indians* make use of their skins for pouches, and ornament them with bits of horn. The finest sort come from the colder parts of that continent: where they are also most numerous. Weſtward of *Carolina* *, there are ſome found of a white color inclining to yellow.

MANNERS.

The otter ſwims and dives with great celerity, and is very deſtructive to fiſh: in rivers it is always obſerved to ſwim againſt the ſtream, to meet its prey. In very hard weather, when its natural ſort of food fails, it will kill lambs, ſucking pigs, and poultry. It is ſaid that two otters will in concert hunt that ſtrong and active fiſh the ſalmon. One ſtations itſelf above, the other below the place where the fiſh lies, and continue chaſing it inceſſantly till the ſalmon quite wearied becomes their prey. To ſuppoſe that they never prey in the ſea is a miſtake: for they have been often ſeen in it both ſwimming and bringing their booty on ſhore, which has been obſerved in the *Orknies* to have been cod, and congers. Its fleſh is exceſſively rank and fiſhy. The *Romiſh* church permits the uſe of it on maigre-days. In the kitchen of the *Carthuſian* convent near *Dijon*, we ſaw one preparing for the dinner of the religious of that rigid order, who, by their rules, are prohibited during their whole lives, the eating of fleſh.

It ſhews great ſagacity in forming its habitation: it burrows under ground on the banks of ſome river or lake; and always makes the entrance of its hole under water; works upwards to the ſurface of the earth, and forms before it reaches the top, ſeveral *holts*, or lodges, that in caſe of high floods,

* *Lawſon's hiſt. Carol.* 119.

it

it may have a retreat, for no animal affects lying drier, and there makes a minute orifice for the admission of air: it is further observed, that this animal, the more effectually to conceal its retreat, contrives to make even this little air hole in the middle of some thick bush.

The otter brings four or five young at a time: as it frequents ponds near gentlemen's houses, there have been instances of litters being found in cellars, sinks, and other drains. It is observable that the male otters never make any noise when taken; but the pregnant females emit a most shrill squeal.

SEA OTTER.

Sir *Robert Sibbald*, in his history of *Fife*, p. 49, mentions a *Sea Otter*, which he says differs from the common sort, in being larger, and having a rougher coat; but probably it does not differ specifically from the kind that frequents fresh waters. Did not *Aristotle* place his *Latax** among the animals which

seek

* Τοιαυτα δε εστιν ὁ τε καλουμενος καστωρ, και το σαθεριον και το σατυριον, και ενυδρις, και ἡ καλουμενη λαταξ. εστι δε τουτο πλατυτερον ενυδριδος, και οδοντας εχει ισχυρους εξιουσα γαρ νυκτωρ πολλακις, τας περι τον ποταμον κερκιδας εκτεμνει τοις οδουσιν. δακνει δε τους ανθρωπους και ἡ ενυδρις, και ουκ αφιησιν, ως λεγουσι, μεχρις αν οστου ψοφον ακουσῃ. το δε τριχωμα εχει ἡ λαταξ σκληρον, και το ειδος μεταξυ του της φωκης τριχωματος, και του της ελαφου. *Aristot. Hist. Anim.* p. 905. A

Sunt etiam in hoc genere (sc. animalium quadrupedum quæ victum ex lacubus et fluviis petunt) fiber, satherium, satyrium, lutris, latax, quæ latior lutre est, dentesque habet robustos,

quippe

feek their food among frefh waters, we fhould imagine we had here recovered this loft animal, which he mentions immediately after the otter, and defcribes as being broader. Though this muft remain a doubt, we may with greater confidence fuppofe the fea otter to be the *Loup marin* of *Belon* *, which from a hearfay account, he fays, is found on the *Englifh* coafts. He compares its form to that of a wolf, and fays, it feeds rather on fifh than fheep. That circumftance alone makes it probable, *Sibbald*'s animal was intended, it being well known, the otter declines flefh when it can get fifh. Little ftrefs ought to be laid on the name, or comparifon of it to a wolf; this variety being of a fize fo fuperior to the common, and its hair fo much more fhaggy, a common obferver might readily catch the idea of the more terrible beaft, and adapt his comparifon to it.

BEAVER. Beavers, which are alfo amphibious animals, were formerly found in *Great Britain*; but the breed has been extirpated many ages ago: the lateft accounts we have of them, is in *Giraldus Cambrenfis* †, who travelled through *Wales* in 1188: he gives a brief hiftory of their manners; and adds, that in his time

quippe quæ noctu plerumque aggrediens, virgulta proxima fuis dentibus, ut ferro præcidat. Lutris etiam hominem mordet, nec defiftit (ut ferunt) nifi fracti offis crepitum fenferit. Lataci pilus durus, fpecie inter pilum vituli marini et cervi.

* *Belon de la Nature des Poifons*, p. 28. pl. 29.

† *Girald. Camb. Itin.* 178, 179.

they

they were found only in the river *Teivi*; two or three waters in that principality, ftill bear the name of *Llyn yr afangc* *, or the beaver lake; which is a further proof, that thefe animals were found in different parts of it: I have feen two of their fuppofed haunts; one in the ftream that runs thro' *Nant Frankon*; the other in the river *Conway* a few miles above *Llanrwft*; and both places in all probability had formerly been croffed by *Beaver* dams. But we imagine they muft have been very fcarce even in earlier times; by the laws of *Hoel dda*, the price of a beaver's fkin (*Croen Lloftlydan* †) was fixed at one hundred and twenty pence, a great fum in thofe days.

* *Raii fyn. quad.* 213.

† *Lloftlydan*, that is, the broad tailed animal. *Leges Wallicæ*, 261.

DIV.

DIV. II. SECT. II.

With only two cutting teeth in each jaw.
Herbivorous, frugivorous.

XII. HARE.

Two cutting teeth in each jaw.
Long ears: ſhort tail.
Five toes before, four behind.

20. COMMON.

Lepus, *Plinii*, lib. viii. c. 55.
The Hare. *Raii ſyn. quad.* 204.
White Hare. *Mort. Northampt.* 445.
Sib. Scot. 11.
Meyer's an. ii. Tab. 32.
Geſner quad. 605.
Lepus caudatus ex cinereo rufus. *Briſſon quad.* 94.
De Buffon, Tom. vi. 246. Tab. 38.
Lepus timidus. *Lin. ſyſt.* 77.
Lepus cauda abrupta pupillis atris. *Faun. Suec.* 35.
Lepus vulgaris cinereus. *Klein quad.* 51.
Br. Zool. 41. *Syn. quad.* No. 184.

Brit. Yſgyfarnog, Ceinach
Fren. Le Lievre
Ital. Lepre, Lievora
Span. Liebre
Port. Lebre
Germ. Has, Haas
Dut. Haas
Swed. Hare
Dan. Hare

TO enter on a minute deſcription of ſo well known an animal, would be to abuſe the reader's patience; yet to neglect pointing out the admirable contrivance of its ſeveral properties and parts, would be fruſtrating the chief deſign of this

this work: that of pointing out the Divine Wiſdom in the animal world.

Being a weak and moſt defenceleſs creature, it is endued, in a very diſtinguiſhed degree, with that preſerving paſſion, fear: this makes it perpetually attentive to every alarm, and keeps it always lean.

To enable it to receive the moſt diſtant notices of dangers, it is provided with very long ears, which (like the tubes made uſe of by the deaf) convey to it the remoteſt ſounds.

EYES.

Its eyes are very large and prominent, adapted to receive the rays of light on all ſides.

To aſſiſt it to eſcape its purſuers by a ſpeedy flight, the hind legs are formed remarkably long, and furniſhed with ſtrong muſcles: their length give the hare ſingular advantages over its enemies in aſcending ſteep places; and ſo ſenſible is the animal of this, as always to make towards the riſing ground when ſtarted.

As it lies always upon the ground, its feet are protected above and below with a thick and warm covering of hair.

The various ſtratagems and doubles it uſes, when hunted, are ſo well known to every ſportſman, as not to deſerve mention; except to awaken their attention to thoſe faculties nature has endowed it with; which ſerve at the ſame time to increaſe their amuſement, as well as to prevent the animal's deſtruction.

It very rarely leaves its form or ſeat in the day; but

but in the night takes a circuit in ſearch of food, always returning through the ſame meuſes, or paſſes.

COLOR. The color approaches very near to that of the ground; which ſecures it more effectually from the ſight of men, and of beaſts and birds of prey. Providence has been ſo careful in reſpect to the preſervation of the ſpecies of animals, as to cauſe in northern countries theſe as well as many others to change color, and become white at the beginning of winter, to render them leſs conſpicuous amidſt the ſnow. Accidental inſtances of white hares are met with in *South Britain*.

Hares differ much in ſize. The ſmalleſt are in the iſle of *Ilay*: the largeſt in that of *Man*, where ſome have been found to weigh twelve pounds.

FOOD. Its food is entirely vegetable; and it does great injury to nurſeries of young trees, by eating the bark off: it is particularly fond of pinks, parſley, and birch.

The hare never pairs; but in the rutting ſeaſon, which begins in *February*, the male purſues and diſcovers the female, by the ſagacity of its noſe. The female goes with young one month, brings uſually two young at a time; ſometimes three, and very rarely four. Sir *Thomas Brown*, in his treatiſe on vulgar errors *, aſſerts the doctrine of ſuperfetation: i. e. a conception upon conception, or an improvement on the firſt fruit before the ſecond is ex-

* P. 118.

cluded;

cluded; and he brings this animal as an inſtance; aſſerting, from his own obſervation, that after the firſt caſt there remain ſucceſſive conceptions, and other younglings very immature, and far from the term of their excluſion; but as the hare breeds very frequently in the year, there is no neceſſity of having recourſe to this accident * to account for their numbers. The antients were acquainted with this circumſtance. *Horace* alludes to it in the ſecond ſatire of the fourth book.

Fæcundi leporis ſapiens ſectabitur armos,

ſays the *bon vivant*, every man of taſte will prefer the wing of the fruitful hare. *Pliny* as a philoſopher is more explicit, and aſſigning a moral reaſon for the great encreaſe of this animal gives the following elegant account of it. Lepus *omnium prædæ naſcens, ſolus præter* Daſypodem *ſuperfœtat, aliud educans, aliud in utero pilis veſtitum, aliud implume, aliud inchoatum gerens pariter.*

Hares are very ſubject to fleas; *Linnæus* tells us, that the *Dalecarlians* make a ſort of cloth of the fur, called *filt*; which, by attracting thoſe inſects, preſerves the wearer from their troubleſome attacks †.

The hair of this creature forms a great article in the hat manufacture; and as this country cannot

* For a farther account of this doctrine, we refer the curious reader to M. *de Buffon*'s works, vol. vi. p. 252, 279, &c.

† *Faun. Suec.* 25.

ſupply

ſupply a ſufficient number, vaſt quantities are annually imported from *Ruſſia* and *Siberia*.

The hare was reckoned a great delicacy among the *Romans* *; the *Britains*, on the contrary, thought it impious even to taſte it †; yet this animal was cultivated by them; either for the pleaſure of the chace; or for the purpoſes of ſuperſtition, as we are informed that *Boadicia*, immediately before her laſt conflict with the *Romans*, let looſe a hare ſhe had concealed in her boſom, which taking what was deemed a fortunate courſe, animated her ſoldiers by the omen of an eaſy victory over a timid enemy ‡.

21. ALPINE. Lepus hieme albus *Forſter hiſt. nat.* Volgæ. *Ph. Tr.* LVII. 343. Alpine Hare. *Syn. quad.* *No.* 184.

THE *Alpine* hare inhabits the ſummits of the *highland* mountains, never deſcends into the vales, or mixes with the common ſpecies which is frequent in the bottoms: it lives among the rocks

* Inter aves turdus, ſi quid me judice verum:
Inter quadrupedes gloria prima Lepus. *Martial.* 13. 92.

† *Leporem et gallinam et anſerem guſtare fas non putant: hæc tamen alunt, animi voluptatiſque cauſa.* Cæſar. Com. lib. v.

‡ Ταυτα ειπȣσα λαγων μεν εκ τȣ κολπȣ, &c. *Xiphilini Epitome Dioniſ.* 173.

with

Pl. X.

ALPINE HARE. *No. 21.* RABBET. *No. 22.*

with *Ptarmigans*, natives of the loftieſt ſituations: does not run faſt; and if purſued is apt to take ſhelter beneath ſtones or in clefts of rocks: is eaſily tamed, and is very ſprightly and full of frolick: is fond of honey, and carraway comfits, and is obſerved to eat its own dung before a ſtorm.

It is leſs than the common hare, weighing only 6 lb. $\frac{1}{2}$. whereas the firſt weighs from eight to twelve pounds. Its hair is ſoft and full; the predominant color grey mixed with a little black and tawny. This is its ſummer's dreſs.

In winter it entirely changes to a ſnowy whiteneſs except the edges and tips of the ears which retain their blackneſs. The alteration of color begins in *September*, and firſt appears about the neck and rump. In *April* it again reſumes its grey coat. This is the caſe in *Styria* *, but in the *polar* tracts ſuch as *Greenland* it never varies from white, the eternal color of the country. In the intermediate climates between temperate and frigid, ſuch as *Scotland* and *Scandinavia* it regularly experiences theſe viciſſitudes of color.

* *Kramer Auſtr.* 315.

Cuniculus

22. RABBET.

Cuniculus. The Rabbet, or Cony. *Raii syn. quad.* 205.
Meyer's an. i. *Tab.* 83.
Gesner quad. 362.
Lepus caudatus, obscure cinereus. *Brisson quad.* 95.
De Buffon Tom. vi. 303. Tab. 50, 51.
Lepus cuniculus. *Lin syst.* 77.
Lepus cauda brevissima papillis rubris. *Faun. Suec.* 26.
Cuniculus terram fodiens; *Klein quad.* 52.
Br. Zool. 43. *Syn. quad. No.* 186.

Brit. Cwningen
Fren. Le Lapin
Ital. Coniglio
Span. Conejo
Port. Coelho
Ger. Koniglein, Kaninchin
Dut. Konyn
Swed. Kanin
Dan. Kanine

IT is well observed by *Pliny*, that nature 'hath 'shewed great kindness, in causing those things 'to be most prolific, that are the most harmless 'and the properest for our food *.'

PROLIFIC.

This excellent observation of his, cannot be better illustrated than in shewing the great fruitfulness of this animal; as it far exceeds that proof, brought by the ingenious author of the œconomy of nature, in support of the same quotation. The instance he produces is the pigeon; whose increase, from one pair, may in four years amount to 14.760 †: but rabbets will breed seven times a year,

* *Benigna circa hoc natura, innocua et esculenta animalia fæcunda generavit.* Lib. viii. c, 55.

† Vide *Swedish* Essays, translated by Mr. *Stillingfleet*, Ed. 1st. p. 75.

year, and bring eight young ones each time: on a ſuppoſition this happens regularly, during four years, their numbers will amount to 1,274,840.

By this account, we might juſtly apprehend being overſtocked with theſe animals, if they had not a large number of enemies which prevents the too great increaſe: not only men, but hawks, and beaſts of prey, make dreadful havoke among the ſpecies. Notwithſtanding theſe different enemies, we are told by *Pliny*, and *Strabo*, that they once proved ſo great a nuiſance to the inhabitants of the *Balearic* iſlands, that they were obliged to implore the aſſiſtance of a military force from the *Romans*, in the time of *Auguſtus*, in order to extirpate them *. Their native country is *Spain*, where they were taken by means of ferrets, as we do at preſent, which animals were firſt introduced there out of *Africa* †: they love a temperate and a warm climate, and are incapable of bearing great cold, ſo that in *Sweden* ‡ they are obliged to be kept in houſes. Our country abounds with them; their furs form a conſiderable article in the hat manufactures; and of late, ſuch part of the fur as is unfit for that purpoſe, has been found as good as feathers for ſtuffing beds and bolſters. Numbers of the ſkins are annually exported into *China*. The *Engliſh* counties that are moſt noted for theſe animals

FUR.

* *Plin.* lib. viii. c. 55. *Strabo*, lib. iii.

† *Strabo*, iii. 144. ‡ *Faun. Suec.* 26.

mals are *Lincolnſhire*, *Norfolk*, and *Cambridgeſhire*. *Methold*, in the laſt county, is famous for the beſt ſort for the table: the ſoil there is ſandy, and full of moſſes and the *Carex* graſs. Rabbets ſwarm in the iſles of *Orkney*, where their ſkins form a conſiderable article of commerce. Excepting otters, brown rats, common mice, and ſhrews, no other quadrupeds are found there. The rabbets of thoſe iſles are in general grey, thoſe which inhabit the hills, grow hoary in winter.

Formerly the ſilver-haired rabbets were in great eſteem for lining of cloaths, and their ſkins ſold at three ſhillings a piece *; but ſince the introduction of the more elegant furs, the price is fallen to ſixpence each. The *Sunk Iſland* † in the *Humber* was once famous for a mouſe-coloured ſpecies, now extirpated by reaſon of the injury it did to the banks by burrowing.

* *Hartlib's Legacy.* † *Ph. Trans.* No. 361.

Two

Two cutting teeth in each jaw.
Four toes before; five behind.
Tufted ears.
Long tail cloathed with long hair.

XIII.
SQUIRREL.

23. COMMON.

Sciurus vulgaris. *Raii ſyn. quad.* 214.
Meyer's an. i. *Tab.* 97.
Geſner quad. 845.
Sciurus rufus, quandoque griſeo admixto. *Briſſon quad.* 104.
De Buffon, Tom. vii. 258. *Tab.* 32.
Sciurus auriculis apice barbatis, palmis 4-dactylis plantis 5-dactylis. *Lin. ſyſt.* 86.
Sciurus palmis ſolis ſaliens. *Faun. Suec.* 37.
Sc. vulgaris rubicundus. *Klein quad.* 53.
Br. Zool. 44. *Syn. quad.* No. 206.

Brit.	Gwiwair
Fren.	L'Ecureuil
Ital.	Scoiattolo, Schiarro, Schiratto
Span.	Harda, Hardilla, Eſquilo
Port.	Ciuro
Germ.	Eichorn, Eichmermlin
Dut.	Inkhoorn
Swed.	Ikorn, graſkin
Dan.	Ekorn

NAME.

THE ſquirrel derives its name from the form of its tail, σκια a ſhade, ουρα a tail, as ſerving this little animal for an umbrella. That part is long enough to cover the whole body, and is clothed with long hairs, diſpoſed on each ſide horizontally, which gives it a great breadth. Theſe ſerve a double purpoſe; when erected, they prove a ſecure protection from the injuries of heat or cold; when extended, they are very inſtrumental in promoting thoſe vaſt leaps the ſquirrel takes from tree to tree. On the authority of *Klein* and *Linnæus*,

I 2 we

we may add a third application of the form of the tail: these naturalists tell us, that when the squirrel is disposed to cross a river, a piece of bark is the boat, the tail the sail.

Manners. This animal is remarkably neat, lively, active, and provident; never leaves its food to chance, but secures in some hollow tree a vast magazine of nuts for winter provision. In the summer it feeds on the buds and young shoots; and is particularly fond of those of the fir and pine, and also of the young cones. It makes its nest of the moss or dry leaves, between the fork of two branches; and brings four or five young at a time. Squirrels are in heat early in the spring, when it is very diverting to see the female feigning an escape from the pursuit of two or three males, to observe the various proofs they give of their agility, which is then exerted in full force.

Descrip. The color of the whole head, body, tail, and legs of this animal, is a bright reddish brown: the belly and breast white: the ears are very beautifully ornamented with long tufts of hair, of a deeper color than those on the body: the eyes are large, black, and lively: the fore teeth, strong, sharp, and well adapted to its food: the legs are short and muscular: the toes long, and divided to their origin; the nails strong and sharp; in short, in all respects fitted for climbing, or clinging to the smallest boughs: on the fore-feet it has only four toes, with a claw a claw in the place of the thumb or interior toe: on the hind feet there are five toes.

When it eats or dreſſes itſelf, it ſits erect, covering the body with its tail, and making uſe of the fore-legs as hands. It is obſerved, that the gullet of this animal is very narrow, to prevent it from diſgorging its food, in deſcending of trees, or in down leaps.

XIV. DORMOUSE.

Two cutting teeth in each jaw.
Four toes before; five behind.
Naked ears.
Long tail covered with hair.

24. DORMOUSE.

Mus avellanarum minor. The Dormouſe or Sleeper. *Raii ſyn. quad.* 220.
The Dormouſe. *Edw.* 266.
Geſner quad. 162.
Glis ſupra rufus infra albicans. *Briſſon quad.* 115.
De Buffon, Tom. viii. 193. *Tab.* 26.
Mus avellanarius. *Lin. ſyſt.* 83.
Mus cauda longa piloſa corpore rufo gula albicante. *Faun. Suec.* 35.
Br. Zool. 45. *Syn. quad.* No. 219.

Brit. Pathew
Fren. Le Muſcardin, Croque-noix, Rat-d'or
Ital. Moſcardino
Span. Liron
Germ. Rothe, Wald-maus
Swed. Skogſmus
Dan. Kaſſel-muus

THIS animal agrees with the ſquirrel in its food, reſidence, and ſome of its actions: on firſt ſight it bears a general reſemblance to it; but on a cloſer inſpection, ſuch a difference may be diſcovered in its ſeveral parts, as vindicates M. *Briſſon* for forming a diſtinct genus of the Dormice, or *Glires*. Theſe want the fifth claw on the interior ſide of their fore-feet; nor are their ears adorned with thoſe elegant tufts of hair that diſtinguiſh the ſquirrel kind. Theſe diſtinctions prevale in the other ſpecies, ſuch as the *Lerot* and *Loir*.

Dormice

MANNERS.

Dormice inhabit woods, or very thick hedges; forming their nests in the hollow of some low tree, or near the bottom of a close shrub: as they want much of the sprightliness of the squirrel, they never aspire to the tops of trees; or, like it, attempt to bound from spray to spray: like the squirrel they form little magazines of nuts, *&c.* for winter provision; and take their food in the same manner, and same upright posture. The consumption of their hoard during the rigor of the season is but small: for they sleep most part of the time; retiring into their holes at the first approach of winter, they roll themselves up, and lie almost torpid the greatest part of that gloomy season. In that space, they sometimes experience a short revival, in a warm sunny day; when they take a little food, and then relapse into their former state.

DESCRIP.

The size of the dormouse is equal to that of a mouse; but has a plumper appearance, and the nose is more blunt; the eyes are large, black, and prominent; the ears are broad, rounded, thin, and semi-transparent: the fore-feet are furnished with four toes; the hind-feet with five; but the interior toes of the hind-feet are destitute of nails: the tail is about two inches and a half long, closely covered on every side with hair: the head, back, sides, belly, and tail, are of a tawny red color; the throat white.

NEST.

These animals seldom appear far from their retreats, or in any open place; for which reason they ſeem leſs common in *England* than they really are. They make their neſts of graſs, moſs, and dead leaves; and bring uſually three or four young at a time.

Two

XV. RAT.

Two cutting teeth in each jaw.
Four toes before, five behind.
Very ſlender tail; naked, or very ſlightly haired.

25. BLACK.

Mus domeſticus major, ſeu Rattus. *Raii ſyn. quad.* 217.
Meyer's an. ii. Tab. 83.
Geſner quad. 731.
Mus cauda longiſſima obſcure cinereus. *Briſſon quad* 118.
De Buffon, Tom. vii. p. 278. *Tab.* 36.

Mus rattus. *Lin. ſyſt.* 83.
Mus cauda longa ſubnuda corpore fuſco cinereſcente. *Faun. Suec.* 33.
Mus Rattus, mus ciſtrinarius. *Klein quad.* 57.
Br. Zool. 46. *Syn. quad.* No. 226.

Brit. Llygoden fferngig
Fren. Le Rat
Ital. Ratto, Sorcio
Span. Raton, Rata
Port. Rato
Germ. Ratz
Dut. Rot
Swed. Rotta
Dan. Rotte

MANNERS.

THE rat is the moſt pernicious of any of our ſmaller quadrupeds: our meat, corn, paper, cloaths, furniture, in ſhort every conveniency of life is a prey to this deſtructive creature: nor does it confine itſelf to theſe; but will make equal havoke among our poultry, rabbets, or young game. Unfortunately for us it is a domeſtic animal, always reſiding in houſes, barns, or granaries; and nature has furniſhed it with fore-teeth of ſuch ſtrength, as enable it to force its way through the hardeſt wood, or oldeſt morter. It makes a lodge, either for its day's reſidence, or for a neſt for its young,

young, near a chimney; and improves the warmth of it, by forming there a magazine of wool, bits of cloth, hay or ſtraw. It breeds frequently in the year, and brings about ſix or ſeven young at a time: this ſpecies increaſes ſo faſt, as to over-ſtock their abode; which often forces them, through deficiency of food, to devour one another: this unnatural diſpoſition happily prevents even the human race from becoming a prey to them: not but that there are inſtances of their gnawing the extremities of infants in their ſleep.

The greateſt enemy the rats have is the weeſel; which makes infinitely more havoke among them than the cat; for the weeſel is not only endowed with ſuperior agility; but, from the form of its body, can purſue them through all their retreats that are impervious to the former. The *Norway* rat has alſo greatly leſſened their numbers, and in many places almoſt extirpated them: this will apologize for a brief deſcription of an animal once ſo well known.

Descrip. Its length from the noſe to the origin of the tail, is ſeven inches: the tail is near eight inches long: the noſe is ſharp-pointed, and furniſhed with long whiſkers: the color of the head and whole upper part of the body is a deep iron-grey, bordering on black; the belly is of a dirty cinereous hue; the legs are of a duſky color, and almoſt naked: the fore-feet want the thumb or interior toe, having only in its place a claw: the hind-feet are furniſhed with five toes.

Among

Among other officers, his *Britiſh* majeſty has a *rat-catcher*, diſtinguiſhed by a particular dreſs, ſcarlet embroidered with yellow worſted, in which are figures of mice deſtroying wheat-ſheaves.

KING'S RAT-CATCHER.

26. BROWN.

Mus ſylveſtris, Rat de bois. *Briſſon quad.* 20.
Le Surmulot. *De Buffon, Tom.* viii. 206. Tab. 27.
Mus norvegicus. *Klein quad.* 56.
Mus ex norvegia. *Seb. Mus. Tom.* ii. 64. *Tab.* 63.
Br. Zool. 47. *Syn. quad. No.* 227.

DESCRIP.

THIS is a very large ſpecies; thicker, and of a ſtronger make than the common rat: the length from the end of the noſe to the beginning of the tail, is nine inches; the length of the tail the ſame; the uſual weight eleven ounces: the ears reſemble thoſe of the rat: the eyes large and black: the color of the head and whole upper part of the body is a light brown, mixed with tawny and aſh-color: the end of the noſe, the throat and belly, are of a dirty white, inclining to grey: the feet and legs almoſt bare; and of a dirty pale fleſh-color: the beginning of the tail is of the ſame color as the back; the reſt of the tail is covered with minute duſky ſcales, mixed with a few hairs.

HIST.

This is the ſpecies well known in this kingdom under the name of the *Norway* rat; but it is an animal quite unknown in *Scandinavia*, as we have been aſſured by ſeveral natives of the countries that

that form that tract: and *Linnæus* * takes no notice of it in his laſt ſyſtem. It is fit here to remark an error that gentleman has in ſpeaking of the common rat, which he ſays was firſt brought from *America* into *Europe* by means of a ſhip bound to *Antwerp*. The fact is, that both rat and mouſe were unknown to the new world before it was diſcovered by the *Europeans*, and the firſt rats it ever knew, were introduced there by a ſhip from *Antwerp* †. This animal never made its appearance in *England* till about forty years ago ‡. It has quite extirpated the common kind wherever it has taken its reſidence; and it is to be feared that we ſhall ſcarce find any benefit by the change; the *Norway* rat having the ſame diſpoſition, with greater abilities for doing miſchief, than the common kind. This ſpecies burrows like the water rat, in the banks of rivers, ponds and ditches; it takes the water very readily, and ſwims and dives with great celerity: like the black ſpecies, it preys on rabbets, poultry, and all kind of game; and on grain and fruits. It increaſes moſt amazingly faſt, producing from fourteen to eighteen young at a time. Its bite is not only ſevere, but dangerous; the wound being immediately attended with a great ſwelling, and is a long time in healing. Theſe rats

* *Lin. ſyſt.* 83.

† *Ovalle's Hiſt. of Chile in Churchill's Voy.* iii. 43.

‡ This ſpecies reached the neighborhood of *Paris*, about ſeventeen years ago.

are

are ſo bold, as ſometimes to turn upon thoſe who purſue them, and faſten on the ſtick or hand of ſuch as offer to ſtrike them.

M. *Briſſon* deſcribes this ſame animal twice under different names, p. 170 under the title of *le rat du bois*; and again, p. 173 under that of *le rat de norvege*. M. *de Buffon* ſtiles it *le Surmulot*; as reſembling the mulots, or field mice, in many reſpects; but exceeding them in bulk.

I ſuſpect that this rat came in ſhips originally from the *Eaſt Indies*; a large brown ſpecies being found there, called *Bandicotes*, which burrow under ground. *Barbot** alſo mentions a ſpecies inhabiting the fields in *Guinea*, and probably the ſame with this.

* *Churchill's Coll. Voy.* 214.

Le

27. WATER. Le Rat d'Eau, *Belon* 30. *pl.* 31.
Mus major aquaticus, ſeu Rattus aquaticus. *Raii ſyn. quad.* 217.
Sorex aquaticus. *Charlton ex.* 25.
Meyer's an. ii. *Tab.* 84.
Mus cauda longa pilis ſupra ex nigro et flaveſcente mixtis, infra cinereis veſtitus. *Briſſon quad.* 124.
De Buffon, Tom. vii. 348. *Tab.* 43.
Mus amphibius. Mus cauda elongata piloſa plantis palmatis. *Lin ſyſt.* 82.
Caſtor cauda lineari tereti. *Faun. Suec.* 25. *Ed.* 1. Mus amphibius 52. *Ed.* 2.
Mus aquatilis. *Klein quad.* 57.
Br. Zool. 48. *Syn. quad. No.* 228.

Brit. Llygoden y dwfr
Fren. Le Rat d'eau
Ital, Sorgo morgange
Span.
Port.
Germ. Waſſer mauſe. W. Ratz
Dut. Water-rot
Swed. Watn-ratta
Dan. Vand-rotte

LINNÆUS, from the external appearance of this animal, has in one of his ſyſtems placed it in the ſame genus with the beaver. The form of the head, the ſhortneſs of the ears, and the thickneſs of the fur and the places it haunts, vindicate in ſome degree the opinion that naturaliſt was at that time of: but the form of the tail is ſo different from that of the beaver, as to oblige him to reſtore the water rat to the claſs in which he found it, in the ſyſtem of our illuſtrious countryman *Ray*.

MANNERS. The water-rat never frequents houſes; but is always found on the banks of rivers, ditches and ponds, where it burrows and breeds. It feeds on ſmall fiſh, or the fry of greater; on frogs, inſects, and

and ſometimes on roots: it has a fiſhy taſte; and in ſome countries is eaten; M. *de Buffon* informing us that the peaſants in *France* eat it on maigre days.

It ſwims and dives admirably well, and continues long under water, though the toes are divided like thoſe of the common rat; not connected by membranes, as Mr. *Ray* imagined; and as *Linnæus*, and other writers, relate after him.

The male weighs about nine ounces; the length ſeven inches from the end of the noſe to the tail; the tail five inches: on each foot are five toes, the inner toe of the fore-foot is very ſmall; the firſt joint of the latter is very flexible, which muſt aſſiſt it greatly in ſwimming, and forming its retreat. The head is large, the ears ſmall, and ſcarce appear through the hair: the noſe blunt, and the eyes little: the teeth large, ſtrong, and yellow: the head and body are covered with thick and pretty long hairs, chiefly black; but mixed with ſome of a reddiſh hue: the belly is of an iron-grey: the tail is covered with ſhort black hairs, the tip of it with white hairs.

A female that we opened had ſix young ones in it.

Mus

28. FIELD.

Mus domeſticus medius. *Raii ſyn. quad.* 218.

Mus cauda longa ſupra e fuſco flaveſcens infra ex albo cinereſcens. *Briſſon quad.* 123.

De Buffon, Tom. vii. 325, Tab. 41.

Mus ſylvaticus, M. cauda longa palmis tetradactylis, plantis pentadactylis, corpore griſeo pilis nigris, abdomine albo. *Lin. ſyſt.* 84.

Faun. Suec. 36.

Brit. Zool. 49. *Syn. quad.* No. 230.

Brit. Llygoden ganolig. Llygoden y maes

Fren. Le Mulot

Dan. Voed

THIS meaſures from the noſe-end to the ſetting on of the tail, four inches and half: the tail is four inches long: the eyes are black, large, and full: the ears prominent: the head and upper part of the body, is of a yellowiſh brown, mixed with ſome duſky hairs: the breaſt is of an ochre color; the reſt of the under ſide is white: the tail is covered with ſhort hair.

MANNERS.

Theſe animals are found only in fields and gardens: in ſome places they are called bean-mice, from the havoke they make among beans when firſt ſown. They feed alſo on nuts, acorns, and corn, forming in their burrows vaſt magazines of winter proviſion.

Sæpe exiguus mus
Sub terris poſuitque domos atque horrea fecit.
Virgil Georg. I. 181.

Often the little mouſe
Illudes our hopes; and ſafely lodged below
Hath

Hath formed his granaries.

Doctor *Derham* takes notice of this wonderful ſagacity of theirs, in providing againſt that ſeaſon when they would find a defect of food abroad: but they provide alſo for other animals· the hog comes in for a ſhare; and the great damage we ſuſtain in our fields, by their rooting up the ground, is chiefly owing to their ſearch after the concealed hoards of the field mice.

They generally make the neſt for their young very near the ſurface, and often in a thick tuft of graſs; they bring from ſeven to ten at a time.

Leſs long-tailed field mouſe, *Br. Zool.* II. *App.* 498. *Syn. quad.* No. 231.

29. HARVEST.

THIS ſpecies is very numerous in *Hampſhire*, particularly during harveſt.

They form their neſt above the ground, between the ſtraws of the ſtanding corn, and ſometimes in thiſtles: it is of a round ſhape, and compoſed of the blades of corn. They bring about eight young at a time.

Theſe never enter houſes: but are often carried in the ſheaves of corn into ricks; and often a hundred of them have been found in a ſingle rick, on pulling it down to be houſed.

Thoſe that are not thus carried away in the ſheaves, ſhelter themſelves during winter under ground, and burrow deep, forming a warm bed for themſelves of dead graſs.

DESCRIP. They are the ſmalleſt of the *Britiſh* quadrupeds: their length from noſe to tail is only two inches and a half: their tail two inches: their weight one ſixth of an ounce. They are more ſlender than the other *long-tailed Field Mouſe*; their eyes leſs prominent; their ears naked, and ſtanding out of the fur; their tail ſlightly covered with hair; their back of a fuller red than the larger ſpecies; inclining to the color of a *Dormouſe*: the belly white; a ſtrait line along the ſides dividing the colors of the back and belly.

30. MOUSE. Mus domeſticus vulgaris ſeu minor. *Raii ſyn. quad.* 218.
Seb. Muſeum, i. *Tab.* 111. f. 6. its ſkeleton. *Tab.* 31.
Geſner quad. 714.
Mus cauda longiſſima, obſcure cinereus, ventre ſubalbeſcente. *Briſſon quad.* 119.
De Buffon, *Tom.* vii. 309. *Tab.* 39.

Mus muſculus. M. cauda elongata, palmis tetradactylis, plantis pentadactylis. *Lin. ſyſt.* 83.
Faun. Suec. 34.
Mus minor, Muſculus vulgaris. *Klein quad.* 57.
Br. Zool. 50. *Syn. quad.* No. 229.

Brit. Llygoden
Fren. La Souris
Ital. Topo, ſorice
Span. Raton
Port. Ratinho
Germ. Maus
Dut. Muys
Swed. Mus
Dan. Muus

THIS timid, cautious, active, little animal, is too well known to require a deſcription: it is

No. 30

MOUSE.

WATER SHREW MOUSE.

No. 33

is entirely domeſtic, being never found in fields; or, as M. *Buffon* obſerves, in any countries uninhabited by mankind: it breeds very frequently in the year, and brings ſix or ſeven young at a time. This ſpecies is often found of a pure white, in which ſtate it makes a moſt beautifull appearance; the fine full eye appearing to great advantage, amidſt the ſnowy color of the fur. The root of *white hellebore* and *ſtaves-acre*, powdered and mixed with meal, is a certain poiſon to them.

31. SHORT TAILED.

Mus agreſtis capite grandi brachiurus. *Raii ſyn. quad.* 218.
Mus cauda brevi pilis e nigricante et ſordide luteo mixtis in dorſo et ſaturate cinereis in ventre veſtitis. *Briſſon quad.* 125.

Mus agreſtis. *Faun. Suec.* 30.
De Buffon, Tom. vii. 369. *Tab.* 47.
Klein quad. 57 *No.* 50.
Br. Zool. 50. *Syn. quad. No.* 233.

Brit. Llygoden gwtta'r maes
Ital. Campagnoli

Fren. Le petit Rat de champs, Le campagnol
Dan. Skier-muus

DESCRIP.

THE length of this ſpecies, from the noſe to the tail, is about ſix inches; the tail only an inch and a half: the head is very large: the eyes prominent: the ears quite hid in the fur: the whole upper part of the body is of a ferruginous color, mixed with black; the belly of a deep aſh-color: the tail is covered with ſhort hair,

K 2 ending

ending with a little buſh, about a quarter of an inch long. The legs, particularly the fore legs, very ſhort.

MANNERS. This animal makes its neſt in moiſt meadows, and brings eight young at a time: it has a ſtrong affection for them: one that was ſeduced into a wire-trap, by placing its brood in it, was ſo intent on foſtering them, that it appeared quite regardleſs of its captivity. The manners of this creature much reſemble the 28th ſpecies: like it, this reſides under ground, and lives on nuts, acorns, but particularly on corn: it differs from the former in the place of its abode: ſeldom infeſting gardens.

It has been obſerved that in houſing a rick of corn, the dogs have devoured all the mice of this ſpecies that they could catch, and rejected the common kind; and that the cats on the contrary would touch none but the laſt.

Two

Two cutting teeth in each jaw pointing forward.
Long ſlender noſe; ſmall ears.
Five toes on each foot.

XVI.
SHREW.

32. FETID.

Mus araneus. Shrew, Shrew Mouſe, or Hardy Shrew. *Raii ſyn. quad.* 239.
Geſner quad. 747.
Mus araneus ſupra ex fuſco rufus infra albicans. *Briſſon quad.* 126.
De Buffon, Tom. viii. 57. *Tab.* 10.
Sorex araneus. S. cauda corpore longiore. *Lin. ſyſt.* 74.
Faun. Suec. 24.
Mus araneus roſtro productiore. *Klein quad.* 58.
Br. Zool. 54. *Syn. quad. No.* 235.

Brit. Llygoden goch, Chwiſtlen, Llyg
Fren. La Muſaraigne
Ital. Toporango
Span. Murganho
Port.
Germ. Spitzmauſe, Ziſſmuſs, Muger
Swed. Nabbmus
Dan. Næbmuus, Muuſeſkier

DESCRIP.

THE length of this little animal, from the end of the noſe to the origin of the tail is two inches and a half: that of the tail, near one inch and a half: the noſe is very long and ſlender; and the upper mandible is much longer than the lower, beſet with long but fine whiſkers: the ears are ſhort, and rounded: the eyes are very ſmall; and, like thoſe of the mole, almoſt concealed in the hair. The color of the head, and upper part of the body, is of a browniſh duſky red: the belly of a dirty white: the tail is covered with ſhort duſky hairs: the legs are very ſhort: the hind legs placed very far back: the feet are divided into five toes.

Above and below are two ſlender cutting teeth pointing forward, and on each a minute proceſs: the reſt of the teeth are ſo cloſely united, as to appear a continued ſerrated bone in every jaw; the whole number is twenty eight.

The ſhrew inhabits old walls, heaps of ſtones, and holes in the earth: is frequently found near hayricks, dunghills, and neceſſary houſes: is often obſerved rooting like a ſwine in ordure: it lives on corn, inſects, and any filth: from its food or the places it frequents, has a diſagreeable ſmell: cats will kill but not eat it: brings four or five young at a time. In *Auguſt* is an annual mortality of them, numbers being in that ſeaſon found dead in the paths. The antients believed them to be injurious to cattle, an error now detected.

33. WATER. Muſaraneus dorſo nigro ventreque albo. *Merret Pinax.* 167.
Sorex fodiens, *Pallas ined.*
La Muſaraigne d'Eau, *de Buffon.* viii. 64.
Water Shrew, *Syn. quad. No.* 256.

THIS ſpecies inhabits the banks of ditches, and other wet ſituations, and is in ſome places

places called the *Blind Mouſe*, from the ſmallneſs of its eyes. The *Germans* call it *Græber* or digger. I imagine it to be the ſame that the inhabitants of *Sutherland* call the water mole, and thoſe of *Cathneſs*, the *Lavellan*, which the laſt imagine poiſons their cattle; and is held by them in great abhorrence. It burrows in banks near the water: and according to *M. de Buffon* brings nine young. It was known to Dr. *Merret* above a century ago; but loſt again till within theſe few years, when it was found to inhabit *Lincolnſhire*, and *Lancaſhire*. Its length from noſe to tail is three inches and three quarters: the tail two inches: the noſe long and ſlender: ears minute: eyes very ſmall and hid in the fur: the color of the head and upper part of the body black: the throat, breaſt, and belly aſh-color; beneath the tail is a triangular duſky ſpot.

XVII. MOLE.

Long ſlender noſe, upper jaw much longer than the lower.

No ears.

Fore-feet very broad, with ſcarce any apparent legs before: hind-feet very ſmall.

34. EUROPEAN.

Talpa. The Mole, Mold-Warp, or Want. *Raii ſyn. quad.* 236.
Meyer's an. i. *Tab.* 2.
Talpa alba noſtras. *Seb. Mus.* i. p. 61. *Tab.* 32. f. 1.
Sib. Scot. 11.
Geſner quad. 931.
Talpa caudata nigricans pedibus anticis et poſticis pentadactylis. *Briſſon quad.* 203.
De Buffon, viii. 81. *Tab.* 12.
Talpa europæus. T. caudata, pedibus pentadactylis. *Lin. ſyſt.* 73.
Faun Suec. 23.
Talpa. *Klein quad.* 60.
Br. Zool. 52. *ſyn. quad.* No. 241.

Brit.	Gwadd, Twrch daear	*Germ.*	Maulwerf
Fren.	La Taupe	*Dut.*	Mol.
Ital.	Talpa.	*Swed.*	Mulvad, Surk
Span.	Topo	*Dan.*	Muldvarp
Port.	Toupeira		

THERE are many animals in which the Divine Wiſdom may be more agreeably illuſtrated; yet the uniformity of its attention to every article of the creation, even the moſt contemptible, by adapting the parts to its deſtined courſe of life, appears more evident in the mole than in any other animal.

A ſubterraneous abode being allotted to it, the ſeeming defects of ſeveral of its parts, vaniſh; which, inſtead of appearing maimed, or unfiniſhed, exhibit

exhibit a moſt ſtriking proof of the fitneſs of their contrivance.

The breadth, ſtrength, and ſhortneſs of the fore-feet, which are inclined ſideways, anſwer the uſe as well as form of hands; to ſcoop out the earth, to form its habitation, or to purſue its prey. Had they been longer, the falling in of the earth would have prevented the quick repetition of its ſtrokes in working, or have impeded its courſe: the oblique poſition of the fore-feet, has alſo this advantage, that it flings all the looſe ſoil behind the animal.

The form of the body is not leſs admirably contrived for its way of life: the fore part is thick and very muſcular, giving great ſtrength to the action of the fore-feet; enabling it to dig its way with amazing force and rapidity, either to purſue its prey, or elude the ſearch of the moſt active enemy. The form of its hind parts, which are ſmall and taper, enables it to paſs with great facility through the earth, that the fore-feet had flung behind; for had each part of the body been of equal thickneſs, its flight would have been impeded, and its ſecurity precarious.

The ſkin is moſt exceſſively compact, and ſo tough as not to be cut but by a very ſharp knife: the hair is very ſhort, and cloſe ſet, and ſofter than the fineſt ſilk: the uſual color is black; not but that there are inſtances of theſe animals being ſpotted*, and a

* *Edw.* 268.

creme

creme colored breed is ſometimes found in my lands near *Downing*.

The ſmallneſs of the eyes (which gave occaſion to the ancients to deny it the ſenſe of ſight*,) is to this animal a peculiar happineſs: a ſmall degree of viſion is ſufficient for an animal ever deſtined to live under ground: had theſe organs been larger, they would have been perpetually liable to injuries, by the earth falling into them; but nature, to prevent that inconvenience, hath not only made them very ſmall, but alſo covered them very cloſely with fur. Anatomiſts mention (beſides theſe) a third very wonderful contrivance for their ſecurity; and inform us that each eye is furniſhed with a certain muſcle, by which the animal has power of withdrawing or exerting them, according to its exigencies.

To make amends for the dimneſs of its ſight, the mole is amply recompenſed, by the great perfection of two other ſenſes, thoſe of hearing and of ſmelling: the firſt gives it notice of the moſt diſtant approach of danger: the other, which is equally exquiſite, directs it in the midſt of darkneſs to its food: the noſe alſo, being very long and ſlender, is well formed for thruſting into ſmall holes, in ſearch of the worms and inſects that inhabit them.

* Aut *oculis capti* fodere cubilia talpæ. *Virg. Georg.* I.
Or *ſightleſs* moles have dug their chamber'd lodge,

Theſe

Theſe gifts may with reaſon be ſaid to compenſate the defect of ſight, as they ſupply in this animal all its wants, and all the purpoſes of that ſenſe. Thus amply ſupplied as it is, with every neceſſary accommodation of life; we muſt avoid aſſenting to an obſervation of a moſt reſpectable writer, and only refer the reader to the note, where he may find the very words of that author; and compare them with thoſe of our illuſtrious countryman, Mr. *Ray* *.

It is ſuppoſed that the verdant circles ſo often ſeen in graſs grounds, called by country people *fairy rings*, are owing to the operations of theſe animals, who at certain ſeaſons perform their burrowings by circumgyrations, which looſening the ſoil, gives the ſurface a greater fertility and rankneſs of graſs than the other parts within or without the ring.

The mole breeds in the ſpring, and brings four or five young at a time: it makes its neſt of moſs, and that always under the largeſt hillock, a little below

* La taupe ſans être aveugle, a les yeux ſi petits ſi couverts, qu'elle ne peut faire grand uſage du ſens de la vûe: *en dedommagement la nature lui a donné avec magnificence l'uſage du ſixième ſens*, &c.

Mr. *Ray* makes the latter obſervation; but forms from it a concluſion much more ſolid and moral. *Teſtes maximos, paraſtatas ampliſſimas, novum corpus ſeminale ab his diverſum et ſeparatum —— penem etiam facile omnium, ni fallor, animalium longiſſimum: ex quibus colligere eſt maximam præ reliquis omnibus animalibus voluptatem in coitu hoc abjectum et vile animalculum percipere, ut habeant quod ipſi invideant, qui in hoc ſupremas vitæ ſuæ delicias collocant.* *Raii ſyn. quad.* 238, 239.

the

the ſurface of the ground. The mole is obſerved to be moſt active, and to caſt up moſt earth, immediately before rain; and in the winter before a thaw; becauſe at thoſe times the worms and inſects begin to be in motion, and approach the ſurface: on the contrary, in very dry weather, this animal ſeldom or never forms any hillocks, as it penetrates deep after its prey, which at ſuch ſeaſons retires far into the ground. During ſummer they run in ſearch of ſnails and worms in the night time among the graſs, which makes them the prey of owls. The mole ſhews great art in ſkinning a worm, which it always does before it eats it; ſtripping the ſkin from end to end, and ſqueezing out all the contents of the body.

Theſe animals do incredible damage in gardens, and meadows; by looſening the roots of plants, flowers, graſs, corn, *&c.* *Mortimer* ſays, that the roots of *Palma chriſti* and *white hellebore*, made into a paſte, and laid in their holes, will deſtroy them. They ſeem not to have many enemies among other animals, except in *Scotland*, where (if we may depend on Sir *Robert Sibbald*) there is a kind of mouſe, with a black back, that deſtroys moles *. We have been aſſured that moles are not found in *Ireland*.

* *Sib. Hiſt. Scot.* Part iii. p. 12. I did not find it was known at preſent.

Five

Five toes on each foot.
Body covered with ſhort ſtrong ſpines.

XVIII. URCHIN.

35. COMMON.

Echinus ſc. erinaceus terreſtris. *Raii ſyn. quad.* 231.
Meyer's an. i. Tab. 95, 96.
Sib. Scot. 11.
Erinaceus parvus noſtras. *Seb. Mus.* i. p. 78. *Tab.* 49. f. 1, 2.
Erinaceus auriculis erectis. *Briſſon quad.* 128.
De Buffon, Tom. viii. 28. *Tab.* 6.
Echinus terreſtris. *Geſner quad.* 368.
Erinaceus europæus. *Lin. ſyſt.* 75.
Erinaceus ſpinoſus auriculatus. *Faun. Suec.* 22.
Acanthion vulgaris noſtras. *Klein quad.* 66.
Br. Zool. 51. *Syn. quad.* No. 247.

Brit. Draenog, Draen y coed
Fren. L'Heriſſon
Ital. Riccio
Span. Erizo
Port. Ourizo
Germ. Igel
Dut. Eegel-varken
Swed. Igelhot
Dan. Pin-ſuin, Pin-ſoe

DESCR.

THE uſual length of this animal, excluſive of the tail, is ten inches: the tail is little more than an inch long; but ſo concealed by the ſpines, as ſcarce to be viſible. The form of the noſe is like that of the hog; the upper mandible being much longer than the lower, and the end flat: the noſtrils are narrow, terminated on each ſide by a thin looſe flap: the color of the noſe is duſky; it is covered by a few ſcattered hairs: the upper part of the head, the ſides, and the rump, are clothed with ſtrong ſtiff hairs, approaching the nature of briſtles, of a yellowiſh and cinereous hue.

The

LEGS. The legs are ſhort, of a duſky color, and almoſt bare: the toes on each foot are five in number, long, and ſeparated the whole way: the thumb, or interior toe, is much ſhorter than the others: the claws long, but weak: the whole upper part of the body and ſides are cloſely covered with ſtrong ſpines, of an inch in length, and very ſharp pointed: their lower part is white, the middle black, the points white. The eyes are ſmall, and placed high in the head: the ears are round, pretty large, and naked. TEETH. The mouth is ſmall, but well furniſhed with teeth: in each jaw are two ſharp pointed cutting teeth: in the upper jaw are on each ſide four tuſhes, and five grinders: in the lower jaw on each ſide are three tuſhes, pointing obliquely forward; and beyond thoſe, four grinders.

The hedge hog is a nocturnal animal, keeping retired in the day; but is in motion the whole night, in ſearch of food. It generally reſides in ſmall thickets, in hedges, or in ditches covered with buſhes; lying well wrapped up in moſs, graſs, or leaves: its food is roots, fruits, worms, and inſects: it lies under the undeſerved reproach of ſucking cattle, and hurting their udders; but the ſmallneſs of its mouth renders that impoſſible.

MANNERS. It is a mild, helpleſs, and patient animal; and would be liable to injury from every enemy, had not Providence guarded it with a ſtrong covering, and a power of rolling itſelf into a ball, by that means ſecuring the defenceleſs parts. The

The barbarity of anatomifts furnifhes us with an amazing inftance of its patience ; one that was diffected alive, and whofe feet were nailed down to the table, endured that, and every ftroke of the operator's knife, without even one groan *.

* *Clavis terebrari fibi pedes et difcindi vifcera patientiffimè ferebat ; omnes cultri ictus fine gemitu plufquam Spartanâ nobilitate concoquens.* Borrich: in Blas; de Echino. 64.

D I V.

DIV. III.

PINNATED QUADRUPEDS,

With fin-like feet: fore legs buried deep in the ſkin: hind legs pointing quite backwards.

XIX. SEAL. Cutting teeth and two canine in each jaw.
Five palmated toes on each foot.
Body thick at the ſhoulders, tapering towards the tail.

36. GREAT. Sea calf, *Ph. Tranſ.* ix. 74. *Tab.* 5.
Le grand Phoque, *de Buffon,* xiii. 345.
Utſuk? *Crantz Greenl.* i. 125.
Great ſeal, *Syn. quad.* No. 266.

A SPECIES not very uncommon on the coaſt of *Scotland*, particularly about the rock *Hiſkyr*, one of the weſtern iſles, which grows to the length of twelve feet.

A young one of this ſpecies was ſome years ago ſhewen in *London*: notwithſtanding it was ſo young as to have ſcarce any teeth, yet it was ſeven feet and a half long.

In my voyage among the *Hebrides* I frequently heard

Pl. XII.

N°

SEALS

heard of this ſpecies, but did not meet with it. Mr. *Thompſon*, our maſter, ſhot one; but it ſunk, and we loſt it.

37 COMMON.

Le Veau marin, ou loup de Mer. *Belon* 25. Pl. 26.
Seal, Seoile, or Sea-calf. Phoca, ſeu vitulus marinus. *Raii ſyn. quad.* 189.
Sea-calf. *Phil. Tranſ.* No. 469. *Tab.* I. *Abridg.* xlvii.
Smith's Kerry, 84, 364.
Borlaſe's Cornw. 284.
Worm. muſe. 289.
Kaſſigiak. *Crantz's hiſt. Greenl.* i. 123.
Le Phoque, *de Buffon*, xiii. 333.
Horr. Icel. 88.
Pontop. Norw. ii. 125.
Briſſon quad. 162.
Phoca vitulina. *Lin. ſyſt.* 56.
Phoca. *Klein quad.* 93.
Phoca dentibus caninis tectis. *Faun. Suec.* 4.
Br. Zool. 34. *Syn. quad.* No. 265.

Brit.	Moelrhon	*Germ.*	Meer wolff, Meer hund
Fren.	Le Veau marin	*Dut.*	Zee hond
Ital.	Vechio marino	*Swed.*	Sial
Span.	Lobo marino	*Dan.*	Sæl hund

DESCRIP.

THE common length of thoſe taken on the *Britiſh* coaſts, is from five to ſix feet.

The ſubject that we took our deſcription from, was a young one; ſo allowance muſt be made for the proportions of the meaſurements of thoſe that have attained their full ſize. Its length, from the end of the noſe to the end of the hind feet, was two feet nine inches; to the end of the tail, two feet three inches: the head was ſeven inches long: the tail two and a half: the fore legs were deeply immerſed in the ſkin of the body; what appeared out, was only eight inches long: the breadth of the fore feet, when extended, was three inches and a half: the hind legs were placed in ſuch a manner, as to point directly backwards; and were ten inches long: each hind foot, when extended, was nine inches and a half broad: every foot was divided into five toes; and each of thoſe connected by a ſtrong and broad web, covered on both ſides with ſhort hair.

The toes were furniſhed with ſtrong claws, well adapted to aſſiſt the animal in climbing the rocks it baſked on: the claws on the hind feet were about an inch long, ſlender, and ſtrait; except at the ends, which were a little incurvated.

The circumference of the body in the thickeſt part, which was near the ſhoulders, was one foot ten inches; but near the hind legs, where it was narroweſt, it meaſured only twelve inches.

The head and noſe were broad and flat, like thoſe of the otter; the neck ſhort and thick; the eyes large and black; it had no external ears, but in lieu of them, two ſmall orifices: the noſtrils were oblong: on each ſide the noſe were ſeveral long ſtiff hairs; and above each eye, were a few of the ſame kind.

TONGUE.

The form of the tongue of this animal is ſo ſingular, that were other notes wanting, that alone would diſtinguiſh it from all other quadrupeds; being forked, or ſlit at the end.

The cutting teeth are ſingular in reſpect to their number, being ſix in the upper jaw, and only four

in

in the lower. It has two canine teeth above and below, and on each ſide of the jaws five grinders; the total thirty-four.

The whole animal was covered with ſhort hair, very cloſely ſet together: the color of that on the head and feet was duſky: on the body duſky, ſpotted irregularly with white: on the back the duſky color predominated; on the belly white: but ſeals vary greatly in their marks and colors, and ſome have been found * entirely white. One that was taken near *Cheſter*, in *May* 1766, had on its firſt capture, the body naked like the ſkin of a porpeſe; and only the head and a ſmall ſpot beneath each fore leg, hairy: it was kept alive ſome time; but before it died, hair began to grow over the whole body †.

The ſeal is common on moſt of the rocky ſhores of *Great Britain* and *Ireland*, eſpecially on the northern coaſts: in *Wales* it frequents the coaſts of *Caernarvonſhire*, and *Angleſey*. It preys entirely on fiſh, and never moleſts the ſea fowl: for I have ſeen numbers of each floating on the waves, as if in company. Seals eat their prey beneath the water; and in caſe they are devouring any very oily fiſh, the place is known by a certain ſmoothneſs

* In the *Aſhmolean Muſeum* at *Oxford*, is a good picture of two white ſeals.

† *Vide* The figure publiſhed in the additional plates of the folio edition of this work.

of the waves immediately above. The power of oil in ſtilling the waves excited by a ſtorm, is mentioned by *Pliny*: the moderns have made the experiment with ſucceſs *; and by that made one advance towards eradicating the vulgar prejudices againſt that great and elegant writer.

We muſt acknowlege the obligations we were under to the Rev. Mr. *Farrington* of *Dinas*, in *Caernarvonſhire*, for ſeveral learned communications; but in particular for the natural hiſtory of this animal, which we ſhall give the public in his own words.

MANNERS. ‘The ſeals are natives of our coaſts; and are ‘found moſt frequently between *Llyn* in *Caernar-* ‘*vonſhire*, and the northern parts of *Angleſey*: they ‘are ſeen often towards *Carrig y moelrhon*, to the ‘weſt of *Bardſey*, or *Ynys Enlli*; and the *Skerries*, ‘commonly called in the *Britiſh* language *Yynys y* ‘*moelrhoniad*, or ſeal iſland. The *Latin* name of ‘this amphibious animal is *Phoca* †: the vulgar ‘name is ſea calf; and on that account, the male is ‘called the bull, and the female the cow; but the ‘*Celtic* appelative is *Moelrhon*, from the word ‘*Moel*, bald, or without ears, and *Rhon*, a ſpear or ‘lance.

* *Phil. Tranſ.* 1774. p. 445.

† Doctor *Charleton* derives the word φωκη ex βωκη, *boatu* quem edit: *vide Exercitationes de dif. An. piſc.* p. 48. But I do not find any authority for his opinion.

‘They

'They are excellent ſwimmers, and ready di-
'vers, and are very bold when in the ſea, ſwim-
'ming careleſsly enough about boats: their dens or
'lodgements are in hollow rocks, or caverns, near
'the ſea; but out of the reach of the tide: in
'the ſummer they will come out of the water, to
'baſk or ſleep in the ſun, on the top of large ſtones,
'or ſhivers of rocks; and that is the opportunity
'our countrymen take of ſhooting them: if they
'chance to eſcape, they haſten towards their pro-
'per element, flinging ſtones and dirt behind them,
'as they ſcramble along; at the ſame time expreſ-
'ſing their fears by piteous moans; but if they
'happen to be overtaken, they will make a vigo-
'rous defence with their feet and teeth, till they
'are killed. They are taken for the ſake of their
'ſkins, and for the oil their fat yields: the former
'ſell for four ſhillings, or four and ſix-pence a
'piece; which, when dreſſed, are very uſeful in
'covering trunks, making waiſtcoats, ſhot pouch-
'es, and ſeveral other conveniencies.'

The fleſh of theſe animals, and even of porpeſes, formerly found a place at the tables of the great; as appears from the bill of fare of that vaſt feaſt that archbiſhop *Nevill* gave in the reign of *Edward* the fourth, in which is ſeen, that ſeveral were provided on the occaſion *. They couple about *April*, on large rocks, or ſmall iſlands, not remote from the

* *Leland's Collectanea*.

L 3 ſhore;

ſhore; and bring forth in thoſe vaſt caverns that are frequent on our coaſts; they commonly bring two at a time, which in their infant ſtate are covered with a whitiſh down, or woolly ſubſtance. The ſeal-hunters in *Cathneſs* have aſſured me that their growth is ſo ſudden, that in nine tides from their birth (fifty-four hours) they will become as active as their parents.

On the coaſt of that county are immenſe caverns opening into the ſea, and running ſome hundreds of yards beneath the land. Theſe are the reſort of ſeals in the breeding time, where they continue till their young are old enough to go to ſea, which is in about ſix or ſeven weeks. The firſt of theſe caves is near the *Ord*, the laſt near *Thrumſter*: their entrance ſo narrow, as only to admit a boat; their inſide very ſpatious and lofty. In the month of *October*, or the beginning of *November*, the ſeal-hunters enter the mouths of the caverns about midnight, and rowing up as far as they can, they land: each of them being provided with a bludgeon, and properly ſtationed, light their torches, and make a great noiſe, which brings down the ſeals from the farther end in a confuſed body with fearfull ſhrieks and cries: at firſt the men are obliged to give way for fear of being over-born; but when the firſt crowd is paſt, they kill as many as ſtraggle behind, chiefly the young, by ſtriking them on the noſe; a very ſlight blow on that part diſpatches them. When the work is over, they drag the ſeals

to

to the boat, which two men are left to guard. This is a moſt hazardous employ; for ſhould their torches go out, or the wind blow hard from ſea during their continuance in the cave, their lives are loſt. The young ſeals of ſix weeks age, yield more oil than their emaciated dams: above eight gallons have been got from a ſingle whelp, which ſells from ſix-pence to nine-pence per gallon; the ſkins from ſix-pence to twelve-pence.

The natural hiſtory of this animal may be further elucidated, by the following extracts from a letter of the Rev. Dr. *William Borlaſe*, dated *October* the 24th, 1763.

'The ſeals are ſeen in the greateſt plenty on the 'ſhores of *Cornwall*, in the months of *May*, *June*, 'and *July*.

'They are of different ſizes; ſome as large as a 'cow, and from that downwards to a ſmall calf.

'They feed on moſt ſorts of fiſh which they can 'maſter, and are ſeen ſearching for their prey near 'ſhore, where the whiſtling fiſh, wraws, and 'polacks reſort.

'They are very ſwift in their proper depth of 'water, dive like a ſhot, and in a trice riſe at fifty 'yards diſtance; ſo that weaker fiſhes cannot avoid 'their tyranny, except in ſhallow water. A per-'ſon of the pariſh of *Sennan*, ſaw not long ſince a 'ſeal in purſuit of a mullet (that ſtrong and ſwift 'fiſh): the ſeal turned it to and fro' in deep water, 'as a gre-hound does a hare: the mullet at laſt 'found it had no way to eſcape, but by running 'into ſhoal water: the ſeal purſued; and the for- 'mer, to get more ſurely out of danger, threw it- 'ſelf on its ſide, by which means it darted into 'ſhoaler water than it could have ſwam in with the 'depth of its paunch and fins, and ſo eſcaped.

'The ſeal brings her young about the begin- 'ning of autumn; our fiſhermen have ſeen two 'ſucking their dam at the ſame time, as ſhe ſtood 'in the ſea in a perpendicular poſition.

'Their head in ſwimming is always above 'water, more ſo than that of a dog.

'They ſleep on rocks ſurrounded by the ſea, or 'on the leſs acceſſible parts of our cliffs, left dry 'by the ebb of the tide; and if diſturbed by any 'thing, take care to tumble over the rocks into 'the ſea. They are extremely watchful, and ne- 'ver ſleep long without moving; ſeldom longer 'than a minute; then raiſe their heads, and if they 'hear or ſee nothing more than ordinary, lie down 'again, and ſo on, raiſing their heads a little, and 'reclining them alternately, in about a minute's 'time. Nature ſeems to have given them this 'precaution, as being unprovided with auricles, 'or external ears; and conſequently not hearing 'very quick, nor from any great diſtance.'

In Sir *R. Sibbald*'s hiſtory of *Scotland*, we find an account of another ſpecies of the ſeal kind, which is copied from *Boethius*. The animal he mentions is the ſea-horſe, *Walrus* or *Morſe*: as this vaſt

vaſt creature is found in the *Norwegian* ſeas, we think it not improbable but that it may have appeared on the *Scottiſh* coaſts; but having no better authority for it, than what is above-mentioned, we dare not give it a place in a *Britiſh Zoology.* The teeth of that animal are as white and hard as ivory; but whether the ελεφάντινα ψάλια, ivory bits, which *Strabo* * mentions among the articles of the *Britiſh* commerce, were made of them, or the tooth of the *Narhwal*, or of ſome of the toothed whales, is not at this time eaſy to be determined. But we may here remark that *Solinus*, in his account of *Britain*, informs us that the fine gentlemen of our iſland adorned the hilts of their ſwords with the teeth of ſea beaſts, which were as white as ivory itſelf †.

* *Strabo, Lib.* iv. 200.

† *Polyhiſt.* c. xxxv.

D I V.

DIV. IV.

WINGED QUADRUPEDS.

XX. BAT. With long extended toes to the fore-feet, connected by thin membranes, extending to the hind-legs.

38. GREAT. La noctule *de Buffon* VIII. *Tab.* xviii. p. 128.
Syn. quad. No. 287.

IS a species less common in *Great-Britain* than the smaller. It ranges high in the air for food, and retires early in the summer.

Is the largest we have: its extent of wing is fifteen inches: its length to the rump two inches eight tenths: of the tail one inch seven tenths.

The nose is slightly bilobated: ears small and rounded: on the chin a minute *verruca*. Hair on the body a reddish ash-color.

They collect under eaves of buildings in vast numbers. The Rev. Doctor *Buckworth* informed me that under those of *Queen's College, Cambridge*, he saw taken in one night, one hundred and eighty-five; the second night sixty-three: the third, two.

La

XIII

No. 40

LONG EARED BAT.

GREAT BAT.

No. 38

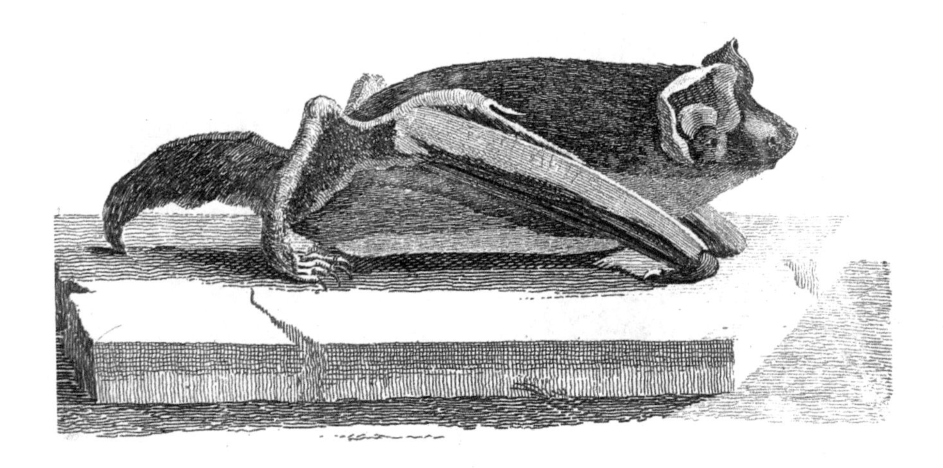

I. Griffiths del

Pl. XIV.

No. 39.

HORSE-SHOE BAT.

La Chauve-souris a fer a cheval. *De Buffon* VIII. 131. *Tab.* xvii. xx.

Horse-shoe Bat. *Syn. quad.* No. 186.

39. HORSE-SHOE.

THIS species was discovered by Mr. *Latham* Surgeon at *Dartford, Kent*; who was so obliging as to communicate it to me. They are found in greatest numbers in the salt-petre houses belonging to the powder mills; and frequent them during the evening for the sake of the gnats which swarm there. They have been also found during winter in a torpid state clinging to the roof. They often feed on *Chafers*, but only eat the body.

The length from the nose to the tip of the tail is three inches and a half: the extent fourteen. At the end of the nose is an upright membrane in form of a horse-shoe. Ears large, broad at their base, inclining backwards; but want the little or internal ear. The color of the upper part of the body is deep cinereous; of the lower whitish.

Edw. av. 201. f. 3.
Alb. iii. Tab. 101.
La petite chauve souris de notre pays. *Brisson quad.* 160.
L'oreillar. *De Buffon, Tom.* viii. 118. 127. Tab. 17. f. 1.

Vespertilio auritus. *Lin. syst.* 47.
V. auritus, naso oreque simplici, auriculis duplicatis, capite majoribus. *Faun. Suec.* 3.
Br. Zool. 56. *Syn. quad. No.* 292.

40. LONG EARED.

THIS species is the lest of the *British* bats: the length being only an inch and three quarters; and the extent of the fore-legs seven inches.

The

The principal diſtinction between this and the common kind, is the ears; which in this are above an inch long, very thin, and almoſt tranſparent: within each of theſe is a leſſer ear, or at leaſt a membrane reſembling one; which, as Mr. *Edwards* obſerves, may poſſibly ſerve as a valve to cloſe the larger, in the ſleeping ſtate of this animal.

41. COMMON. Veſpertilio. Bat, Flitter, or Flutter Mouſe. *Raii ſyn. quad.* 243.
Short-eared *Engliſh* Bat. *Edw. av.* 201. f. 2.
Seb. Mus. i.
The Rear Mouſe. *Charlton ex.* 80.
Meyer's an i. Tab. 3.
Geſner av. 766.
Veſpertilio murini coloris, pedibus omnibus pentadactylis. *Briſſon quad.* 158.
La chauve ſouris. *De Buffon, Tom.* viii. 113. Tab. 16.
Veſpertilio murinus. *Lin. ſyſt.* 47.
V. caudatus naſo oreque ſimplici. *Faun. Suec.* 2.
V. major. *Klein quad.* 61.
Veſpertilio. *Plinii Lib.* x. 6. 61.
Br. Zool. 55. *Syn. quad.* No. 291.

Brit. Yſtlum
Fren. La Chauve ſouris
Ital. Nottola, Notula, Sporteglione, Vilpiſtrello, Vilpiſtrello
Span. Murcielago, Morciegalo
Port. Morcego
Germ. Speckmaus, Fledermaus
Dut. Vledermuys
Swed. Laderlap, Fladermus
Dan. Flagermuus, Aftenbakke

THIS ſingular animal was placed by *Pliny*, *Geſner*, *Aldrovandus*, and ſome other naturaliſts, among the birds: they did not conſider, that it wanted every character of that order of animals, except the power of flying: if the irregular,

gular, uncertain, and jerking motion * of the bat in the air, can merit the name of flight. No birds whatſoever are furniſhed with teeth, or bring forth their young alive, and ſuckle them: were other notes wanting, theſe would be ſufficient to determine that the bat is a quadruped.

The ſpecies now deſcribed, is the moſt common: the uſual length of it is about two inches and a half: the extent of the fore-legs nine inches.

The members that are uſually called the wings, are nothing more than the four interior toes of the fore-feet, produced to a great length, and connected by a thin membrane; which extends alſo to the hind legs; and from them to the tail: the firſt toe is quite looſe, and ſerves as a heel, when the bat walks; or as a hook, when it would adhere to any thing. The hind-feet are diſengaged from the membrane, and divided into five toes, furniſhed with pretty ſtrong claws. The membranes are of a duſky color: the body is covered with ſhort fur, of a mouſe-color, tinged with red. The eyes are very ſmall: the ears like thoſe of the mouſe.

This ſpecies of bat is very common in *England*: it makes its firſt appearance early in the ſummer, and begins its flight in the duſk of the evening: it principally frequents the ſides of woods, glades, and ſhady walks; and is alſo frequently obſerved

* The *Engliſh* ſynonym of this animal, *Flitter*, or *Flutter mouſe*, is very expreſſive of its action in the air.

to

to ſkim along the ſurface of pieces of water, in queſt of gnats and inſects: theſe are not its only food; for it will eat meat of any kind that it happens to find hanging up in a larder.

The bat brings only two young at a time; which it ſuckles from two teats placed on the breaſt, like thoſe of the human race. Theſe animals are capable of being brought to ſome degree of familiarity. The Rev. Mr. *White* of *Selborne* has ſeen a bat ſo far tamed as to eat inſects out of a perſon's hand; and while it was feeding would bring its wings round before its mouth, hovering in the manner of birds of prey.

Towards the latter end of ſummer, the bat retires into caves, ruined buildings, the roofs of houſes, or hollow trees; where it remains the whole winter, in a ſtate of inaction; ſuſpended by the hind-feet, and cloſely wrapped up in the membranes of the fore-feet.

The voice of the bat is ſomewhat like that of the mouſe; but very low, and weak. *Ovid* takes notice both of that, and the derivation of its *Latin* name,

Lucemque peroſæ
Nocte volant, ſeroque tenent a veſpere nomen.
Minimam pro corpore vocem
Emittunt, peraguntque levi ſtridore querelas.
Met. lib. iv. 10.

Their little bodies found
No words, but murmur'd in a fainting ſound.

In

In towns, not woods, the ſooty bats delight,
And never till the duſk begin their flight;
Till *Veſper* riſes with his evening flame;
From whom the *Romans* have derived their name.

Euſden.

CLASS

CLASS II.

BIRDS.

AVES INTERNUNCIÆ JOVIS.

CLASS II.

BIRDS.

Div. I. LAND BIRDS.

II. WATER BIRDS.

Div. I. Order I. RAPACIOUS.

Genus.

I. FALCON.
II. FOWL.

II. PIES.

III. SHRIKE.
IV. CROW.
V. CUCKOO.
VI. WRYNECK.
VII. WOODPECKER.
VIII. KINGFISHER.
IX. NUTHATCH.
X. HOOPOE.
XI. CREEPER.

III. GALLINACEOUS.

COCK.
TURKEY.

M 2 PINTADO.

Genus.	
	PINTADO.
	PEACOCK.
	PHEASANT.
XII.	GROUS.
XIII.	BUSTARD.

IV. COLUMBINE.

XIV.	PIGEON.

V. PASSERINE.

XV.	STARE.
XVI.	THRUSH.
XVII.	CHATTERER.
XVIII.	GROSBEAK.
XIX.	BUNTING.
XX.	FINCH.
XXI.	FLY-CATCHER.
XXII.	LARK.
XXIII.	WAGTAIL.
XXIV.	WARBLERS.
XXV.	TITMOUSE
XXVI.	SWALLOW.
XXVII.	GOATSUCKER.

Div. II. WATER BIRDS.

VI. CLOVEN FOOTED.

XXVIII.	HERON.
XXIX.	CURLEW.
XXX.	SNIPE.

SAND-

Genus.

XXXI. SANDPIPER.
XXXII. PLOVER.
XXXIII. OYSTER-CATCHER.
XXXIV. RAIL.
XXXV. GALLINULE.

VII. FIN FOOTED.

XXXVI. PHALAROPE.
XXXVII. COOT.
XXXVIII. GREBE.

VIII. WEB FOOTED.

XXXIX. AVOSET.
XL. AUK.
XLI. GUILLEMOT.
XLII. DIVER.
XLIII. TERN.
XLIV. GULL.
XLV. PETREL.
XLVI. MERGANSER.
XLVII. DUCK.
XLVIII. CORVORANT.

EXPLANATION OF SOME TECHNICAL TERMS IN ORNITHOLOGY USED IN THIS WORK, AND BY LINNÆUS.

Fig.

1. *Cere. Cera* — THE naked ſkin that covers the baſe of the bill in the *Hawk* kind.

2. *Capiſtrum* — A word uſed by *Linnæus* to expreſs the ſhort feathers on the forehead juſt above the bill. In *Crows* theſe fall forwards over the noſtrils.

3. *Lorum* — The ſpace between the bill and the eye generally covered with feathers, but in ſome birds naked, as in the black and white *Grebe*.

4. *Orbits. Orbita* — The ſkin that ſurrounds the eye, which is generally bare, particularly in the *Heron* and *Parrot*.

5. *Emarginatum* — A bill is called *roſtrum emarginatum* when there is a ſmall notch near the end: this is conſpicuous in that of *Butcher-birds* and *Thruſhes*.

6. *Vibriſſæ* — *Vibriſſæ pectinatæ*, ſtiff hairs that grow on each ſide the mouth, formed like a double comb, to be ſeen in the *Goatſucker*, *Flycatcher*, &c.

7. *Baſtard wing. Alula ſpuria* — A ſmall joint riſing at the end of the middle part of the wing, or the

cubitus ;

cubitus; on which are three or five feathers.

8. *Leſſer coverts of the wings. Tectrices primæ* — The ſmall feathers that lie in ſeveral rows on the bones of the wings. The *under coverts* are thoſe that line the inſide of the wings.

9. *Greater coverts. Tectrices ſecundæ* — The feathers that lie immediately over the quil-feathers and ſecondary feathers.

10. *Quil-feathers. Primores* — The largeſt feathers of the wings, or thoſe that riſe from the firſt bone.

11. *Secondary feathers. Secondariæ* — Thoſe that riſe from the ſecond.

12. *Coverts of the tail. Uropygium* — Thoſe that cover the baſe of the tail.

13. *Vent-feathers.* — Thoſe that lie from the vent to the tail. *Criſſum Linnæi.*

14. *The tail. Rectrices*

15. *Scapular feathers* — That riſe from the ſhoulders and cover the ſides of the back.

16. *Nucha* — The hind part of the head.

17. *Roſtrum ſubulatum* — A term *Linnæus* uſes for a ſtrait and ſlender bill.

18. — To ſhew the ſtructure of the feet of the *Kingfiſher*.

19. *Pes ſcanſorius* — The foot of the *Woodpecker* formed for climbing. Climbing feet.

20. *Finned foot. Pes lobatus, pinnatus* — Such as thoſe of the *Grebes*, &c. Such as are indented, as *fig.* 21. are

M 4 called

called ſcalloped, ſuch are thoſe of *Coots* and ſcallop-toed *Sandpipers.*

22. *Pes tridactylus*	Such as want the back toe.
23. *Semi-palmated. Pes ſemi-palmatus*	When the webs only reach half way of the toes.
24. *Ungue poſtico ſeſſili*	When the hind claw adheres to the leg without any toe, as in the *Petrels.*
25. *Digitis* 4 *omnibus palmatis*	All the four toes connected by webs as in the *Corvorants.*

EXPLANATION OF OTHER LINNÆAN TERMS.

Roſtrum cultratum	WHEN the edges of the bill are very ſharp, ſuch as in that of the *Crow.*
Unguiculatum	A bill with a nail at the end, as in thoſe of the *Gooſanders* and *Ducks.*
Lingua ciliata	When the tongue is edged with fine briſtles, as in *Ducks.*
Integra	When quite plain or even.
Lumbriciformis	When the tongue is long, round and ſlender like a worm, as that of the *Woodpecker.*
Pedes compedes	When the legs are placed ſo far behind as to make the bird walk with difficulty, or as if *in fetters*; as is the caſe with the *Auks, Grebes* and *Divers.*
Nares Lineares	When the noſtrils are very narrow, as in *Sea Gulls.*
Marginatæ	With a rim round the noſtrils, as in the *Stare.*

CLASS

EXPLANATION of TECHNICAL TERMS.

P. Mazell fn.

[XVI.

BRITISH ZOOLOGY.

CLASS II. BIRDS.

DIV. I. LAND-BIRDS.

LONDON,
Printed for Benj. White,
MDCCLXXVI.

CLASS II.

BIRDS.

DIV. I. LAND BIRDS.

ORDER I. RAPACIOUS.

I. FALCONS.

Strong hooked BILL, the baſe covered with a CERE or naked ſkin. The firſt joint of the middle toe connected to that of the outmoſt by a membrane.

42. GOLDEN-EAGLE.

Grand aigle royal. *Belon av.* 89.
Aquila Germana. *Geſn. av.* 168.
Aquila, aguglia, Chryſaetos. *Aldr.* I. 62.
Gneſios. *Plinii lib.* 10. *c.* 3.
The golden eagle. *Wil. orn.* 8.
Aquila aurea, ſeu fulva. *Raii ſyn. av.* 6.
Falco Chryſaetos. *Lin. ſyſt.* 125.
Orn. *Faun. Suec. ſp.* 54.
L' Aigle doré. *Briſſon av.* I. 431.
Golden eagle. *Br. Zool.* 61. *Tab.* A. *Pl. Enl.* 410.
Stein adler. *Kram.* 325. *Scopoli.* No. 1.
Le grand Aigle. *Hiſt. D'Oys.* 1. 76.

THIS ſpecies is found in the mountanous parts of *Ireland* where it breeds in the lofti-eſt cliffs: it lays three, and ſometimes four eggs, of which ſeldom more than two are prolific; pro-vidence

vidence denying a large increaſe to rapacious birds*, becauſe they are noxious to mankind; but graciouſly beſtowing an almoſt boundleſs one on ſuch as are of uſe to us. This kind of eagle ſometimes migrates into *Caernarvonſhire*, and there are inſtances, though rare, of their having bred in *Snowdon* hills; from whence ſome writers give that tract the name of *Creigiau'r eryrau*, or the eagle rocks; others that of *Creigiau'r eira*, or the ſnowy rocks; the latter ſeems the more natural epithet; it being more reaſonable to imagine that thoſe mountains, like *Niphates* in *Armenia*, and *Imaus* † in *Tartary*, derived their name from the circumſtance of being covered with ſnow, which is ſure to befal them near the half of every year, than from the accidental appearance of a bird on them, once only in ſeveral years.

Descrip. The golden eagle weighs about twelve pounds; its length is three feet; the extent of its wings ſeven feet four inches; the bill is three inches long, and of a deep blue color; the cere is yellow; the irides of a hazel color: the ſight and ſenſe of ſmelling are very acute: *her eyes behold afar off* ‡: the head and neck are cloathed with narrow ſharp pointed feathers, and of a deep brown color, bor-

* Τῶν γαμψωνύχων ὀλιγοτόκα πάντα. Ariſt. hiſt. an.

† *Imaus*——incolarum lingua nivoſum ſignificante. *Plin. lib.* 6. *c.* 21.

‡ *Job* 39, 27. Where the natural hiſtory of the eagle is finely drawn up.

dered

dered with tawny; the hind part of the head in particular is of a bright ruſt-color.

The whole body, above as well as beneath, is of a dark brown; and the feathers on the back are finely clouded with a deeper ſhade of the ſame: the wings, when cloſed, reach to the end of the tail: the quil feathers are of a chocolate color, the ſhafts white: the tail is of a deep brown, irregularly barred and blotched with an obſcure aſh color, and uſually white at the roots of the feathers: the legs are yellow, ſhort, and very ſtrong, being three inches in circumference, and are feathered to the very feet: the toes are covered with large ſcales, and armed with moſt formidable claws, the middle of which are two inches long.

Eagles in general are very deſtructive to fawns, lambs, kids, and all kind of game; particularly in the breeding ſeaſon, when they bring a vaſt quantity of prey to their young. *Smith*, in his hiſtory of *Kerry*, relates that a poor man in that county got a comfortable ſubſiſtence for his family, during a ſummer of famine, out of an eagle's neſt, by robbing the eaglets of the food the old ones brought, whoſe attendance he protracted beyond the natural time, by clipping the wings and retarding the flight of the former. It is very unſafe to leave infants in places where eagles frequent; there being inſtances in *Scotland** of two being carried off by them, but fortunately,

* *Martin's hiſt. Weſt. Iſles*, 299. *Sib. hiſt, Scot.* 14.

Illæſum

Illæſum unguibus hæſit onus.

the theft was diſcovered in time, and the children reſtored unhurt out of the eagles neſts, to the affrighted parents. In order to extirpate theſe pernicious birds, there is a law in the *Orkney* iſles, which entitles any perſon that kills an eagle to an hen out of every houſe in the pariſh, in which it was killed *.

Eagles ſeem to give the preference to the carcaſſes of dogs or cats. Perſons, who make it their buſineſs to kill theſe birds, lay that of one or other by way of bait; and then conceal themſelves within gun-ſhot. They fire the inſtant the eagle alights, for ſhe that moment looks about before ſhe begins to prey. Yet quick as her ſight may be, her ſenſe of hearing ſeems ſtill more exquiſite. If hooded crows or ravens happen to be nearer the carrion and reſort to it firſt, and give a ſingle croak, the eagle is certain of inſtantly repairing to the ſpot, if there is one in any part of the neighborhood.

LONGEVITY. Eagles are remarkable for their longevity; and for their power of ſuſtaining a long abſtinence from food. Mr. *Keyſler* relates that an eagle died at *Vienna* after a confinement of 104 years. This preeminent length of days probably gave occaſion

* *Camden's Brit.* I. 1474. The impreſſion of an eagle and child on the coin of the *Iſle of Man*, was probably owing to ſome accident of this kind.

to

to the ſaying of the PSALMIST, *thy youth is renewed like the eagle's.* One of this ſpecies, which was nine years in the poſſeſſion of *Owen Holland*, Eſq; of *Conway*, lived thirty-two years with the gentleman who made him a preſent of it; but what its age was when the latter received it from *Ireland* is unknown. The ſame bird alſo furniſhes a proof of the truth of the other remark, having once, through the neglect of ſervants, endured hunger for twenty-one days, without any ſuſtenance whatſoever.

43. BLACK EAGLE.

Golden eagle, with a white ring about its tail. *Wil. orn.* 59. *Raii ſyn. av.* 6.
White tailed eagle. *Edw.* 1.

Falco fulvus. *Lin. ſyſt.* 125. *Briſſon av.* I. 420. *Hiſt. d'oys.* I. 86.
Ring-tail Eagle. *Br. Zool.* 62. *Pl. Enl.* 409.

DESCRIP.

THIS bird is common to the northern parts of *Europe* and *America*; that figured by Mr. *Edwards*, differing only in ſome white ſpots on the breaſt, from our ſpecies. It is frequent in *Scotland*, where it is called the *Black Eagle*, from the dark color of the plumage. It is very deſtructive to deer, which it will ſeize between the horns, and by inceſſantly beating it about the eyes with its wings, ſoon makes a prey of the haraſſed animal. The eagles in the iſle of *Rum* have nearly extirpated the ſtags that uſed to abound there. This ſpecies generally builds in clefts of rocks near the deer foreſts;

forests; and makes great havoke not only among them, but also the white hares and *Ptarmigans*.

It is equal in size to the preceding: the bill is of a blackish horn color; the cere yellow; the whole body is of a deep brown, slightly tinged with rust color; but what makes a long description of this kind unnecessary, is the remarkable band of white on the upper part of the tail; the end only being of a deep brown: which character it maintains through every stage of life, and in all countries where it is found. The legs are feathered to the feet: the toes yellow, the claws black. Mr. *Willughby* gives the following very curious account of the nest of this species, p. 21.

NEST.

'In the year of our Lord 1668, in the woodlands 'near the river *Derwent*, in the *Peak* of *Derbyshire*, 'was found an eagle's nest made of great sticks, 'resting one end on the edge of a rock, the other 'on two birch trees; upon which was a layer of 'rushes, and over them a layer of heath, and up'on the heath rushes again; upon which lay one 'young one, and an addle egg; and by them a 'lamb, a hare, and three heath poults. The 'nest was about two yards square, and had no 'hollow in it. The young eagle was black as a 'hobby, of the shape of a goshawk, of almost the 'weight of a goose, rough footed, or feathered 'down to the foot: having a white ring about the 'tail.'

Mr. *Willughby* imagines, his first *pygargus*, or white

SEA EAGLE.

white tailed eagle, p. 61. to be but a variety of this, having the ſame characteriſtic mark, and differing only in the pale color of the head.

The antients believed, that the pebble, commonly called the *ætites**, or eagle ſtone, was formed in the eagle's neſt; and that the eggs could not be hatched without its aſſiſtance. Many abſurd ſtories have been raiſed about this foſſil, which (as it bears but an imaginary relation to the eagle) muſt be omitted in a zoologic work.

44. SEA EAGLE.

Bein-brecher, Oſſifraga, Meer-adler, Fiſch-arn, Haliæetos. *Geſner av.* 201. 203.
Haliætos. *Turneri.*
Auguiſta barbata, Oſſifraga. *Aldr. av.* i. 118.
Haliæetos. *Plinii lib.* 10. *c.* 3.
Sib. hiſt. Scot. 14.
Sea eagle, or oſprey. *Wil. orn.* 59.
Raii ſyn. av. 7.
Sea eagle. *Dale's Harwich.* 396.
Martin's hiſt. Weſt. iſles 70.
Le grand aigle de mer. *Briſſon av.* i. 437.
Sea eagle. *Br. Zool.* 63. *Pl. Enl.* 112. 415.
Falco oſſifragus. *Lin. ſyſt.* 124.
Gaaſe orn. *Brunnich* 13.
L'Orfrair. *Hiſt. d'oys.* I. 112.

THIS ſpecies is found in *Ireland*, and ſeveral parts of *Great Britain*; the ſpecimen we took our deſcription from, was ſhot in the county of *Galway*: Mr. *Willughby* tells us there was an aery

* If the reader's curioſity ſhould be excited, we refer him for information to *Pliny*, lib. x. c. 3. lib. xxx. c. 21. to *Boetius de gemmis*, p. 375. to Dr. *Woodward's* catalogue of foſſils, vol. i. p. 53. c. 268, 269. and *Grew's Rarities*, p. 297.

of

of them in *Whinfield-park*, *Weſtmoreland*; and the eagle ſoaring in the air, with a cat in its talons, which *Barlow* drew from the very fact which he ſaw in *Scotland**, is of this kind. The cat's reſiſtance brought both animals to the ground, when *Barlow* took them up; and afterwards cauſed the event to be engraved in the thirty-ſixth plate of his collection of prints. *Turner* ſays, that in his days, it was too well known in *England*, for it made horrible deſtruction among the fiſh; he adds, the fiſhermen were fond of anointing their baits with the fat of this bird, imagining that it had a peculiar alluring quality: they were ſuperſtitious enough to believe that whenever the *ſea eagle* hovered over a piece of water, the fiſh, (as if charmed) would riſe to the ſurface with their bellies upwards; and in that manner preſent themſelves to him. No writer ſince *Cluſius* has deſcribed the ſea eagle; though no uncommon ſpecies, it ſeems at preſent to be but little known; being generally confounded with the golden eagle, to which it bears ſome reſemblance.

Descrip. The color of the head, neck and body, are the ſame with the latter; but much lighter, the tawny part in this predominating: in ſize it is far ſuperior; the extent of wings in ſome being nine or ten feet. The bill is larger, more hooked, and more arched; underneath grow ſeveral ſhort, but ſtrong hairs or briſtles, forming a ſort of beard. This

* Mr. *Walpole's* catalogue of engravers, p. 49.

gave

gave occaſion to ſome writers to ſuppoſe it to be the *aquila barbata* or bearded eagle of *Pliny*. The interior ſides, and the tips of the feathers of the tail, are of a deep brown; the exterior ſides of ſome are ferruginous, in others blotched with white. The legs are yellow, ſtrong and thick; and feathered but little below the knees; which is an invariable ſpecific difference between this and our firſt ſpecies. This nakedneſs of the legs is beſides no ſmall convenience to a bird who preys among the waters. The claws are of a deep and ſhining black, exceedingly large and ſtrong, and hooked into a perfect ſemicircle; thoſe of the hind and firſt toe are an inch and a half long.

FOOD.

All writers agree, that this eagle feeds principally on fiſh; which it takes as they are ſwimming near the ſurface *, by darting itſelf down on them; not by diving or ſwimming, as ſeveral authors have invented, who furniſh it for that purpoſe with one webbed foot to ſwim with, and another divided foot to take its prey with. *Pliny*, with his uſual elegance, deſcribes the manner of its fiſhing. *Supereſt haliæetos, clariſſima oculorum acie, librans ex alto ſeſe, viſoque in mari piſce, præceps in eum ruens, et diſcuſſis pectore aquis rapiens.*

* *Martin*, ſpeaking of what he calls the great eagles in the weſtern iſles, ſays, that they faſten their talons in the back of the fiſh, commonly of ſalmon, which are often above water, or on the ſurface. Thoſe of *Greenland* will even take a young ſeal out of the water.

It also preys on water fowl. The same writer prettily describes the chace, an amusement the inhabitants near the large lakes formed by the *Shannon* frequently enjoy.

It is strange that authors should give the name of *Nisus* to the sparrow hawk, when *Ovid* expressly mentions this as the bird to which the father of *Scylla* was transformed.

Quam pater ut vidit (nam jam pendebat in auras
Et modo factus erat fulvis HALIÆETOS alis)
Ibat, ut hærentem rostro laniaret adunco.

A hawk from upper air came pouring down,
('Twas *Nisus* cleft the air with wings new grown.)
At *Scylla*'s head his horny bill he aims.

Croxal.

45. CINEREOUS.

Pygargus, or white tailed eagle. *Wil. orn.* 61.
Raii syn. av. 7. *Brisson* I. 427.
Ern. *Br. Zool. Pl. Enl.* 411.
Hist. d'Oys. I. 99.

Pygargus hinnularius, an *Erne. Sib. Scot.*
Vultur albiulla. *Lin. syst.* 123.
Braunfahle Adler. *Frisch* I. 70.
Gamsen geyer. *Kram.* 326.
Postoina. *Scopoli.* No. 2.

DESCRIP. IS inferior in size to the golden eagle: the beak, cere and irides are of a very pale yellow; the space between that and the eyes bare, and of a bluish color. The head and neck are of a pale ash-color: the body and wings cinereous clouded with brown,

CINEREOUS EAGLE.

brown, the quil feathers very dark: the tail white: the legs feathered but little below the knees, and of a very light yellow. The male is of a darker color than the female.

The bill of this is rather ſtraiter than is uſual in the eagle, which ſeems to have induced *Linnæus* to place it among the *vultures*; but it can have no clame to be ranked with that genus, for the *pygargus* is wholly feathered; whereas, the characteriſtical mark of the vulture is, that the head and neck are either quite bare, or only covered with down.

Inhabits *Scotland*, and the *Orknies*, and feeds on fiſh, as well as on land animals.

FALCONRY.

Falconry was the principal amuſement of our anceſtors: a perſon of rank ſcarce ſtirred out without his hawk on his hand; which, in old paintings, is the criterion of nobility. *Harold*, afterwards king of *England*, when he went on a moſt important embaſſy into *Normandy*, is painted embarking with a bird on his fiſt, and a dog under his arm *: and in an antient picture of the nuptials of *Henry* VI. a nobleman is repreſented in much the ſame manner †; for in thoſe days, *It was thought ſufficient for noblemen's ſons to winde their*

* *Monfaucon monumens de la monarchie françoiſe*, I. 372.

† *Mr. Walpole's anecdotes of painting*, I. 33.

born and to carry their hawk fair, and leave ſtudy and learning to the children of mean people *. The former were the accompliſhments of the times; *Spenſer* makes his gallant Sir *Triſtram* boaſt,

> Ne is there hauke which mantleth her on pearch,
> Whether high towring, or accoaſting low,
> But I the meaſure of her flight doe ſearch,
> And all her pray, and all her diet know †.

In ſhort, this diverſion was, among the old *Engliſh*, the pride of the rich, and the privilege of the poor, no rank of men ſeems to have been excluded the amuſement: we learn from the *book* of *St. Albans* ‡, that every degree had its peculiar hawk, from the *emperor* down to the *holy water clerk*. Vaſt was the expence that ſometimes attended this ſport; in the reign of *James* I. Sir *Thomas Monſon* ‖ is ſaid to have given a thouſand pounds for a caſt of hawks: we are not then to wonder at the rigor of the laws that tended to preſerve a pleaſure that was carried to ſuch an extravagant pitch. In the 34th of *Edward* III. it was made felony to ſteal a hawk: to take its eggs, even in a perſon's own ground, was puniſhable with impriſonment for a year and a day; beſides a fine at the king's pleaſure: in queen *Elizabeth*'s reign the impriſonment was reduced to three months; but the offender was to

* *Biog. Brit.* article *Caxton.*

† *Book* VI. *Canto* 2.

‡ A treatiſe on hunting, hawking and heraldry, printed at *St. Albans* by *Caxton*, and attributed to *Dame Julian Barnes.*

‖ Sir *Ant. Weldon*'s court of K. *James*. 105.

find

find ſecurity for his good behaviour for ſeven years, or lie in priſon till he did. Such was the enviable ſtate of the times of *old England:* during the whole day our gentry were given to the fowls of the air, and the beaſts of the field: in the evening they celebrated their exploits with the moſt abandoned and brutiſh ſottiſhneſs: at the ſame time the inferior rank of people, by the moſt unjuſt and arbitrary laws, were liable to capital puniſhments, to fines, and loſs of liberty, for deſtroying the moſt noxious of the feathered tribe.

According to *Olearius*, the diverſion of falconry is more followed by the *Tartars* and *Perſians*, than ever it was in any part of *Europe*. *Il n'y avoit point de hutte qui n'euſt ſon aigle ou ſon faucon* *.

Our anceſtors made uſe of ſeveral kinds of native hawks; though that penetrating and faithful naturaliſt Mr. *Ray*, has left us only the bare name of a falcon in his liſt of the *Engliſh* birds, without mentioning the ſpecies.

The falcons or hawks that were in uſe in theſe kingdoms, are now found to breed in *Wales*, and in *North-Britain*, and its iſles. The peregrine falcon inhabits the rocks of *Caernarvonſhire*. The ſame ſpecies, with the *gyrfalcon*, the *gentil*, and the *goſhawk* are found in *Scotland*, and the *lanner* in *Ireland*.

We may here take notice that the *Norwegian* breed was, in old times, in high eſteem with our coun-

* *Tom.* I. 217. 328.

trymen: they were thought bribes worthy a king. *Jeoffrey Fitzpierre* gave two good *Norway* hawks to king *John*, to obtain for his friend the liberty of exporting 100 weight of cheese: and *Nicholas* the *Dane* was to give the king a hawk every time he came into *England*, that he might have free liberty to traffick throughout the king's dominions *.

They were also made the tenures that some of our nobility held their estates by, from the crown. Thus Sir *John Stanley* had a grant of the *Isle of Man* from *Henry* IV. to be held of the king, his heirs and successors, by homage and the service of two falcons, payable on the day of his or their coronation †. And *Philipp* de *Hastang*, held his manour of *Combertoun*, in *Cambridgeshire*, by the service of keeping the king's falcons ‡.

46. OSPREY.

Une Orfraye. *Belon. av.* 96.
Fisch-adler, Masswy, Aquila anataria, Clanga, Planga, Percnos, Morphnos. *Gesner. av.* 196.
Haliætus, seu aquila marina. *Gesner av.* 804.
Balbushardus. *Turneri.*
Auguista piumbina, Aquilastro, Haliætus, seu Morphnos. *Aldr. av.* I. 105. 114.
Haliætus. *Caii opusc.* 85.
Bald Buzzard. *Wil. orn.* 69.
Bald Buzzard, or sea eagle. *Raii syn. av.* 16.
Fishing hawk. *Catesby's Carol.* I. *Tab.* 2.
Falco cyanopus. *Klein Stem. Tab.* 8.
Falco Haliætus. *Lin. syst.* 129.
Blafot, Fisk-orn. *Faun. Suec. sp.* 63.
Aigle de mer. *Brisson av.* I. 440. *Tab.* 34. *Hist. d'Oys.* I. 103.
The Osprey. *Br. Zool.* 63. *Tab.* A. 1. *Pl. Enl.* 414.
Fisk-oern. *Brunnich*, p. 5.

MR. *Ray* places this bird among the hawks, instead of the eagles, on a supposition that

* *Madox* antiq. exchequer. I. 469, 470.

† *Blunt*'s antient tenures. 20.

‡ *Madox* I. 652.

Mr.

Mr. *Willughby* had exceeded in his account of its weight; but as we had an opportunity of confirming the words of the latter, from one of this ſpecies juſt taken, we here reſtore it to the aquiline rank, under the name of the Oſprey: which was the name it was known by in *England* above one hundred and ſixty years ago; as appears by Dr. *Kay*, or *Caius*'s deſcription of it, who alſo calls it an eagle.

NEST.

This bird haunts rivers, lakes, and the ſea-ſhores. It builds its neſt on the ground among reeds, and lays three or four white eggs of an elliptical form; rather leſs than thoſe of a hen.

FOOD.

It feeds chiefly on fiſh *, taking them in the ſame manner as the ſea eagle does, by precipitating itſelf on them, not by ſwimming; its feet being formed like thoſe of other birds of prey, for the left is not at all palmated, as ſome copying the errors of antient writers, aſſert it to be. The *Italians* compare the violent deſcent of this bird on its prey, to the fall of lead into water, and call it, *Auguiſta piumbina*, or the *leaden eagle*.

DESCRIP.

The bird here deſcribed was a female; its weight was ſixty-two ounces: the length twenty-three inches: the breadth five feet four inches: the wing when cloſed reached beyond the end of the tail: that, as in all the hawk kind, conſiſts of twelve feathers: the two middle feathers were duſky;

* *Turner* ſays it preys alſo on coots, and other water fowl.

the others barred alternately on their inner webs with brown and white: on the joint of the wing next the body was a ſpot of white: the quil feathers of the wings were black; the ſecondary feathers and the coverts duſky, the former having their interior webs varied with brown and white. The inner coverts white ſpotted with brown. The head ſmall and flat, the crown white marked with oblong duſky ſpots. The cheeks, chin, belly and breaſt white, the laſt ſpotted with a dull yellow: from the corner of each eye is a bar of brown that extends along the ſides of the neck pointing towards the wing. The legs very ſhort, thick and ſtrong: their length being only two inches and a quarter; their circumference two inches: their color a pale blue: the outward toe turns eaſily backwards, and what merits attention, the claw belonging to it is larger than that of the inner toe; in which it differs from all other birds of prey; but ſeems peculiarly neceſſary to this kind, for better ſecuring its ſlippery prey: the roughneſs of the ſoles of the feet contributes to the ſame end. The difference in weight, and other trifling particulars, makes us imagine that the bird Mr. *Willughby* ſaw was a male; as the females of all the hawk kind, are larger, ſtronger, and fiercer than the males; the defence of their young, and the providing them food, reſting chiefly on them.

HEAD.

LEGS.

Le

Paillou pinx.
Mazell sc.

Le Gerfault. *Belon av.* 94
Gyrfalco. *Aldr. av.* I. 243.
Jer-falcon. *Wil. orn.* 78.
Gyrfalco. *Raii syn. av.* 13.
White Falcon. *Wil. orn.* 80.

F. Islandus albus. *Brunnich* 7. 8.
Le Gerfault. *Brisson av.* I. 370.
Sib. Scot. 14.
Charlton Ex. 317.

47. GYRFALCON.

DESCRIP.

THIS elegant species is not much inferior in size to the *Osprey*. The *irides* dusky: the bill is very much hooked and yellow. The throat is of a pure white: the whole plumage of the same color, but marked with dusky lines, spots or bars: the head, breast and belly with narrow lines, thinly scattered and pointing down: the wings with large heart-shaped spots: the middle feathers of the tail with a few bars: the feathers on the thighs are very long, and of a pure white; the legs of a pale blue, and feathered a little below the knees. This kind is sometimes found quite white: it was a bird in high esteem when falconry was in vogue, and used for the noblest game, such as cranes and herons.

This is the *Gyrfalco* of all the ornithologists except *Linnæus*, whose bird we are totally unacquainted with: though he gives several of their synonyms, his description differs entirely from each of them. Inhabits the north of *Scotland*; shot near *Aberdeen*.

Belon

48. PERE-GRINE. *Belon av.* 116.
Falco peregrinus niger. *Aldr. av.* I. 239.
Blue backed falcon. *Charl. Ex.* 73.
Ditto. *Br. Zool. Tab.* A*. 5.
Sparviere pellegrino femmina. *Lorenzi av. Tab.* 24.
Le Faucon pelerin. *Briſſon av.* I. 341. *Hiſt. D' Oys.* I. 249.

DESCRIP. IN ſize equal to the moor-buzzard: the bill ſtrong, ſhort, and very much hooked, armed near the end of the upper mandible with a very ſharp proceſs: blue at the baſe, black at the point: the irides duſky.

The feathers on the forehead whitiſh: the crown of the head black mixed with blue: the hind part of the neck black: the back, ſcapulars, and coverts of the wings, elegantly barred with deep blue and black. The quil feathers duſky, marked with elliptical white ſpots placed tranſverſe: the inner coverts croſſed with black and white bars: the throat white: the fore part of the neck, and upper part of the breaſt white ſlightly tinged with yellow, the laſt marked with a few ſmall duſky lines pointing downwards. The reſt of the breaſt, the belly, thighs and vent feathers, white inclining to grey, and croſſed with duſky ſtrokes pointed in their middle. The tail conſiſts of feathers of equal length, finely and frequently barred with blue and black. The legs ſhort and yellow: the toes very long.

This ſpecies ſeems to vary: we have ſeen one that

PEREGRINE FALCON.

that was ſhot in *Hampſhire*, juſt as it had ſtruck down a *Rook* and was tearing it to pieces. The whole under ſide of the body was of a deep dirty yellow, but the black bars were the ſame as in that above deſcribed. The weight of this was two pounds eight ounces; the breadth thirty eight inches.

This ſpecies breeds on the rocks of *Llandidno* in *Caernarvonſhire*. That promontory has been long famed for producing a generous kind, as appears by a letter extant in *Gloddaeth* library, from the lord treaſurer *Burleigh* to an anceſtor of Sir *Roger Moſtyn*, in which his lordſhip thanks him for a preſent of a fine caſt of hawks taken on thoſe rocks, which belong to the family. They are alſo very common in the north of *Scotland*; and are ſometimes trained for falconry by ſome few gentlemen who ſtill take delight in this amuſement in that part of *Great Britain*. Their flight is amazingly rapid: one that was reclamed by a gentleman in the ſhire of *Angus*, a county on the eaſt ſide of *Scotland*, eloped from his maſter with two heavy bells to each foot, on the twenty-fourth of *September* 1772, and was killed in the morning of the twenty-ſixth, near *Moſtyn*, *Flintſhire*.

Grey

49. GREY. Grey Falcon. *Br. Zoology*, 65. *Octavo* I. 137.

DESCRIP.

THIS kind was ſhot near *Halifax* 1762, and the following account tranſmitted to us by Mr. *Bolton*, of *Worly-clough*. This bird was about the ſize of a raven: the bill was ſtrong, ſhort, much hooked, and of a bluiſh color: the cere, and edges of the eye-lids yellow: the irides red: the head was ſmall, flatted at the top; the fore part of a deep brown; the hind part white: the ſides of the head and throat were creme colored: the belly white, marked with oblong black ſpots: the hind part of the neck, and the back were of a deep grey: the wings were very long, and when cloſed reached beyond the train: the firſt of the quil feathers were black, with a white tip; the others were of a bluiſh grey, and their inner webs irregularly ſpotted with white: the tail was long, and wedge ſhaped; the two middle feathers being the longeſt, were plain, (the color not mentioned) the reſt ſpotted: the legs were long, naked, and yellow.

Gentil

Pl. XXI

Nº

FALCON GENTIL.

FALCON GENTIL.

a Variety

Gentil Falcon. *Wil. orn.* 80.
Raii ſyn. av.
Falco gentilis. F. cere pedibusque flavis, corpore cinereo maculis fuſcis, cauda fuſcis quatuor nigricantibus.
Lin. ſyſt. 126.
Falk. *Faun. Suec.* ſp. 58.
Kram. Auſtr. 328.
Falco gentilis. *Brun.* No. 6.
Scopoli, No. 3.
L'Autour. *Hiſt. d'Oys.* I. 230.

50. GENTIL.

DESCRIP.

THIS ſpecies of an elegant make is larger than the goſhawk. Cere, and legs yellow; irides light yellow: pupil large and of a full black: head light ruſt color, with oblong black ſpots: whole under ſide from chin to tail white, tinged with yellow: each feather marked with heartſhaped duſky ſpots pointing down: back brown: quil feathers duſky; barred on the out-moſt web with black, on the lower part of the inner with white. Coverts of the wings, and the ſcapulars, brown edged with ruſt color: wings reach only one half the length of the tail. The tail barred with four or five bars of black, and the ſame of cinereous: the firſt edged above and below with a line of dull white. The very tips, all the tail feathers white.

The young birds vary in having on their breaſts tranſverſe bars inſtead of cordated ſpots, as in the ſpecimen, *Plate*

This ſpecies inhabits the north of *Scotland*; and was in high eſteem as a bold and ſpirited bird in the days of falconry. It makes its neſt in rocks.

The

51. LANNER. The Lanner. *Wil. orn.* 82. Falco Lanarius. *Lin. ſyſt.* 129. Lanarius. *Raii ſyn. av.* 15. *Faun. Suec. ſp.* 62.

THIS ſpecies breeds in *Ireland*: the bird our deſcription is taken from, was caught in a decoy in *Lincolnſhire*, purſuing ſome wild ducks under the nets, and communicated to us by *Taylor White* Eſq; under the name of the *Lanner*.

DESCRIP. It was leſs than the buzzard. The cere was of a pale greeniſh blue; the crown of the head of a brown and yellow clay color: above each eye, to the hind part of the head, paſſed a broad white line; and beneath each, a black mark pointing down: the throat white: the breaſt tinged with dull yellow, and marked with brown ſpots pointing downwards: the thighs and vent ſpotted in the ſame manner: back and coverts of the wings deep brown, edged with a paler: quil feathers duſky: the inner webs marked with oval ruſt colored ſpots: the tail was ſpotted like the wings.

The legs ſhort and ſtrong, and of a bluiſh caſt, which Mr. *Willughby* ſays, are the characters of that bird. We are here to obſerve, that much caution is to be uſed in deſcribing the hawk kind, no birds being ſo liable to change their colors the two or three firſt years of their lives: inattention to this has

LANNER.

has caused the number of hawks to be multiplied far beyond the reality. The marks to be attended to as forming the characters of the species, are those on the quil feathers and the tail, which do not change. Another reason for this needless increase of the species of this tribe of birds, is owing to the names given to the same kinds in different periods of their lives, by the writers on falconry, which ornithologists have adopted and described as distinct kinds: even Mr. *Ray* has been obliged to copy them. The falcon, the falcon gentil, and the haggard, are made distinct species, whereas they form only one: this is explained by a *French* author, who wrote in the beginning of the last century, and effectually clears up this point; speaking of the falcon, he tells us, " S'il est prins " en *Juin*, *Juillet* & *Aoust*, vous le nommerez " *Gentil:* si en *Septembre*, *Octobre*, *Novembre* ou " *Decembre*, vous le nommerez *Pellerin* ou *Passa-* " *ger:* s'il est prins en *Janvier*, *Feburier* et *Mars*, " il sera nommé *Antenere:* et apres estre muë une " fois et avoir changé son cerceau, non aupara- " vant, vous le dires *Hagar*, mot *Hebrieu*, qui sig- " nifie estranger *."

* *La fauconnerie de Charles d'Arcussia seigneur d'Esparron, p.* 14. 5*me edit. Paris* 1607.

Autour.

52. GOS-HAWK.

Autour. *Belon av.* 112.
Gesner av. 5.
Aldr. av. i. 181.
Sib. Scot. 15.
Goshawk, accipiter palumbarius. *Wil. av.* 85.
Raii syn. av. 18.

L'Atour, Astur. *Brisson av.* i. 317.
Grosser gepfeilter Falck. *Frisch.* I. 82.
Astore. *Zinan* 87.
Falco palumbarius. *Lin. syst.* 130.

DESCRIP.

THE goshawk is larger than the common buzzard, but of a longer and more elegant form. The bill is blue towards the base, black at the tip: the cere a yellowish green: over each eye is a white line; and on the side of the neck is a bed of broken white: the head, hind part of the neck, back and wings are of a deep brown color: the breast and belly white, beautifully marked with numerous transverse bars of black and white: the tail is long, of a brownish ash-color, marked with four or five dusky bars placed remote from each other.

The legs are yellow: the claw of the back, and that of the inner toe very large and strong.

This species and the sparrow hawk, are distinguished by Mr. *Willughby* by the name of short winged hawks, because their wings, when closed, fall short of the end of the tail.

The goshawk was in high esteem among falconers, and flown at cranes, geese, pheasants and partridges. It breeds in *Scotland*, and builds its nest

in

XXIV. **The GOSHAWK.** No. 52.

in trees; is very deſtructive to game, and daſhes through the woods after its quarry with vaſt impetuoſity; but if it cannot catch the object of its purſuit almoſt immediately, deſiſts, and perches on a bough till ſome new game preſents itſelf.

53. KITE.

Le Milan royal. *Belon av.* 129.
Milvus. *Geſn. av.* 609.
Glede, Puttok, Kyte *Turneri.*
Milvio, Nichio. *Ald. av.* i. 201.
Kite, or Glead. *Wil. orn.* 74.
Milvus. *Plinii lib.* x. *c.* 10.
Raii ſyn. av. 17.
Rother Milon. *Kram.* 326.
Falco milvus. *Lin. ſyſt.* 126.
Glada. *Faun. Suec. ſp.* 57.
Le Milan royal. *Briſſon av.* i. 414. *Tab.* 33. *Hiſt. d'Oys.* 1. 197.
Nibbio. *Zinan.* 82.
The Kite. *Br. Zool.* 66. *Tab.* A. 2. *Pl. Enl.* 422.
Glente. *Brunnich* 3.

THE kite generally breeds in large foreſts, or wooded mountanous countries: its neſt is made externally with ſticks, lined with ſeveral odd materials, ſuch as rags, bits of flannel, rope, and paper. It lays two, or at moſt three eggs: which, like thoſe of other birds of prey, are much rounded, and blunt at the ſmaller end; they are white, ſpotted with a dirty yellow. Its motion in the air diſtinguiſhes it from all other birds; being ſo ſmooth and even, as to be ſcarce perceptible; ſometimes it will remain quite motionleſs for a conſiderable ſpace; at others glides through the ſky, without the leaſt apparent action of its wings: from thence is derived the old name of Glead, or Glede, from the *Saxon* Glida. Lord *Bacon* obſerves, that when kites fly high, it portends fair and dry weather. Some have ſuppoſed theſe to be birds of paſſage; but in *England* they certainly continue the whole year. *Cluſius* relates * that when he was in *London*, he obſerved a moſt amazing number of kites that flocked there for the ſake of the offals, &c. which were flung in the ſtreets. They were ſo tame as to take their prey in the midſt of the greateſt crowds; and it was forbidden to kill them.

The tail of this kind is ſufficient to diſtinguiſh it from all other *Britiſh* birds of prey, being forked. *Pliny* thinks that the invention of the rudder aroſe from the obſervation men made of the various motions of that part, when the kite was ſteering through the air †. Certain it is that the moſt uſeful arts were originally copied from animals; however we may now have improved upon them. Still in thoſe nations which are in a ſtate of nature, (ſuch as the *Samoieds* and *Eſquimaux*) their dwellings are inferior to thoſe of the beavers, which thoſe ſcarcely human beings but poorly copy.

DESCRIP. The weight of this ſpecies is forty-four ounces: the length twenty-ſeven: the breadth five feet one inch. The bill is two inches long, and very much in

* *Belon obſ. ad finem Clus. exot.* 108.

† Iidem videntur artem gubernandi docuiſſe caudæ flexibus. *Lib.* x. *c.* 10.

hooked at the end: the cere yellow: *irides* of a ſtraw-color. The head and chin are of a light grey, in ſome, white, marked with oblong ſtreaks of black: the neck and breaſt are of a tawny red, but the middle of the feathers black. On the belly and thighs, the ſpots are fewer, and under the tail they almoſt vaniſh. The upper part of the back is brown, the middle covered with very ſoft white down. The five firſt quil feathers are black; the inner webs of the others duſky barred with black, and the lower edges white. The coverts of the wings are varied with tawny black and white: the tail is forked, and of a tawny red: the outmoſt feather on each ſide of a darker hue than the reſt; and marked with a few obſcure duſky ſpots: the thighs are covered with very long feathers: the legs are yellow and ſtrong.

Theſe birds differ in their colors. We have ſeen a beautiful variety ſhot in *Lincolnſhire* that was entirely of a tawny color.

54. BUZZARD.

Le Buſe, ou Buſard. *Belon av.* 100.
Buteo. *Geſner. av.* 46.
Buſharda *Turneri.*
Buteo, ſeu Triorches. *Ald. av.* I. 190.
Triorches, Buteo. *Plinii lib.* x. c. 7.
Raii ſyn. av. 16.
Common Buzzard, or Puttock. *Wil. orn.* 70.
Wald Geyer. *Kram.* 329.
Falco buteo. *Lin. ſyſt.* 127.
Quidfogel. *Faun. Suec. ſp.* 60.
La Buſe. *Briſſon av.* I. 406. *Hiſt. d'Oys.* I. 206.
Pojana. *Zinan.* 85. *Scopoli.* No. 4.
Br. Zool. 66. *Tab.* A. 3. *Pl. Enl.* 419.
Oerne Falk. *Brunnich* p. 5.

THIS bird is the commoneſt of the hawk kind we have in *England.* It breeds in large woods, and uſually builds on an old crow's neſt, which it enlarges and lines with wool, and other ſoft materials: it lays two or three eggs, which are ſometimes wholly white; ſometimes ſpotted with yellow. The cock buzzard will hatch and bring up the young, if the hen is killed*. The young conſort with the old ones for ſome little time after they quit the neſt; which is not uſual with other birds of prey, who always drive away their brood as ſoon as they can fly. This ſpecies is very ſluggiſh and inactive; and is much leſs in motion than other hawks, remaining perched on the ſame bough for the greateſt part of the day, and is found at moſt times near the ſame place. It feeds on birds, rabbets, moles and mice; it will alſo eat frogs, earth-worms and inſects.

* *Ray's Letters.* 352.

This

BUZZARD.

SPOTTED FALCON.

This bird is ſubject to ſome variety in its colors: we have ſeen ſome whoſe breaſt and belly were brown, and only marked croſs the craw with a large white creſcent: uſually the breaſt is of a yellowiſh white, ſpotted with oblong ruſt-colored ſpots, pointing downwards: the chin ferruginous: the back of the head and neck, and the coverts of the wings are of a deep brown, edged with a pale ruſt color: the ſcapular feathers brown; with white towards their roots: the middle of the back is covered only with a thick white down: the ends of the quil feathers are duſky: their lower exterior ſides aſh-colored: their interior ſides blotched with darker and lighter ſhades of the ſame: the tail is barred with black and aſh-color, and ſometimes with ferruginous: the bar next the very tip is black, and the broadeſt of all; the tip itſelf of a duſky white. The *irides* are white, tinged with red. The weight of this ſpecies is thirty-two ounces: the length twenty-two inches; the breadth fifty-two. DESCRIP. SIZE.

Spotted Falcon, *Br. Zool.* iv. *tab.* 11. 55. SPOTTED.

TWO of theſe birds have been ſhot near *Longnor*, *Shropſhire*.

O 3 Size

Size of a buzzard: bill black; cere and legs yellow: irides pale yellow: crown, and hind part of the neck white, ſpotted with light reddiſh brown: back and ſcapulars of the ſame color edged with white. Quil feathers duſky barred with aſh color.

Under ſide of the neck, breaſt, belly, and thighs, white; the firſt, alſo the beginning of the breaſt marked with a few ruſty ſpots: rump white: middle feathers of the tail barred with white, and a deep brown: the others with a lighter and darker brown. The legs very ſtrong.

56. Honey Buzzard.

Le Goiran, ou Bondrèe. *Belon av.* 101.
Ald. av. I. 191.
Honey-Buzzard. *Wil. orn.* 72.
Raii ſyn. av. 16.
Froſch-geyerl. *Kram.* 331.
Falco Apivorus. *Lin. ſyſt.* 130.
Slag-hok. *Faun. Suec. ſp.* 65.
La Bondrèe. *Briſſon av.* i. 410. *Hiſt. d'Oys.* I. 208.
Zinan. 84.
Br. Zool. 67. *Tab.* A. 4. A*. 4. *Pl. Enl.* 420.
Muſe-Hoeg, Muſe-Baage, *Brunnich* p. 5.

Descrip.

THE weight of this ſpecies is thirty ounces: the length twenty-three inches: the breadth fifty-two: the bill and cere are black; the latter much wrinkled: the irides of a fine yellow: the crown of the head aſh-colored: the neck, back, ſcapulars, and covert feathers of the wings, are of a deep brown: the chin is white; the breaſt and belly of the ſame color, marked with duſky ſpots pointing downwards. The tail is long, of a dull

Tail.

dull brown color, marked with three broad duſky bars; between each of which are two or three of the ſame color, but narrower: the legs are ſhort, ſtrong, and thick: the claws large and black.

After the publication of the *folio Zoology*, Mr. *Plymly* favored us with a variety of this ſpecies, engraved in the additional plates, ſuppoſed to be a female, being ſhot on the neſt: it was entirely of a deep brown color, but had much the ſame marks on the wings and tail as the male; and the head was tinged with aſh color. There were two eggs in the neſt, blotched over with two reds ſomething darker than thoſe of the keſtril; though Mr. *Willughby* ſays they are of a different color: that naturaliſt informs us, that this bird builds its neſt with ſmall twigs, which it covers with wool; that its eggs are cinereous, marked with darker ſpots: as he found the combs of waſps in the neſt, he gave this ſpecies the name of the honey-buzzard: he adds, that it feeds on the erucæ of thoſe inſects, on frogs, lizards, *&c.* and that it runs very ſwiftly like a hen.

EGGS.

57. MOOR BUZZARD.

Le ſau-Perdrieux. *Belon av.* 114.
Circus Accipiter. *Geſner av.* 49.
Milvus æruginoſus. *Ald. av.* i. 203.
Moor Buzzard. *Wil. orn.* 75. *Raii ſyn. av.* 17.
Brauner rohr Geyer. *Kram.* 328.
Falco æruginoſus. *Lin. ſyſt.* 91.
Hoenſ-tjuf. *Faun. Suec. ſp.* 66.
Pojana roſſa. *Zinan.* 83.
Le Buſard de marais. *Briſſon av.* i. 401. *Hiſt. d'Oys.* 1. 218.
Schwartz-brauner Fiſch-Geyer mit dem gelben Kopf. *Friſch.* I. 77.
Hoenſe Hoeg. *Brunnich* p. 5.
Br. Zool. 67. *Tab.* A. 5.

THIS ſpecies frequents moors, marſhy places, and heaths; it never ſoars like other hawks; but commonly ſits on the ground, or on ſmall buſhes: it makes its neſt in the midſt of a tuft of graſs or ruſhes: we have found three young ones in it, but never happened to meet with the eggs: it is a very fierce and voracious bird, and is a great deſtroyer of rabbets, young wild ducks *, and other water fowl. It alſo preys, like the oſprey, upon fiſh.

DESCRIP. Its uſual weight is twenty ounces: the length twenty-one inches: the breadth four feet three inches: the bill is black; cere yellow; irides of the ſame color: the whole bird, head excepted, is of a chocolate brown, tinged with ruſt color: on the head is a large yellowiſh ſpot; we have ſeen

* In ſome places it is called the *duck hawk.*

ſome

Pl. XXVII

No. 67

MOOR BUZZARD.

HENHARRIER.

ſome birds of this kind with their head and chin entirely white; and others again have a whitiſh ſpot on the coverts of their wings; but theſe are only to be deemed varieties. The uniform color of its plumage, and the great length and ſlenderneſs of its legs, diſtinguiſhes it from all other hawks.

58. HEN-HARRIER.

Lanarius albus. *Aldr. av.* i. 197.
Rubetarius *Turneri.*
Wil. orn. 70.
Raii ſyn. av. 17.
Blue Hawk. *Edw.* 225. *the male.*
Falco Cyaneus. *Lin. ſyſt.* 126.
Le Lanier cendrè. *Briſſon av.* i. 365. *the male. Hiſt. d'Oys.* i. 212.
Br. Zool. 68. *Tab.* A. 6. *Pl. Enl.* 459.
Grau-weiſſe Geyer. *Friſch.* I. 79, 80.
Brunnich 14.

DESCRIP.

THE HEN-HARRIER weighs about twelve ounces: the length is ſeventeen inches; the breadth three feet three inches: the bill is black: cere, irides, and edges of the eye-lids yellow: the head, neck, back, and coverts of the wings, are of a bluiſh grey: the back of the head white, ſpotted with a pale brown: the breaſt, belly, and thighs, are white: the former marked with a few ſmall duſky ſtreaks: the ſcapular feathers are of a deep grey, inclining to duſky: the two middle feathers of the tail are entirely grey; the others only on their exterior webs; the interior being white, marked with duſky bars: the legs yellow, long and ſlender.

Theſe

These birds are extremely destructive to young poultry, and to the feathered game: they fly near the ground, skimming the surface in search of prey. They breed on the ground, and never are observed to settle on trees.

59. Ring-tail.

Subbuteo. *Gesner. av.* 48.
Ringtail. Pygargus accipiter. *Raii syn. av.* 17. *Wil. orn.* 70.
Le faucon a collier. *Brisson av.* 1. 345. *Pl. Enl.* 443, and 480.
Une autre oyseau St. *Martin. Belon av.* 104.
Rubetarius *Turneri.* La soubuse. *Hist. d'Oys.* I. 215.
Brunnich No. 14. *Br. Zool.* 68. *Tab.* 4. 7.

Descrip.

THE ringtail weighs sixteen ounces: is twenty inches long; and three feet nine inches broad: the *cere* and *irides* yellow: on the hind part of the head, round the ears to the chin, is a wreath of short stiff feathers of a dusky hue, tipt with a reddish white: on the top of the head, and the cheeks, the feathers are dusky, bordered with rust color; under each eye is a white spot: the back is dusky, the rump white, with oblong yellowish spots on each shaft: the tail is long; the two middle feathers marked with four dusky, and four broad cinereous bars; the others with three black, and three tawny bars; but the tips of all, white: the breast and belly are of a yellowish brown, with a cast of red, and marked with oblong dusky spots, but

but they are ſubject to vary, for we have met with one ſpecimen that had theſe parts entirely plain. The legs in color and ſhape reſemble thoſe of the preceding.

This has generally been ſuppoſed to be the female of the former: but from ſome late obſervations by the infallible rule of diſſection, males have been found of this ſpecies. *Willughby* ſays, that the eggs are white, much beſmeared with red. Theſe birds fly higher than the *hen-barrier*; and I have ſeen them perch on trees.

60. KESTREL

La Creſſerelle. *Belon av.* 125.
Geſner av. 54.
Kiſtrel, Kaſtrel, or Steingal, *Turneri.*
Aldr. av. 188.
The Keſtril, Stannel, Stonegall, Windhover. *Wil. orn.* 84.
Raii ſyn. av. 16.
La Creſſerelle. *Briſſon av.* I. 393.
Hiſt. d'Oys. I. 280.
Windwachl, Rittlweyer, Wannenweher. *Kram,* 331.
Roethel-Geyer. *Friſch.* I. 84. *fœm.* Mauſe-Falck. *Friſch.* I. 88.
Falco tinnunculus. *Lin. ſyſt.* 127.
Kyrko-Falk. *Faun. Suec. ſp.* 61.
Kirke-Falk. *Brunnich* 4. 5.
Gheppio, Acertello, Gavinello. *Zinan.* 88.
Br. Zool. 68. plate A. *Pl. Enl.* 401. 471.
Poſtoka, Splintza, Skoltſch. *Scopoli.* No. 5.

DESCRIP.

THE male of this beautiful ſpecies weighs only ſix ounces and a half: its length fourteen inches: the breadth two feet three inches: cere and legs yellow: *irides* dark. Its colors at once diſtinguiſh it from all other hawks: the crown

crown of the head, and the greater part of the tail, are of a fine light grey, the lower end of the latter is marked with a broad black bar: the inner webs of the three feathers next the two middle barred with black: the tips white: the back and coverts of the wings are of a brick red, elegantly ſpotted with black: the interior ſides of the quil feathers are duſky, deeply indented with white. The whole under ſide of the bird, of a pale ruſt color, ſpotted with black; the thighs and vent only, plain.

FEMALE. The female weighs eleven ounces: the color of the back and wings are far leſs bright than thoſe of the male: it differs too in the colors of the head and tail; the former being of a pale reddiſh brown, ſtreaked with black; the latter of the ſame color, marked with numerous tranſverſe black bars: the breaſt is of a dirty yellowiſh white; and the middle of each feather has an oblong duſky ſtreak, pointing downwards.

The keſtrel breeds in the hollows of trees, in the holes of high rocks, towers and ruined buildings: it lays four eggs, of the ſame color with thoſe of the preceding ſpecies: its food is field mice, ſmall birds and inſects; which it will diſcover at a great diſtance. This is the hawk that we ſo frequently ſee in the air fixed in one place, and as it were fanning it with its wings; at which time it is watching for its prey. It flings up the indigeſted fur and feathers in form of a round ball. When falconry

was

was in uſe in *Great Britain*, this kind was trained for catching ſmall birds and young partridges.

61. HOBBY.

Le Hobreau. *Belon av.* 118.
Geſner av. 75. *fæm.*
Hobbia. *Turneri.*
Æſalon. *Aldr. av.* I. 187.
The Hobby. *Wil. orn.* 83.
Le Hobreau, Dendro-falco. *Briſſon av.* I. 375. *Hiſt. d'Oys.* I. 277.
Raii ſyn. av. 15.
Falco ſubbuteo. *Lin. ſyſt.* 127.
Faun. Suec. ſp. 59.
Barletta. *Lorenzi av.* 45.
Stein-Falck. *Friſch.* I. 86.
Laerke-Falk. *Brunnich* 10. 11.
Br. Zool. 69. plate A. 9. *Pl. Enl.* 431.

THIS bird was alſo uſed in the humbler kind of falconry; particularly in what was called daring of larks: the hawk was caſt off; the larks aware of their moſt inveterate enemy, are fixed to the ground through fear; which makes them a ready prey to the fowler, by drawing a net over them. The hobby is a bird of paſſage; but breeds in *England*, and migrates in *October*.

DESCRIP. The male weighs ſeven ounces: the length is one foot; the breadth two feet three inches: cere and orbits yellow: irides hazel: upper mandible furniſhed with a proceſs: above each eye a white line: the crown of the head and back are of a deep bluiſh black: the hind part of the head is marked with two pale yellow ſpots; each cheek with a large black one pointing downwards: the coverts of the wings are of the ſame color with the back, but ſlightly edged with ruſt color: the interior

rior webs of the ſecondary and quil feathers, are varied with oval tranſverſe reddiſh ſpots: the breaſt white, marked with oblong ſpots of black: thighs and vent feathers, pale orange: the two middle feathers of the tail are entirely of a deep dove color: the others are barred on their interior ſides with ruſt color, and tipt with a dirty white.

FEMALE. The ſpots on the breaſt of the female are of a higher color than thoſe of the male: it is greatly ſuperior in ſize, its legs have a tinge of green, in other reſpects it reſembles the former.

62. SPARROW HAWK.

L'Eſpervier. *Belon av.* 121. *Geſner av.* 51.
Sparhauc *Turneri.*
Accipiter fringillarius, ſparviero. *Aldr. av.* i. 183.
Wil. orn. 86.
L'Epervier, accipiter. *Briſſon av.* I. 310. *Hiſt. d'oys.* I. 225.
Raii ſyn. av. 18.
Sperber *Friſch.* I. 90. 91. *Kram.* 332.
Falco niſus. *Lin. ſyſt.* 130.
Sparfhoek. *Faun. Suec. ſp.* 69.
Spurre-hoeg. *Brunnich p.* 5.
Scopoli. No. 6.
Br. Zool. 69. plate A. 10. A. 11. *Pl. Enl.* 466, 467. 412.

THE difference between the ſize of the male and female ſparrow hawks, is more diſproportionate than in moſt other birds of prey; the former ſometimes ſcarce weighing five ounces, the latter nine ounces.

DESCRIP. The length of the male is about twelve inches, the breadth twenty-three: the female

female is fifteen inches long; in breadth twenty-ſix.

Theſe birds, as well as the hawk kind in general, vary greatly in their colors; in ſome, the back, head, coverts of the wings and tail, are of a deep bluiſh grey; in others of a deep brown, edged with a ruſty red: the quil feathers are dusky, barred with black on their exterior webs, and ſpotted with white on the lower part of their inner webs: the tail is of a deep aſh color marked with fine broad black bars, the tip white: the breaſt and belly are of a whitiſh yellow, adorned with tranſverſe waved bars; in ſome of a deep brown color, in others orange: the cere, irides, and legs yellow. The colors of the female differ from thoſe of the male: the head is of a deep brown; the back, and coverts of the wings, are duſky mixed with dove color; the coverts of the tail of a brighter dove color; the waved lines that croſs the breaſt, are more numerous than thoſe on that of the male; and the breaſt itſelf of a purer white.

Manners.

This is the moſt pernicious hawk we have; and makes great havoke among pigeons, as well as partridges. It builds in hollow trees, in old neſts of crows, large ruins, and high rocks: lays four white eggs, encircled near the blunter end with red ſpecks. Mr. *Willughby* places this among the ſhort-winged hawks; or ſuch whoſe wings, when cloſed, fall ſhort of the end of the tail.

L'Eſmerillon.

63. MERLIN.*

L'Efmerillon. *Belon av.* 118.
Æfalon. *Gefner av.* 44.
Merlina. *Turneri.*
Smerlus, Smerillus. *Aldr. av.* I. 187.
Wil. orn. 85.
Raii fyn. av. 15.
L'Emerillon. *Briffon av.* I. 382.
Smerlio, o Smeriglio. *Lorenzi av. tab.* 18. 19.
Br. Zool. 70. plate A. 12.
Pl. Enl. 468.
Hift. D'Oys. I. 288.

DESCRIP.

THE Merlin weighs near five ounces and a half: its length is twelve inches, its breadth twenty five. The bill is of a bluifh lead color: the cere of a lemon color: the irides very dark, almoft black: the head is ferruginous, and each feather is marked with a bluifh black ftreak along the fhaft: the back and wings are of a deep bluifh afh color, adorned with ferruginous ftreaks and fpots, and edged with the fame: the quil feathers are almoft black, marked with reddifh oval fpots: the under coverts of the wings brown, beautifully marked with round white fpots: the tail is five inches long, croffed with alternate bars of dufky and reddifh clay color; on fome of the feathers of the fame bird are thirteen, on fome fifteen, but in one bird I examined, were no more than eight: the breaft and belly are of a yellowifh white, marked with oblong brown fpots pointing downwards: the legs yellow: the wings when clofed reach within an inch and a half of the end of the tail. This and

MANNERS.

* Merularius; quia merulas infectatur. *Skinner.*

the

the preceding kind were often trained for hawking: and this ſpecies, ſmall as it is, was inferior to none in point of ſpirit: it was uſed for taking partridg-es, which it would kill by a ſingle ſtroke on the neck. The Merlin flies low, and is often ſeen along roads' ſides, ſkimming from one ſide of the hedges to the other, in ſearch of prey.

It does not breed in *England*, but migrates here in *October*, about the time that the *Hobby* diſappears; for the Lark-catchers obſerve that in *September* they take no *Merlins* but abundance of *Hobbies*: but in the following month, *Merlins* only.

It was known to our *Britiſh* anceſtors by the name of *Llamyſden*; was uſed in hawking; and its neſt was valued at twenty-four pence. They made uſe of four other ſpecies, but have left us only their names; the *Hebog* or *Hawk*, whoſe neſt was eſtimated at a pound; the *Gwalch*'s or *Falcon*'s at one hundred and twenty pence; the *Hwyedig*'s or *long winged*, at twenty-four pence; and a ſpecies call-ed *Cammin* or *crooked bill*, at four pence. The *Penhebogyd* or *chief falconer*, held the fourth place at the court of the *Welch* prince: but notwith-ſtanding the hoſpitality of the times, this officer was allowed only three draughts out of his horn, leaſt he ſhould be fuddled and neglect his birds *.

* *Leges Wallicæ*, 253. 25.

II. OWL. Large round HEAD, ſtrong hooked BILL, no CERE. Feathers round the face diſpoſed in a circular form. Outmoſt TOE capable of being turned back, and doing the office of a hind toe.

64. EAGLE. Bubo maximus nigri et fuſci coloris. *Sib. Scot.* 14.
Great Owl, or Eagle Owl. *Wil. orn.* 99. *Raii ſyn. av.*
Strix Bubo. *Lin. ſyſt.* 131.
Uff. *Faun. Suec.* No. 69.
Berg Uggle, Katugl hane. *Strom. Sondm.* 222.
Buhu. *Kram. Auſtr.* 323.
Sova. *Scopoli.* No. 7.
Le grand duc. *Briſſon.* I. 477. *De Buffon,* I. 332.
Eagle Owl. *Br. Zool.* IV. *Tab.* VI. *Pl. Enl.* 385. 435.

THE eagle owl has been ſhot in *Scotland* and in *Yorkſhire.* It inhabits inacceſſible rocks and deſert places; and preys on hares and feathered game. Its appearance in cities was deemed an unlucky omen; *Rome* itſelf once underwent a luſtration, becauſe one of them ſtrayed into the *capitol.* The antients had them in the utmoſt abhorrence, and thought them, like the ſcreech owls, the meſſengers of death. *Pliny* ſtyles it *Bubo funebris* & *noctis monſtrum.*

Solaque culminibus ferali *carmine* Bubo
Sæpe queri et longas in fletum ducere voces.

VIRGIL.

Perch'd

PL. XXIX

N.° 64

Pl. XXX. №6

LONG-EARED OWL.

Perch'd on the roof the bird of night complains,
In lengthen'd ſhrieks, and dire funereal ſtrains.

In ſize it is almoſt equal to an eagle. *Irides* bright yellow: head and whole body finely varied with lines, ſpots and ſpecks of black, brown, cinereous, and ferruginous. Wings long: tail ſhort, marked with duſky bars. Legs thick, covered to the very end of the toes with a cloſe and full down of a teſtaceous color. Claws great, much hooked and duſky.

EARED OWLS.

65. LONG EARED.

L'Hibou cornu. *Belon av.* 136. *Geſner av.* 635.
Aſio, ſeu Otus. *Aldr. av.* I. 265.
The Horn Owl. *Wil. orn.* 100. *Raii ſyn. av.* 25.
Noctua aurita. *Sib. Scot.* 14.
Strix otus. *Lin. ſyſt.* 132.
Le moyen Duc ou le Hibou. *Briſſon av.* I. 486. *Hiſt. d'Oys.* I. 342.
Horn-uggla. *Faun. Suec. ſp.* 71.
Haſſelquiſt itin. 233.
Horn Ugle. *Brunnich* 16.
Horn-eule. *Kram.* 323.
Br. Zool. Plate 4. f. 1. *Pl. Enl.* 29. 473.
Mala Sova. *Scopoli* No. 9.
Rothe Kautzlein. *Friſch* I. 99.

THIS ſpecies is found, though not frequently, in the north of *England*, in *Cheſhire* and in *Wales*. The weight of the female, according to Mr. *Willughby* (for we never had opportunity of weighing it) is ten ounces: the length fourteen inches and a half: the breadth three feet four

DESCRIP.

P 2 inches:

inches: the irides are of a bright yellow: the bill black: the circle of feathers ſurrounding the eyes is white tipt with reddiſh and duſky ſpots, and the part next the bill black: the breaſt and belly are of a dull yellow, marked with ſlender brown ſtrokes pointing downwards: the thighs and vent feathers of the ſame color, but unſpotted. The back and coverts of the wings are varied with deep brown and yellow: the quil feathers of the ſame color, but near the ends of the outmoſt is a broad bar of red: the tail is marked with duſky and reddiſh bars, but beneath appears aſh colored: the horns or ears are about an inch long, and conſiſt of ſix feathers variegated with yellow and black: the feet are feathered down to the claws.

66. SHORT EARED.

Br. Zool. 71. *Tab.* B. 3. and B. 4. Fig. 2.

THE horns of this ſpecies are very ſmall, and each conſiſts of only a ſingle feather; theſe it can raiſe or depreſs at pleaſure; and in a dead bird they are with difficulty diſcovered. This kind is ſcarcer than the former; both are ſolitary birds, avoiding inhabited places. Theſe ſpecies may be called long winged owls; the wings when cloſed reaching beyond the end of the tail; whereas in the common kinds, they fall ſhort of it.

This

SHORT EARED OWL.

This is a bird of paſſage, and has been obſerved to viſit *Lincolnſhire* the beginning of *October*, and to retire early in the ſpring; ſo probably, as it performs its migrations with the woodcock, its ſummer retreat is *Norway*. During day it lies hid in long old graſs; when diſturbed, it ſeldom flies far, but will light and ſit looking at one, at which time the horns may be ſeen very diſtinctly. It has not been obſerved to perch on trees, like other owls: it will alſo fly in ſearch of prey in cloudy hazy weather. Farmers are fond of ſeeing theſe birds in their fields, as they clear them from mice. It is found frequently on the hill of *Hoy* in the *Orknies*, where it flies about and preys by day like a hawk. I have alſo received this ſpecies from *Lancaſhire*, which is a hilly and wooded country: and my friends have alſo ſent it from *New England* and *Newfoundland*.

DESCRIP.

The length of the ſhort eared owl is fourteen inches: extent three feet: the head is ſmall and hawk-like: the bill is duſky: weight fourteen ounces: the circle of feathers that immediately ſurrounds the eyes is black: the larger circle white, terminated with tawny and black: the feathers on the head, back, and coverts of the wings are brown edged with pale dull yellow: the breaſt and belly are of the ſame color, marked with a few long narrow ſtreaks of brown pointing downwards;

wards: the thighs, legs and toes, are covered with plain yellow feathers; the quill-feathers are dusky, barred with red: the tail is of a very deep brown, adorned on each side the shaft of the four middle feathers with a yellow circle which contains a brown spot: the tip of the tail is white.

The other *European* horn owl, the little horn owl, *Scops* or *Petit Duc* of *M. de Buffon*, I. 353, is unknown in *Great Britain*.

OWLS WITH SMOOTH HEADS.

67. WHITE. *Belon av.* 143 *.
Aluco minor. *Aldr. av.* I. 272.
Common barn, white, or church Owl, Howlet, madge Howlet, Gillihowter. *Wil. orn.* 104.
Raii syn. av. 25.
Le petit Chat-huant. *Brisson av.* I. 503.
Allocco. *Zinan.* 99.
Strix flammea. *Lin. syst.* 133.
Faun. Suec. 73.
Br. Zool. 71. plate B. *Pl. Enl.* 474.
L'Effraie. *Hist. d'Ois.* I. 366.
Perl-Eule. *Frisch.* I. 97.

THIS species is almost domestic: inhabiting for the greatest part of the year, barns,

* This refers only to the figure, for his description means the *Goatsucker*.

haylofts,

haylofts, and other outhoufes; and is as ufeful in clearing thofe places from mice, as the congenial cat: towards twilight it quits its perch, and takes a regular circuit round the fields, fkimming along the ground in queft of field mice, and then returns to its ufual refidence: in the breeding feafon it takes to the eaves of churches, holes in lofty buildings, or hollows of trees. During the time the young are in the neft, the male and female alternately fally out in queft of food, make their circuit, beat the fields with the regularity of a fpaniel, and drop inftantly on their prey in the grafs. They very feldom ftay out above five minutes; return with their prey in their claws; but as it is neceffary to fhift it into their bill, they always alight for that purpofe on the roof, before they attempt to enter their neft.

This fpecies I believe does not hoot; but fnores and hiffes in a violent manner; and while it flies along, will often fcream moft tremendoufly. Its only food is mice: as the young of thefe birds keep their neft for a great length of time, and are fed even long after they can fly, many hundreds of mice will fcarcely fuffice to fupply them with food.

Owls caft up the bones, fur or feathers of their prey in form of fmall pellets, after they have devoured it, in the fame manner as hawks do. A gentleman, on grubbing up an old pollard afh that had been the habitation of owls for many generations, found at the bottom many bufhels of this rejected ſtuff. Some owls will, when they are ſatisfied, like dogs, hide the remainder of their meat.

The elegant plumage of this bird makes amends for the uncouthneſs of its form: a circle of ſoft white feathers ſurround the eyes. The upper part of the body, the coverts and ſecondary feathers of the wings are of a fine pale yellow: on each ſide the ſhafts are two grey and two white ſpots placed alternate: the exterior ſides of the quil feathers are yellow; the interior white, marked on each ſide with four black ſpots: the lower ſide of the body is wholly white: the interior ſides of the feathers of the tail are white; the exterior marked with ſome obſcure duſky bars: the legs are feathered to the feet: the feet are covered with ſhort hairs: the edge of the middle claw is ſerrated. The uſual weight of this ſpecies is eleven ounces: its length fourteen inches: its breadth three feet.

SIZE.

68. TAWNY OWL.

Ulula. *Geſner av.* 773.
Strix. *Aldr. av.* I. 285.
Common brown or ivy Owl. *Wil. orn.* 102.
Raii ſyn. av. 25.
Le Chat huant. *Briſſon av.* I. 500. *Hiſt. d'Oys.* I. 362.
Strige. *Zinan*, 100. *Scopoli*, No. 12.
Strix ſtridula. *Lin. ſyſt.* 133.
Skrik uggla. *Faun. Suec.* 77.
Strix Orientalis. *Haſſelquiſt itin.* 233.
Nacht Eule, Gemeine. *Kram.* 324.
Braune-Eule, or Stock-Eule? *Friſch*, I. 96.
Nat Ugle. *Brunnich*, 18.
Br. Zool. 72. plate B. 3. *Pl. Enl.* 437.

THIS is the *Strix* of *Aldrovandus*, what we call the *Screech Owl*; to which the folly of ſuperſtition perftition had given the power of prefaging death by its cries. The antients believed that it fucked the blood of young children; a fact not incredible, for *Haffelquift** defcribes a fpecies found in *Syria*, which frequently in the evening flies in at the windows, and deftroys the helplefs infant.

Nocte volant puerofque petunt nutricis egentes,
Et vitiant cuneis corpora rapta fuis.
Carpere dicuntur lactentia vifcera roftris,
Et plenum poto fanguine guttur habent.
Eft illis ftrigibus nomen, fed nominis hujus
Caufa quod horrenda ftridere nocte folent.

Ovid. Faft. VI. 135.

DESCRIP.

The female of this fpecies weighs nineteen ounces: the length is fourteen inches: the breadth two feet eight inches: the irides are dufky: the ears in this, as in all owls, very large; and their fenfe of hearing very exquifite. The color of this kind is fufficient to diftinguifh it from every other: that of the back, head, coverts of the wings, and on the fcapular feathers, being a fine tawny red, elegantly fpotted and powdered with black or dufky fpots of various fizes: on the coverts of the wings, and on the fcapulars, are feveral large white fpots: the coverts of the tail are tawny, and quite free from any marks: the tail is varioufly blotched, barred and fpotted with pale red and black; in

* *Itin.* 255.

the

the two middle feathers the red predominates: the breaſt and belly are yellowiſh, mixed with white, and marked with narrow black ſtrokes pointing downwards: the legs are covered with feathers down to the toes.

This is a hardier ſpecies than the former; and the young will feed on any dead thing, whereas thoſe of the white owl muſt have a conſtant ſupply of freſh meat.

69. BROWN.

The grey Owl. *Wil. orn.* 103.
Raii ſyn. av. 26.
La Hulote. *Briſſon av.* I. 507.
Strix Ulula. *Lin. ſyſt.* 133.
Faun. Suec. 78.
Ugle. *Brunnich,* 19.
Graue Eule? *Friſch,* I. 94.
Br. Zool. 72. Plate B. 1.

AS the names this and the precedent ſpecies bear do by no means ſuit their colors, we have taken the liberty of changing them to others more congruous. Both theſe kinds agree entirely in their marks; and diffre only in the colors: in this the head, wings and back are of a deep brown, ſpotted with black in the ſame manner as the former: the coverts of the wings and the ſcapulars are adorned with ſimilar white ſpots: the exterior edges of the four firſt quil feathers in both are ſerrated: the breaſt in this is of a very pale aſh color mixed with tawny, and marked with oblong jagged ſpots: the feet too are feathered down to the very

DESCRIP.

BROWN OWL.

very claws: the circle round the face is aſh-colored, ſpotted with brown.

Both theſe ſpecies inhabit woods, where they reſide the whole day; in the night they are very clamorous; and when they hoot, their throats are inflated to the ſize of an hen's egg. In the duſk they approach our dwellings; and will frequently enter pigeon houſes, and make great havoke in them. They deſtroy numbers of little leverets, as appears by the legs frequently found in their neſts. They alſo kill abundance of moles, and ſkin them with as much dexterity as a cook does a rabbet. Theſe breed in hollow trees, or ruined edifices; lay four eggs of an elliptic form, and of a whitiſh color.

LITTLE.

La Cheveche. *Belon av.* 140.
Noctua. *Geſner av.* 620.
Little Owl. *Wil. orn.* 105.
Raii ſyn. av. 26.
Edw. 228.
Tſchiavitl. *Kram.* 324.
Faun. Suec. 79.
La petite Chouette, ou la Cheveche. *Briſſon av.* I. 514.
Strix paſſerina. *Lin. ſyſt.* 133.
La Civetta. *Olina,* 65. *Scopoli,* No. 17.
Krak-Ugle. *Brunnich,* 20.
Kleinſte Kautzlein. *Friſch,* I. 100.
Br. Zool. 73. plate B. 5.

DESCRIP.

THIS elegant ſpecies is very rare in *England*; it is ſometimes found in *Yorkſhire*, *Flintſhire*, and alſo near *London*: in ſize it ſcarcely exceeds a thruſh, though the fullneſs of its plumage makes it it appear larger: the irides are of a light yellow: the bill of a paper color: the feathers that encircle the face are white tipt with black: the head brown, ſpotted with white: the back, and coverts of the wings of a deep olive brown; the latter ſpotted with white: on the breaſt is a mixture of white and brown: the belly is white, marked with a few brown ſpots: the tail of the ſame color with the back: in each feather barred with white: in each adorned with circular white ſpots, placed oppoſite one another on both ſides the ſhaft: the legs and feet are covered with feathers down to the claws.

The *Italians* made uſe of this owl to decoy ſmall birds to the limed twig: the method of which is exhibited in *Olina*'s *uccelliera*, p. 65.

Mr. *Steuart*, the admirable author of the Antiquities of *Athens*, informed me that this ſpecies of owl was very common in *Attica*; that they were birds of paſſage, and appeared there the beginning of *April* in great numbers; that they bred there; and that they retired at the ſame time as the *Storks*, whoſe arrival they a little preceded.

ORDER

GREAT FEMALE SHRIKE.

ORDER II. PIES.

III. SHRIKE.

Strong bill, ſtrait at the baſe, and hooked at the end. Each ſide of the upper mandible marked with one notch. Outmoſt toe cloſely joined to the middlemoſt as far as the firſt joint.

71. GREAT.

Le grande Pie grieſche. *Belon av.* 126.
Lanius cinereus. *Geſner av.* 579.
Skrike, nyn murder *Turneri.*
Lanius cinereus, Collurio major. *Aldr av.* I. 199.
Caſtrica, Ragaſtola. *Olina*, 41.
Greater Butcher Bird, or Mattageſs; in the *North of England*, Wierangle. *Wil. orn.* 87.
Raii ſyn. av. 18.
Speralſter, Grigelalſter, Neuntodter. *Kram.* 364.
Butcher Bird, Murdering Bird or Skreek. *Mer. Pinax*, 170.
Cat. Carol. app. 36.
Night Jar. *Mort. Northampt.* 424.
La Pie-grieſche griſe. *Briſſon av.* II. 141. *Hiſt. d'Oys.* I. 296.
Pl. Enl. 32. f. 1.
Lanius excubitor. *Lin. ſyſt.* 135.
Warfogel. *Faun. Suec.* 80.
Daniſh Torn-Skade. *Norvegis* Klavert. *Br.* 21. 22.
Br. Zool. 73. plate C. *Pl. Enl.* 445.
Velch Skrakoper. *Scopoli*, No. 18.
Berg-Aelſter (Mountain Magpie) or groſſer Neuntocdter. *Friſch*, I. 59.

SIZE.

THIS bird weighs three ounces: its length is ten inches: its breadth fourteen: its bill is black, one inch long, and hooked at the end; the upper mandible furniſhed with a ſharp proceſs: the noſtrils are oval, covered with black briſtles pointing downwards: the muſcles that move the bill are very

very thick and ftrong; which makes the head very large. This apparatus is quite requifite in a fpecies whofe method of killing its prey is fo fingular, and whofe manner of devouring it is not lefs extraordinary: fmall birds it will feize by the throat, and ftrangle *; which probably is the reafon the *Germans* call this bird *Wurchangel* †, or the fuffocating angel. It feeds on fmall birds, young neftlings, beetles and caterpillars. When it has killed the prey, it fixes them on fome thorn, and when thus fpitted pulls them to pieces with its bill: on this account the *Germans* call it *Thorntráer* and *Thornfreker*. We have feen them, when confined in a cage, treat their food in much the fame manner, fticking it againft the wires before they would devour it. Mr. *Edwards* very juftly imagines that as nature has not given thefe birds ftrength fufficient to tear their prey to pieces with their feet, as the hawks do, they are obliged to have recourfe to this artifice.

MANNERS.

It makes its neft with heath and mofs, lining it with wool and goffamer; and lays fix eggs, of a dull olive green, fpotted at the thickeft end with black.

DESCRIP.

The crown of the head, the back, and the coverts that lie immediately on the joints of the wings are afh-colored; the reft of the coverts black: the quil feathers are black, marked in their middle

* *Edw. Gl.* III. 233. † *Wil. orn.* 87.

with

with a broad white bar; and except the four firſt feathers, and the ſame number of thoſe next the body, are tipt with white: the tail conſiſts of twelve feathers of unequal lengths, the middle being the longeſt; the two middlemoſt are black, the next on each ſide tipt with white, and in the reſt the white gradually increaſes to the outmoſt, where that color has either entire poſſeſſion, or there remains only a ſpot of black: the cheeks are white, but croſſed from the bill to the hind part of the head with a broad black ſtroke: the throat, breaſt and belly are of a dirty white: the legs are black. The female is of the ſame color with the male, the breaſt and belly excepted, which are marked tranſverſely with numerous ſemicircular brown lines.

TAIL,

CHEEKS.

72. RED-BACKED.

La petite Pie grieſche griſe. *Belon av.* 128.
Lanius tertius. *Aldr. av.* I. 199.
Leſſer Butcher Bird, called in *Yorkſhire* Fluſher. *Wil. orn.* 88. *ſp.* 2. the *male.* 89. *ſp.* 3. the *female.*
Raii ſyn. av. 18.
Daniſh Tornſkade. *Norv.* Hantvark. *Br.* 23.
Mort. Northampt. 424.
L'Ecorcheur. *Briſſon av.* II. 151.
Pl. Enl. 31. f. 2. *Hiſt. d' Oys,* I. 304
Lanius collurio. *Lin. ſyſt.* 136.
Faun. Suec. 81. Tab. II. f. 81.
Dorngreul, Dornheher. *Kram.* 363.
Bufferola, Ferlotta roſſa. *Zinan,* 91.
Br. Zool. 74. plate C. 1.
Mali Skrakoper. *Scopoli,* No. 19.

DESCRIP.

THE male weighs two ounces; the female two ounces two drams. The length of the former

former is ſeven inches and a half; the breadth eleven inches. The irides are hazel; the bill reſembles that of the preceding ſpecies: the head and lower part of the back are of a fine light grey: acroſs the eyes from the bill runs a broad black ſtroke: the upper part of the back, and coverts of the wings, are of a bright ferruginous color; the breaſt, belly and ſides are of an elegant bloſſom color; the two middle feathers of the tail are longeſt, and entirely black; the lower part of the others white, and the exterior webs of the outmoſt feather on each ſide wholly ſo.

In the female the ſtroke acroſs the eyes is of a reddiſh brown: the head of a dull ruſt color mixed with grey: the breaſt, belly and ſides of a dirty white, marked with ſemicircular duſky lines: the tail is of a deep brown; the outward feather on each ſide excepted, whoſe exterior webs are white.

Theſe birds build their neſts in low buſhes, and lay ſix eggs of a white color, but encircled at the bigger end with a ring of browniſh red.

Lanius

Lanius minor primus. *Aldr. av.* I. 200.
Another ſort of Butcher Bird. *Wil. orn.* 89. *ſp.* 4.
The Wood-chat. *Raii ſyn. av.* 19. *ſp.* 6.
Br. Zool. 74. plate C. 2.
Dorngreul mit rother platten. *Kram.* 363.
La Pie grieſche rouſſe. *Briſſon av.* II. 147. *Hiſt. d'Oys.* I. 301.
Pl. Enl. 9. f. 2.
Buferola, Ferlotta bianca. *Zinan.* 89.
Kleiner Neuntoedter. *Friſch*, I. 61.

73. WOOD-CHAT.

DESCRIP.

IN ſize it ſeems equal to the preceding: the bill is horn colored: the feathers that ſurround the baſe are whitiſh; above is a black line drawn croſs the eyes, and then downwards each ſide the neck: the head and hind part of the neck are of a bright bay: the upper part of the back duſky: the coverts of the tail grey: the ſcapulars white: the coverts of the wings duſky: the quil feathers black, marked towards the bottom with a white ſpot: the throat, breaſt and belly of a yellowiſh white. The two middle feathers appear by the drawing to be entirely black: the exterior edges and tips of the reſt white: the legs black.

FEMALE.

The female differs: the upper part of head, neck and body are reddiſh, ſtriated tranſverſely with brown: the lower parts of the body are of a dirty white, rayed with brown: the tail is of a reddiſh brown, marked near the end with duſky, and tipt with red.

IV. CROW. Strait ſtrong BILL: NOSTRILS covered with briſtles reflected down. Outmoſt TOE cloſely connected to the middle toe as far as the firſt joint.

74. RAVEN. Le Corbeau. *Belon av.* 279.
Corvus. *Geſner av.* 334.
Corvo, Corbo. *Aldr. av.* I. 343.
Wil. orn. 121.
Raii ſyn. av. 39.
Le Corbeau. *Briſſon av.* II. 8.
Velch oru. *Scopoli*, No. 35.
Corvus corax. *Lin. ſyſt.* 155.
Korp. *Faun. Suec.* 85.
Daniſh Raun. *Norv.* Korp. *Br.* 27.
Rab. *Kram.* 333. *Friſch*, I. 63.
Br. Zool. 75. *Hiſt. d'Oys.* III. 13.

DESCRIP. THIS ſpecies weighs three pounds: its length is two feet two inches: its breadth four feet: the bill is ſtrong and thick; and the upper mandible convex. The color of the whole bird is black, finely gloſſed with a rich blue; the belly excepted, which is duſky.

Ravens build in trees, and lay five or ſix eggs of a pale green color marked with ſmall browniſh ſpots. They frequent in numbers the neighborhood of great towns; and are held in the ſame ſort of veneration as the vultures are in *Egypt* *,

* *Haſſelquiſt itin.* 23.

and

Pl. XXXIV JACKDAW. No. 81

CROW. No. 75

and for the ſame reaſon; for devouring the carcaſes and filth, that would otherwiſe prove a nuſance. A vulgar reſpect is alſo paid to the raven, as being the bird appointed by Heaven to feed the prophet *Elijah*, when he fled from the rage of *Ahab* *. The raven is a very docil bird, may be taught to ſpeak, and fetch and carry. In clear weather they fly in pairs a great height, making a deep loud noiſe, different from the common croaking. Their ſcent is remarkably good; and their life prolonged to a great ſpace.

The quils of ravens ſell for twelve ſhillings the hundred, being of great uſe in tuning the lower notes of a harpſichord, when the wires are ſet at a conſiderable diſtance from the ſticks.

75. CARRION.

La Corneille. *Belon av.* 281.
Cornix (Krae). *Geſner av.* 320.
Cornice, Cornacchio. *Aldr. av.* I. 369.
Wil. orn. 122.
Raii ſyn. av. 39.
La Corbine. *Hiſt. d'Oys.* III. 45.
La Corneille. *Briſſon av.* 12.
Corvus corone. *Lin. ſyſt.* 155.
Faun. Suec. 86.
Krage. *Br.* 30.
Br. Zool. 75.
Oru. *Scopoli,* No. 36.

THE crow in the form of its body agrees with the raven; alſo in its food, which is

* 1. *Kings* 17.

carrion and other filth. It will alſo eat grain and inſects; and like the raven will pick out the eyes of young lambs when juſt dropped: for which reaſon it was formerly diſtinguiſhed from the rook, which feeds entirely on grain and inſects, by the name of the *gor* or *gorecrow*; thus *Ben Johnſon* in his *Fox*, *act* I. *ſcene* 2.

Vulture, kite,
Raven and *gor-crow*, all my birds of prey.

Virgil ſays that its croaking foreboded rain:

Tum Cornix *plena pluviam vocat improba voce.*

It was alſo thought a bird of bad omen, eſpecially if it happened to be ſeen on the left hand:

Sæpe ſiniſtra cava prædixit ab illice Cornix.

England breeds more birds of this tribe than any other country in *Europe*. In the twenty-fourth of *Henry* VIII. they were grown ſo numerous, and thought ſo prejudicial to the farmer, as to be conſidered an evil worthy parlementary redreſs: an act was paſſed for their deſtruction, in which rooks and choughs were included. Every hamlet was to provide crow nets for ten years; and all the inhabitants were obliged at certain times to aſſemble during that ſpace, to conſult the propereſt method of extirpating them.

Though the crow abounds in our country, yet in *Sweden*

Sweden it is ſo rare, that *Linnæus* mentions it only as a bird that he once knew killed there.

It lays the ſame number of eggs as the raven, and of the ſame color: immediately after deſerting their young, they go in pairs. Both theſe birds are often found white, or pied; an accident that befals black birds more frequently than any others: I have alſo ſeen one entirely of a pale brown color, not only in its plumage, but even in its bill and feet. The crow weighs about twenty ounces. Its length eighteen inches: its breadth two feet two inches.

76. ROOK.

La Graye, Grolle ou Freux. *Belon av.* 283.
Cornix frugivora (Roeck). *Geſner av.* 332.
Aldr. av. I. 378.
Wil. orn. 123.
Raii ſyn. av. 39.
Corvus frugilegus. *Lin. ſyſt.* 156.
Le Freux, ou la Frayonne. *Hiſt. d'Oys.* III. 55.
La Corneille Moiſſoneuſe. *Briſſon av.* II. 16.
Roka. *Faun. Suec.* 87.
Spermologus, ſeu frugilega. *Caii opuſc.* 100.
Schwartze krau, Schwartze krahe. *Kram.* 333. *Friſch*, I. 64.
Br. Zool. 76.

THE Rook is the *Corvus* of *Virgil*, no other ſpecies of this kind being gregarious.

E paſtu decedens agmine magno
Corvorum increpuit denſis exercitus alis.

A very natural deſcription of the evening return of theſe birds to their neſts.

Q 3 This

This bird differs not greatly in its form from the carrion crow: the ſize of the rook is ſuperior; but the colors in each are the ſame, the plumage of both being gloſſed with a rich purple. But what diſtinguiſhes the rook from the crow is the bill; the noſtrils, chin, and ſides of that and the mouth being in old birds white and bared of feathers, by often thruſting the bill into the ground in ſearch of the *erucæ* of the Dor-beetle *; the rook then, inſtead of being proſcribed, ſhould be treated as the farmer's friend; as it clears his ground from caterpillars, that do incredible damage by eating the roots of the corn. Rooks are ſociable birds, living in vaſt flocks: crows go only in pairs. They begin to build their neſts in *March*; one bringing materials, while the other watches the neſt, leſt it ſhould be plundered by its brethren: they lay the ſame number of eggs as the crow, and of the ſame color, but leſs. After the breeding ſeaſon rooks forſake their neſt-trees, and for ſome time go and rooſt elſewhere, but return to them in *Auguſt*: in *October* they repair their neſts †.

* Scarabæus melolantha. *Lin. ſyſt.* 351. *Roſel,* II. *Tab.* 1. *Liſt. Goed.* 265.

† *Calendar of Flora.*

La

77. HOODED.

La Corneille emantelée. *Belon. av.* 285.
Cornix varia, Marina, Hyberna, (Nabelfrae.) *Gesner av.* 332.
Cornix cinerea. *Aldr. av.* I. 379.
Raii syn. av. 39.
Martin's West. Isles, 376.
Hooded Crow. *Sib. Scot.* 15.
Pl. Enl. 76.
La Corneille mantelée. *Brisson av.* II. 19. *Hist. d' Ois.* III. 61.
Mulacchia cinerizia, Monacchia. *Zinan.* 70.
Corvus cornix. *Lin. syst.* 156.
Kraka. *Faun. Suec. sp.* 88.
Grave Kran, Kranveitl. *Kram.* 333.
Graue-Krœhe (grey-Cow), Nebel-Krœhe (mist Crow). *Frisch*, I. 65.
Br. Zool. 76. plate D. 1.
Urana. *Scopoli*, No. 37.

THE bill of this species agrees in shape with that of the rook, to which it bears great similitude in its manners, flying in flocks, and feeding on insects. In *England* it is a bird of passage; it visits us in the beginning of winter, and leaves us with the woodcocks. They are found in the inland as well as maritime parts of our country; in the latter they feed on crabs and shelfish.

It is very common in *Scotland*: in many parts of the *Highlands*, and in all the *Hebrides*, *Orknies*, and *Shetlands*, is the only species of genuine crow; the *Carrion* and the *Rook* being unknown there. It breeds and continues in those parts, the whole year round. Perhaps those that inhabit the northern parts of *Europe*, are they which migrate here. In the *Highlands* they build indifferently in all kinds

kinds of trees: lay ſix eggs: have a ſhriller note than the common crows, are much more miſchievous, pick out the eyes of lambs, and even of horſes when engaged in bogs: are therefore in many places proſcribed, and rewards given for killing them. For want of other food, they will eat cran-berries and other mountain berries.

Belon, *Geſner*, and *Aldrovand*, agree that this is a bird of paſſage in their reſpective countries: that it reſorts in the breeding ſeaſon to high mountains, and deſcends into the plains on the approach of winter. It breeds alſo in the ſouthern parts of *Germany*, on the banks of the *Danube* *.

DESCRIP. The weight of this ſpecies is twenty-two ounces: the length twenty-two inches; the breadth twenty-three. The head, under ſide of the neck, and wings are black, gloſſed over with a fine blue: the breaſt, belly, back, and upper part of the neck, are of a pale aſh color: the irides hazel: the legs black, and weaker than thoſe of the Rook. The bottom of the toes are very broad and flat, to enable them to walk without ſinking on marſhy and muddy grounds, where they are converſant.

* *Kram.* 333.

La

78. MAGPIE.

La Pie. *Belon av.* 291.
Pica varia et caudata. *Gesner av.* 695.
Aldr. av. I. 392.
The Magpie, or Pianet. *Wil. orn.* 127.
Raii syn. av. 41.
La Pie. *Brisson*, II. 35. *Hist. d'Oys*, III. 85.
Gazza, Putta. *Zinan.* 66.

Corvus Pica. *Lin. syst.* 157.
Skata, Skiura, Skara. *Faun. Suec. sp.* 92.
Danish Skade, Huus Skade. *Norv.* Skior, Tunfugl. *Brunnich*, 32.
Aelster. *Frisch*, I. 58.
Alster. *Kram.* 335.
Br. Zool. 77. plate D. 2.
Praka. *Scopoli*, No. 38.

THE great beauty of this very common bird was so little attended to, that the editors of the *British Zoology* thought fit to publish a print of it after a painting by the celebrated *Barlow*. The marks of this species are so well known, that it would be impertinent to detain the reader with the particulars.

We shall only observe the colors of this bird: its black, its white, its green, and purple, and the rich and gilded combination of glosses on the tail, are at lest equal to those that adorn the plumage of any other. It bears a great resemblance to the butcher-bird in its bill, which has a sharp process near the end of the upper mandible; in the shortness of its wings, and the form of the tail, each feather shortening from the two middlemost: it agrees also in its food; which are worms, insects, and small birds. It will destroy young chick-

ens;

ens; is a crafty, reſtleſs, noiſy bird: *Ovid* therefore with great juſtice ſtyles it,

---------- Nemorum convicia *Pica.*

Is eaſily tamed; may be taught to imitate the human voice: it builds its neſt with great art, covering it entirely with thorns, except one ſmall hole for admittance; and lays ſix or ſeven eggs, of a pale green color ſpotted with brown. The magpie weighs near nine ounces: the length is eighteen inches; the breadth only twenty four.

79. JAY.

Le Jay. *Belon av.* 289.
Pica glandaria. *Geſner av.* 700.
Aldr. av. I. 393.
Olina, 35.
Wil. orn. 130.
Raii ſyn. av. 41.
Ghiandaia. *Zinan.* 67.
Corvus glandarius. *Lin. ſyſt.* 156.
Le Geay, Garrulus. *Briſſon av.* II. 47. *Hiſt. d'Oyſ.* III. 107.
Allonſkrika, Kornſkrika. *Faun. Suec. ſp.* 90.
Skov-ſkade. *Br.* 33.
Nuſſ-heher. *Kram.* 335.
Eichen-Heher (Oak-Jay), or Holtz-Schreyer (Wood-Cryer). *Friſch,* I. 55.
Br. Zool. 77. plate D.
Skoia, Schoga. *Scopoli,* No. 39.

DESCRIP.

THIS is one of the moſt beautifull of the *Britiſh* birds. The weight is between ſix and ſeven ounces: the length thirteen inches; the breadth twenty and a half.

The bill is ſtrong, thick and black; about an inch and a quarter long. The tongue black, thin, and cloven at the tip: the irides white. The chin is white: at the angle of the mouth are two large

large black ſpots. The forehead is white, ſtreak-ed with black: the head is covered with very long feathers, which at pleaſure it can erect into the form of a creſt: the whole neck, back, breaſt and belly are of a faint purple daſhed with grey; the covert feathers of the wings are of the ſame color.

The firſt quil feather is black; the exterior webs of the nine next are aſh-colored, the interior webs duſky: the ſix next black; but the lower ſides of their exterior webs are white tinged with blue; the two next wholly black; the laſt of a fine bay color tipt with black.

The leſſer coverts are of a light bay: the great-er covert feathers moſt beautifully barred with a lovely blue, black and white: the reſt black: the rump is white. The tail conſiſts of twelve black feathers. The feet are of a pale brown: the claws large and hooked. It lays five or ſix eggs, of a dull whitiſh olive, mottled very obſcurely with pale brown. The neſt is made entirely of the fine fi-bres of roots of trees; but has for a foundation ſome coarſe ſticks: it is generally placed on the top of the underwood, ſuch as hazels, thorns, or low birch. The young follow their parents till the ſpring: in the ſummer they are very injurious to gardens, being great devourers of peaſe and cher-ries; in the autumn and winter they feed on acorns, from whence the Latin name. Dr. *Kramer* *

* *Kram. elench.* 335.

obſerves,

obſerves, that they will kill ſmall birds. Jays are very docil, and may be taught to imitate the human voice: their native note is very loud and diſagreeable. When they are enticing their fledged young to follow them, they emit a noiſe like the mewing of a cat.

80. RED LEGGED.

Scurapola. *Belon. obſ.* 12.
La Chouette ou Chouca rouge. *Belon av.* 286.
Pyrrhocorax gracculus ſaxatilis (Stein-tahen, Stein-frae). *Geſner av.* 522, 527.
Spelvier, Taccola. *Aldr. av.* I. 386.
Wil. orn. 126.
Raii ſyn. av. 40.
Le Crave. *Hiſt. d' Oys.* III. 1.
The Killegrew. *Charlton ex.* 75.
Cornwall Kae. *Sib. Scot.* 15.
Borlaſe Cornw. 249. *Tab.* 24.
Camden, Vol. I. 14.
Le Coracias. *Briſſon av.* II. 4. *Tab.* 1.
Corvus gracculus. *Lin. ſyſt.* 158.
Monedula pyrrhocorax. *Haſſelquiſt itin.* 238.
Br. Zool. 83. plate L *.
Gracula pyrrhocorax. *Scopoli*, No. 46.

THIS ſpecies is but thinly ſcattered over the northern world: no mention is made of it by any of the *Fauniſts*; nor do we find it in other parts of *Europe*, except *England*, and the *Alps* *. In *Aſia*, the iſland of *Candia* produces it †. In *Africa*, *Ægypt*: which laſt place it viſits towards the end of the inundations of the *Nile* ‡. Except

* *Plin. nat. hiſt. lib.* X. *c.* 48. *Briſſon*, II. 5.
† *Belon obſ.* 17.
‡ *Haſſelquiſt itin.* 240.

Ægypt

Pl. XXXV

No. 80

RED LEGGED CROW.

Ægypt it affects mountanous and rocky ſituations; and builds its neſt in high cliffs, or ruined towers, and lays four or five eggs, white ſpotted with a dirty yellow. It feeds on inſects, and alſo on new ſown corn: they commonly fly high, make a ſhriller noiſe than the jackdaw, and may be taught to ſpeak. It is a very tender bird, and unable to bear very ſevere weather; is of an elegant, ſlender make; active, reſtleſs, and thieving; much taken with glitter, and ſo meddling as not to be truſted where things of conſequence lie. It is very apt to catch up bits of lighted ſticks; ſo that there are inſtances of houſes being ſet on fire by its means; which is the reaſon that *Camden* calls it *incendiaria avis*. Several of the *Welſh* and *Corniſh* families bear this bird in their coat of arms. It is found in *Cornwall*, *Flintſhire*, *Caernarvonſhire*, and *Angleſea*, in the cliffs and caſtles along the ſhores; and in different parts of *Scotland* as far as *Straithnavern*; and in ſome of the *Hebrides*. They are alſo found in ſmall numbers on *Dover* cliff, where they came by accident: A gentleman in that neighborhood had a pair ſent as a preſent from *Cornwall*, which eſcaped, and ſtocked thoſe rocks. They ſometimes deſert the place for a week or ten days at a time, and repeat it ſeveral times in the year.

DESCRIP.

Its weight is thirteen ounces: the breadth thirty-three inches: the length ſixteen: its color is wholly black, beautifully gloſſed over with blue and

and purple: the legs and bill are of a bright orange, inclining to red: the tongue almoſt as long as the bill, and a little cloven: the claws large, hooked, and black. *Scopoli* ſays that in *Carniola* the feet of ſome, during autumn, turn black.

81. JACK-DAW.

Chouca, Chouchette, ou Chouette. *Belon av.* 286.
Gracculus, ſeu monedula. *Geſner av.* 521.
Aldr. av. I. 387.
Wil. orn. 125.
Raii ſyn. av. 40.
Le Choucas. *Briſſon av.* 24.
Scopoli, No. 38.
Mulacchia nera. *Zinan.* 70.
Corvus monedula. *Lin. ſyſt.* 156.
Kaja. *Faun. Suec. ſp.* 89.
Daniſh Alike. *Norv.* Kaae, Kaye, Raun Kaate, Raage. *Br.* 31.
Tagerl, Dohle, Tſchockerl. *Kram.* 334.
Graue-Dohle. *Friſch*, I. 67.
Br. Zool. 78.
Hiſt. d'Oys. III. 69.

DESCRIP.

THE jack-daw weighs nine ounces: the length thirteen inches: the breadth twenty-eight. The head is large in proportion to its body; which Mr. *Willughby* ſays argues him to be ingenious and crafty. The irides are white: the forehead is black: the hind part of the head a fine light grey: the breaſt and belly of duſky hue, inclining to aſh-color: the reſt of the plumage is black, ſlightly gloſſed with blue: the feet and bill black: the claws very ſtrong, and hooked. It is a docil loquacious bird.

Jack-daws breed in ſteeples, old caſtles, and in high

high rocks; laying five or ſix eggs. I have known them ſometimes to breed in hollow trees near a rookery, and join thoſe birds in their foraging parties. In ſome parts of *Hampſhire* they make their neſts in rabbet holes: they alſo build in the interſtices between the upright and tranſome *ſtones* of *Stone-Henge*; a proof of the prodigious height of that ſtupendous antiquity; for their neſts are placed beyond the reach of the ſhepherd-boys, who are always idling about the ſpot. They are gregarious birds; and feed on inſects, grain, and ſeeds *.

* The Caryocatactes, *Wil. orn.* 132. *Edw. tab.* 240. a bird of this genus, was ſhot near *Moſtyn*, *Flintſhire*, in *October*, 1753; ſuppoſed to have ſtraggled from *Germany*, where they are common: and the Roller, another bird of this claſs, was killed near *Helſtone bridge*, *Cornwall*, in the autumn, 1766. It is alſo a native of *Germany*; and is far the moſt beautifull of the *European* birds. As an acquaintance with theſe wanderers may be agreeable to our readers, we have given its figure, as well as that of the former. The one is copied from Mr. *Edwards*; the other from a drawing by *Paillou*. *Vide Appendix*.

Bill

V.
CUCKOO.

Bill a little arched.
Short tongue.
Ten feathers in the tail.
Climbing feet.

82. CUCKOO. Le Coqu. *Belon av.* 132.
Cuculus. *Gesner av.* 362.
Aldr. av. I. 20.
Cuculo. *Olina* 38.
Wil. orn. 97.
Raii syn. av. 23.
Le Coucou. *Brisson av.* 105.
Cuculus canorus *Lin. syst.* 168.
Gjok. *Faun. Suec. sp.* 96.
Danish Gjoeg v. Kuk. *Norv.* Gouk. *Br.* 36.
Kuckuck. *Frisch.* I. 40, 41, 42.
Kuctuct. *Kram.* 337.
Br. Zool. 80. plate G. G. 1.
Kukautza. *Scopoli.* No. 48.

THIS singular bird appears in our country early in the spring, and makes the shortest stay with us of any bird of passage; it is compelled here, as Mr. *Stillingfleet* observes, by that constitution of the air which causes the fig-tree to put forth its fruit *. From the coincidence of the first appearance of the summer birds of passage, and the leafing and fruiting of certain plants; this ingenious writer would establish a natural calendar in our rural œconomy; to instruct us in the time of sowing our most useful seeds, or of doing such work as depends on a certain temperament of the air. As the fallibility of human calendars need not be

* *Calendar of Flora. vid. Preface throughout.*

insisted

F. CUCKOO.

WRYNECK. Nº 83.

insisted on, we must recommend to our country-men some attention to these feathered guides, who come heaven-taught, and point out the true commencement of the season *; their food being the insects of those seasons they continue with us.

It is very probable, that these birds, or at lest part of them do not entirely quit this island during winter; but that they seek shelter in hollow trees, and lie torpid, unless animated by unusually warm weather. I have two evidences of their being heard to sing as early as *February*: one was in the latter end of that month 1771, the other on the fourth 1769: the weather in the last was uncommonly warm; but after that they were heard no more, chilled again as I suppose into torpidity. There is an instance of their being heard in the summer time to sing at midnight.

There is a remarkable coincidence between their song, and the season of the mackerel's continuance in full roe; that is from about the middle of *April*, to the latter end of *June*.

The cuckoo is silent for some little time after his arrival: his note is a call to love, and used only by the male, who sits perched generally on some dead tree, or bare bough, and repeats his song, which he loses as soon as the amorous season is over. In a trap, which we placed on a tree frequented by

* In *Sweden*, which is a much colder climate than our own, the cuckoo does not appear so early by a month.

cuckoos, we caught not fewer than five male birds in one ſeaſon. His note is ſo uniform, that his name in all languages ſeems to have been derived from it; and in all other countries it is uſed in the ſame reproachful ſenſe.

> The plain ſong *cuckoo* grey,
> Whoſe note full many a man doth mark,
> And dares not anſwer nay.
>
> *Shakeſpear.*

The reproach ſeems to ariſe from this bird making uſe of the bed or neſt of another to depoſit its eggs in; leaving the care of its young to a wrong parent; but *Juvenal* with more juſtice gives the infamy to the bird in whoſe neſt the ſuppoſititious eggs were layed,

Tu tibi tunc curruca *places* *.

A water-wagtail, a yellow hammer, or hedge-ſparrow †, is generally the nurſe of the young cuckoos; who, if they happen to be hatched at the ſame time with the genuine off-ſpring, quickly deſtroy them, by overlaying them as their growth is

* *Sat.* VI. 275.

‡ I have been eye-witneſs to two inſtances: when a boy I ſaw a young cuckoo taken out of the neſt of a hedge ſparrow: and in 1773 took another out of that of a yellow hammer: the old yellow hammer ſeemed as anxious about the loſs as if it had been its proper offspring.

ſoon

ſoon ſo ſuperior. This want in the cuckoo of the common attention other birds have to their young, ſeems to ariſe from ſome defect in its make, that diſables it from incubation; but what that is, we confeſs ourſelves ignorant, referring the inquiry to ſome ſkilful anatomiſt. A friend tells me that the ſtomach is uncommonly large, even ſo as to reach almoſt to the vent: may not the preſſure of that in a ſitting poſture, prevent incubation?

This bird has been ridiculouſly believed to change into a hawk, and to devour its nurſe on quitting the neſt, whence the *French* proverb *ingrat comme un coucou.* But it is not carnivorous, feeding only on worms and inſects: it grows very fat, and is ſaid to be as good eating as a land rail. The *French* and *Italians* eat them to this day. The *Romans* admired them greatly as a food: *Pliny** ſays, that there is no bird to compare with them for delicacy.

DESCRIP.

The weight of the cuckoo is a little more than five ounces; the length is fourteen inches; breadth twenty-five. The bill black, very ſtrong, a little incurvated, and about two-thirds of an inch long. The irides yellow. The head, hind part of the neck, the coverts of the wings, and the rump are of a dove color; darker on the head and paler on the rump. The throat and upper part of the

* *Lib.* X. c. 9.

neck are of a pale grey; the breaſt and belly white, croſſed elegantly with undulated lines of black. The vent feathers of a buff color, marked with a few duſky ſpots. The wings are very long, reaching within an inch and a half of the end of the tail; the firſt quil feather is three inches ſhorter than the others; they are duſky, and their inner webs are barred with large oval white ſpots. The tail conſiſts of ten feathers of unequal lengths like thoſe of the butcher bird: the two middle are black tipt with white; the others are marked with white ſpots on each ſide their ſhafts. The legs are ſhort; and the toes diſpoſed two backwards and two forwards like the woodpecker, though it is never obſerved to run up the ſides of trees. The female differs in ſome reſpects. The neck before and behind is of a browniſh red: the tail barred with the ſame color and black, and ſpotted on each ſide the ſhaft with white. The young birds are brown mixed with ferruginous and black, and in that ſtate have been deſcribed by ſome authors as old ones.

Weak

VI. WRYNECK.

Weak BILL, ſlightly incurvated.
NOSTRILS bare.
TONGUE long, ſlender, armed at the point.
Ten flexible feathers in the TAIL.
Climbing FEET.

83. WRYNECK.

Le Tercou, Torcou, ou Turcot. *Belon av.* 306.
Jynx. *Geſner av.* 573.
Aldr. av. I. 421.
The Wryneck. *Wil. orn.* 138.
Raii ſyn. av. 44.
Le Torcol, Torquilla. *Briſſon av.* IV. 4. *Tab.* 1. *fig.* 1.
Collotorto, Verticella. *Zinan.* 72.
The Emmet Hunter. *Charlton ex.* 93.
Jynx torquilla. *Lin. ſyſt.* 172.
Gjoktyta. *Faun. Suec. ſp.* 97.
Bende-Hals. *Br.* 37.
Natterwindl, Wendhalſs. *Kram.* 336.
Dreh-Hals. *Friſch*, I. 38.
Br. Zool. 80. plate F.
Iſhudeſch. *Scopoli*, No. 50.

DESCRIP.

NATURE, by the elegance of its penciling the colors of this bird, hath made ample amends for their want of ſplendor. Its plumage is marked with the plaineſt kinds. A liſt of black and ferruginous ſtrokes divides the top of the head and back. The ſides of the head and neck are aſh colored, beautifully traverſed with fine lines of black and reddiſh brown. The quil feathers are duſky; but each web is marked with ruſt colored ſpots. The chin and breaſt are of a light yellowiſh brown, adorned with ſharp pointed bars of black. The tail conſiſts of ten feathers, broad at their ends and

R 3 weak;

weak; of a pale aſh color, powdered with black and red, and marked with four equidiſtant bars of black. The tongue is long and cylindric; for the ſame uſe as that of the woodpecker. The toes are alſo diſpoſed the ſame way. The bill is ſhort, weak, and a little arcuate. The irides are of a yellowiſh hazel.

The Wryneck we believe to be a bird of paſſage, appearing here in the ſpring before the cuckoo. The *Welſh* conſider it as the forerunner or ſervant of that bird, and call it *Gwâs y gog*, or the cuckoo's attendant: the *Swedes* regard it in the ſame light*.

The food of this ſpecies is inſects, but chiefly ants, for on examination we found the ſtomach of one filled with their remains. As the tongue of this bird, like that of the *Ant-bear* or *Tamandria*, is of an enormous length; it poſſibly not only makes uſe of it to pick thoſe inſects out of their retreat, but like that quadruped may lay it acroſs their path, and when covered with ants draw it into its mouth.

Its weight is one ounce and a quarter: the length ſeven inches; the breadth eleven. It takes its name from a manner it has of turning its head back to the ſhoulders; eſpecially when terrified: it has alſo the faculty of erecting the feathers of the

* Jynx hieme non apparet, vere autem remigrans, cuculi, poſt quatuordecem dies, adventum ruricolis annuntiat. *Amœn. acad.* IV. 584.

head

head like thoſe of the jay. Its note is like that of the *Keſtril*, a quick repeated ſqueak. Its eggs are white, and have ſo thin a ſhell that the yolk may be ſeen through it. This bird builds in the hollows of trees, making its neſt of dry graſs, in which we have counted nine young.

VII. WOOD-PECKER. Strait ſtrong angular BILL. Noſtrils covered with briſtles.

TONGUE very long, ſlender, and armed at the end with a ſharp bony point.

TEN ſtiff feathers in the tail.

Climbing feet.

84. GREEN. Le Pic mart, Pic verd, Pic jaulne. *Belon av.* 299.
Geſner av. 710.
Pico verde. *Aldr. av.* I. 416.
Green Woodpecker, or Woodſpite; called alſo the Rain Fowl, High Hoe, and Hewhole. *Wil. orn.* 135.
Raii ſyn. av. 42.
Le Pic verd. *Briſſon av.* 4. 9.
Picus viridis. *Lin. ſyſt.* 175.
Wedknar, Gronſpik, Grongjoling. *Faun. Suec. ſp.* 99.
Haſſelquiſt itin. Ter. Sanct. 291.
Girald. Cambrens. 191.
Daniſh & Norw. Groenſpet. *Br.* 39.
Grunſpecht. *Kram.* 334. *Friſch*, I. 35.
Br. Zool. 78. plate E.
Deteu, Detela. *Scopoli*, No. 52.

THE wiſdom of Providence in the admirable contrivance of the fitneſs of the parts of animals to their reſpective nature, cannot be better illuſtrated than from this genus: which we ſhall give from the obſervations of our illuſtrious countryman Mr. *Ray**.

Theſe birds feed entirely on inſects: and their principal action is that of climbing up and down the bodies or boughs of trees: for the firſt purpoſe

* *Ray* on the Creation, p. 143.

they

they are provided with a long ſlender tongue, armed with a ſharp bony end barbed on each ſide, which by the means of a curious apparatus of muſcles * they can exert at pleaſure, darting it to a great length into the clifts of the bark, transfixing and drawing out the inſects that lurk there.

Nest.

They make their neſts in the hollows of trees: in order therefore to force their way to thoſe cavities, their bills are formed ſtrong, very hard, and wedge-like at the end; Dr. *Derham* obſerves, that a neat ridge runs along the top, as if an artiſt had deſigned it for ſtrength and beauty. Yet it has not power to penetrate a ſound tree: their perforation of any tree is a warning to the owner to throw it down.

Their legs are ſhort, but ſtrong; their thighs very muſcular: their toes diſpoſed, two backwards, two forward: the feathers of the tail are very ſtiff, ſharp pointed and bending downwards. The three firſt circumſtances do admirably concur to enable them to run up and down the ſides of trees with great ſecurity; and the ſtrength of the tail ſupports them firmly when they continue long in one place, either where they find plenty of food, or while they are forming an acceſs to the interior part of the timber. This form of the tail makes their flight very awkward, as it inclines their body down, and forces them to fly with ſhort and frequent jerks

* *Phil. Tranſ. Martin's abridg.* V. p. 55. plate 2.

when

when they would afcend, or even keep in a line.

This fpecies feeds oftener on the ground than any other of the genus: all of them make their nefts in the hollows of trees; and lay five or fix eggs, of a beautifull femitranfparent white.

EGGS.

DESCRIP.

This kind weighs fix ounces and a half. Its length is thirteen inches; the breadth twenty and a half: the bill is dufky, triangular, and near two inches long: the crown of the head is crimfon, fpotted with black. The eyes are furrounded with black, beneath which (in the males only) is a rich crimfon mark. The back, neck, and leffer coverts of the wings are green. The rump of a pale yellow. The greater quil feathers are dufky, fpotted on each fide with white. The tail confifts of ten ftiff feathers, whofe ends are generally broken as the bird refts on them in climbing; their tips are black: the reft of each is alternately barred with dufky and deep green. The whole under part of the body is of a very pale green; and the thighs and vent marked with dufky lines. The legs and feet are of a cinereous green.

L'epeiche,

L'epeiche, Cul rouge, Pic rouge. *Belon av.* 300.
Picus varius, feu albus. *Gefner av.* 709.
Greater fpotted Woodpecker, or Witwal. *Wil. orn.* 137.
Raii fyn. av. 43.
Picchio. *Zinan.* 73.
Le grand Pic varié. *Briffon av.* IV. 34.
Picus major. *Lin. fyft.* 176.
Gyllenrenna. *Faun. Suec. fp.* 100.
Hakke-fpeet. *Brunnich*, 40.
Groffes Baumhackl. *Kram.* 336.
Bunt Specht. *Frifch*, I. 36.
Br. Zool. 79. plate E.
Kobilar. *Scopoli*, No. 53.

85. GREAT SPOTTED.

DESCRIP.

THIS fpecies weighs two ounces three quarters: the length is nine inches; the breadth is fixteen. The bill is one and a quarter long, of a black horn color. The *irides* are red. The forehead is of a pale buff color. The crown of the head a gloffy black. The hind part marked with a rich deep crimfon fpot: the cheeks white; bounded beneath by a black line that paffes from the corner of the mouth, and furrounds the hind part of the head. The neck is encircled with a black color. The throat and breaft are of a yellowifh white. The vent feathers of a fine light crimfon. The back, rump, and coverts of the tail, and leffer coverts of the wings are black; the fcapular feathers and coverts adjoining to them are white. The quil feathers black, elegantly marked on each web with round white fpots.

TAIL.

The four middle feathers of the tail are black, the next tipt with dirty yellow; the bottoms of the

the two outmoſt black; the upper parts a dirty white. The exterior feather marked on each web with two black ſpots; the next with two on the inner web, and only one on the other. The legs are of a lead color. The female wants that beautiful crimſon ſpot on the head; in other reſpects the colors of both agree. This ſpecies is much more uncommon than the preceding; and keeps altogether in the woods.

FEMALE.

86. MIDDLE. Picus medius. P. albo nigroque varius, criſſo pileoque rubris. *Lin. ſyſt.* 176. *Faun. Suec. ſp.* 82. *Scopoli*, No. 54.
Le Pic variè. *Briſſon av.* IV. 38.

THIS ſpecies agrees with the preceding in colors and ſize, excepting that the crown of the head in this is of a rich crimſon; the crown of the head in the male of the former black; and the crimſon is in form of a bar on the hind part.

Birds thus marked have been ſhot in *Lancaſhire*, and other parts of *England*; but I am doubtfull whether they are varieties, or diſtinct ſpecies.

Geſner

MIDDLE & LITTLE SPOTTED WOODPECKERS.

87. LEST SPOTTED.

Gesner av. 709.
Aldr. av. I. 416.
Lesser spotted Woodpecker, or Hickwall. *Wil. orn.* 138.
Raii syn. av. 43.
Picus minor. *Lin. syst.* 176.
Le petit Pic variè. *Brisson av.* iv. 41.
Faun. Suec. sp. 192. *Scopoli.* No. 55.
Hasselquist itin. 242.
Kleiner Bunt-Specht. *Frisch.* I. 37.
Kleiner Baumhackl. *Kram.* 336.
Br. Zool. 79. plate E.

DESCRIP.

THIS species is the least of the genus, scarce weighing an ounce; the length is six inches; the breadth eleven. The forehead is of a dirty white: the crown of the head (in the male) of a beautiful crimson: the cheeks and sides of the neck are white, bounded by a bed of black beneath the former. The hind part of the head and neck, and the coverts of the wings are black: the back is barred with black and white: the scapulars and quil feathers spotted with black and white: the four middle feathers of the tail are black; the others varied with black and white: the breast and belly are of a dirty white: the crown of the head (in the female) is white; the feet are of a lead color.

IT has all the characters and actions of the greater kind, but is not so often met with.

Strait

VIII. KING-FISHER.

Strait, ſtrong, ſharp pointed BILL.
Tongue ſhort and pointed.
Three loweſt joints of the outmoſt TOE connected to the middle toe.

88. KING-FISHER

Le Martinet peſcheur. *Belon av.* 218.
Iſpida (Isfogel) *Geſner av.* 571.
Aldr. av. III. 200.
Olina 39, 40.
Wil. orn. 146.
Raii ſyn. av. 48.
Pl. Enl. 77.
Alcedo iſpida. *Lin. ſyſt.* 179.
Le Martin-pêcheur. *Briſſon av.* iv. 471.
Piombino, Martino peſcatore, Peſcatore del re. *Zinan* 116.
Isfogel. *Muſ. Fr. ad.* 16.
Scopoli. No. 64.
Jis-fugl. *Brunnich in Append.*
Eisvogel. *Friſch.* II. 223.
Meerſchwalbe. *Kram.* 337.
Br. Zool. 82. plate I.

DESCRIP.

THIS bird weighs an ounce and a quarter: its length is ſeven inches; its breadth eleven: its ſhape is very clumſy, the head and bill being very large, and the legs diſproportionably ſmall: the bill is two inches long; the upper mandible black, the lower yellow: the irides are red: the colors of this bird atone for its inelegant form: the crown of the head, and the coverts of the wings are of a deep blackiſh green, ſpotted with bright azure: the ſcapular feathers, and coverts of the tail are alſo of a moſt ſplendent azure: the whole underſide of the body is orange colored; a broad mark of the ſame

Pl. XXXVIII

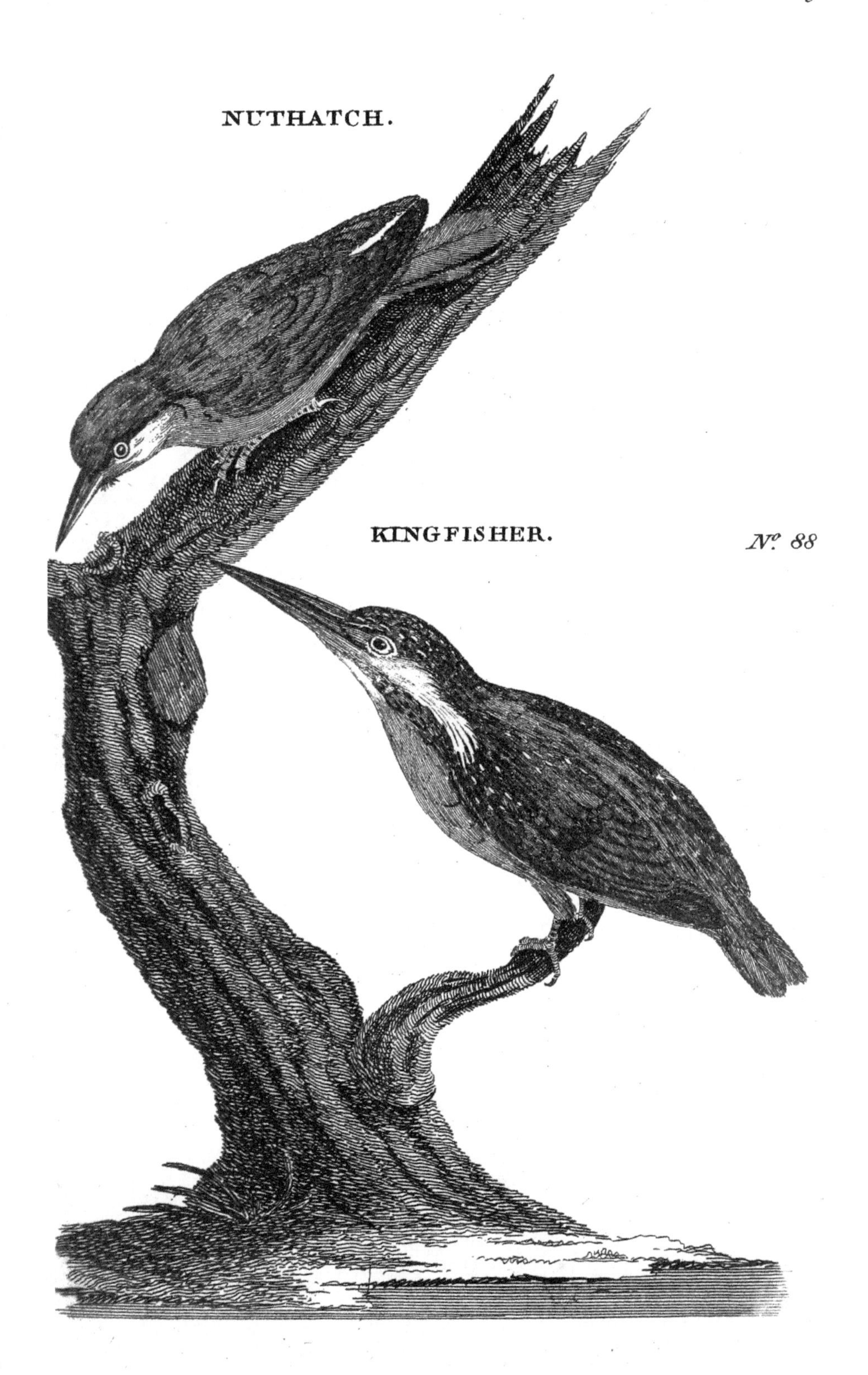

ſame paſſes from the bill beyond the eyes; beyond that is a large white ſpot: the tail is ſhort, and conſiſts of twelve feathers of a rich deep blue: the feet are of a reddiſh yellow: the three lower joints of the outmoſt toe adhere to the middle toe: the inner toe adheres to it by one joint.

The kingfiſher frequents the banks of rivers, and feeds on fiſh. To compare ſmall things to great, it takes its prey after the manner of the *oſprey*, balancing itſelf at a certain diſtance over the water for a conſiderable ſpace, then darting below the ſurface, brings the prey up in its feet. While it remains ſuſpended in the air, in a bright day, the plumage exhibits a moſt beautiful variety of the moſt dazzling and brilliant colors. This ſtriking attitude did not eſcape the notice of the antients, for *Ibycus*, as quoted by *Athenæus*, ſtyles theſe birds ἀλκυονες τανυσιπτεροι *, the *halcyons* with expanded wings. It makes its neſt in holes in the ſides of the cliffs, which it ſcoops to the depth of three feet; and lays from five to nine eggs †, of a moſt beautiful ſemi-tranſparent white. The neſt is very fetid, by reaſon of the remains of the fiſh brought to feed the young.

This ſpecies is the ἀλκυωναφωνος, or mute *halcyon* of *Ariſtotle* ‡, which he deſcribes with more preci-

* P. 388.

† *Geſner* ſays he found nine young in one neſt.

‡ *Hiſt. an.* 892, 1050.

ſion

ſion than is uſual with that great philoſopher: after his deſcription of the bird, follows that of its neſt, than which the moſt inventive of the antients have delivered nothing that appears at firſt ſight more fabulous and extravagant. He relates, that it reſembled thoſe concretions that are formed by the ſea-water; that it reſembled the long necked gourd, that it was hollow within, that the entrance was very narrow, ſo that ſhould it overſet the water could not enter; that it reſiſted any violence from iron, but could be broke with a blow of the hand; and that it was compoſed of the bones of the Βελονη or ſea-needle *.

NEST.

The neſt had medical virtues aſcribed to it; and from the bird was called *Halcyoneum*. In a fabulous age every odd ſubſtance that was flung aſhore received that name; a ſpecies of tubular coral, a ſponge, a zoophyte, and a miſcellaneous concrete having by the antients been dignified with that title from their imaginary origin †. Yet much of this ſeems to be founded on truth. The form of the neſt agrees moſt exactly with the curious account of it that Count *Zinanni* has favored us with ‡. The

* 1050. See alſo *Ælian*. lib. ix. c. 17. *Plin.* lib. x. c. 32.

† *Plin.* lib. xxxii. c. 8. *Dioſc.* lib. v. c. 94.

‡ Nidifica egli nelle ripe degli acquidotti, o de piccoli torrenti vicino al mare, formando però il nido nei ſiti più alti di dette ripe, acciocchè l'eſcreſcenza delle acque non poſſa inſinuarſi nel di lui foro; e fa egli detto nido incavando internamente il terreno in tondo per la lunghezza di tre piedi, e riducendo

The materials which *Ariſtotle* ſays it was compoſed of, are not entirely of his own invention. Whoever has ſeen the neſt of the kingfiſher, will obſerve it ſtrewed with the bones and ſcales of fiſh; the fragments of the food of the owner and its young: and thoſe who deny that it is a bird that frequents the ſea, muſt not confine their ideas to our northern ſhores; but reflect, that birds that inhabit a ſheltered place in the more rigorous latitudes, may endure expoſed ones in a milder clime. *Ariſtotle* made his obſervations in the eaſt: and allows, that the *halcyon* ſometimes aſcended rivers *; poſſibly to breed: for we learn from *Zinanni*, that in his ſoft climate, *Italy*, it breeds in *May*, in banks of ſtreams that are near the ſea; and having brought up the firſt hatch, returns to the ſame place to lay a ſecond time.

On the foundation laid by the philoſopher, ſucceeding writers formed other tales extremely abſurd; and the poets, indulging the powers of imagination, dreſſed the ſtory in all the robes of romance. This neſt was a floating one;

> Incubat *halcyone* pendentibus æquore nidis†.

it was therefore neceſſary to place it in a tranquil ſea, and to ſupply the bird with charms to allay the

riducendo il fine di detto foro a foggia di batello, tutto coperto di ſcaglie di peſci, che reſtano vagamente intrecciate; ma forſe non ſono così diſpoſte ad arte, bensì per accidente.

* Ἀναβαίνει δὲ τε ἐπὶ τοὺς ποταμοὺς Hiſt. an. 1050.

† *Ovid Met.* lib. xi.

fury of a turbulent element during the time of its incubation; for it had, at that ſeaſon, power over the ſeas and the winds.

> Χ' ἀλκυόνες ϛορεσεῦντι τὰ κύματα, την τε θάλασσαν,
> Τόν τε νότον, τον τ' εὖρον, ὃς ἔσχατα φυκία κινεῖ·
> Ἀλκυόνες, γλαυκαῖς Νηρηίσι ταί τε μάλιϛα
> Ορνίθων ἐφίλαθεν. *Theocrit. Idyl.* vii. l. 57.

May *Halcyons* ſmooth the waves, and calm the ſeas,
And the rough ſouth-eaſt ſink into a breeze;
Halcyons of all the birds that haunt the main,
Moſt lov'd and honor'd by the *Nereid* train.

Fawkes.

These birds were equally favourites with *Thetis* as with the *Nereids*;

> Dilectæ *Thetidi* Halcyones. *Virg. Georg.* I. 399.

As if to their influence theſe deities owed a repoſe in the midſt of the ſtorms of winter, and by their means were ſecured from thoſe winds that diſturb their ſubmarine retreats, and agitated even the plants at the bottom of the ocean.

Such are the accounts given by the *Roman* and *Sicilian* poets. *Ariſtotle* and *Pliny* tell us, that this bird is moſt common in the ſeas of *Sicily*: that it ſat only a few days, and thoſe in the depth of winter; and during that period the mariner might ſail in full ſecurity;

ſecurity; for which reaſon they were ſtiled, *Halcyon days**.

> Perque dies placidos hiberno tempore ſeptem
> Incubat *Halcyone* pendentibus æquore nidis:
> Tum via tuta maris: ventos cuſtodit, et arcet
> Æolus egreſſu. *Ovid. Met.* lib. XI.

> *Alcyone* compreſs'd,
> Seven days ſits brooding on her watery neſt
> A wintry queen; her ſire at length is kind,
> Calms every ſtorm and huſhes every wind. *Dryden.*

In after times, theſe words expreſſed any ſeaſon of proſperity: theſe were the *Halcyon days* of the poets; the brief tranquillity; the *ſeptem placidi dies* of human life.

The poets alſo made it a bird of ſong: *Virgil* ſeems to place it in the ſame rank with the *Linnet:*

> Littoraque *Halcyonem* reſonant, & *Acanthida* dumi.
> *Georg.* III. 338.

And *Silius Italicus* celebrates its muſic, and its floating neſt:

> Cum ſonat *Halcyone* cantu, nidoſque natantes
> Immotâ geſtat ſopitis fluctibus undâ.
> *Lib.* XIV. 275.

But we ſuſpect that theſe writers have transferred to our ſpecies, the harmony that belongs to

* *Ariſt. hiſt. an.* 541. *Plin.* lib. x. c. 32. lib. xviii. c. 24. Αλκυονειαι ημεραι of the former; and *dies halcyonides* of the latter.

S 2 the

the *vocal alcedo* of the philosopher, καὶ ἡ μὲν φθέγγεται, καθιζάνουσα επι τῶν δονάκων *, *which was vocal and perched upon reeds*. *Aristotle* says, it is the lest of the two, but that both of them have a cyanean back †. *Belon* labors to prove the *vocal alcedo* to be the *rousserole*, or the *greater reed sparrow* ‡, a bird found in *France* and some other parts of *Europe*, and of a very fine note: it is true that it is conversant among reeds, like the bird described by *Aristotle*; but as its colors are very plain, and that striking character of the fine blue back is wanting, we cannot assent to the opinion of *Belon*; but rather imagine it to be one of the lost birds of the antients.

Those who think we have said too much on this subject, should consider how incumbent it is on every lover of science, to attempt placing the labors of the antients in a just light: to clear their works from those errors, that owe their origin to the darkness of the times; and to evince, that many of their accounts are strictly true; many founded on truth; and others contain a mixture of fable and reality, which certainly merit the trouble of separation. It is much to be lamented that travel-

* *Hist. an.* 892.

† Νῶτον κυανεον, the color of the *cyanus*, or *lapis lazuli*.

‡ Le Roufferolle, *Belon av.* 221. Le Roucherolle, *Brisson av.* II. 218. Greater reed sparrow, *Wil. orn.* 143. Turdus arundinaceus, *Lin. syst. sp.* 296.

lers

lers, either on claſſic or any other ground, have not been more aſſiduous in noting the zoology of thoſe countries, which the antients have celebrated for their productions: for, from thoſe who have attended to that branch of natural knowledge, we have been able to develope the meaning of the old naturaliſts; and ſettle with preciſion ſome few of the animals of the antients.

Italy, a country crowded with travellers of all nations, hath not furniſhed a ſingle writer on claſſical zoology. The *Eaſt* has been more fortunate: *Belon*, the firſt voyager who made remarks in natural hiſtory during his travels, mentions many of the animals of the places he viſited, and may be very uſeful to aſcertain thoſe of *Ariſtotle*, eſpecially as he has given their modern *Greek* names. Our countryman, Dr. *Ruſſel*, enumerates thoſe of *Syria*. Dr. *Haſſelquiſt* has made ſome additions to the ornithology of *Egypt*: but all theſe fall ſhort of the merits of that moſt learned and inquiſitive traveller, Dr. *Shaw*; who with unparalleled learning and ingenuity, has left behind him the moſt ſatisfactory, and the moſt beautiful comments on the animals of the antients, particularly thoſe mentioned in HOLY WRIT, or what relates to the *Ægyptian* mythology: ſuch as do honor to our country, and we flatter ourſelves will prove incentives to other travellers, to complete what muſt prove ſuperior to any one genius, be it ever ſo great: from ſuch we may be ſupplied with the means of illuſtrating the works of the antient naturalifts; whilft commentators, after loading whole pages with unenlightening learning, leave us as much in the dark, as the age their authors wrote in.

Strait

Strait triangular BILL.
Short TONGUE, horny at the end, and jagged.

IX. NUT-HATCH.

89. NUT-HATCH

Le grand Grimpereau, le Torchepot. *Belon av.* 304.
Picus cinereus, seu Sitta. *Gesner av.* 711.
Ziolo. *Aldr. av.* I. 417.
The Nuthatch, or Nut-jobber. *Wil. orn.* 142.
Raii syn. av. 47.
The Woodcracker. *Plott's hist. Ox.* 175.
Sitta Europæa. *Lin. syst.* 177.
Le Torchepot, Sitta. *Brisson av.* III. 588. *tab.* 29. *fig.* 3.
Picchio grigio, Raparino. *Zinan.* 74.
Notwacka, Notpacka. *Faun. Suec. sp.* 104.
Danish Spœtt-meise. *Norv.* Nat-Bake. *Br.* 42.
Klener, Nusszhacker. *Kram.* 362.
Blau-specht. *Frisch*, I. 39.
Br. Zool. 81. plate H.
Barless. *Scopoli*, No. 57.

DESCRIP.

THE nuthatch weighs near an ounce; its length is near five inches three-quarters; breadth nine inches; the bill is strong and strait, about three quarters of an inch long; the upper mandible black, the lower white: the irides hazel; the crown of the head, back, and coverts of the wings are of a fine bluish grey: a black stroke passes over the eye from the mouth: the cheeks and chin are white: the breast and belly of a dull orange color; the quil feathers dusky; the wings underneath are marked with two spots, one white at the root of the exterior quils; the other black at the joint of the bastard wing; the tail consists of twelve feathers; the two middle are grey: the two

S 4 exterior

exterior feathers tipt with grey, then succeeds a transverse white spot; beneath that the rest is black; the legs are of a pale yellow; the back toe very strong, and the claws large.

This bird runs up and down the bodies of trees, like the woodpecker tribe; and feeds not only on insects, but nuts, of which it lays up a considerable provision in the hollows of trees: it is a pretty sight, says Mr. *Willughby*, to see her fetch a nut out of her hoard, place it fast in a chink, and then standing above it with its head downwards, striking it with all its force, breaks the shell, and catches up the kernel: it breeds in the hollows of trees; if the entrance to its nest be too large, it stops up part of it with clay, leaving only room enough for admission: in autumn it begins to make a chattering noise, being silent for the greatest part of the year. *Doctor Plott* tells us, that this bird, by putting its bill into a crack in the bough of a tree, can make such a violent sound as if it was rending asunder, so that the noise may be heard at lest twelve score yards.

Slender

Pl. XXXIX
No. 9
CREEPER.
HOOPO.
No. 9

Slender incurvated BILL.
Very ſhort TONGUE.
Ten feathers in the TAIL.

X.
HOOPOE.

La Huppe. *Belon av.* 293.
Upupa. *Geſner av.* 776.
Aldr. av. II. 314.
Bubbola. *Olina,* 36.
The Hoop, or Hoopoe. *Wil. orn.* 145.
Raii ſyn. av. 48.
The Dung Bird. *Charlton ex.* 98. *tab.* 99.
Plott's Oxf. 177.
Edw. 345.
Pl. enl. 52.
La Hupe ou Puput. *Briſſon av.* III. 455. *tab.* 43.
Upupa epops. *Lin. ſyſt.* 183.
Harfogel, Pop. *Faun. Suec. ſp.* 105.
Ter Chaous *Pococke Trav.* I. 209.
Her-fugl. *Brunnich,* 43.
Widhopf. *Kram.* 337.
Upupa; arquata ſtercoraria; gallus lutoſus. *Klein Stem. av.* 24. *tab.* 25.
Br. Zool. 83. plate L.
Smerda kaura. *Scopoli,* No. 62.

90. HOOPOE.

DESCRIP.

THIS bird may be readily diſtinguiſhed from all others that viſit theſe iſlands by its beautiful creſt, which it can erect or depreſs at pleaſure: it weighs three ounces: its length is twelve inches: its breadth nineteen: the bill is black, two inches and a half long, ſlender, and incurvated: the tongue triangular, ſmall, and placed low in the mouth: the irides are hazel: the creſt conſiſts of a double row of feathers; the higheſt about two inches long: the tips are black, their lower part of a pale orange color: the neck is of a pale reddiſh brown: the breaſt and belly white; but in young birds

birds marked with narrow duſky lines pointing down: the leſſer coverts of the wings are of a light brown: the back, ſcapulars and wings croſſed with broad bars of white and black: the rump is white: the tail conſiſts of only ten feathers, white marked with black, in form of a creſcent, the horns pointing towards the end of the feathers. The legs are ſhort and black: the exterior toe is cloſely united at the bottom to the middle toe.

According to *Linnæus* it takes its name from its note*, which has a ſound ſimilar to the word; or it may be derived from the *French huppè*, or creſted: it breeds in hollow trees, and lays two aſh-colored eggs: it feeds on inſects which it picks out of ordure of all kinds: the antients believed that it made its neſt of human excrement; ſo far is certain, that its hole is exceſſively fœtid from the tainted food it brings to its young. The country people in *Sweden* look on the appearance of this bird as a preſage of war;

——— Facies armata videtur.

and formerly the vulgar in our country eſteemed it a forerunner of ſome calamity. It viſits theſe iſlands frequently; but not at ſtated ſeaſons, neither does it breed with us. It is found in many parts of *Europe*, in *Egypt*, and even as remote as *Ceylon*. The *Turks* call it *Tir Chaous* or the meſſenger bird,

* *Faun. Suec.* 2d *edit.* 37.

from

from the refemblance its creft has to the plumes worn by the Chaous or *Turkifh* couriers.

Ovid fays that *Tereus* was changed into this bird:

> Vertitur in volucrem, cui ftant in vertice criftæ,
> Prominet immodicum pro longa cufpide roftrum:
> Nomen *Epops* volucri. *Metam.* lib. vi. l. 672.

> *Tereus*, through grief, and hafte to be reveng'd,
> Shares the like fate and to a bird is chang'd.
> Fix'd on his head the crefted plumes appear,
> Long is his beak and fharpen'd as a fpear. *Croxall*

Very

XI.
CREEPER.

Very ſlender BILL, very much incurvated.
Twelve feathers in the TAIL.

91. CREEPER.

Le petit Grimpereau. *Belon av.* 375.
Certhia. *Geſner av.* 251.
Aldr. av. I. 424.
Wil. orn. 144.
Raii ſyn. av. 47.
The Oxeye Creeper. *Charlton ex.* 93.
Picchio piccolo. *Zinan.* 75.
Le Grimpereau. *Briſſon* III. 603.
Cat. Carol. app. 37.
Certhia familiaris. *Lin. ſyſt.* 184.
Krypare. *Faun. Suec. ſp.* 106.
Træe-Pikke v. Lie-Heſten. *Br.* p. 12. *Scopoli*, No. 59.
Grau-Specht. *Friſch*, I. 39.
Baumlaufferl. *Kram.* 337.
Br. Zool. 82. plate K.

DESCRIP.

THE creeper weighs only five drams: and next to the creſted wren is the leſt of the *Britiſh* birds: the manner it has of ruffling its feathers, and their length give it a much larger appearance than is real. The length of this bird is five inches and a half: the breadth ſeven and a half: the bill is hooked like a ſickle: the irides hazel: the legs ſlender: the toes and claws very long, to enable it to creep up and down the bodies of trees in ſearch of inſects, which are its food: it breeds in hollow trees; and lays ſometimes twenty eggs: the head and upper part of the neck are brown, ſtreaked with black: the rump is tawny: the coverts of the wings are variegated with brown and black: the quil-feathers duſky, tipt with white, and

and edged and barred with tawny marks: the breaſt and belly are of a ſilvery white: the tail is very long, and conſiſts of twelve ſtiff feathers; notwithſtanding Mr. *Willughby* and other ornithologiſts give it but ten: they are of a tawny hue, and the interior ends of each ſlope off to a point.

ORDER

ORDER III. GALLINACEOUS.

XII. GROUS.

Short arched BILL.

Outmoſt, and inner TOES connected to the firſt joint of the middle toe by a ſmall membrane.

* With legs feathered to the feet: broad ſcarlet eye-brows.

** With naked legs.

92. WOOD.

Le Coc de bois ou Faiſan bruyant. *Belon av.* 249.
Urogallus major (the Male). *Geſner av.* 490.
Grygallus major (the Female). 495.
Gallo cedrone, Urogallus ſive Tetrao. *Aldr. av.* II. 29.
Gallo alpeſtre, Tetrax *Nemeſiani* (fem.) *Aldr. av.* II. 33.
Pavo ſylveſtris. *Girald. Topogr. Hibern.* 706.
Cock of the Mountain, or Wood. *Wil. orn.* 172.
Raii ſyn. av. 53.
Pl. Enl. 73. 74.
Capricalca. *Sib. Scot.* 16. *tab.* 14, 18.
Le cocque de Bruyeres. *Briſſon av.* I. 182. *Hiſt. d' Oys.* II. 191.
Tetrao urogallus. *Lin. ſyſt.* 273.
Kjader. *Faun. Suec. ſp.* 200.
Pontop. II. 101.
Tjader-hona. *Haſſelquiſt itin.* † 571.
Klein Stem. tab. 27.
Mas *Norvegis* Tiur, Teer, Toedder. Foemina *Norv.* Roey. *Brunnich,* 194.
Aurhan. *Kram.* 356.
Auerhahn. *Friſch,* I. 107, 108.
Br. Zool. 84. plates M. M*.
Pl. Enl. 73, 74.
Devi peteln. *Scopoli,* No. 169.

† *Swediſh* edition. This bird was ſhot in the iſle of *Milo,* on a *palm tree. Belon* tells us, it is often found in *Crete; Obſ.* p. 11. The *Engliſh* tranſlator of *Haſſelquiſt* gives a falſe name to the bird, calling it the *Black Game.*

THIS

WOOD GROUS.

aillon pinx.t

P. Mazell sculp.

THIS ſpecies is found in no other part of *Great Britain* than the Highlands of *Scotland*, North of *Inverneſs*; and is very rare even in thoſe parts. It is there known by the name of *Capercalze*, *Auer-calze*, and in the old law books *Caperkally*: the laſt ſignifying the horſe of the woods; this ſpecies being, in compariſon of others of the genus, pre-eminently large *. We believe that the breed is extinct in *Ireland*, where it was formerly found.

Giraldus Cambrenſis † deſcribes it under the title of *Peacock* of the wood, from the rich green that ſhines on the breaſt of the male. *Boethius* ‡ alſo mentions it under the name of *Capercalze*; and truely deſcribes its food, the extreme ſhoots of the pine. He afterwards gives an exact deſcription of the *black cock*, but gives it the name of the cock of the wood, a name now confined to this ſpecies. Biſhop *Leſsly* || is a third of our hiſtorians who makes mention of this bird along with two others of the genus, the black cock and common grous; but the *Ptarmigan* is overlooked by them. None of theſe writers were converſant in the ſtudy of natural hiſtory, therefore are very excuſable for their inaccuracy.

* For the ſame reaſon the *Germans* call it *Aur-han*, or the *Urus* or wild ox cock.

† *Topogr. Hibern.* 706. ‡ Deſcr. Regni *Scotiæ*. 7.

|| *Scotiæ* Deſcr. 24.

It

It inhabits wooded and mountanous countries; in particular, forests of pines, birch trees and junipers; feeding on the tops of the former, and berries of the latter; the first infects often the flesh with such a taste, as to render it scarcely eatable. In the spring it calls the females to its haunts with a loud and shrill voice; and is at that time so very inattentive to its safety, as to be very easily shot. It stands perched on a tree, and descends to the females on their first appearance. They lay from eight to sixteen eggs; eight at the first, and more as they advance in age *.

These birds are common to *Scandinavia*, *Germany*, *France*, and several parts of the *Alps*. In our country I have seen one specimen at *Inverness*, a male, killed in the woods of Mr. *Chisolme*, North of that place.

Descrip. The length of the male is two feet eight inches; the breadth three feet ten: its weight sometimes fourteen pounds. The female is much less, the length being only twenty-six inches; the breadth forty. The sexes differ also greatly in colors. The bill of the male is of a pale yellow: the nostrils are covered with dusky feathers: the head, neck and back are elegantly marked, slender lines of grey and black running transversely. The feathers on the hind part of the head are long, and beneath the throat is a large tuft of long feathers.

* *Schwenckfelt* Aviarium *Silesiæ*. 372.

The

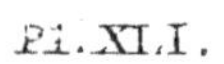
Pl. XLI.

N
F. WOOD GROUS
P. Mazell

The upper part of the breaſt is of a rich gloſſy green; the reſt of the breaſt and the belly black, mixed with ſome white feathers: the ſides are marked like the neck: the coverts of the wings croſſed with undulated lines of black and reddiſh brown: the exterior webs of the greater quil feathers are black: at the ſetting on of the wings in both ſexes is a white ſpot; the inner coverts are of the ſame color: the tail conſiſts of eighteen feathers, the middle of which is the longeſt; are black, marked on each ſide with a few white ſpots: the vent feathers black mixed with white. The legs very ſtrong, covered with brown feathers: the edges of the toes pectinated.

The female differs greatly from the male: the bill is duſky: the throat red: the head, neck and back are marked with tranſverſe bars of red and black: the breaſt has ſome white ſpots on it, and the lower part is of a plain orange color: the belly barred with pale orange and black; the tips of the feathers white. The feathers of the back and ſcapulars black, the edges mottled with black and pale reddiſh brown; the ſcapulars tipt with white. The inner webs of the quil feathers duſky: the exterior mottled with duſky and pale brown. The tail is of a deep ruſt color barred with black, tipt with white, and conſiſts of ſixteen feathers.

Geſner, as Mr. *Willughby* * has long ſince ob-

* *Wil. orn.* 173. *Geſner av.* 490. 495.

ſerved, deceived by the very different plumage of the male and female of this kind, has formed of them two ſpecies.

93. BLACK. Urogallus minor (the Male). *Geſner av.* 493. Grygallus minor (the Female). 496.
Faſan negro, Faſiano alpeſtre, Urogallus ſive Tetrao minor Gallus Scoticus ſylveſris. *Aldr. av.* II. 32. 160.
Raii ſyn. av. 53.
Heath-cock, black Game, or Grous. *Wil. orn.* 173.
Tetrao tetrix. *Lin. ſyſt.* 274.
Orre. *Faun. Suec. ſp.* 102.
Le Coq-de-bruyeres a queue fourchue. *Briſſon av.* I. 186. *Hiſt. d'Oys.* II. 210.
Cimbris mas Urhane, *fæmina* Urhoene. *Norvegis* Orrfugl. *Brunnich*, 196.
Berkhan, Schildhan. *Kram.* 356.
Birckhahn. *Friſch*, I. 109.
Br. Zool. 85. *tab.* M. 1. 2.
Pl. Enl. 172, 173.
Gallo sforcello *Italis. Scopoli*, No. 169.

MANNERS. THESE birds, like the former, are fond of wooded and mountanous ſituations; they feed on bilberries, and other mountain fruits; and in the winter on the tops of the heath. They are often found in woods; this and the preceding ſpecies perching like the pheaſant: in the ſummer they frequently deſcend from the hills to feed on corn: they never pair; but in the ſpring the male gets upon ſome eminence, crows and claps his wings *; on which ſignal all the females within hearing

* The ruffed heathcock of *America*, a bird of this genus, does the ſame. *Edw. Gl.* p. 80. The cock of the wood agrees too in this exultation during the amorous ſeaſon; at which time the peaſants in the *Alps*, directed by the ſound, have an opportunity of killing them.

BLACK COCK.

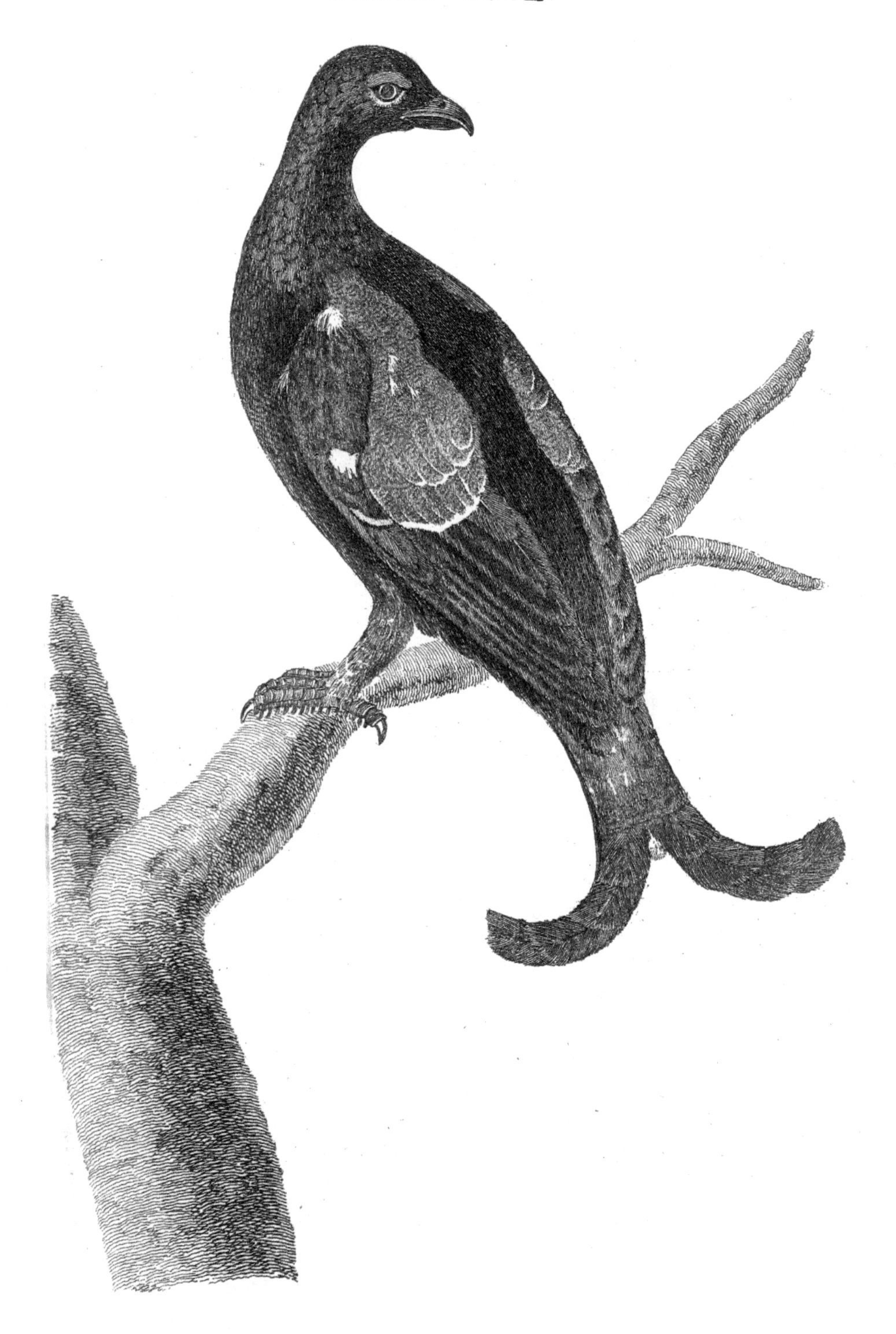

hearing resort to him: the hen lays seldom more than six or seven eggs. The young males quit their mother in the beginning of winter; and keep in flocks of seven or eight till spring; during that time they inhabit the woods: they are very quarrelsome, and will fight together like game cocks; and at that time are so inattentive to their own safety, that it has often happened that two or three have been killed at one shot.

DESCRIP.

An old black cock weighs near four pounds: its length is one foot ten inches; its breadth two feet nine: the bill dusky: the plumage of the whole body black, glossed over the neck and rump with a shining blue. The coverts of the wings are of a dusky brown: the four first quil feathers are black, the next white at the bottom; the lower half of the secondary feathers white, and the tips are of the same color: the inner coverts of the wings white: the thighs and legs are covered with dark brown feathers; on the former are some white spots: the toes resemble those of the former species. The tail consists of sixteen black feathers, and is much forked; the exterior feathers bend greatly outwards, and their ends seem as if cut off. The feathers under the tail and inner coverts of the wings are of a pure white.

FEMALE.

The female weighs only two pounds: its length is one foot six inches; its breadth two feet six. The head and neck are marked with alternate bars of dull red and black: the breast with dusky, black and white; but the laſt predominates. The back, coverts of the wings and tail are of the ſame colors as the neck, but the red is deeper: the inner webs of the quil feathers are mottled with black and white: the inner coverts of the wings are white; and in both ſexes form a white ſpot on the ſhoulder. The tail is ſlightly forked; it conſiſts of eighteen feathers variegated with red and black. The feathers under the tail are white, marked with a few bars of black and orange. This bird hatches its young late in the ſummer. It lays from ſix to eight eggs, of a dull yellowiſh white color, marked with numbers of very ſmall ferruginous ſpecks; and towards the ſmaller end with ſome blotches of the ſame hue.

MIXED BREED.

Beſides the common ſpecies of black cock, M. *Briſſon* mentions a variety found in *Scotland*, under the name of *le coq de bruyere piqueté*, or ſpotted black cock. It differs from the common ſort in being ſpotted on the neck, breaſt, wings and thighs with red. This I ſuppoſe to have been a ſpurious breed between this and the former ſpecies, as the *Tetrao Hybridus* of *Linnæus* is. I could not learn that this mixed race was found at preſent in *North Britain*, perhaps becauſe the cock of the wood is now become ſo very rare. It is alſo found in *Sweden*, and deſcribed by *Linnæus* in his *Faun. Suec.* *ſp.* 201. by the title of *Tetrao caudâ bifurcâ ſubtus albo punctata*, in *Swediſh*, *Racklehane* or *Roſlare:* the legs of this and the preceding kind are feathered only

F. GROUS. No. 94 PTARMIGAN. No.

only to the feet: they both inhabit woods in the winter; therefore nature hath not given them the ſame kind protection againſt the cold, as ſhe has the grous and ptarmigan, who muſt undergo all the rigor of the ſeaſon beneath the ſnow, or on the bare ground.

94. RED.

Gallina campeſtris. *Girald. topogr. Hibern.* 706.
Red Game, Gorcock, or Moor-cock, *Wil. orn.* 177.
Lagopus altera Plinii. *Raii ſyn. av.* 54.
Moor-cock, or Moor-fowl. *Sib. Scot.* 16.
La Gelinote Hupée. *Briſſon av.* I. 209. *Hiſt. d'Oys.* II. 252.
La Gelinote d'Ecoſſe, Bonaſa Scotica. *Idem* 199. *tab.* 22. f. 1. *Hiſt. d'Oys.* II. 242.
Br. Zool. 85. plate M. 3.

DESCRIP.

THE male weighs about nineteen ounces. The length is fifteen inches and a half: the breadth twenty-ſix. The bill is black: the noſtrils covered with red and black feathers: the irides hazel colored. At the baſe of the lower mandible, on each ſide, is a white ſpot: the throat is red. The plumage on the head and neck is of a light tawny red; each feather is marked with ſeveral tranſverſe bars of black.

The back and ſcapular feathers are of a deeper red, and on the middle of each feather is a large black ſpot: the breaſt and belly are of a dull purpliſh brown, croſſed with numerous narrow duſky lines: the quil feathers are duſky: the tail conſiſts

T 3 of

of ſixteen feathers of an equal length, all of them (except the four middlemoſt) are black, and the middle feathers are barred with red: the thighs are of a pale red, barred obſcurely with black, the legs and feet cloathed to the very claws with thick ſoft white feathers*; the claws are whitiſh, very broad and ſtrong.

The female weighs only fifteen ounces. The colors in general are duller than thoſe of the male: the breaſt and belly are ſpotted with white: and the tips of ſome of the coverts of the wings are of the ſame color. The red naked part that lies above the eyes is leſs prominent than in the male, and the edges not ſo deeply fringed.

We believe this ſpecies to be peculiar to the *Britiſh* iſlands; not having met with any account of it, except in the writings of our countrymen Mr. *Ray* and *Willughby*, and in M. *Briſſon* under the name of *Bonaſa Scotica*; the ſame writer deſcribes it again by the title of *Attagen*, but his references are either to authors who have copied our naturaliſts, or to ſuch who mean quite another kind. Mr. *Ray* ſeems to think his bird, the other *Lagopus* of *Pliny*†, or the *Francolino* of the

* The feet in the figure given by *M. Briſſon* are engraven naked, or bare of feathers. The ſpecimen probably came to that gentleman in that condition: his deſcription in other reſpects is very accurate.

† Eſt et alia nomine eodem, a coturnicibus magnitudine tantum differens, croceo tinctu cibis gratiſſima. lib. x. c. 48.

modern

modern *Italians:* but the account left us by *Pliny* ſeems too brief and uncertain to determine at this time what ſpecies he intended; and that the *Francolino* is not the ſame with our grous, is evident from the figure of it exhibited by our accurate friend Mr. *Edwards* *.

Theſe birds pair in the ſpring, and lay from ſix to ten eggs: the young brood or packs follow the hen the whole ſummer; in the winter they join in flocks of forty or fifty, and become remarkably ſhy and wild: they always keep on the tops of the hills, are ſcarce ever found on the ſides, and never deſcend into the vallies; their food is the mountain berries, and the tops of heath.

95. PTARMIGAN.

La perdris blanche. *Belon av.* 259.
Lagopus. *Geſner av.* 576.
Perdrix alba ſeu Lagopus, Perdice alpeſtre. *Aldr. av.* II. 66.
Lagopus. *Plinii* lib. x. c. 48.
Tetrao Lagopus. *Lin. ſyſt.* 274.
Snoripa. *Faun. Suec. ſp.* 203.
La Gelinote blanche. *Briſſon av.* I. 216.
Raii ſyn. av. 55.
White Game, erroneouſly called the white Partridge. *Wil. orn.* 176.
The Ptarmigan. *Sib. Scot.* 16.
Pl. Enl. 129. *Hiſt. d'Oys.* II. 264.
Norv. Rype. *Mas Iſlandis,* Riupkarre, *Fæm.* Riupa. *Brunnich* 199.
Schneehuhn. *Friſch,* I. 110.
Schneehun. *Kram.* 356.
Br. Zool. 86 plates M. 4. 5.
Scopoli. No. 118.

THIS bird is well deſcribed by Mr. *Willughby,* under the name of the white game.

* Plate 246.

T 4 M. *Briſſon*

M. *Briſſon** joins it with the white partridge of Mr. *Edwards*, plate 72. I have received both ſpecies at the ſame time from *Norway*, and am convinced that they are not the ſame.

Theſe two birds differ greatly; the former being above twice the ſize of the *Ptarmigan*; and the color of its ſummer plumage quite different; that of Mr. *Edwards'* bird being marked with large ſpots of white, and dull orange; that of the *Ptarmigan* is of a pale brown or aſh-color, elegantly

DESCRIP. croſſed or motled with ſmall duſky ſpots, and minute bars: the head and neck with broad bars of black, ruſt-color, and white: the wings are white, but the ſhafts of the greater quil-feathers black: the belly white. In the male, the grey color predominates, except on the head and neck where there is a great mixture of red, with bars of white: but the whole plumage in this ſex is extremely elegant. The females and young birds have a great deal of ruſt-color in them: both agree in their winter dreſs, being intirely white, except as follows: in the male a black line occurs between the bill and the eyes; the ſhaft of the ſeven firſt quil feathers are black: the tail of the *Ptarmigan* conſiſts of ſixteen feathers; the two middle of which are aſh-colored, motled with black, and tipt with white; the two next black ſlightly marked with white at their ends, the reſt wholly black; the fea-

* *Tom.* I. *p.* 216.

thers

thers incumbent on the tail white, and almoſt entirely cover it.

The length of theſe birds is near fifteen inches; the extent twenty three: the weight nineteen ounces.

Ptarmigans are found in theſe kingdoms only on the ſummits of the higheſt hills of the highlands of *Scotland* and of the *Hebrides*; and a few ſtill inhabit the lofty hills near *Keſwick* in *Cumberland.* They live amidſt the rocks perching on the grey ſtones, the general color of the ſtrata in thoſe exalted ſituations: they are very ſilly birds, ſo tame as to bear driving like poultry; and if provoked to riſe take very ſhort flights, taking a ſmall circuit like pigeons: they taſte ſo like a grous as to be ſcarcely diſtinguiſhed; like the grous they keep in ſmall packs; but never like thoſe birds take ſhelter in the heath; but beneath looſe ſtones.

Theſe birds are called by *Pliny*, *Lagopi*, their feet being cloathed with feathers to the claws, as the hare's are with fur: the nails are long, broad and hollow: the firſt circumſtance guards them from the rigor of the winter; the latter enables them to form a lodge under the ſnow, where they lie in heaps to protect them from the cold: the feet of the grous are cloathed in the ſame manner, but thoſe of the two firſt ſpecies here deſcribed, which perch upon trees, are naked, the legs only being feathered, not being in want of ſuch a protection.

In *Scotland* they inhabit from the hill of *Benlomond*

mond to the naked mountain of *Scaroben* in *Cathnefs*, the iſle of *Arran*, many of the *Hebrides*, and the *Orknies*.

** With naked legs.

96. PARTRIDGE.

La Perdris griſe ou Gouache. *Belon av.* 257.
Perdix (Waldhun) *Geſner av.* 669.
Perdix minor ſive cinerea. *Aldr. av.* II. 66.
Wil. orn. 166.
Raii ſyn. av. 57.
Tetrao Perdrix. *Lin. ſyſt.* 276.
Rapphona. *Faun. Suec. ſp.* 205.
La Perdrix griſe. *Briſſon av.* I. 219.
Pl. Enl. 27. *Hiſt. d'Oys.* II. 401.
Starna. *Zinan.* 30.
Agerhoene. *Br.* 201.
Rebhun. *Kram.* 357.
Rebhuhn. *Friſch*, I. 114.
Br. Zool. 86. plate M.
Serebitza *Scopoli*. No. 175.

DESCRIP.

THE male partridge weighs near fifteen ounces; the female near two ounces leſs: the length to the end of the tail thirteen inches; the breadth twenty. The bill is whitiſh: the crown of the head is brown ſpotted with reddiſh white: behind each eye is a naked red ſkin. The chin, cheeks and forehead of a deep orange color, but in the females much paler than in the other ſex. The neck and breaſt are prettily marked with narrow undulated lines of aſh-color and black; and in the hind part of the neck is a ſtrong mixture of ruſt color: on the breaſt of the male is a broad a broad mark in form of a horſe-ſhoe, of a deep orange hue; in the female it is leſs diſtinct.

Each feather on the back is finely marked with ſeveral ſemicircular lines of reddiſh brown and black: the ſcapulars with a narrow white line along their ſhafts, and with black and cinereous undulated lines on the webs; whoſe ſides are marked with a large ſpot of ruſt color. The greater quil-feathers are duſky, ſpotted on each web with pale red: it has eighteen feathers in the tail; the ſix outmoſt on each ſide are of a bright ruſt color tipt with white; the others marked tranſverſely with irregular lines of pale reddiſh brown and black: the legs are of a whitiſh caſt.

SALACIOUS.

The nature of this bird is ſo well known, that it will be unneceſſary to detain the readers with any account of it: all writers agree, that its paſſion for venery exceeds that of any bird of the genus; ſhould the reader's curioſity be excited to ſee a more particular account, we beg leave to refer them to thoſe authors who have recorded this part of its natural hiſtory *.

The *Britiſh* name of this bird is *Kor-iâr*, a word now obſolete; that now in uſe is *Pertriſen*, borrowed from the *Normans*. *Sâr* is the generic name for the tribe.

* *Pliny* lib. 10. c. 23. *Wil. orn.* 168. *Edw. preface to Gleanings, part* 2.

L

97. QUAIL. La Caille. *Belon av.* 263.
Gesner av. 334.
Coturnix Latinorum. *Aldr. av.* II. 69.
Wil. orn. 169.
Raii syn. av. 58.
La Caille. *Brisson av.* I. 247.
Hist. d'Oys. II. 449.
Quaglia. *Zinan.* 36.
Tetrao coturnix. *Lin. syst.* 278.
Wachtel. *Faun. Suec. sp.* 206.
Vagtel. *Brunnich,* 202.
Wachtel. *Kram.* 357. *Frisch,* I. 117.
Br. Zool. 87. plate M. 6.
Perpelitza *Scopoli,* No. 176.

DESCRIP. THE length of the quail is seven inches and a half; the breadth fourteen: the bill is of a dusky color: the feathers of the head are black, edged with rusty brown: the crown of the head divided by a whitish yellow line, beginning at the bill and running along the hind part of the neck to the back: above each eye is another line of the same color: the chin and throat of a dirty white: the cheeks spotted with brown and white: the breast is of a pale yellowish red spotted with black: the scapular feathers and those on the back are marked in their middles with a long pale yellow line, and on their sides with ferruginous and black bars: the coverts of the wings are reddish brown, elegantly barred with paler lines bounded on each side with black. The exterior side of the first quil feathers is white, of the others dusky spotted with red: the tail consists of twelve short feathers barred with black and very pale brownish red: the legs are of a pale hue.

Quails

Quails are found in moſt parts of *Great-Britain*; but not in any quantity: they are birds of paſſage: ſome entirely quitting our iſland, others ſhifting their quarters. A gentleman, to whom this work lies under great obligations for his frequent aſſiſtance, has aſſured us, that theſe birds migrate out of the neighbouring inland countries, into the hundreds of *Eſſex*, in *October*, and continue there all the winter: if froſt or ſnow drive them out of the ſtubble fields and marſhes, they retreat to the ſea-ſide; ſhelter themſelves among the weeds, and live upon what they can pick up from the *algæ*, &c. between high and low water mark. Our friend remarks, that the time of their appearance in *Eſſex*, coincides with that of their leaving the inland counties; the ſame obſervation has been made in *Hampſhire*.

Theſe birds are much leſs prolific than the partridge, ſeldom laying more than ſix or ſeven whitiſh eggs, marked with ragged ruſt colored ſpots: yet Mr. *Holland* of *Conway*, once found a neſt with twelve eggs, eleven of which were hatched: they are very eaſily taken, and may be enticed any where by a call.

They are birds of great ſpirit; inſomuch that quail fighting among the *Athenians* was as great an entertainment as cock fighting is in this country: it is at this time a faſhionable diverſion in *China*, and large ſums are betted there on the event*. The

* *Bell's Travels.* I. 371.

bodies

bodies of theſe birds are extremely hot; the *Chineſe* on that account hold them in their hands in cold weather in order to warm themſelves*. *Chaude comme une Caille*, is a common proverb.

The antients never eat this bird, ſuppoſing them to have been unwholeſome, as they were ſaid to feed on *Hellebore*.

To the birds of this genus we ſhould add the whole tribe of domeſtic land fowl, ſuch as *Peacocks*, *Pheaſants*, &c. but theſe cannot clame even an *European* origin.

PEACOCKS. *India* gave us *Peacocks*; and we are aſſured † they are ſtill found in the wild ſtate, in vaſt flocks, in the iſlands of *Ceylon* and *Java*. So beautiful a bird could not long be permitted to be a ſtranger in the more diſtant parts; for ſo early as the days of *Solomon* ‡, we find among the articles imported in his *Tarſhiſh* navies, *Apes* and *Peacocks*. A monarch ſo converſant in all branches of natural hiſtory, *who ſpoke of trees, from the cedar of Libanon, even unto the hyſſop that ſpringeth out of the wall: who ſpoke alſo of beaſts and of fowl*, would certainly not neglect furniſhing his officers with inſtructions for collecting every curioſity in the countries they voyaged to, which gave him a knowledge that

* *Oſbeck's Voyage*. I. 269.

† *Knox's hiſt. of Ceylon*. 28.

‡ *Kings*, I. 10.

the

diſtinguiſhed him from all the princes of his time. *Ælian* * relates, that they were brought into *Greece* from ſome barbarous country; and that they were held in ſuch high eſteem, that a male and female were valued at *Athens* at 1000 *drachmæ*, or 32*l.* 5*s.* 10*d.* Their next ſtep might be to *Samos*; where they were preſerved about the temple of *Juno*, being the birds ſacred to the goddeſs †: and *Gellius* in his *noctes Atticæ*, c. 16. commends the excellency of the *Samian* peacocks. It is therefore probable that they were brought here originally for the purpoſes of ſuperſtition, and afterwards cultivated for the uſes of luxury. We are alſo told, when *Alexander* was in *India* ‡, he found vaſt numbers of wild ones on the banks of the *Hyarotis*, and was ſo ſtruck with their beauty, as to appoint a ſevere puniſhment on any perſon that killed them.

Peacocks' creſts, in antient times, were among the ornaments of the Kings of *England*. *Ernald de Aclent* fined to King *John* in a hundred and forty palfries, with ſackbuts, lorains, gilt ſpurs and peacocks' creſts, ſuch as would be for his credit. *Maddox Antiq. Exch* .I. 273.

POULTRY.

Our common *poultry* came originally from *Perſia* and *India*. *Ariſtophanes* || calls the cock περσικός ὄρνις, the *Perſian* bird; and tells us, it enjoyed

* *Ælian de nat. an.* lib. v. 21.

† *Athenæus. lib.* xiv. p. 655.

‡ *Q. Curtius.* lib. ix. || *Aves, lin.* 483.

that

that kingdom before *Darius* and *Megabyzus*: at this time we know that theſe birds are found in a ſtate of nature in the iſles of *Tinian* *, and others of the *Indian* ocean; and that in their wild condition their plumage is black and yellow, and their combs and wattles purple and yellow †. They were early introduced into the weſtern parts of the world; and have been very long naturalized in this country; long before the arrival of the *Romans* in this iſland, *Cæſar* informing us, they were one of the forbidden foods of the *old Britains*. Theſe were in all probability imported here by the *Phœnicians*, who traded to *Britain*, about five hundred years before *Chriſt*. For all other domeſtic fowls, turkies, geeſe, and ducks excepted, we ſeem to be indebted to our conquerors, the *Romans*. The wild fowl were all our own from the period they could be ſuppoſed to have reached us after the great event of the flood.

PHEASANTS.

Pheaſants were firſt brought into *Europe* from the banks of the *Phaſis*, a river of *Colchis*.

> Argiva primùm ſum tranſportata carina,
> Ante mihi notum nil, niſi *Phaſis* erat.
>
> *Martial*. lib. xiii. ep. 72.

GUINEA HENS.

Guinea hens, the *Meleagrides* or *Gallinæ numidicæ*

* *Dampier's voy*. I. 392. Lord *Anſon's voy*. 309.

† For this information we are indebted to governor *Loten*.

of

of the antients, came originally from *Africa**. We are much furprized how *Belon* and other learned ornithologifts could poffibly imagine them to have been the fame with our *Turkies*; fince the defcriptions of the *meleagri* left us by *Athenæus* and other antient writers, agree fo exactly with the *Guinea hen*, as to take away (as we fhould imagine) all power of miftake. *Athenæus* (after *Clytus Milefius*, a difciple of *Ariftotle*) defcribes their nature, form and colors: he tells us, "They want natural affection "towards their young; that their head is naked, "and that on the top of it is a hard round body "like a peg or nail; that from the cheeks hangs a "red piece of flefh like a beard; that it has no wat-"tles like the common poultry; that the feathers "are black fpotted with white; that they have no "fpurs; and that both fexes are fo like, as not to "be diftinguifhed by the fight †". *Varro* and *Pli-*

* *Bofman's hiftory of Guinea.* 248. *Voyages de Marchais* III. 323. *Barbot's defcr. Guinea. Churchill's coll. voy.* v. 29.

† Ἔςι δὲ ἄςοργον πρὸς τα ἔκγονα τὸ ὄρνεον, καὶ ὀλιγωρεῖ τῶν νεωτέρων, — ἐπ' αὐτῆς δὲ λόφον σάρκινον σληκρὸν, ςρογγύλον ἐξέκοντα τῆς κεφαλῆς ὥσπερ πάτταλον —— πρὸς δὲ ταῖς γνάθοις από τȣ σώματος αρξαμένην ἀντὶ πώγωνος μακραν σάρκα, καὶ ἐρυθροτέραν των ορνιθων την δὲ τοῖς ὄρνισιν ἐπί τῶ ῥύγχει γινομένην, ην ἔνιοι πώγωνα καλȣσιν, ȣκ ἔχει, διο καὶ ταύτη κολοϐόν ἐςι —— σῶμα ἅπαν ποικίλον, μέλανος ὄντος τȣ χρώματος ὀλεπτίλοις λευκοῖς —— σκέλη και ἄκεντρα —— παραπλήσιαι δὲ εἰσὶν αἱ θήλειαι τοῖς ἄῤῥεσιν· διὸ και δυςδιάκριτόν ἐςι τὸ των μελεαγρίδων γένος. Athenæus, 655.

ny * take notice of their ſpotted plumage, and the gibbous ſubſtance on their head: ſo that from theſe citations we find every character of the *Guinea hen*, but none that agrees with the *Turky*.

Barbot † informs us that very few *turkies* are to be met with in *Guinea*; and thoſe only in the hands of the chiefs of the *European* forts; the negroes declining to breed any on account of their tenderneſs which ſufficiently proves them not to be natives of that climate. On the contrary the ſame writer ſays, that the *Guinea* hens, or as he calls them *Pintadas*, are found there in flocks of two or three hundred, that perch in trees, feed on worms and graſshoppers; that they are run down and taken by dogs, and that their fleſh is tender and ſweet, generally white, though ſometimes black.

He alſo remarks that neither the common poultry or ducks are natural to *Guinea*, any more than the *Turky*.

Neither is that bird a native of *Aſia*: the firſt that were ſeen in *Perſia* were brought from *Venice* by ſome *Armenian* merchants ‡. They are alſo cultivated in *Ceylon*, but not found wild.

In fact the *Turky* was unknown to the antient naturaliſts, and even to the *old world* before the diſcovery of *America*. It was a bird peculiar to

* *Varro*. lib. 3. c. 9. *Pliny*. lib. 10. c. 26. † *Barbot* 217.
‡ *Tavernier*. 146.

the

the new continent, and is now the commoneſt wild fowl of the northern parts of that country. It was firſt ſeen in *France*, in the reign of *Francis* I. and in *England*, in that of *Henry* VIII. By the date of the reign of theſe monarchs, the firſt birds of this kind muſt have been brought from *Mexico*, whoſe conqueſt was completed, A. D. 1521. the ſhort lived colony of the *French* in *Florida* not being attempted before 1562; nor our more ſucceſsful one in *Virginia*, effected till 1585; when both thoſe monarchs were in their graves.

Ælian, indeed, mentions a bird found in *India** that ſome writers have ſuſpected to be the *Turky*, but we conclude with *Geſner*, that it was either the *Peacock*, or ſome bird of that genus. On conſulting ſome gentlemen who have long reſided in the *Indies*, we find, that though the *Turky* is bred there, it is only conſidered as a domeſtic bird, and not a native of the country.

* *Æliani hiſt. an.* lib. xvi. c. 2.

XIII. BUSTARD.

Strong BILL, a little incurvated.
No back TOE.

98. GREAT.

Tetrax. *Athenæi*, lib. IX. 398.
L'Oſtarde. *Belon av.* 235.
Otis, vel Biſtarda. *Geſner av.* 484, 486.
Otis ſive Tarda. *Aldr. av.* II. 39.
Wil. orn. 178.
Raii ſyn. av. 58.
Guſtard. *Boethii*, 7. and *Sib. Scot.* 16.
Edw. Tab. 73, 74.
L'Outarde. *Briſſon av.* V. 18. *Hiſt. d'Oys.* II.
Otis tarda. *Lin. ſyſt.* 264.
Faun. Suec. ſp. 196.
Trap. *Kram.* 355.
Acker-Trappe. *Friſch*, I. 106. *Scopoli*, No. 160.
Br. Zool. 87. plate N. *Pl Enl.* 245.

DESCRIP.

THE buſtard is the largeſt of the *Britiſh* land fowl; the male at a medium weighing twenty-five pounds; there are inſtances of ſome very old ones weighing twenty-ſeven. The breadth nine feet; the length near four. Beſides the ſize and difference of color, the male is diſtinguiſhed from the female by a tuft of feathers about five inches long on each ſide the lower mandible. Its head and neck are aſh colored: the back is barred tranſverſely with black and bright ruſt color: the greater quil feathers are black: the belly white: the tail is marked with broad red and black bars, and conſiſts of twenty feathers: the legs duſky.

FEMALE.

The female is about half the ſize of the male: the

Pl.XLIV

No 98

BUSTARD.

the crown of the head is of a deep orange, traverſed with black lines; the reſt of the head is brown. The lower part of the fore-ſide of the neck is aſh-colored: in other reſpects it reſembles the male, only the colors of the back and wings are far more dull.

PLACE.

Theſe birds inhabit moſt of the open countries of the ſouth and eaſt parts of this iſland, from *Dorſetſhire*, as far as the *Wolds* in *Yorkſhire* *. They are exceeding ſhy, and difficult to be ſhot; run very faſt, and when on the wing can fly, though ſlowly, many miles without reſting. It is ſaid that they take flight with difficulty, and are ſometimes run down with grehounds. They keep near their old haunts, ſeldom wandering above twenty or thirty miles. Their food is corn and other vegetables, and thoſe large earth worms that appear in great quantities on the *Downs*, before ſun-riſing in the ſummer. Theſe are replete with moiſture, anſwer the purpoſe of liquids, and enable them to live long without drinking on thoſe extenſive and dry tracts. Beſides this, nature hath given the males an admirable magazine for their ſecurity againſt drought, being a pouch †, whoſe entrance lies immediately under the tongue, and which is capable

* In Sir *Robert Sibbald*'s time they were found in the *Mers*, but I believe that they are now extinct in *Scotland*.

† The world is obliged to the late Dr. *Douglas* for this diſcovery; and to Mr. *Edwards* for communicating it.

of holding near ſeven quarts; and this they probably fill with water, to ſupply the hen when ſitting, or the young before they can fly. Buſtards lay only two eggs, of the ſize of thoſe of a gooſe, of a pale olive brown, marked with ſpots of a darker color; they make no neſt, only ſcrape a hole in the ground. In autumn they are (in *Wiltſhire)* generally found in large turnep fields near the Downs, and in flocks of fifty or more.

99. LESSER. The *French* Canne-patiere. V. 24. *de Buffon,* II. 40.
Wil. orn. 179. *Pl. Enl.* 10. 25.
La petite outard. *Briſſon av.* Otis Tetrax. *Lin. ſyſt.* 264.

THERE have been three or four inſtances of this ſpecies being ſhot in *England,* but the ſpecimens I have ſeen have been all female. Whether they were accidental ſtragglers from the continent; or whether they breed here, and the male has eſcaped the ſportſman's notice, is not yet aſcertained.

This bird is about the ſize of a pheaſant. The male, which I have ſeen in *France,* varies much in the colors of the neck from the female, being black, marked tranſverſely above and below with a band of white. The crown of the head black and ferruginous. The back, ſcapulars, and coverts of the wings varied with black and ferruginous lines.

lines. The quil feathers black at their ends, white at their bottoms; the white predominating to the secondaries, which are quite white. The breast, belly, and thighs white. The middle feathers of the tail, tawny barred with black: the rest white. Legs cinereous.

The neck of the female agrees in colors with the back: in other respects the marks pretty nearly agree.

They inhabit open countries; feed on grain, seeds, and insects.

100. THICK-KNEED.

Norfolk Plover. *Br. Zool.* II. 378.
Un Ostardeau, Oedicnemus. *Belon av.* 239.
Charadrius (Triel vel Griel). *Gesner av.* 256.
The Stone Curlew. *Wil. orn.* 306.
Raii syn. av. 108.
Le grand Pluvier, Courly de terre. *Brisson av.* V. 76. *Tab.* 7. *fig.* 1.
Charadrius oedicnemus. *Lin. syst.* 255.
Br. Zool. 127.
Kervari. *Hasselquist Itin.* 210? *Engl. Ed.* 200.

DESCRIP.

THE weight of this species is eighteen ounces. The length to the tail eighteen inches: the breadth thirty six. The head is remarkably round: the space beneath the eyes is bare of feathers, and of a yellowish green: the irides yellow: the feathers of the head, neck, back, and scapulars, and coverts of the wings are black, edged deeply with pale reddish brown: the belly and thighs are of a pure white: the two firſt quil feathers are black, marked on the middle of each web with a large white ſpot.

The tail conſiſts of twelve feathers; the tips of the two outmoſt are black, beneath is a broad white bar, the remaining part barred with white and duſky brown: in the next feathers the white leſſens; in the middle it almoſt diſappears, changing it to a pale reddiſh brown, mottled with a darker: its mouth very wide: the legs are of a fine yellow: the toes very ſhort, bordered with a ſtrong membrane: the knees thick, as if ſwelled, like thoſe of a gouty man; from whence *Belon* gives it the name of *Oedicnemus* *.

This bird ſeems unknown in the weſtern parts of this kingdom; but is found in *Hampſhire*, *Norfolk*, and on *Lincoln heath*, where, from a ſimilarity of colors to the curlew, it is called the *Stone Curlew*. It breeds in ſome places in rabbet boroughs; alſo among ſtones on the bare ground, laying two eggs of a copper color, ſpotted with a darker red. The young run ſoon after they are hatched. Theſe birds feed in the night on worms and caterpillars: they will alſo eat toads; and *Geſner* ſays they will catch mice, which is confirmed by *Haſſelquiſt*.

They make a moſt piercing ſhrill noiſe, which they begin in the evening; and are ſo loud, as to

* From οιδεω, and κνήμη.

be

be heard near a mile in a ſtill night. They inhabit fallow lands and downs; affect dry places, never being ſeen near any waters. When they fly, they extend their legs ſtrait out behind: are very ſhy birds; run far before they take to wing; and often ſquat: are generally ſeen ſingle; and are eſteemed very delicate food.

In habit, make, and manners, theſe birds approach near to the *Buſtard.* We have therefore removed them into that genus, from that of *Plovers.*

They are migratory: appear in *England* about the middle of *April,* and retire in autumn.

ORDER

ORDER IV. COLUMBINE.

XIV. PIGEON.

Soft ſtrait BILL.
NOSTRILS lodged in a tuberous naked ſkin.
TOES divided to their origin.

101. COMMON.

La Pigeon privè. *Belon av.* 313.
Columba vulgaris. *Geſner av.* 279. Livia. 307.
Columba domeſtica. *Aldr. av.* II. 225.
Common wild Dove, or Pigeon. *Wil. orn.* 180. and the Stock Dove, or Wood Pigeon *. 185.
Raii ſyn. av. 59, 62.
Golob. *Scopoli,* No. 177.
Le Pigeon domeſtique. *Briſſon av.* I. 68. *Hiſt. d' Oys.* II. 491.
Le Biſet. 498.
Columba Oenas. *Lin. ſyſt.* 279.
Skogs dufwa, Dufwa, Hemdufwa. *Faun. Suec. ſp.* 207.
Kirke-Due, Skov-Due. *Brunnich,* 203.
Feldtaube, Hauſtaube, Hohltaube. *Kram.* 358.
Blau-Taube, or Holtz-Taube. *Friſch,* I. 139.
Br. Zool. 88. plate 88.

THE tame pigeon, and all its beautifull varieties, derive their origin from one ſpecies, the *Stock Dove:* the *Engliſh* name implying its being the *ſtock* or *ſtem* from whence the other domeſtic kinds ſprung. Theſe birds, as *Varro* † obſerves, take their (Latin) name, *Columba,* from their voice

* Columba livia. *Aldr. av.* II. 234. et Oenas, ſeu vinago. 233.
† De *Ling. Lat.* lib. IV.

or

№ 103

ROCK PIGEON.

№ 101

or cooing; and had he known it, he might have added the *Britiſh*, &c. for *K'lommen*, *Kylobman*, *Kulm* and *Kolm* ſignify the ſame bird. They were, and ſtill are in moſt parts of our iſland, in a ſtate of nature; but probably the *Romans* taught us the method of making them domeſtic, and conſtructing pigeon houſes. Its characters in the ſtate neareſt that of its origin, is a deep bluiſh aſh color; the breaſt daſhed with a fine changeable green and purple; the ſides of the neck with ſhining copper color; its wings marked with two black bars, one on the coverts of the wings, the other on the quil feathers. The back white, and the tail barred near the end with black. The weight fourteen ounces.

DESCRIP.

In the wild ſtate it breeds in holes of rocks, and hollows of trees, for which reaſon ſome writers ſtile it *columba cavernalis* *, in oppoſition to the Ring Dove, which makes its neſt on the boughs of trees. Nature ever preſerves ſome agreement in the manners, characters, and colors of birds reclamed from their wild ſtate. This ſpecies of pigeon ſoon takes to build in artificial cavities, and from the temptation of a ready proviſion becomes eaſily domeſticated. The drakes of the tame duck, however they may vary in color, ever retain the mark of their origin from our *Eng-*

* The *Columba ſaxatilis*, a ſmall ſort, that is frequent on moſt of our cliffs, is only a variety of the wild pigeon.

liſh

liſh mallard, by the curled feathers of the tail: and the tame gooſe betrays its deſcent from the wild kind, by the invariable whiteneſs of its rump, which they always retain in both ſtates.

Multitudes of theſe birds are obſerved to migrate into the ſouth of *England:* and while the beech woods were ſuffered to cover large tracts of ground, they uſed to haunt them in *myriads*, reaching in ſtrings of a mile in length, as they went out in the morning to feed. They viſit us the lateſt of any bird of paſſage, not appearing till *November*; and retire in the ſpring. I imagine that the ſummer haunts of theſe are in *Sweden*, for Mr. *Eckmark* makes their retreat thence coincide with their arrival here *. But many breed here, as I have obſerved, on the cliffs of the coaſt of *Wales*, and of the *Hebrides*.

VARIETIES.

The varieties produced from the domeſtic pigeon are very numerous, and extremely elegant; theſe are diſtinguiſhed by names expreſſive of their ſeveral properties, ſuch as *Tumblers*, *Carriers*, *Jacobines*, *Croppers*, *Powters*, *Runts*, *Turbits*, *Owls*, *Nuns*, &c. † The moſt celebrated of theſe is the

CARRIER.

Carrier, which from the ſuperior attachment that

* *Amœn. Acad.* IV. 593.

† Vide *Wil. orn. Moore's Columbarium*, and a treatiſe on domeſtic pigeons, publiſhed in 1765. The laſt illuſtrates the names of the birds, with ſeveral neat figures.

pigeon

pigeon ſhews to its native place, is employed in many countries as the moſt expeditious courier: the letters are tied under its wing, it is let looſe, and in a very ſhort ſpace returns to the home it was brought from, with its advices*. This practice was much in vogue in the *Eaſt*; and at *Scanderoon*, till of late years †, uſed on the arrival of a ſhip, to give the merchants at *Aleppo* a more expeditious notice than could be done by any other means. In our own country, theſe aerial meſſengers have been employed for a very ſingular purpoſe, being let looſe at *Tyburn* at the moment the

* This cuſtom was obſerved by that legendary traveller, Sir *John Maundevile*, knight, warrior and pilgrim; who, with the true ſpirit of religious chivalry, voyaged into the *Eaſt*, and penetrated as far as the borders of *China*, during the reigns of *Edward* II. and III.

" In that contree," ſays he, " and other contrees bezonde, thei han a cuſtom, whan thei ſchulle uſen werre, and whan men holden ſege abouten cytee or caſtelle, and thei withinnen dur not ſenden out meſſagers with lettere, fro lord to lord, for to aſke ſokour, thei maken here letters and bynden hem to the nekke of a *Colver*, and leten the *Colver* flee; and the *Colveren* ben ſo taughte, that thei fleen with tho letters to the verry place, that men wolde ſend hem to. For the *Colveres* ben noryſſcht in tho places, where thei ben ſent to; and thei ſenden hem thus, for to beren here letters. And the *Colveres* retournen azen, where as thei ben noriſſcht and ſo they don comounly." The voiage and travaile of Sir *J. Maundevile*, knight, ed. 1727.

† Dr. *Ruſſel* informs us, that the practice is left off. *Hiſt. Aleppo*, 66.

fetal

fatal cart is drawn away, to notify to diſtant friends, the departure of the unhappy criminal.

In the *Eaſt*, the uſe of theſe birds ſeems to have been improved greatly, by having, if we may uſe the expreſſion, relays of them ready to ſpread intelligence to all parts of the country. Thus the governor of *Damiata* circulated the news of the death of *Orrilo:*

> Toſto che'l Caſtellan di *Damiata*
> Certificoſſi, ch'era morto *Orrilo,*
> La *Colomba* laſciò, ch'avea legata
> Sotto l'ala la lettera col filo.
> Quelle andò al *Cairo,* ed indi fu laſciata
> Un' altra altrove, come quivi e ſtilo:
> Si, che in pochiſſime ore andò l'avviſo
> Per tutto *Egitto,* ch'era *Orrilo* ucciſo *.

But the ſimple uſe of them was known in very early times: *Anacreon* tells us, he conveyed his billet-doux, to his beautifull *Bathyllus*, by a dove.

> Εγὼ δ' Ἀνακρέοντι
> Διακονῶ τοσαῦτα·
> Και νῦν οἵας ἐκείνου
> Ἐπιστολας κομίζω †.

* 'As ſoon as the commandant of *Damiata* heard that *Orrilo* was dead, he let looſe a pigeon, under whoſe wing he 'had tied a letter; this fled to *Cairo*, from whence a ſecond 'was diſpatched to another place, as is uſual; ſo that in a 'very few hours, all *Egypt* was acquainted with the death 'of *Orrilo*.' *Arioſto, canto* 15.

† *Anacreon, ode* 9. εἰς περιστεράν.

I am

I am now *Anacreon*'s ſlave,
And to me entruſted have
All the o'erflowings of his heart
To *Bathyllus* to impart;
Each ſoft line, with nimble wing,
To the lovely boy I bring.

Tauroſthenes alſo, by means of a pigeon he had decked with purple, ſent advice to his father, who lived in the iſle of *Ægina*, of his victory in the *Olympic* games, on the very day he had obtained it *. And, at the ſiege of *Modena*, *Hirtius* without, and *Brutus* within the walls, kept, by the help of pigeons, a conſtant correſpondence; baffling every ſtratagem of the beſieger *Antony* †, to intercept their couriers. In the times of the *Cruſades*, there are many more inſtances of theſe birds of peace being employed in the ſervice of war: *Joinville* relates one during the cruſade of *Saint Louis* ‡; and *Taſſo* another, during the ſiege of *Jeruſalem* ||.

The nature of pigeons is to be gregarious; to lay only two eggs; to breed many times in the year §;

* *Ælian var. hiſt.* lib. IX. 2. *Pliny, lib.* X. *c.* 24. ſays, that ſwallows have been made uſe of for the ſame purpoſe.

† *Pliny*, lib. X. c. 37. *Exclames*, Quid vallum et vigil obſidio atque etiam retia amne pretenta profuere *Antonio, per cœlum eunte nuncio?*

‡ *Joinville*, 638. *app.* 35. || *Taſſo*, Book XVIII.

§ So quick is their produce, that the author of the *Oeconomy of nature* obſerves, that in the ſpace of four years, 14,760 may come from a ſingle pair. *Stillingfleet's tracts*, 75.

to

to bill in their courtſhip; for the male and female to ſit by turns, and alſo to feed their young; to caſt their proviſion out of their craw into the young ones' mouths; to drink, not like other birds by ſipping, but by continual draughts like quadrupeds; and to have notes mournful, or plaintive.

102. RING.

Le Ramier. *Belon av.* 307.
Phaſſa. *Belon obſ.* 13.
Palumbus. *Geſner av.* 310.
Palumbus major ſive torquatus. *Aldr. av.* II. 227.
Colombaccio. *Olina,* 54.
Ring-dove, Queeſt, or Cuſhat. *Wil. orn.* 185.
Le Pigeon Ramier. *Briſſon av.* I. 89. *Hiſt. d'Oys.* II. 531.
Griunik. *Scopoli,* No. 178.
Raii ſyn. av. 62.
Columba palumbus. *Lin. ſyſt. ſp.* 282.
Ringdufwa, Siutut. *Faun. Suec. ſp.* 208.
Wildtaube, Ringltaube. *Kram.* 359.
Ringel-Taube. *Friſch,* I. 138.
Dan. Ringel-due *Bornholmis,* Skude. *Brunnich,* 204.
Br. Zool. 89. plate O.

THIS ſpecies forms its neſt of a few dry ſticks in the boughs of trees: attempts have been made to domeſticate them, by hatching their eggs under the common pigeon in dove houſes; but as ſoon as they could fly, they always took to their proper haunts. In the beginning of the winter they aſſemble in great flocks, and leave off cooing; which they begin in *March,* when they pair. The ring dove is the largeſt pigeon we have; and may be at once diſtinguiſhed from all

DESCRIP.

others by the ſize. Its weight is about twenty ounces;

ounces: its length eighteen inches; its breadth thirty. The head, back, and coverts of the wings are of a bluiſh aſh color: the lower ſide of the neck and the breaſt are of a purpliſh red, daſhed with aſh color: on the hind part of the neck is a ſemicircular line of white; above and beneath that the feathers are gloſſy, and of changeable colors as oppoſed to the light. The belly is of a dirty white: the greater quil feathers are duſky; the reſt aſh colored: underneath the baſtard wing is a white ſtroke pointing downwards.

103. TURTLE.

La Turtrelle. *Belon av.* 309.
Turtur. *Geſner av.* 316.
Turtur. *Aldr. av.* II. 235.
Tortora. *Olina*, 34.
The Turtle-dove. *Wil. orn.* 183.
Raii ſyn. av. 61.
Wilde Turtel taube. *Kram.* 359.
Turtel-Taube. *Friſch*, I. 140.
Le Tourterelle. *Briſſon av.* I. 92. *Scopoli*, No. 181.
Br. Zool. 89. plate O. 1.
Hiſt. d'Oys. II. 545.

THIS ſpecies is found in *Buckinghamſhire*, *Glouceſterſhire*, *Shropſhire*, and in the *Weſt* of *England*. They are very ſhy and retired birds, breeding in thick woods, generally of oak: we believe that they reſide in *Buckinghamſhire* during the breeding ſeaſon, migrating into the other countries in autumn.

DESCRIP.

The length is twelve inches and a half; its breadth twenty-one: the weight four ounces. The irides are of a fine yellow: a beautifull crimſon circle encompaſſes the eye lids. The chin and forehead are whitiſh: the top of the head aſh colored mixed with olive: on each ſide of the neck is a ſpot of black feathers prettily tipt with white: the back aſh colored, bordered with olive brown: the ſcapulars and coverts of a reddiſh brown ſpotted with black: the quil feathers of a duſky brown, the tips and outward edges of a yellowiſh brown: the breaſt of a light purpliſh red, having the verge of each feather yellow: the belly white: the ſides and inner coverts of the wings bluiſh. The tail is three inches and a half long; the two middlemoſt feathers are of a duſky brown; the others black, with white tips: the end and exterior ſide of the outmoſt feathers wholly white.

Order

Pl. XLVI

No 110

ORDER V. PASSERINE.

Strait BILL; depreſſed: the NOSTRILS ſurrounded with a prominent rim.

XV. STARE.

104. STARE.

L'Eſtourneau. *Belon av.* 321.
Sturnus. *Geſner av.* 746.
Aldr. av. II. 284.
Stare, or Starling. *Wil. orn.* 196.
Raii ſyn. av. 67.
L'Etourneau. *Briſſon av.* II. 439. *Hiſt. d'Oys.* III. 176.
Sanſonet. *Pl. Enl.* 75.
Starl. *Scopoli*, No. 189.
Storno. *Zinan.* 69.
Olina, 18.
Sturnus vulgaris. *Lin. ſyſt.* 290.
Stare. *Faun. Suec. ſp.* 213.
Haſſelquiſt, Itin. 284.
Danis & Norvegis, Stær. *Br.* 229.
Staar. *Friſch*, II. 217.
Starl. *Kram.* 362.
Br. Zool. 93. plate P. 2. f. 1.

THE Stare breeds in hollow trees, eaves of houſes, towers, ruins, cliffs, and often in high rocks over the ſea, ſuch as thoſe of the *Iſle of Wight*. It lays four or five eggs, of a pale greeniſh aſh color; and makes its neſt of ſtraw, ſmall fibres of roots, and the like. In winter, ſtares aſſemble in vaſt flocks: they collect in myriads in the fens of *Lincolnſhire*, and do great damage to the fen men, by rooſting on the reeds, and breaking them down by their weight; for reeds are the thatch of the country, and are harvested with great care.

These birds feed on worms, and insects; and it is said that they will get into pigeon houses, for the sake of sucking the eggs. Their flesh is so bitter, as to be scarce eatable. They are very docil, and may be taught to speak.

DESCRIP. The weight of the male of this species is about three ounces; that of the female rather less. The length is eight inches three quarters: the breadth fourteen inches. Bill, in old birds, yellow. The whole plumage is black, very resplendent with changeable blue, purple, and copper: each feather marked with a pale yellow spot. The lesser coverts are edged with yellow, and slightly glossed with green. The quil feathers and tail dusky: the former edged with yellow on the exterior side; the last with dirty white. The legs of a reddish brown.

Strait-

Strait BILL, a little bending at the point, with a small notch near the end of the upper mandible.
Outmost TOE adhering as far as the first joint to the middle toe.

XVI. THRUSH.

105. MISSEL.

La Grive ou Siserre. *Belon av.* 324.
Turdus viscivorus. *Gesner av.* 759.
Aldr. av. II. 273.
Tordo. *Olina*, 25.
Missel-bird, or Shrite. *Wil. orn.* 187.
Raii syn. av. 64.
Misseltoe-thrush, or Shreitch. *Charlton ex.* 89.
Turdus viscivorus. *Lin. syst.* 291.
Tordo viscada, Zicchio. *Zinan.* 39.
La Draine. *Hist. d'Oys.* III. 295.
La grosse grive, Turdus major. *Brisson av.* II. 200.
Scopoli, No. 193.
Biork-Trast. *Faun. Suec. sp.* 216.
Dobbelt-Kramsfugl. *Brunnich*, 231.
Zariker, Mistler, Zerrer. *Kram.* 361.
Mistel-Drossel, or Schnarre. *Frisch*, I. 25.
Br. Zool. 90. plate P. f. 1.

THIS is the largest of the genus, and weighs near five ounces. Its length is eleven inches: its breadth sixteen and a half. The bill is shorter and thicker than that of other thrushes; dusky, except the base of the lower mandible, which is yellow. The *irides* hazel.

Head, back, and lesser coverts of the wings are of a deep olive brown. The lower part of the

X 3 back

back tinged with yellow. The loweſt order of leſſer coverts, and the great coverts brown: the firſt tipt with white; the laſt both tipt and edged with the ſame color. The quil feathers, and ſecondaries duſky; but the lower part of the inner webs white. The inner coverts of the wings white. Tail brown; the three outmoſt feathers tipt with white. Cheeks and throat mottled with brown and white: breaſt and belly whitiſh yellow, marked with large ſpots of black: the legs yellow.

Theſe birds build their neſts in buſhes, or on the ſide of ſome tree, generally an aſh, and lay four or five eggs: their note of anger or fear is very harſh, between a chatter and ſkreek; from whence ſome of its *Engliſh* names; its ſong though is very fine, which it begins, ſitting on the ſummit of a high tree, very early in the ſpring, often with the new year, in blowing ſhowery weather, which makes the inhabitants of *Hampſhire* to call it the *Storm-cock.* It feeds on inſects, holly and miſſeltoe berries, which are the food of all the thruſh kind: in ſevere ſnowy weather, when there is a failure of their uſual diet, they are obſerved to ſcratch out of the banks of hedges, the root of *Arum*, or the cuckoo pint: this is remarkably warm and pungent, and a proviſion ſuitable to the ſeaſon.

This bird migrates into *Burgundy* in the months of *October* and *November*: in *Great-Britain*, continues the whole year. The *Welſh* call this bird *Pen*

Pen y llwyn, or the maſter of the coppice, as it will drive all the leſſer ſpecies of thruſhes from it. The antients believed that the *miſſeltoe* (the baſis of bird-lime) could not be propagated but by the berries that had paſt through the body of this bird; and on that is founded the proverb of *Turdus malum ſibi cacat*.

It may be obſerved, that this is the largeſt bird, *Britiſh or foreign* (within our knowledge) that ſings or has any melody in its note: the notes of all ſuperior being either ſcreaming, croaking, chattering, &c. the pigeon kind excepted, whoſe ſlow plaintive continued monotone has ſomething ſweetly ſoothing in it. *Thompſon* (the naturaliſt's poet) in the concert he has formed among the feathered tribe, allows the imperfection of voice in the larger birds, yet introduces them as uſeful as the baſe in chorus, notwithſtanding it is unpleaſing by itſelf.

> The jay, the rook, the daw,
> And each harſh pipe (diſcordant heard alone)
> Aid the full concert: while the ſtock-dove breaths
> A melancholy murmur thro' the whole *.

* *Seaſons. Spring.* l. 606.

106. FIELD-FARE.

La Litorne. *Belon av.* 328.
Turdus pilaris *Gesner av.* 753.
Aldr. av. II 274.
Wil. orn. 188.
Raii syn. av. 64.
La Litorne, ou Tourdelle. *Brisson av.* II. 214. *Hist. d'Oys.* III. 301.
Lin. syst. 291.
Kramsfogel, snoskata. *Faun. Suec.* No. 215
Brinauka. *Scopoli*, No. 194.
Dan. Dobbelt Kramsfugl. *Cimbris.* Snarrer. *Norvegis*, Graae Trost, Field-Trost, Nordenvinds Pibe, *Bornholmis*, Simmeren. *Br.* 232.
Kranabets vogel, Kranabeter. *Kram.* 361.
Wacholder-Drossel, Juniper Thrush), or Ziemer. *Frisch*, I. 26.
Br. Zool. 90. plate P. 2. f. 1.

THIS bird passes the summer in the northern parts of *Europe*; also in lower *Austria**. It breeds in the largest trees †; feeds on berries of all kinds, and is very fond of those of the juniper. Fieldfares visit our islands in great flocks about *Michaelmas*, and leave us the latter end of *February*, or the beginning of *March*. We suspect that the birds that migrate here, come from *Norway*, &c. forced by the excessive rigor of the season in those cold regions; as we find that they winter as well as breed in *Prussia*, *Austria* ‡, and the moderate climates.

These birds and the *Redwings* were the *Turdi* of the *Romans*, which they fattened with crums of

* *Kramer elench.* 361. † *Faun. Suec. sp.* 78.

‡ *Klein hist. av.* 178.

figs

figs and bread mixed together. *Varro* informs us that they were birds of paſſage, coming in autumn, and departing in the ſpring. They muſt have been taken in great numbers, for they were kept by thouſands together in their fattening aviaries *. They do not arrive in *France* till the beginning of *December*.

DESCRIP.

Theſe birds weigh generally about four ounces; their length is ten inches, their breadth ſeventeen. The head is aſh-colored inclining to olive, and ſpotted with black; the back and greater coverts of the wings of a fine deep cheſnut; the rump aſh-colored: the tail is black: the lower parts of the two middlemoſt feathers, and the interior upper ſides of the outmoſt feathers excepted; the firſt being aſh-colored, the latter white. The legs are black; the talons very ſtrong.

* *Varro*, lib. III. c. 5.

La

107. THROSTLE.

La petite Grive. *Belon av.* 226.
Turdus minor alter. *Gesner av.* 762.
Aldr. av. II. 275.
Storno. *Olina,* 18.
Mavis, Throstle, or Song Thrush. *Wil. orn.* 188.
Raii syn. av. 64.
La petite Grive, Turdus minor. *Brisson av.* II. 205.
Hist. d'Oys. III. 280.
Turdus musicus. *Lin. syst.* 292.
Faun. Suec. sp. 217.
Turdus in altissimis. *Klein stem. av. tab.* 13.
Weindroschl, Weissdroschl, Sommer-droschl. *Kram.* 361.
Sing-Drossel, or Weiss-drossel. *Frisch,* I. 27.
Cimbris & Bornholmis, Vün-drossel. *Norvegis,* Tale Trast. *Br.* 236.
Br. Zool. 91. plate P. f. 2.
Drasich. *Scopoli,* No. 195.

DESCRIP.

THE weight of this species is three ounces: the length nine inches: the breadth thirteen and a half. In colors it so nearly resembles the missel thrush, that no other remark need be added, but that it is lesser, and that the inner coverts of the wings are yellow.

The throstle is the finest of our singing birds, not only for the sweetness and variety of its notes, but for long continuance of its harmony; for it obliges us with its song for near three parts of the year. Like the missel bird, it delivers its music from the top of some high tree; but to form its nest descends to some low bush or thicket: the nest is made of earth, moss, and straws, and the inside is curiously plaistered with clay. It lays five or six eggs,

eggs, of a pale bluiſh green, marked with duſky ſpots.

In *France* theſe birds are migratory: in *Burgundy*, they appear juſt before vintage, in order to feed on the ripe grapes, are therefore called there *la Grive de vigne.*

108. REDWING.

Le Mauvis. *Belon av.* 327.
Turdus minor. *Geſner av.* 761.
T. Illas ſeu Tylas. *Aldr. av.* II. 275.
Redwing, Swinepipe, or Wind Thruſh. *Wil. orn.* 189.
Raii ſyn. av. 54.
Le Mauvis. *Briſſon av.* II. 208. *tab.* 20. *fig.* 1. *Hiſt. d'Oys.* III. 309.
Scopoli, No. 196.
Pl. Enl. 51.
Turdus iliacus. *Lin. ſyſt.* 292.
Klera, Kladra, Tall-Traſt. *Faun. Suec. ſp.* 218.
Rothdroſchl, Walddroſchl, Winterdroſchl. *Kram.* 361.
Wein-Droſſel. Roth-Droſſel. *Friſch,* I. 28.
Br. Zool. 91. plate P. f. 2.

THESE birds appear in *Great-Britain* a few days before the fieldfare; they come in vaſt flocks, and from the ſame countries as the latter. With us they have only a diſagreeable piping note; but in *Sweden* during the ſpring ſing very finely, perching on the top of ſome tree among the foreſts of maples. They build their neſts in hedges, and lay ſix bluiſh green eggs ſpotted with black *.

DESCRIP.

They have a very near reſemblance to the throſtle;

* *Faun. Suec. ſp.* 218.

tle; but are leſs, only weighing two ounces and a quarter: their colors are much the ſame; only the ſides under the wings and the inner coverts in this are of a reddiſh orange; in the throſtle yellow: above each eye is a line of yellowiſh white, beginning at the bill and paſſing towards the hind part of the head. The vent feathers are white.

Beſides theſe three ſorts of throſtles, the author of the *epitome of the art of huſbandry* *, mentions a fourth kind under the name of the *heath throſtle*, which he commends as far ſuperior to the others in its ſong: he ſays it is the leſt of any, and may be known by its dark breaſt; that it builds its neſt by ſome heath-ſide, is very ſcarce, and will ſing nine months in the year.

109. Blackbird.

Le Merle noir. *Belon av.* 320.
Merula. *Geſner av.* 602.
Aldr. av. II. 276.
Merlo. *Zinan.* 39. *Olina*, 29.
Wil. orn. 190.
Raii ſyn. av. 65.
La Merle. *Briſſon av.* II. 227.
Hiſt. d'Oys. III. 330.
Pl. Enl 2.
Turdus merula. *Lin. ſyſt.* 295.
Kohl-Traſt. *Faun. Suec. ſp.* 220.
Dan. & *Norvegis* Solfort. *Br.* 234.
Amſel, Amarl. *Kram.* 360.
Schwartze Amſel. *Friſch*, I. 29.
Br. Zool. 92.
Koſs. *Scopoli*, No. 197.

THIS bird is of a very retired and ſolitary nature: frequents hedges and thickets, in

* *By* J. B. *gent. third edit.* 1685.

which

M. & F. BLACKBIRD.

which it builds earlier than any other bird: the neſt is formed of moſs, dead graſs, fibres, &c. lined or plaiſtered with clay, and that again covered with hay or ſmall ſtraw. It lays four or five eggs of a bluiſh green color, marked with irregular duſky ſpots. The note of the male is extremely fine, but too loud for any place except the woods: it begins to ſing early in the ſpring, continues its muſic part of the ſummer, deſiſts in the moulting ſeaſon; but reſumes it for ſome time in *September*, and the firſt winter months.

DESCRIP.

The color of the male, when it has attained its full age, is of a fine deep black, and the bill of a bright yellow: the edges of the eye-lids yellow. When young the bill is duſky, and the plumage of a ruſty black, ſo that they are not to be diſtinguiſhed from the females; but at the age of one year they attain their proper color.

Le

110. RING-OUZEL.

Le Merle ou Collier. *Belon av.* 318.
Merula torquata. *Gesner av.* 607.
Merlo alpestre. *Aldr. av.* II. 282.
Wil. orn. 194. Rock or Mountain-Ouzel, 195.
Mwyalchen y graig. *Camden Brit.* 795.
Le Merle a plastron blanc. *Hist. d'Oys.* III. 340.
Raii syn. av. 65.
Morton Northampt. 425.
Le Merle a Collier. *Brisson av.* II. 235.
Turdus torquatus. *Lin. syst.* 296.
Faun. Suec. sp. 221. *Scopoli,* No. 198.
Dan. Ringdrossel. *Norvegis* Ring Trost. *Br.* 237.
Ringlamsel. *Kram.* 360.
Ringel-Amsel. *Frisch,* I. 30.
Br. Zool. 92. plate P. 1. f. 1.

DESCRIP.

THESE birds are superior in size to the black bird: their length is eleven inches; their breadth seventeen. The bill in some is wholly black, in others the upper half is yellow: on each side the mouth are a few bristles: the head and whole upper part of the body are dusky, edged with pale brown: the quil-feathers, and the tail are black. The coverts of the wings, the upper part of the breast, and the belly are dusky, slightly edged with ash-color. The middle of the breast is adorned with a white crescent, the horns of which point to the hind part of the neck. In some birds this is of a pure white, in others of a dirty hue. In the females and in young birds this mark is wanting, which gave occasion to some naturalists to form two species of them.

The *Ring-Ouzel* inhabits the *Highland* hills, the north of *England*, and the mountains of *Wales*. They

They are alſo found to breed in *Dartmoor*, in *Devonſhire*, in banks on the ſides of ſtreams. I have ſeen them in the ſame ſituation in *Wales*, very clamorous when diſturbed.

They are obſerved by the Rev. Mr. *White*, of *Selborn*, near *Alton*, *Hants*, to viſit his neighbourhood regularly twice a year, in flocks of twenty or thirty, about the middle of *April*, and again about *Michaelmas*. They make it only a reſting place in their way to ſome other country; in their ſpring migration they only ſtay a week, in their autumnal a fortnight. They feed there on haws, and for want of them on yew berries. On diſſection, the females were found full of the ſmall rudiments of eggs, which prove them to be later breeders than any others of this genus, which generally have fledged young about that time. The places of their retreat is not known: thoſe that breed in *Wales* and *Scotland* never quitting thoſe countries. In the laſt they breed in the hills, but deſcend to the lower parts to feed on the berries of the mountain aſh.

They migrate in *France* at the latter ſeaſon: and appear in ſmall flocks about *Montbard*, in *Burgundy*, in the beginning of *October*, but ſeldom ſtay above two or three weeks. Notwithſtanding this, they are ſaid, to breed in *Sologne* and the foreſt of *Orleans*.

Merula

111. WATER-OUZEL.

Merula aquatica. *Gefner av.* 608.
Lerlichirollo. *Aldr. av.* III. 186.
Water-craw. *Turner.*
The Water-Ouzel, or Water-Crake. *Wil. orn.* 149.
Raii fyn. av. 66.
Sturnus cinclus. *Lin. fyft.* 290.
Watnftare. *Faun. Suec. fp.* 214.
Povodni Kofs. *Scopoli,* No. 223.
Le Merle d'eau. *Briffon av.* v. 252.
Merlo aquatico. *Zinan.* 109.
Norvegis, Foffe Fald, Foffe Kald, Quærn Kald, Stroem-Stær, Bække Eugl. *Brunnich.* 230.
Waffer-amfel, Bach-amfel. *Kra.* 374.
Br. Zool. 92. plate. P. 1. f. 2.

THIS bird frequents fmall brooks, particularly thofe with fteep banks, or that run through a rocky country. It is of a very retired nature, and never feen but fingle, or with its mate. It breeds in holes in the banks, and lays five white eggs adorned with a fine blufh of red. It feeds on infects and fmall fifh; and as Mr. *Willughby* obferves, though not web-footed, will dart itfelf after them quite under water. The neft is conftructed in a curious manner, of hay and fibres of roots, and lined with dead oak leaves, having a portico, or grand entrance made with green mofs.

NEST.

DESCRIP.

Its weight is two ounces and a half: the length feven inches one quarter: the breadth eleven: the bill is narrow, and compreffed fideways: the eyelids are white: the head, cheeks, and hind part of of the neck are duſky, mixed with ruſt color: the back, coverts of the wings, and of the tail alſo duſky, edged with bluiſh aſh color: the throat and breaſt white: the belly ferruginous, vent feathers a deep aſh color: the legs are of a pale blue before, black behind: the tail ſhort and black, which it often flirts up, as it is ſitting.

Theſe are all the birds of this genus that can clame a place in this work. The roſe colored ouzel, *Wil. orn.* 194. *Edw.* 20. a foreign bird, has been ſhot at *Norwood* near *London*; for its hiſtory we refer our readers to the appendix.

XVII. CHATTERER.

BILL ſtrait, a little convex above, and bending towards the point. Near the end of the lower mandible a ſmall notch on each ſide.

NOSTRILS hid in briſtles.

Middle TOE connected at the baſe to the outmoſt.

112. WAXEN.

Garrulus Bohemicus. *Geſner av.* 703.
Aldr. av. I. 395.
Bohemian Chatterer. *Wil. orn.* 133.
Bell's Travels. I. 198.
Silk Tail. *Raii ſyn. av.* 85.
Ray's Letters, 198. 200.
Le Jaſeur de Boheme, Bombycilla Bohemica. *Briſſon av.* II. 333. *Scopoli.* No. 20.
Phil. Tranſ. No. 175.
Ampelis garrulus. *Lin. ſyſt.* 297.
Siden-Suantz, Snotuppa. *Faun. Suec. ſp.* 82.
Sieden vel Sieben Suands. *Brunnich* 25.
Zuſerl, Geidenſchweiffl. *Kram.* 363.
Seiden-ſchwantz. *Friſch*, I. 32.
Br. Zool. 77. plate C. 1.

THESE birds appear but by accident in *South Britain*: about *Edinburgh* in *February*, they come annually and feed on the berries of the mountain aſh: they alſo appear as far ſouth as *Northumberland*, and like the fieldfare make the berries of the white thorn their food. Their native country is *Bohemia*, from whence they wander over *Europe*, and were once ſuperſtitiouſly conſidered as preſages of a peſtilence. They are gregarious: feed

CHATTERER.

feed on grapes where vineyards are cultivated; are esteemed delicious food: easily tamed.

DESCRIP.

The length of the bird I saw was eight inches: the bill short, thick, and black; the base covered with black bristles; from thence passes to the hind part of the head over each eye a bar of black: on the head is a sharp pointed crest reclining backwards: the *irides* are of a bright ruby colour: the cheeks tawny: the throat black, with a small bristly tuft in the middle.

The head, crest, and back ash colored mixed with red: the rump a fine cinereous: breast and belly, pale chesnut dashed with a vinaceous cast: the vent feathers bright bay: the lower part of the tail black; the end of a rich yellow: the lesser coverts of the wings brown, the greater black tipt with white: the quil-feathers black, the three first tipt with white; the six next have half an inch of their exterior margin edged with fine yellow, the interior with white. But what distinguishes this from all other birds are the horny appendages from the tips of seven of the secondary feathers of the color and gloss of the best red wax; some have one more or one less: The legs are black.

I think that the females want the yellow marks in the wings.

XVIII. GROSBEAK

BILL ſtrong, thick, convex above and below.
NOSTRILS ſmall and round.
TONGUE as if cut off at the end.

113. HAW.

Le Groſbec ou Pinſon royal. *Belon av.* 373.
Coccothrauſtes (ſteinbeiſſer) *Geſner av.* 276.
Aldr. av. II. 289.
Froſone. *Olina* 37.
Groſbeak, or Hawfinch. *Wil. orn.* 244.
Raii ſyn. av. 85.
Charlton ex. 91.
Dleſchk. *Scopoli*, No. 201.
Edw. av. 188. The male.
Le Groſbec. *Briſſon av.* III. 219.
Pl. enl. 99, 100.
Loxia coccothrauſtes. *Lin. ſyſt.* 299.
Stenkneck. *Faun. Suec. ſp.* 222.
Kernbeis, Nuſbeiſſer. *Kram.* 365.
Kirſchfinch (Cherry-finch). *Friſch*, I. 4.
Brunnich. in append.
Br. Zool. 105. plate U. F. 1.

THE birds we deſcribe were ſhot in *Shropſhire:* they viſit us only at uncertain times, and are not regularly migrant. They feed on berries; and even on the kernels of the ſtrongeſt ſtones, ſuch as thoſe of cherries and almonds, which they crack with the greateſt facility: their bills are well adapted to that work, being remarkably thick and ſtrong. Mr. *Willughby* tells us, they are common in *Germany* and *Italy*; that in the ſummer they live in woods, and breed in hollow trees, laying five or ſix eggs; but in the winter they come down into the plains.

DESCRIP. This ſpecies weighs near two ounces: its length is ſeven inches; the breadth thirteen: the bill is of a fun-

Pl. XLIX.
F. CROSS BILL.
PINE GROSBEAK.

a funnel ſhape; ſtrong, thick, and of a dull pale pink color; at the baſe are ſome orange colored feathers: the irides are grey: the crown of the head and cheeks of a fine deep bay: the chin black: from the bill to the eyes is a black line: the breaſt and whole under ſide is of a dirty fleſh color: the neck aſh-colored: the back and coverts of the wings of a deep brown, thoſe of the tail of a yellowiſh bay: the greater quil-feathers are black, marked with white on their inner webs. The tail is ſhort, ſpotted with white on the inner ſides. The legs fleſh color.

The great particularity of this bird, and what diſtinguiſhes it from all others, is the form of the ends of the middle quil-feathers; which Mr. *Edwards* juſtly compares to the figure of ſome of the antient *battle-axes*: theſe feathers are gloſſed over with a rich blue; but are leſs conſpicuous in the female: the head in that ſex is of dull olive, tinged with brown; it alſo wants the black ſpot under the chin.

Loxia enucleator. *Lin. ſyſt.* 299.
Tallbit, Natt-waka. *Faun. Suec.* No. 223.
Greateſt Bulfinch. *Edw.* 123, 124.
Coccothrauſtes Canadenſis. *Briſſon*, III. 250.

114. PINE.

THESE are common to *Hudſon's Bay*, *Sweden* and *Scotland*. I have ſeen them flying above the great pine foreſts of *Invercauld*, *Aberdeenſhire*.

Y 3 I ima-

I imagine they breed there, for I ſaw them *Auguſt* 5th. They feed on the ſeeds of the pine. *Linnæus* ſays, they ſing in the night.

It is near twice the ſize of the bulfinch. The bill ſtrong, duſky, forked at the end; leſs thick than that of the common bulfinch: head, back, neck, and breaſt of a rich crimſon: the bottoms of the feathers aſh-color; the middle of thoſe on the back and head black: the lower belly and vent aſh-color: the leſſer coverts of the wings duſky, edged with orange; the next with a broad ſtripe of white: the loweſt order of greater coverts with another; exterior edges of the ſame color: the quil-feathers and tail duſky; their exterior edges of a dirty white: legs black: length nine inches and a half. There ſeems an agreement in colors, as well as food, between this ſpecies and the croſs-bill; one that I ſaw in *Scotland*, and believe to be the female, was (like the female croſs-bill) of a dirty green; the tail and quil-feathers duſky.

115. CROSS-BILLED.

Loxia. *Gesner av.* 591.
Aldr. av. I. 426.
Shell-apple, or Crofs-bill. *Wil. orn.* 248.
Raii syn. av. 86.
Charlton ex. 77.
Edw. av. 303.
Cat. Carol. app. 37.
Le Bec-croife. *Briffon av.* III. 329. *tab.* 17. *fig.* 3.

Loxia curviroftra. *Lin. syft.* 299.
Korffnaff, Kinlgelrifvare. *Faun. Suec. sp.* 224. *Scopoli*, No. 200.
Krumbfchnabl, Kreutzvogel. *Kram.* 365.
Kreutz-Schnabel. *Frifch*, I. 11.
Norveg. Kors-Næb. Kors-fugl. *Br.* 238.
Br. Zool. 106. plate U. f. 2.

THERE are two varieties of this bird: Mr. *Edwards* has very accurately figured the leffer, which we have feen frequently: the other is very rare. We received a male and female out of *Shropfhire*, which were fuperior in fize to the former, the bill remarkably thick and fhort, more encurvated than that of the common kind, and the ends more blunt.

Thefe birds, like the former, are inconftant vifitants of this ifland: in *Germany* and *Switzerland** they inhabit the pine forefts, and breed in thofe trees fo early as the months of *January* and *February*. They feed on the feeds of the cones of pines and firs; and are very dexterous in fcaling them, for which purpofe the crofs ftructure of the lower mandible of their bill is admirably adapted; they feed alfo on hemp feed, and the pips or kernels

* *Gefner* 59. *Kramer Elench.* 365.

of apples, and are ſaid to divide an apple with one ſtroke of the bill to get at the contents. *Linnæus* * ſays, that the upper mandible of this bird is moveable; but on examination we could not diſcover its ſtructure to differ from that of others of the genus.

It is an undoubted fact, that theſe birds change their colors; or rather the ſhades of their colors: that is, the males which are red, vary at certain ſeaſons to deep red, to orange, or to a ſort of a yellow: the females which are green, alter to different varieties of the ſame color.

116. BULFINCH.

Le Pivoine. *Belon av.* 359.
Aſprocolos, *obſ.* 13.
Rubicilla, ſive pyrrhula. *Geſner av.* 733.
Aldr. av. II. 326.
Ciufolotto. *Olina*, 40.
Bulfinch, Alp, or Nope. *Wil. orn.* 247.
Raii. ſyn. av. 86.
Blutfinck, *Friſch*, I. 2.
Le Bouvreuil. *Briſſon av.* III. 308.
Pl. enl. 145.
Monachino, Sufolotto. *Zinan.* 58.
Loxia pyrrhula. *Lin. ſyſt.* 300.
Domherre. *Faun. Suec. ſp.* 225.
Gumpl. *Kram.* 365. Gimpl. *Scopoli*, No. 202.
Danis & Norvegis Dom-pape, *quibuſdam* Dom-Herre. *Br.* 240.
Br. Zool. 106. plate U. f. 3. 4.

THE wild note of this bird is not in the leſt muſical; but when tamed it becomes remarkably docil, and may be taught any tune after a pipe, or to whiſtle any notes in the juſteſt manner: it ſeldom forgets what it has learned; and

* *Faun. Suec. ſp.* 224.

will

will become ſo tame as to come at call, perch on its maſter's ſhoulders, and (at command) go through a difficult muſical leſſon. They may be taught to ſpeak, and ſome thus inſtructed are annually brought to *London* from *Germany*.

Descrip.

The male is diſtinguiſhed from the female by the ſuperior blackneſs of its crown, and by the rich crimſon that adorns the cheeks, breaſt, belly, and throat of the male; thoſe of the female being of a dirty color: the bill is black, ſhort, and very thick: the head large: the hind part of the neck and the back are grey: the coverts of the wings are black; the lower croſſed with a white line: the quil-feathers duſky, but part of their inner webs white: the coverts of the tail and vent feathers white: the tail black.

In the ſpring theſe birds frequent our gardens, and are very deſtructive to our fruit-trees, by eating the tender buds. They breed about the latter end of *May*, or beginning of *June*, and are ſeldom ſeen at that time near houſes, as they chuſe ſome very retired place to breed in. Theſe birds are ſometimes wholly black; I have heard of a male bulfinch which had changed its colors after it had been taken in full feather, and with all its fine teints. The firſt year it began to aſſume a dull hue, blackening every year, till in the fourth it attained the deepeſt degree of that color. This was communicated to me by the Reverend Mr. *White* of *Selborne*. Mr. *Morton*, in his Hiſtory of *Northampton-*

Northamptonſhire * gives another inſtance of ſuch a change, with this addition, that the year following, after moulting, the bird recovered its native colors. Bulfinches fed entirely on hemp-ſeed are apteſt to undergo this change.

117. GREEN.

Belon av. 365.
Aſſarandos, *obſ.* 13.
Chloris. *Geſner av.* 258.
Aldr. av. II. 371.
Olina, 26.
Wil. orn. 246.
Raii ſyn. av. 85.
Le Verdier. *Briſſon av.* III. 190.
Grindling. *Scopoli*, No. 206.
Verdone, Verdero, Antone. *Zinan.* 63.
Loxia chloris. *Lin. ſyſt.* 304.
Swenſka. *Faun. Suec. ſp.* 226.
Svenſke. *Br.* 242.
Grunling. *Kram.* 368.
Grünfinck (Greenfinch) *Friſch* I. 2.
Br. Zool. 107.

DESCRIP.

THE head and back of this bird are of a yellowiſh green; the edges of the feathers are grey; the rump more yellow: the breaſt of the ſame color; the lower belly white: the edges of the outmoſt quil-feathers are yellow, the next green, the fartheſt grey: the tail is a little forked: the two middle feathers are wholly duſky: the exterior webs of the four outmoſt feathers on both ſides the tail are yellow. The colors in the female are much leſs vivid than in the male.

NEST.

Theſe birds are very common in this iſland: they make their neſt in hedges; the outſide is

* Page 437.

compoſed

composed of hay or stubble, the middle part of moss, the inside of feathers, wool, and hair. During breeding-time, that bird which is not engaged in incubation, or nutrition, has a pretty way of sporting on wing over the bush. They lay five or six eggs of a pale green color, marked with blood colored spots. Their native note has nothing musical in it; but a late writer on singing-birds says, they may be taught to pipe or whistle in imitation of other birds.

This bird is so easily tamed, that it frequently eats out of one's hand five minutes after it is taken, if you have an opportunity of carrying it into the dark; the bird should be then put upon your finger, which it does not attempt to move from (as being in darkness it does not know where to fly) you then introduce the finger of your other hand under its breast, which, making it inconvenient to stay upon the first finger on which it was before placed, it climbs upon the second, where it likewise continues, and for the same reason. When this hath been nine or ten times repeated, and the bird stroked and caressed, it finds that you do not mean to do it any harm; and if the light is let in by degrees, it will very frequently eat any bruised seed out of your hand, and afterwards continue tame.

BILL

XIX. BUNTING.

BILL ſtrong and conic, the ſides of each mandible bending inwards: in the roof of the upper, a hard knob, of uſe to break and comminute hard ſeeds.

118. COMMON.

Le Proyer, Prier, ou Pruyer. *Belon av.* 266.
Emberiza alba. *Geſner av.* 654.
Aldr. av. II. 264.
Strillozzo. *Olina*, 44.
Wil. orn. 267.
Raii ſyn. av. 93.
Le Proyer, Cynchramus. *Briſſon av.* III. 292.
Pl. Enl. 30.
Petrone, Capparone, Stardacchio. *Zinan.* 63.
Emberiza Miliaria. *Lin. ſyſt.* 308.
Faun. Suec. ſp. 228.
Korn Larkor. *Lin. it. ſcan.* 292. *tab.* 4.
Cimbris Korn-Lærke. *Norveg.* Knotter. *Brunnich* 247.
Graue Ammer. *Friſch*, I. 6.
Braſler. *Kram.* 372.
Br. Zool. 111. plate W. f. 7.

DESCRIP.

THE bill of this bird, and the other ſpecies of this genus, is ſingularly conſtructed; the ſides of the upper mandible form a ſharp angle, bending inwards towards the lower; and in the roof of the former is a hard knob, adapted to bruiſe corn or other hard ſeeds.

The throat, breaſt, ſides, and belly are of a yellowiſh white: the head and upper part of the body of a pale brown, tinged with olive; each of which (except the belly) are marked with oblong black ſpots; towards the rump the ſpots grow fainter. The quil-feathers are duſky, their exterior edges

Pl. L

edges of a pale yellow. The tail is a little forked, of a duſky hue, edged with white; the legs are of a pale yellow.

This bird reſides with us the whole year, and during winter collects in flocks.

Belon av. 366.
Emberiza flava. *Geſner av.* 653.
Cia pagglia riccia, Luteæ alterum genus. *Aldr. av.* II. 372.
Wil. orn. 268.
Yellow Hammer, *Raii ſyn. av.* 93.
Le Bruant. *Briſſon av.* III. 258.
Pl. Enl. 30. f. 2.
Sternardt. *Scopoli*, No. 209.
Emberiza citrinella. *Lin. ſyſt.* 309.
Groning, Golſpink. *Faun. Suec. ſp.* 230.
Ammering, Goldammering. *Kram.* 370. *Friſch*, I. 5.

119. YELLOW.

NEST.

THIS ſpecies makes a large flat neſt on the ground, near or under a buſh or hedge; the materials are moſs, dried roots, and horſe hair interwoven. It lays ſix eggs of a white color, veined with a dark purple: is one of our commoneſt birds, and in winter frequents our farm yards with other ſmall birds.

DESCRIP.

The bill is of a duſky hue: the crown of the head is of a pleaſant pale yellow; in ſome almoſt plain, in others ſpotted with brown: the hind part of the neck is tinged with green: the chin and throat are yellow: the breaſt is marked with an orange red: the belly yellow: the leſſer coverts of the wings are green; the others duſky, edged with ruſt color:

color: the back of the ſame colors: the rump of a ruſty red: the quil-feathers duſky, their exterior ſides edged with yellowiſh green: the tail is a little forked; the middle feathers are brown; the two middlemoſt edged on both ſides with green; the others on their exterior ſides only: the interior ſides of the two outmoſt feathers are marked obliquely near their ends with white.

120. REED.

Schœniclus. *Geſner av.* 573, 652.
Wil. orn. 269.
Reed Sparrow. *Raii ſyn. av.* 95.
The Nettle-monger. *Morton Northampt.* 423.
Ror-Spurv. *Brunnich* 251.
L'Ortulan de Roſeaux, Hortulanus arundinaceus. *Briſſon av.* III. 274.
Emberiza ſchœniclus. *Lin. ſyſt.* 311.
Saf-ſparf. *Faun. Suec. ſp.* 231.
Rohrammering, Meerſpatz. *Kram.* 371.
Rohrammer (Reed-hammer) *Friſch*, I. 7.
Br. Zool. 112. plate W.

NEST.

THE reed ſparrow inhabits marſhy places, moſt commonly among reeds; from which it takes its name. Its neſt is worthy notice for the artful contrivance of it, being faſtened to four reeds, and ſuſpended by them like a hammock, about three feet above the water; the cavity of the neſt is deep, but narrow, and the materials are ruſhes, fine bents and hairs. It lays four or five eggs, of a bluiſh white, marked with irregular purpliſh veins, eſpecially on the larger end. It is a bird much admired for its ſong, and like the nightingale it ſings in the night.

In

In the male, the head, chin, and throat are black: the tongue livid: at each corner of the mouth commences a white ring, which encircles the head. At approach of winter the head changes to hoary, but on the return of ſpring reſumes its priſtine jettyneſs. The whole under ſide of the body is white. The back, coverts of the wings, and the ſcapular feathers are black, deeply bordered with red. The two middle feathers of the tail are of the ſame colors; the three next black. The exterior web, and part of the interior of the outmoſt feather is white. The head of the female is ruſt-coloured, ſpotted with black; it wants the white ring round the neck: but in moſt other reſpects reſembles the male.

DESCRIP.

Great pied Mountain Finch, or Brambling. *Wil. orn.* 255.
Raii. ſyn. av. 88.
L'Ortolan de Neige, Hortulanus nivalis. *Briſſon av.* III. 285.
Schnee-ammer (Snow-hammer). *Friſch,* I. 6.
Br. Zool. 112. plate f. 6.

121. TAWNY.

THE weight of this bird is rather more than an ounce: the length is ſix inches three quarters: the breadth twelve inches three quarters. The bill is very ſhort; yellow, except the point, which is black. The crown of the head is of a tawny color, darkeſt near the forehead: the whole neck is of the ſame color, but paler: the throat almoſt

DESCRIP.

almoſt white: the upper part of the breaſt is of a dull yellow; the breaſt and whole under part of the body white, daſhed with a yellowiſh tinge. The back and ſcapular feathers are black, edged with a pale reddiſh brown: the rump and covert feathers of the tail are white on their lower half; on their upper, yellow.

The tail conſiſts of twelve feathers, and is a little forked: the three exterior feathers are white: the two outmoſt marked with a duſky ſpot on the exterior ſide; the third is marked with the ſame color on both ſides the tip: the reſt of the tail feathers are entirely duſky. The wings, when cloſed, reach about the middle of the tail: the color, of as much of the ſix firſt quil-feathers as appears in view, is duſky, ſlightly tipt with a reddiſh white: their lower part on both ſides white: in the ſeven ſucceeding feathers the duſky color gradually gives place to the white; which in the ſeventh of theſe poſſeſſes the whole feather, except a ſmall ſpot on the exterior upper ſide of each; the two next are wholly white: the reſt of the quil-feathers and the ſcapular feathers are black, edged with a pale red: the baſtard wing, and the outmoſt ſecondary feathers are of the ſame color with the quil-feathers: the reſt of them, together with the coverts, are entirely white, forming one large bed of white. The legs, feet and claws are black: the hind toe is very long, like that of a lark, but not ſo ſtrait.

Theſe

Theſe birds are ſometimes found in different parts of *England*; but are not common. I am unacquainted with their breeding places, or their hiſtory: are ſometimes found white, and then miſtaken for white larks.

122. SNOW.

Emberiza nivalis. *Lin. ſyſt.* 308.
Snoſparf. *Faun. Suec.* No. 227.
Le Pincon de neige ou la niverolle. *Briſſon*, III. 162.
Cimbris, ſneekok, vinter fugl. *Norvegis*. Sneefugl, Fialſter. *Brunnich*, 245.
Avis ignota a *Piperino* miſſa. *Geſner av.* 798.
Scopoli, No. 214.
Snow-bird, *Edw.* 126. *Egede Greenl.* 64. *Marten's Spitzbergen*, 73.
Forſter in Ph. Tr. vol. LXII. p. 403.

THE weight of this ſpecies is one ounce and a quarter: the bill and legs black: the forehead and crown white, with ſome mixture of black on the hind part of the head: the back of a full black, the rump white: the baſtard wing and ends of the greater coverts black, the others white: the quil-feathers black, their baſe white: the ſecondaries white, with a black ſpot on their inner webs. The middle feathers of their tail black; the three outmoſt white, with a duſky ſpot near their ends: from chin to tail of a pure white.

Theſe birds are called in *Scotland*, *Snow-flakes*, from their appearance in hard weather and in deep ſnows. They arrive in that ſeaſon among the *Cheviot* hills, and in the *Highlands* in amazing flocks. A few breed in the laſt on the ſummit of the higheſt hills in the ſame places with the *Ptarmigans*; but the greateſt numbers migrate from the extreme north. They appear in the *Shetland* iſlands, then in the *Orknies*, and multitudes of them often fall, wearied with their flight, on veſſels in the *Pentland Firth*. Their appearance is a certain fore-runner of hard weather, and ſtorms of ſnow, being driven by the cold from their common retreats. Their progreſs ſouthward is probably thus; *Spitzbergen* and *Greenland*, *Hudſon's Bay*, the *Lapland Alps*, *Scandinavia*, *Iceland*, the *Ferroe* iſles, *Shetland*, *Orknies*, *Scotland*, and the *Cheviot* hills. They viſit at that ſeaſon all parts of the northern hemiſphere, *Pruſſia*, *Auſtria*, and *Siberia** They arrive lean and return fat. In *Auſtria* they are caught and fed with millet, and like the *Ortolan*, grow exceſſively fat. In their flights, they keep very cloſe to each other, mingle moſt confuſedly together; and fling themſelves collectively into the form of a ball, at which inſtant the fowler makes great havoke among them.

* *Kram. Auſtria*, 372. *Bell's Travels*, I. 198.

Leſſer

Leſſer Mountain-finch, or Brambling. *Wil. orn.* 255. *Morton Northampt.* 423. *tab.* 13. *fig.* 3. *Br. Zool.* 113.

123. MOUNTAIN.

DESCRIP.

WE are obliged to borrow the following deſcription from the account of Mr. *Johnſon* tranſmitted to Mr. *Ray*; having never ſeen the bird. Mr. *Ray* ſuſpected that it was only a variety of the former, but Mr. *Morton*, having frequent opportunity of examining this ſpecies, proves it to be a diſtinct kind.

According to Mr. *Johnſon*, its bill is ſhort, thick, and ſtrong; black at the point, the reſt yellow. The forehead is of a dark cheſtnut; the hind part of the head and cheeks of a lighter; the hind part of the neck, and the back are aſh-colored; the latter more ſpotted with black; the throat is white: the breaſt and belly waved with flame color; at the ſetting on of the wing grey; the five firſt feathers are of a blackiſh brown, the reſt white with the point of each daſhed with brown: the three outmoſt feathers of the tail are white; the reſt dark brown; the feet black; the hind claw as long again as any of the reſt. The breaſt of the female is of a darker color than that of the male. The ſpecies, by the above-mentioned writer's account, is found in *Yorkſhire* and *Northamptonſhire*.

XX. FINCH. BILL perfectly conic, ſlender towards the end, and ſharp-pointed.

124. GOLDFINCH.

Belon av. 353.
Carduelis. *Geſner av.* 242.
Aldr. av. II. 349.
Cardelli. *Olina,* 10.
Goldfinch, or Thiſtlefinch. *Wil. orn.* 256.
Raii ſyn. av. 89.
Le Chardonneret. *Briſſon av.* III. 53.
Pl. Enl. 4. f. 1.
Cardellino. *Zinan.* 59.
Fringilla carduelis. *Lin. ſyſt.* 318.
Stiglitza. *Faun. Suec. ſp.* 236.
Stiglitz. *Br.* 257. *Scopoli,* No. 211.
Stiglitz. *Kram.* 365. Diſtelfinck. *Friſch,* I. 1.
Br. Zool. 108. plate V. f. 1.

DESCRIP. THIS is the moſt beautifull of our hard billed ſmall birds; whether we conſider its colors, the elegance of its form, or the muſic of its note. The bill is white, tipt with black, the baſe is ſurrounded with a ring of rich ſcarlet feathers: from the corners of the mouth to the eyes is a black line: the cheeks are white: the top of the head is black; and the white on the cheeks is bounded almoſt to the forepart of the neck with black: the hind part of the head is white: the back, rump, and breaſt, are of a fine pale tawny brown, lighteſt on the two laſt: the belly is white: the covert feathers of the wings, in the male, are black: the quil-feathers black, marked in their middle with a beautifull yellow; the tips white: the

the tail is black, but moſt of the feathers marked near their ends with a white ſpot: the legs are white.

The female is diſtinguiſhed from the male by theſe notes; the feathers at the end of the bill in the former are brown; in the male black: the leſſer coverts of the wings are brown: and the black and yellow in the wings of the female are leſs brilliant. The young bird, before it moults, is grey on the head; and hence it is termed by the bird-catchers a *grey pate.*

There is another variety of goldfinch, which is, perhaps, not taken above once in two or three years, which is called by the *London* bird-catchers a *cheverel*, from the manner in which it concludes its *jerk:* when this ſort is taken, it ſells at a very high price: it is diſtinguiſhed from the common ſort by a white ſtreak, or by two, and ſometimes three white ſpots under the throat.

Their note is very ſweet, and they are much eſteemed on that account, as well as for their great docility. Towards winter they aſſemble in flocks, and feed on ſeeds of different kinds, particularly thoſe of the thiſtle. It is fond of orchards; and frequently builds in an apple or pear tree: its neſt is very elegantly formed of fine moſs, liverworts, and bents on the outſide; lined firſt with wool and hair, and then with the goſlin or cotton of the ſallow. It lays five white eggs, marked with deep purple ſpots on the upper end.

This bird ſeems to have been the χρυσομῖτρις* of *Ariſtotle*; being the only one that we know of, that could be diſtinguiſhed by a *golden fillet* round its head, feeding on the ſeeds of *prickly* plants. The very ingenious tranſlator † of *Virgil*'s eclogues and georgics, gives the name of this bird to the *acalanthis* or *acanthis*:

Littoraque *alcyonen* reſonant, *acanthida* dumi.

In our account of the *Halcyon* of the antients, p. 191 of the former edition, we followed his opinion; but having ſince met with a paſſage in *Ariſtotle* that clearly proves that *acanthis* could not be uſed in that ſenſe, we beg, that, till we can diſcover what it really is, the word may be rendered *linnet*; ſince it is impoſſible the philoſopher could diſtinguiſh a bird of ſuch ſtriking and brilliant colors as the *goldfinch*, by the epithet κακοχροος, or bad colored; and as he celebrates his *acanthis* for a fine note, φωνην μέν τοι λιγυράν ἔχουσι ‡, both characters will ſuit the linnet, being a bird as remarkable for the ſweetneſs of its note, as for the plaineſs of its plumage.

* Which he places among the ἀκανθοφάγα. *Scaliger* reads the word ῥυσομίτρις, which has no meaning; neither does the critic ſupport his alteration with any reaſons. *Hiſt. an.* 887.

† Dr. *Martyn.*

‡ *Hiſt. an.* 1055.

Le

125. CHAFFINCH.

Le Pinſon. *Belon av.* 371.
Fringilla. *Geſner av.* 337.
Aldr. av. II. 356.
Olina 31.
Wil. orn. 253.
Raii ſyn. av. 88.
Fringuello. *Zinan.* 61.
Le Pinçon. *Briſſon av.* 148.
Schinkovitz. *Scopoli*, No. 217.
Pl. enl. 54. f. 1.
Fringilla cœlebs. *Lin. ſyſt.* 318.
Fincke, Bofincke. *Faun. Suec. ſp.* 232.
Buchfinck (Beachfinch) *Friſch*, I. 1.
Finke. *Kram.* 367.
Bofinke. *Br.* 253.
Br. Zool. 108. plate V. f. 2. 3.

THIS ſpecies entertains us agreeably with its ſong very early in the year; but towards the latter end of ſummer aſſumes a chirping note: both ſexes continue with us the whole year. What is very ſingular in *Sweden*, the females quit that country in *September*, migrating in flocks into *Holland*, leaving their mates behind; in the ſpring they return.* In *Hampſhire* Mr. *White* has obſerved ſomething of this kind; vaſt flocks of females with ſcarcely any males among them. Their neſt is almoſt as elegantly conſtructed as that of the goldfinch, and of much the ſame materials, only the inſide has the addition of ſome large feathers. They lay four or five eggs, of a dull white color, tinged and ſpotted with deep purple.

DESCRIP.

The bill is of a pale blue, the tip black: the feathers on the forehead black: the crown of the head, the hind part and the ſides of the neck are

* *Amœn. acad.* II. 42. IV. 595.

of a bluifh grey: the fpace above the eyes, the cheeks, throat, and forepart of the neck, are red: the fides and belly white, tinged with red: the upper part of the back of a deep tawny color; the lower part and rump green: the coverts on the very ridge of the wing black and grey; beneath them is a large white fpot: the baftard wing and firft greater coverts black, the reft tipt with white: the quil-feathers black; their exterior fides edged with pale yellow: their inner and outward webs white on their lower part, fo as to form a third white line acrofs the wing: the tail is black, except the outmoft feather, which is marked obliquely with a white line from top to bottom; and the next which has a white fpot on the end of the inner web: the legs are dufky: the colors of the female are very dull: it entirely wants the red on the breaft and other parts: the head and upper part of the body are of a dirty green: the belly and breaft of a dirty white: the wings and tail marked much like thofe of the male.

Le

Le Montain. *Belon av.* 372.
Montifringilla montana. *Gesner av.* 388.
Aldr. av. II. 358.
Fringuello montanina. *Olina* 32.
Bramble, or Brambling. *Wil. orn.* 254.
Mountain-finch. *Raii syn. av.* 88.
Le Pinçon d'ardennes. *Brisson av.* III. 155.
Pl. enl. 54. f. 2.

Fringilla montifringilla. *Lin. syst.* 318.
Pinosch. *Scopoli,* No. 218.
Norquint. *Faun. Suec sp.* 233.
Quæker, Bofinkens Hore-Unge, Akerlan. *Brunnich* 255.
Nicowitz, Mecker, Piencken. *Kram.* 367.
Bergfinck (Mountainfinch). *Frisch,* I. 3.
Br. Zool. 108. plate V. f. 4.

126. BRAMBLING.

DESCRIP.

THIS bird is not very common in these islands. It is superior in size to the chaffinch: the top of the head is of a glossy black, slightly edged with a yellowish-brown: the feathers of the back are of the same colors, but the edges more deeply bordered with brown: the chin, throat, and breast are of an orange color: the lesser coverts of the wings of the same color; but those incumbent on the quil-feathers barred with black, tipt with orange: the inner coverts at the base of the wings are of a fine yellow: the quil-feathers are dusky; but their exterior sides edged with yellow; the tail a little forked: the exterior web of the outmost feather is white, the others black, except the two middle, which are edged and tipt with ash color.

Lo

127. SPARROW.

Le Moineau, Paiſſe, ou Moiſſon. *Belon av.* 361.
Paſſer. *Geſner av.* 643.
Aldr. av. II. 246.
Paſſera noſtrale. *Olina*, 42.
The Houſe-ſparrow. *Wil. orn.* 249.
Raii ſyn. av. 86.
Le Moineau franc. *Briſſon av.* III. 72.
Pl. enl. 6. f. 1. 55. f. 1.
Fringilla domeſtica. *Lin. ſyſt.* 323.
Tatting, Graſparf. *Faun. Suec. ſp.* 242.
Danis Graae-Spurre. *Norveg.* Huus-Kald. *Br.* 264.
Hauſſpatz. *Kram.* 369.
Br. Zool. II. 300.

DESCRIP.

THE bill of the male is black: the crown of the head is grey: under each eye is a black ſpot; above the corner of each is a broad bright bay mark, which ſurrounds the hind part of the head. The cheeks are white: the chin and under ſide of the neck are black; the latter edged with white: the belly of a dirty white: the leſſer coverts of the wings are of a bright bay: the laſt row black, tipt with white: the great coverts black, outwardly edged with red; the quil-feathers the ſame: the back ſpotted with red and black: tail duſky.

The lower mandible of the bill of the female is white: beyond each eye is a line of white: the head and whole upper part are brown, only on the back are a few black ſpots: the black and white marks on the wings are obſcure; the lower ſide of the body is a dirty white.

Sparrows are proverbially ſalacious: they breed early

M. & F. SPARROWS.

Pl. LII
No 128
TREE SPARROW.
SEDGE BIRD.

early in the ſpring, make their neſts under the eaves of houſes, in holes of walls, and very often in the neſts of the martin, after expelling the owner. *Linnæus* tells us (a tale from *Albertus Magnus)* that this inſult does not paſs unrevenged; the injured martin aſſembles its companions, who aſſiſt in plaiſtering up the entrance with dirt; then fly away, twittering in triumph, and leave the invader to periſh miſerably.

They will often breed in plumb-trees and apple-trees, in old rooks's neſts, and in the forks of boughs beneath them.

128. TREE SPARROW.

Paſſerinus. *Geſner av.* 656.
Aldr. av. II. 261.
Olina, 48.
Wil. orn. 252.
Raii ſyn. av. 87.
Edw. av. 269.
Le Moineau de Montagne, Paſſer montanus. *Briſſon av.* III. 79.
Paſſere Montano. *Zinan.* 81.
Fringilla montana. *Lin. ſyſt.* 324.
Faun. Suec. ſp. 243. *Scopoli*, No. 221.
Skov-Spurre. *Brunnich*, 267.
Feldſpatz, Rohrſpatz. *Kram.* 370. *Friſch*, I. 1.
Br Zool. 109.
Grabetz. *Scopoli*, No. 220.

THIS ſpecies is inferior in ſize to the common ſparrow. The bill is thick and black: the crown of the head; hind part of the neck; and the leſſer coverts of the wings, of a bright bay: the two firſt plain; the laſt ſpotted with black: the chin black; the cheeks and ſides of the head white, marked with a great black ſpot beneath

beneath each ear: the breaſt and belly of a dirty white. Juſt above the greater coverts is a row of feathers black edged with white; the greater coverts are black edged with ruſt color: quil-feathers duſky, edged with pale red: lower part of the back of an olive brown: tail brown: legs ſtraw color.

Theſe birds are very common in *Lincolnſhire*; are converſant among trees, and collect like the common kind in great flocks.

129. SISKIN. *Belon av.* 354.
Acanthis, ſpinus, ligurinus. *Geſner av.* 1.
Aldr. av. II. 352.
Lucarino. *Olina,* 17.
Wil. orn. 261.
Raii ſyn. av. 91.
Le Serin. *Briſſon av.* III. 65.
Fringilla ſpinus. *Lin. ſyſt.* 322.
Siſka, Gronſiſka. *Faun. Suec. ſp.* 237.
Siſgen. *Brunnich,* 261.
Zeiſel, Zeiſerl. *Kram.* 366.
Friſch, I. 2. *Scopoli,* No. 212.
Br. Zool. 109. plate V.

DESCRIP. THE head of the male is black: the neck and back green; but the ſhafts on the latter are black: the rump is of a greeniſh yellow; the throat and breaſt the ſame: the belly white: the vent-feathers yellowiſh, marked with oblong duſky ſpots in their middle: the pinion quil is duſky edged with green: the outward webs of the nine next quil-feathers are green; the green part is widened by degrees in every feather, till in the laſt it takes up half the length: from the tenth almoſt the

Pl. LIII

No. 129

SISKIN. M. & F.

TWITE. M. & F. No. 133

the lower half of each feather is yellow, the upper black: the exterior coverts of the wings are black: the two middle feathers of the tail are black; the reſt above half way are of a moſt lovely yellow, with black tips. The colors of the female are paler: her throat and ſides are white ſpotted with brown; the head and back are of a greeniſh aſh color, marked alſo with brown.

Mr. *Willughby* tells us, that this is a ſong bird: that in *Suſſex* it is called the *barley-bird*, becauſe it comes to them in *barley-ſeed* time. We are informed that it viſits theſe iſlands at very uncertain times, like the groſbeak, &c. It is to be met with in the bird ſhops in *London*, and being rather a ſcarce bird, ſells at a higher price than the merit of its ſong deſerves: it is known there by the name of the *Aberdavine.* The bird catchers have a notion of its coming out of *Ruſſia.* Dr. *Kramer* * informs us, that this bird conceals its neſt with great art; though there are infinite numbers of young birds in the woods on the banks of the *Danube*, that ſeem juſt to have taken flight, yet no one could diſcover it.

* *Kramer elench.* 366.

Belon

130. LINNET. *Belon av.* 356.
Linaria, Henfling, Schofzling, Flacklin. *Gesner av.* 590.
Haenfling. *Frisch*, I. 9.
Aldr. av. II. 359.
Wil. orn. 258.
Raii syn. av. 90.
Fanello. *Zinan.* 61.
La Linotte. *Brisson av.* III. 131.
Pl. enl. 151. f. 1.
Br. Zool. 110.

DESCRIP. THE bill of this species is dusky, but in the spring assumes a bluish cast: the feathers on the head are black edged with ash color: the sides of the neck deep ash color: the throat marked in the middle with a brown line; bounded on each side with a white one: the back black bordered with reddish brown: the bottom of the breast is of a fine blood red, which heighthens in color as the spring advances: the belly white: the vent feathers yellowish: the sides under the wings spotted with brown: the quil-feathers are dusky; the lower part of the nine first white: the coverts incumbent on them black; the others of a reddish brown; the lowest order tipt with a paler color: the tail is a little forked, of a brown color, edged with white; the two middle feathers excepted, which are bordered with dull red. The females and young birds want the red spot on the breast; in lieu of that, their breasts are marked with short streaks of brown pointing downwards: the females have also less white in their wings.

These

GREATER AND LESSER RED POLLS.

These birds are much esteemed for their song: they feed on seeds of different kinds, which they peel before they eat: the seed of the *linum* or *flax* is their favorite food; from whence the name of the linnet tribe.

They breed among furze and white thorn: the outside of their nest is made with moss and bents; and lined with wool and hair. They lay five whitish eggs, spotted like those of the goldfinch.

131. RED HEADED LINNET.

Linaria rubra. *Gesner av.* 591.
Fanello marino. *Aldr. av.* II. 360.
Wil. orn. 260.
Raii syn. av. 91.
La grande Linotte des vignes. *Brisson av.* III. 135.
Fringilla cannabina. *Lin. syst.* 322. *Scopoli*, No. 219.
Hampling. *Faun. Suec. sp.* 240.
Torn-Irisk. *Brunnich*, 263.
Hauefferl, Hampfling. *Kram.* 368.
Blut Hänfling (Bloody Linnet). *Frisch*, I. 9.
Br. Zool. 110.

DESCRIP.

THIS bird is less than the former: on the forehead is a blood colored spot; the rest of the head and the neck are of an ash color: the breast is tinged with a fine rose color: the back, scapular feathers, and coverts of the wings, are of a bright reddish brown: the first quil-feather is entirely black; the exterior and interior edges of the eight following are white, which forms a bar of that color on the wing, even when closed: the sides are yellow; the middle of the belly white: the

the tail, like that of the former, is forked, of a duſky color, edged on both ſides with white, which is broadeſt on the inner webs. The head of the female is aſh color, ſpotted with black: the back and ſcapulars are of a dull browniſh red: and the breaſt and ſides of a dirty yellow, ſtreaked with duſky lines. It is a common fraud in the bird ſhops in *London*, when a male bird is diſtinguiſhed from the female by a red breaſt, as in the caſe of this bird, to ſtain or paint the feathers, ſo that the deceit is not eaſily diſcovered, without at leſt cloſe inſpection.

Theſe birds are frequent on our ſea-coaſts; and are often taken in *flight* time near *London*: it is a familiar bird; and is chearful in five minutes after it is caught.

132. LESS RED HEADED LINNET.

Wil. orn. 260.
Raii ſyn. av. 91.
La petite Linotte des vignes. *Briſſon av.* III. 138.
Pl. enl. 151. f. 2.
Fringilla linaria. *Lin. ſyſt.* 322.
Grafiſka. *Faun. Suec. ſp.* 241.
Graſel, Meerzeiſel, Tſchotſcherl. *Kram.* 369.
Rothplattige Staenfling. *Friſ.* I. 10.
Br. Zool. 111.

DESCRIP.

THIS is the leſt of the linnets, being ſcarce half the ſize of the preceding. Its bill is duſky, but the baſe of the lower mandible yellow: the forehead ornamented with a rich ſhining ſpot of a purpliſh red: the breaſt is of the ſame color, but

but not ſo bright; yet in the breaſts of ſome we have found the red wanting: the belly is white: the back duſky, edged with reddiſh brown: the ſides in ſome yellowiſh, in others aſh color, but both marked with narrow duſky lines: the quil-feathers, and thoſe of the tail, are duſky, border-ed with dirty white: the coverts duſky, edged with white, ſo as to form two tranſverſe lines of that color. The ſpot on the forehead of the female is of a ſaffron color. The legs are duſky.

We have ſeen the neſt of this ſpecies on an *alder* ſtump near a brook, between two or three feet from the ground: it was made on the outſide with dried ſtalks of graſs and other plants, and here and there a little wool, the lining was hair and a few feathers: the bird was *ſitting* on four eggs of a pale bluiſh green, thickly ſprinkled near the blunt end with ſmall reddiſh ſpots. The bird was ſo tena-cious of her neſt, as to ſuffer us to take her off with our hand, and we found that after we had re-leaſed her ſhe would not forſake it.

This ſeems to be the ſpecies known about *Lon-don* under the name of *ſtone redpoll:* is gregarious.

133. TWITE. Le Picaveret? *Belon av.* 358.
Wil. orn. 261.
Raii ſyn. av. 91.
Linaria montana. Linaria minima.
La petite Linotte, ou le Cabaret. *Briſſon av.* III. 142, 145.
Linaria fera ſaxatilis. *Klein. hiſt. av.* 93.
Br. Zool. 111.

THIS is an inhabitant of the hilly parts of our country, as Mr. *Willughby* informs us. He ſays it is twice the ſize of the laſt ſpecies: that the color of the head and back is the ſame with that of the common linnet: that the feathers on the throat and breaſt are black edged with white: the rump is of a rich ſcarlet or orange tawny color. The edges of the middle quil-feathers are white, as are the tips of thoſe of the ſecond row: the two middle feathers of the tail are of a uniform duſky color; the others edged with white. This ſpecies is taken in the flight ſeaſon near *London* with the linnets; it is there called a *Twite.* The birds we examined differed in ſome particulars from Mr. *Willughby*'s deſcription. In ſize they are rather inferior to the common linnet, and of a more taper make: their bills ſhort and entirely yellow: above and below each eye is a pale brown ſpot: the edges of the greater coverts of the wings white; in other reſpects both agree. The female wants the red mark on the rump.

DESCRIP.

These

These birds take their name from their note, which has no music in it: it is a familiar bird, and more easily tamed than the common *linnet*.

We believe it breeds only in the *Northern* parts of our island.

CANARY BIRD.

Here it may not be improper to mention the *Canary bird* *, which is of the finch tribe. It was originally peculiar to those isles, to which it owes its name; the same that were known to the antients by the addition of the *fortunate*. The happy temperament of the air, the spontaneous productions of the ground in the varieties of fruits; the sprightly and chearful disposition of the inhabitants †; and the harmony arising from the number of the birds found there ‡, procured them that romantic distinction. Though the antients celebrate the isle of *Canaria* for the multitude of birds, they have not mentioned any in particular. It is pro-

* *Wil. orn.* 262. *Raii syn. av.* 91. *Vide* Serin des Canaries. *Brisson av.* III. 184. Fringilla Canaria. *Lin. syst.* 321.

† Fortunatæ *insulæ abundant sua sponte genitis, et subinde aliis super aliis innascentibus nihil solicitos alunt; beatius quam aliæ urbes excultæ. Mela de sit. orb.* III. 17. He then relates the vast flow of mirth among this happy people, by a figurative sort of expression, that alludes to their tempering discretion with their jollity, and never suffering it to exceed the bounds of prudence. This he delivers under the notion of two fountains found among them, *alterum qui gustavere risu solvuntur in mortem; ita affectis remedium est ex altero bibere.*

‡ *Omnes copia pomorum, et avium omnes generis abundant,* &c. *Plin.* lib. VI. C. 32.

A a 2 bable

bable then, that our ſpecies was not introduced into *Europe* till after the ſecond diſcovery of theſe iſles, which was between the thirteenth and fourteenth centuries. We are uncertain when it firſt made its appearance in this quarter of the globe. *Belon*, who wrote in 1555, is ſilent in reſpect to theſe birds: *Geſner** is the firſt who mentions them; and *Aldrovand*† ſpeaks of them as rarities; that they were very dear on account of the difficulty attending the bringing them from ſo diſtant a country, and that they were purchaſed by people of rank alone. *Olina*‡ ſays, that in his time there was a degenerate ſort found on the iſle of *Elba*, off the coaſt of *Italy*, which came there originally by means of a ſhip bound from the *Canaries* to *Leghorn*, and was wrecked on that iſland. We once ſaw ſome ſmall birds brought directly from the *Canary Iſlands*, that we ſuſpect to be the genuine ſort; they were of a dull green color, but as they did not ſing, we ſuppoſed them to be hens. Theſe birds will produce with the goldfinch and linnet, and the offspring is called a mule-bird, becauſe, like that animal, it proves barren.

* *Geſner av.* 240.

† *Aldr. av.* II. 355.

‡ *Olina uccel.* 7.

They

They are ſtill found * on the ſame ſpot to which we were firſt indebted for the production of ſuch charming ſongſters; but they are now become ſo numerous in our country, that we are under no neceſſity of croſſing the ocean for them.

* *Glas's hiſt. Canary Iſles*, 199.

XXI. FLY-CATCHER.

BILL flatted at the base; almoſt triangular: notched near the end of the upper mandible, and beſet with briſtles.

TOES divided to their origin.

134. SPOTTED.

Stoparola. *Aldr. av.* II. 324.
A ſmall bird without a name, like the *Stopparola* of *Aldrovand. Wil. orn.* 217.
Zinan. 45.
The Cobweb. *Morton Northampt.* 426.
Raii ſyn. av. 77.
Le Gobe-mouche, Muſcicapa. *Briſſon av.* II. 357, *tab.* 35. f. 3.
Muſcicapa griſola. *Lin. ſyſt.* 328.
Br. Zool. 99. plate P. 2. f. 4.

THE fly-catcher is a bird of paſſage, appears in the ſpring, breeds with us, and retires in *Auguſt*. It builds its neſt on the ſides of trees, towards the middle: *Morton* ſays in the corners of walls where ſpiders weave their webs. We have ſeen them followed by four or five young, but never ſaw their eggs. When the young can fly the old ones withdraw with them into thick woods, where they frolick among the top branches; dropping from the boughs frequently quite perpendicular on the flies that ſport beneath, and riſe again in the ſame direction. It will alſo take its ſtand on the top of ſome ſtake or poſt, from whence it ſprings forth on its prey, returning ſtill to the ſame ſtand for many times together. They feed

feed alſo on cherries, of which they ſeem very fond.

DESCRIP.

The head is large, of a browniſh hue ſpotted obſcurely with black: the back of a mouſe color: the wings and tail duſky; the interior edges of the quil-feathers edged with pale yellow: the breaſt and belly white; the ſhafts of the feathers on the former duſky; the throat and ſides under the wings are daſhed with red: the bill is very broad at the baſe, is ridged in the middle, and round the baſe are ſeveral ſhort briſtles: the inſide of the mouth is yellow: the legs and feet ſhort and black.

135. PIED.

Atri capilla ſive ficedula. *Aldr. av.* II. 331.
Cold finch. *Wil. orn.* 236. *Raii ſyn. av.* 77. *Edw.* 30. *Friſche*, I. 22.
Le Traquet d' Angleterre. Rubetra anglicana. *Briſſon*, III. 436.
Meerſchwartz pluffle. *Kramer Auſt.* 377.
Cold-finch. *Br. Zool.*
Muſcicapa atricapilla. *Lin. ſyſt.* 326. *Faun. Suec.* No. 256. Tab. 1.

MALE.

THIS is leſſer than a hedge ſparrow. The bill and legs black: the forehead white: head, cheeks, and back black: the coverts of the tail ſpotted with white: coverts of the wings duſky, traverſed with white bar: quil feathers duſky: the exterior ſides of the ſecondaries white; the interior duſky: the middle feathers of the tail black;

the exterior marked with white: the whole under side of the body white.

FEMALE. The female wants the white spot on the forehead: the whole head, and upper part of the body dusky brown: the white in the wings less conspicuous: the under side of the body of a dirty white.

Found in different parts of *England*: but is a rare species.

Weak

Weak BILL, ftrait, bending towards the point. NOSTRILS covered with feathers or briftles. TOES divided to their origin. BACK TOE armed with a long and ftrait claw.

XXII. LARK.

L' Alouette. *Belon av.* 269.
Chamochilada. *Obf.* 12.
Alauda fine crifta. *Gefner av.* 78.
Aldr. av. II. 369.
Lodola. *Olina,* 12.
Common Field Lark, or Sky Lark. *Wil. orn.* 203.
Raii fyn. av. 69.
L' Alouette. *Briffon av.* III. 335.
Allodola, Panterana. *Zinan.* 55.
Alauda arvenfis. *Lin. fyft.* 287.
Larka. *Faun. Suec. fp.* 209.
Alauda cœlipeta. *Klein ftem. Tab.* 15. f. 1.
Sang-Lœrke. *Br.* 221.
Feldlerche. *Kram.* 362. *Frifch,* I. 15.
Br. Zool. 93. plate S. 2. f. 7.
Lauditza. *Scopoli,* No. 184.

136. SKY.

DESCRIP.

THE length of this fpecies is feven inches one-fourth: the breadth twelve and a half: the weight one ounce and a half: the tongue broad and cloven: the bill flender: the upper mandible dufky, the lower yellow: above the eyes is a yellow fpot: the crown of the head a reddifh brown fpotted with deep black: the hind part of the head afh-color: chin white. It has the faculty of erecting the feathers of the head. The feathers on the back, and coverts of the wings dufky edged with reddifh brown, which is paler on the latter: the quil-feathers dufky: the exterior web edged

edged with white, that of the others with reddiſh brown: the upper part of the breaſt yellow ſpotted with black: the lower part of the body of a pale yellow: the exterior web, and half of the interior web next to the ſhaft of the firſt feather of the tail are white; of the ſecond only the exterior web; the reſt of thoſe feathers duſky; the others are duſky edged with red; thoſe in the middle deeply ſo, the reſt very ſlightly: the legs duſky: ſoles of the feet yellow: the hind claw very long and ſtrait.

This and the wood lark are the only birds that ſing as they fly; this raiſing its note as it ſoars, and lowering it till it quite dies away as it deſcends. It will often ſoar to ſuch a height, that we are charmed with the muſic when we loſe ſight of the ſongſter; it alſo begins its ſong before the earlieſt dawn. *Milton*, in his *Allegro*, moſt beautifully expreſſes theſe circumſtances: and Bp. *Newton* obſerves, that the beautifull ſcene that *Milton* exhibits of rural chearfulneſs, at the ſame time gives us a fine picture of the regularity of his life, and the innocency of his own mind; thus he deſcribes himſelf as in a ſituation

> To hear the lark begin his flight,
> And ſinging ſtartle the dull night,
> From his watch tower in the ſkies,
> 'Till the dappled dawn doth riſe.

It

It continues its harmony ſeveral months, beginning early in the ſpring, on pairing. In the winter they aſſemble in vaſt flocks, grow very fat, and are taken in great numbers for our tables. They build their neſt on the ground, beneath ſome clod; forming it of hay, dry fibres, &c. and lay four or five eggs.

The place theſe birds are taken in the greateſt quantity, is the neighbourhood of *Dunſtable*: the ſeaſon begins about the fourteenth of *September*, and ends the twenty-fifth of *February*; and during that ſpace, about 4000 dozen are caught, which ſupply the markets of the metropolis. Thoſe caught in the day are taken in clap-nets of fiveteen yards length, and two and a half in breadth; and are enticed within their reach by means of bits of looking-glaſs, fixed in a piece of wood, and placed in the middle of the nets, which are put in a quick whirling motion, by a ſtring the larker commands; he alſo makes uſe of a decoy lark. Theſe nets are uſed only till the fourteenth of *November*, for the larks will not *dare*, or frolick in the air except in fine ſunny weather; and of courſe cannot be inviegled into the ſnare. When the weather grows gloomy, the larker changes his engine, and makes uſe of a trammel net twenty-ſeven or twenty-eight feet long, and five broad; which is put on two poles eighteen feet long, and carried by men under each arm, who paſs over the fields and quarter the ground as a ſetting dog; when they

they hear or feel a lark hit the net, they drop it down, and ſo the birds are taken.

137. WOOD. Tottavilla. *Olina*, 27.
Wil orn. 204.
Raii ſyn. av. 69.
L' Alouette de Bois ou le Cujelier. *Briſſon av.* III. 340. *Tab.* 20. *fig.* 1.
Alauda arborea. *Lin. ſyſt.* 287.
Faun. Suec. ſp. 211.
Ludllerche, Waldlerche *Kram.* 362.
Danis Skov-Lerke, *Cimbris* Heede-Leker, Lyng-Lreke.
Br. 224.
Br. Zool. 94. plate Q. f. 3.
Zippa. *Scopoli*, No. 186.

THIS bird is inferior in ſize to the ſky lark, and is of a ſhorter thicker form; the colors are paler, and its note leſs ſonorous and leſs varied, though not leſs ſweet. Theſe and the following characters, may ſerve at once to diſtinguiſh it from the common kind: it perches on trees; it whiſtles like the black-bird. The crown of the head, and the back, are marked with large black ſpots edged with pale reddiſh brown: the head is ſurrounded with a whitiſh coronet of feathers, reaching from eye to eye: the throat is of yellowiſh white, ſpotted with black: the breaſt is tinged with red: the belly white: the coverts of the wings are brown, edged with white and dull yellow: the quil-feathers duſky; the exterior edges of the three firſt white; of the others yellow, and their tips blunt and white: the firſt feather of the wing is ſhorter than the ſecond; in the common lark it is is near equal: the tail is black, the outmoſt feather is tipt with white: the exterior web, and inner ſide of the interior are alſo white; in the ſecond feather, the exterior web only: the legs are of a dull yellow; the hind claw very long. The wood lark will ſing in the night; and, like the common lark, will ſing as it flies. It builds on the ground, and makes its neſt on the outſide with moſs, within of dried bents lined with a few hairs. It lays five eggs, duſky and blotched with deep brown, marks darkeſt at the thicker end.

The males of this and the laſt are known from the females by their ſuperior ſize. But this ſpecies is not near ſo numerous as that of the common kind.

138. TIT.

La Farlouſe, Fallope ou L'Alouette de pre. *Belon av.* 272.
Aldr. av. II. 370.
Lodolo di Prato. *Olina,* 27.
Wil. orn. 206.
Raii ſyn. av. 69.
L'Alouette de prez ou la Farlouſe. *Briſſon av.* III. 343.
Mattolina, Petragnola, Corriera. *Zinan.* 55.
Alauda pratenſis. *Lin. ſyſt.* 287.
Faun. Suec. ſp. 210.
Wieſen Lerche (Meadows Lark) *Friſch,* I. 16.
Englerke. *Br.* 223.
Br. Zool 94. plates Q. f. 6. P. 1. f. 3.

THIS bird is found frequently in low marſhy grounds: like other larks it builds its neſt among the graſs, and lays five or ſix eggs. Like the

the woodlark it ſits on trees; and has a moſt remarkable fine note, ſinging in all ſituations, on trees, on the ground, while it is ſporting in the air, and particularly in its deſcent. This bird with many others, ſuch as the thruſh, blackbird, willow wren, &c. become ſilent about midſummer, and reſume their notes in *September*: hence the interval is the moſt mute of the year's three vocal ſeaſons, ſpring, ſummer, and autumn. Perhaps the birds are induced to ſing again as the autumnal tempe-

DESCRIP. rament reſembles the vernal. It is a bird of an elegant ſlender ſhape: the length is five inches and a half: the breadth nine inches: the bill is black: the back and head is of a greeniſh brown, ſpotted with black: the throat and lower part of the belly are white: the breaſt yellow, marked with oblong ſpots of black: the tail is duſky; the exterior feather is varied by a bar of white, which runs acroſs the end and takes in the whole outmoſt web. The claw on the hind toe is very long, the feet yellowiſh: the ſubject figured in plate P. 1. of the *folio* edition, is a variety with duſky legs, ſhot on the rocks on the coaſt of *Caernarvonſhire*.

139. FIELD.

The Leſſer Field Lark. *Wil. orn.* 207.

DESCRIP. THIS ſpecies we received from Mr. *Plymly*. It is larger than the *tit lark*; the head and hind part

part of the neck are of a pale brown, ſpotted with duſky lines, which on the neck are very faint. The back and rump are of a dirty green; the former marked in the middle of each feather with black, the latter plain. The coverts of the wings duſky, deeply edged with white. The quil-feathers duſky; the exterior web of the firſt edged with white, of the others with a yellowiſh green.

The throat is yellow: the breaſt of the ſame color, marked with large black ſpots: the belly and vent-feathers white: on the thighs are a few dusky oblong lines: the tail is dusky: half the exterior and interior web of the outmoſt feather is white; the next is marked near the end with a ſhort white ſtripe pointing downwards. The legs are of a very pale brown; and the claw on the hind toe very ſhort for one of the lark kind, which ſtrongly diſtinguiſhes it from the *tit lark.*

Edw. 297. *Br. Zool.* II. 239. *Briſſon Suppl.* 94. 140. RED.

I MET with this ſpecies in the magnificent and elegant *Muſeum* of ASHTON LEVER, Eſq; where the lover of *Britiſh* or exotic ornithology, may find delight and inſtruction equally intermixed.

This ſpecies is equal in ſize to the common lark. A white line croſſes each eye, and another paſſes beneath.

beneath. The bill is thick: the chin and throat whitiſh: the head, neck, back, and coverts of the wings of a ruſty brown, ſpotted with black: breaſt whitiſh, with dusky ſpots: belly of a dirty white: the middle feather of the tail black edged with brown: the two exterior white: legs of a pale brown.

This bird is common to the neighbourhood of *London*, to *North America*, and to the South of *Europe*; but with us is rare. Mr. *Edwards* firſt diſcovered it: he remarks, that when the wing is gathered up, the third primary feather reaches to the tip of the firſt.

141. CREST-ED.

Alauda criſtata minor. *Aldr. av.* II. 371.
Wil. orn. 209.
Raii ſyn. av. 69.

La petite alouette hupée. *Briſſon av.* III. 361.
Br. Zool. 95.

THIS ſpecies we find in Mr. *Ray*'s hiſtory of *Engliſh* birds; who ſays it is found in *Yorkſhire*, and gives us only this brief deſcription of it, from *Aldrovandus*: it is like the greater creſted lark, but much leſs, and not ſo brown; that it hath a conſiderable tuft on its head for the ſmallneſs of its body; and that its legs are red. We never ſaw this kind; but by Mr. *Bolton*'s liſt of *Yorkſhire* birds, which he favored us with, we are informed it is in plenty in that country.

Slender

Pl. LV

No 1.

WHITE WAGTAIL.

YELLOW WAGTAIL.

No

SKY LARK.

No

Slender BILL, with a ſmall tooth near the end of the upper mandible.
Lacerated TONGUE.
Long TAIL.

XXIII. WAGTAIL.

Belon av. 349.
Motacilla alba. *Geſner av.* 618.
Aldr. av. II. 323.
Ballarina, Cutrettoſa. *Olina,* 43.
Wil. orn. 237.
Raii ſyn. av. 75.
La Lavandiere. *Briſſon av.* III. 461.
Monachina. *Zinan.* 51.
Pliſka, Paſtaritra. *Scopoli,* No. 224.
M. alba. *Lin. ſyſt.* 331.
Arla, Sadeſarla. *Faun. Suec. ſp.* 252.
Danis Vip-Stiert, Havre-Sæer. *Norvegis* Erle, Lin-Erle. *Brunnich,* 271.
Weiſs und ſchwartze Bachſteltze. *Friſch,* I. 23.
Graue Bachſtelze. *Kram.* 374.
Br. Zool. 104.

142. WHITE WAGTAIL.

THIS bird frequents the ſides of ponds, and ſmall ſtreams; and feeds on inſects and worms, as do all the reſt of this genus. Mr. *Willughby* juſtly obſerves, that this ſpecies ſhifts its quarters in the winter; moving from the north to the ſouth of *England*, during that ſeaſon. In ſpring and autumn it is a conſtant attendant of the plough, for the ſake of the worms thrown up by that inſtrument.

The head, back, and upper and lower ſide of the neck as far as the breaſt are black: in ſome the chin is white, and the throat marked with a black crescent: the breast and belly are white: the quil-feathers are dusky: the coverts black tipt and edged with white. The tail is very long, and always in motion. The exterior feather on each side is white: the lower part of the inner web excepted, which is dusky; the others black: the bill, inside of the mouth, and the legs, are black. The back claw very long.

143. Yellow Wagtail.

Susurada. *Belon obs.* II.
Motacilla flava (Gale Wassarsteltz). *Gesner av.* 618.
Aldr. av. II. 323.
Wil. orn. 238.
Raii syn. av. 75.
Edw. av. 258. The Male.
Codatremola. *Zinan.* 51.
La Bergeronette du Printemps, Motacilla verna. *Brisson av.* III. 468. *Pl. enl.* 28. f. 1.
Motacilla flava. *Lin. syst.* 331.
Gelb-brüstige Bachsteltze. *Frisch*, I. 23.
Faun. Suec. sp. 253.
Gulspink. *Brunnich.* 273.
Gelbe Bachstelze. *Kram.* 374.
Scopoli, No. 225.
Br. Zool. 105.

Descrip. THE male is a bird of great beauty: the breast, belly, thighs, and vent-feathers, being of a most vivid and lovely yellow: the throat is marked with some large black spots: above the eye is a bright yellow line: beneath that, from the bill cross the eye is another of a dusky hue; and beneath the eye is a third of the same color: the head and whole upper part of the body is of an olive green, which brightens in the coverts of the tail; the quil-feathers are dusky: the coverts of the wings olive colored,

colored, but the lower rows dusky, tipt with yellowiſh white: the two outmoſt feathers of the tail half white; the others black, as in the former.

The colors of the female are far more obſcure than thoſe of the male: it wants alſo thoſe black ſpots on the throat.

It makes its neſt on the ground, in corn fields: the outſide is compoſed of decayed ſtems of plants, and ſmall fibrous roots; the inſide is lined with hair: it lays five eggs. This ſpecies migrates in the North of *England*, but in *Hampſhire* continues the whole year.

144. GREY WAGTAIL.

La Bergerette. *Belon av.* 351.
Motacilla flava alia. *Aldr. av.* II. 323.
Wil. orn. 238.
Raii ſyn. av. 75.
Edw. av. 259. The Male.

La Bergeronette jaune, Motacilla flava. *Briſſon av.* III. 471. *tab.* 23. *fig.* 3. The Male.
Br. Zool. 105.

DESCRIP.

THE top of the head, upper part of the neck, and the back of this ſpecies are aſh colored; ſlightly edged with yellowiſh green: the ſpace round each eye is aſh colored: beneath and above which is a line of white: in the male, the chin and throat are black: the feathers incumbent on the tail are yellow: the tail is longer, in proportion to its ſize, than that of the other kinds: the two exterior feathers are white; the reſt black: the breaſt, and whole under ſide of the body are yellow: the quil-feathers are dusky; thoſe next the back edged with yellow. The colors of the female are uſually more obſcure; and the black ſpot on the throat is wanting in that ſex.

The birds of this genus are much in motion: ſeldom perch: perpetually flirting their tails: ſcream when they fly: frequent waters: feed on inſects; and make their neſts on the ground.

BILL

BILL ſlender and weak.
NOSTRILS ſmall and ſunk.
Exterior TOE joined at the under part of the laſt joint to the middle toe.

XXIV. WARBLERS.

* Thoſe with tails of one color.

** Thoſe with particolored tails.

Le Roſſignol. *Belon av.* 335.
Adoni, Aidoni. *Obſ.* 12.
Luſcinia. *Geſner av.* 592.
Aldr. av. II. 336.
Wil. orn. 220.
Raii ſyn. av. 78.
Le Roſſignol. *Briſſon av.* III. 397.
Slauz. *Scopoli*, No. 227.
Ruſignulo. *Zinan.* 54.
Motacilla luſcinia. *Lin. ſyſt.* 328.
Nachtergahl. *Faun. Suec. ſp.* 244.
Haſſelquiſt Itin. Ter. Sanct. 291.
Nattergale. *Brunnich in append.*
Au-vogel, Auen-nachtigall. *Kram.* 376.
Nachtigall. *Friſch*, I. 21.
Br. Zool. 100. plate S. 1. f. 2.

145. NIGHTINGALE.

THE nightingale takes its name from *night*, and the *Saxon* word *galan* to ſing; expreſſive of the time of its melody. In ſize it is equal to the *redſtart*; but longer bodied, and more elegantly made. The colors are very plain. The head and back are of a pale tawny, daſhed with olive: the tail is of a deep tawny red: the throat, breaſt,

DESCRIP.

 and

and upper part of the belly of a light gloſſy aſh-color: the lower belly almoſt white: the exterior webs of the quil-feathers are of a dull reddiſh brown; the interior of browniſh aſh-color: the irides are hazel, and the eyes remarkably large and piercing: the legs and feet a deep aſh-color.

This bird, the moſt famed of the feathered tribe, for the variety *, length, and ſweetneſs of its notes, viſits *England* the beginning of *April*, and leaves us in *Auguſt*. It is a ſpecies that does not ſpread itſelf over the iſland. It is not found in *North Wales*; or in any of the *Engliſh* counties north of it, except *Yorkſhire*, where they are met with in great plenty about *Doncaſter*. They have been alſo heard, but rarely, near *Shrewſbury*. It is alſo remarkable, that this bird does not migrate ſo far weſt as *Devonſhire* and *Cornwall*; counties where the ſeaſons are ſo very mild, that myrtles flouriſh in the open air during the whole year: neither are they found in *Ireland*. *Sibbald* places them in his liſt of *Scotch* birds; but they certainly are unknown in that part of *Great Britain*, probably from the ſcarcity and the recent introduction of hedges there. Yet they viſit *Sweden*, a much more ſevere climate. With us they frequent thick

* For this reaſon, *Oppian*, in his *halieutics*, l. I. 728. gives the *nightingale* the epithet of αἰολοφώνη, or *various voiced*; and *Heſiod*, (figuratively) of ποικιλόδειρα, or *various throated*. Εργα και ἡμέραι, l. 201.

hedges,

hedges, and low coppices; and generally keep in the middle of the bufh, fo that they are very rarely feen. They form their neft of oak leaves, a few bents and reeds. The eggs are of a deep brown. When the young firft come abroad, and are helplefs, the old birds make a plaintive and jarring noife with a fort of fnapping as if in menace, purfuing along the hedge the paffengers.

They begin their fong in the evening, and continue it the whole night. Thefe, their vigils, did not pafs unnoticed by the antients: the flumbers of thefe birds were proverbial; and not to reft as much as the *nightingale*, expreffed a very bad fleeper *. This was the favorite bird of the *Britifh* poet, who omits no opportunity of introducing it, and almoft conftantly noting its love of folitude and night. How finely does it ferve to compofe part of the folemn fcenery of his *Penferofo*; when he defcribes it

> In her faddeft fweeteft plight,
> Smoothing the rugged brow of night;
> While *Cynthia* checks her dragon yoke,
> Gently o'er th' accuftom'd oak;
> Sweet bird, that fhunn'ft the noife of folly,
> Moft mufical, moft melancholy!
> Thee, chauntrefs, oft the woods among,
> I woo to hear thy evening fong.

* *Ælian var. hift.* 577. both in the text and note. It muft be remarked, that nightingales fing alfo in the day.

In another place he ftyles it the *folemn bird*; and again fpeaks of it,

As the wakeful bird
Sings darkling, and in fhadieft covert hid,
Tunes her nocturnal note.

The reader muft excufe a few more quotations from the fame poet, on the fame fubject; the firft defcribes the approach of evening, and the retiring of all animals to their repofe.

Silence accompanied; for beaft and bird,
They to their graffy couch, thefe to their nefts
Were flunk; all but the wakeful *nightingale*,
She all night long her amorous defcant fung.

When *Eve* paffed the irkfome night preceding her fall, fhe, in a dream, imagines herfelf thus reproached with lofing the beauties of the night by indulging too long a repofe:

Why fleep'ft thou, *Eve*? now is the pleafant time,
The cool, the filent, fave where filence yields
To the night-warbling bird, that now awake
Tunes fweeteft his love-labor'd fong.

The fame birds fing their nuptial fong, and lull them to reft. How rapturous are the following lines! how expreffive of the delicate fenfibility of our *Milton*'s tender ideas!

The Earth
Gave fign of gratulation, and each hill;
Joyous the birds; frefh gales and gentle airs

Whifper'd

Whiſper'd it to the woods, and from their wings
Flung roſe, flung odors from the ſpicy ſhrub,
Diſporting, till the amorous bird of night
Sung ſpouſal, and bid haſte the evening ſtar
On his hill-top to light the bridal lamp.

These, lull'd by *nightingales*, embracing ſlept;
And on their naked limbs the flowery roof
Shower'd roſes, which the morn repair'd.

Theſe quotations from the beſt judge of melody, we thought due to the ſweeteſt of our feathered choiriſters; and we believe no reader of taſte will think them tedious.

Virgil ſeems to be the only poet among the antients, who hath attended to the circumſtance of this bird's ſinging in the night time.

Qualis populeâ mœrens *Philomela* ſub umbrâ
Amiſſos queritur fœtus, quos durus arator
Obſervans nido implumes detraxit: at illa
Flet noctem, ramoque ſedens miſerabile carmen
Integrat, et mœſtis late loca queſtibus implet.

Georg. IV. l. 511.

As *Philomel* in poplar ſhades, alone,
For her loſt offspring pours a mother's moan,
Which ſome rough ploughman marking for his prey,
From the warm neſt, unfledg'd hath dragg'd away;
Percht on a bough, ſhe all night long complains,
And fills the grove with ſad repeated ſtrains.

F. Warton.

Pliny has deſcribed the warbling notes of this bird, with an elegance that beſpeaks an exquiſite ſenſibility

ſenſibility of taſte: notwithſtanding that his words have been cited by moſt other writers on natural hiſtory, yet ſuch is the beauty, and in general the truth of his expreſſions, that they cannot be too much ſtudied by lovers of natural hiſtory, therefore clame a place in a work of this kind. We muſt obſerve notwithſtanding, that a few of his thoughts are more to be admired for their vivacity than for ſtrict philoſophical reaſoning; but theſe few are eaſily diſtinguiſhable.

" *Luſciniis* diebus ac noctibus continuis xv. garrulus ſine " intermiſſu cantus, denſante ſe frondium germine, non in " noviſſimum digna miratu ave. Primum tanta vox tam parvo " in corpuſculo, tam pertinax ſpiritus. Deinde in una per- " fecta muſicæ ſcientia modulatus editur ſonus: & nunc con- " tinuo ſpiritu trahitur in longum, nunc variatur inflexo, " nunc diſtinguitur conciſo, copulatur in torto: promittitur " revocato, infuſcatur ex inopinato: interdum & ſecum ipſe " murmurat: plenus, gravis, acutus, creber, extentus, ubi " viſum eſt, vibrans, ſummus, medius, imus. Breviterque " omnia tam parvulis in faucibus, quæ tot exquiſitis tibi- " arum tormentis ars hominum excogitavit: ut non ſit dubi- " um hanc ſuavitatem præmonſtratam efficaci auſpicio, cum " in ore *Steſichori* cecinit infantis. Ac ne quis dubitet artis " eſſe, plures ſingulis ſunt cantus, nec iidem omnibus, ſed " ſui cuique. Certant inter ſe, palamque animoſa conten- " tio eſt. Victa morte finit ſæpe vitam, ſpiritu prius defici- " ente, quam cantu. Meditantur aliæ juniores, verſuſque " quos imitentur accipiunt. Audit diſcipula intentione mag- " na & reddit, vicibuſque reticent. Intelligitur emendatæ " correctio & in docente quædam reprehenſio " *.

* *Plin.* lib. 10. c. 29.

Le

146. RED-START.

Le Roffignol de Muraille. *Belon av.* 347.
Ruticilla, five Phœnicurus (Sommerotele) *Gefner av.* 731.
Aldr. av. II. 327.
Codoroffo. *Olina,* 47.
Wil. orn. 218.
Raii fyn. av. 78.
Ruticilla. *Briffon av.* III. 403.
Culo ranzo, Culo roffo. *Zinan.* 53. *Scopoli,* No. 232.
Motacilla Phœnicurus. *Lin. fyft.* 335.
Rodftjert. *Faun. Suec. fp.* 257.
Norvegis Blod-fugl. *Danis* Roed-ftiert. *Brunnich,* 280.
Schwartzkehlein (Blackthroat) *Frifch,* I. 19.
Waldrothfchweiffl. *Kram.* 376.
Br. Zool. 99. plate S. f. 6. 7.

THIS alfo appears among us only in the fpring and fummer, and is obferved to come over nearly at the fame time with the nightingale. It makes its neft in hollow trees, and holes in walls and other buildings; which it forms with mofs on the outfide, and lines with hair and feathers. It lays four or five eggs, very like thofe of the hedge-fparrow, but rather paler, and more taper at the lefs end. This bird is fo remarkably fhy, that it will forfake its neft, if the eggs are only touched. It has a very fine foft note; but being a fullen bird, is with difficulty kept alive in confinement. It is remarkable in fhaking its tail, it moves it horizontally as a dog does when fawning.

DESCRIP.

The bill and legs of the male are black: the forehead white: the crown of the head, hind part of the neck, and the back are of a deep blue grey: the cheeks and throat black: the breaft, rump and

and ſides are red: the two middle feathers of the tail brown, the others red: the wings brown. In the female, the top of the head and back are of a deep aſh-color: the rump and tail of a duller red than thoſe of the male: the chin white; the lower ſide of the neck cinereous; the breaſt of a paler red.

147. RED-BREAST.

Rubeline. *Belon av.* 348.
Rubecula. *Geſner av.* 730.
Erithacus. *Aldr. av.* II. 325.
Olina, 16.
Robin Red-breaſt, or Ruddock. *Wil. orn.* 219.
Raii ſyn. av. 78.
Le Rouge-gorge. *Briſſon av.* III. 418.
Pettoroſſo. *Zinan.* 46.
Motacilla rubecula. *Lin. ſyſt.* 337.
Rotgel. *Faun. Suec. ſp.* 260.
Roed-Finke, Roed-Kielke. *Br.* 283.
Rothkehlein. *Friſch*, I. 19.
Rothkropfl. *Kram.* 376.
Br. Zool. 100. plate S. 2.
Smarnza, Taſchtza. *Scopoli*, No. 231.

THIS bird, though ſo very petulant as to be at conſtant war with its own tribe, yet is remarkably ſociable with mankind: in the winter it frequently makes one of the family; and takes refuge from the inclemency of the ſeaſon even by our fire ſides. *Thomſon** has prettily deſcribed the annual viſits of this gueſt.

> The RED-BREAST, ſacred to the houſhold gods,
> Wiſely regardful of th' embroiling ſky,

* In his Seaſons, *vide Winter*, line 246.

In

In joyleſs fields, and thorny thickets, leaves
His ſhivering mates, and pays to truſted Man
His annual viſit. Half afraid, he firſt
Againſt the window beats; then, briſk, alights
On the warm hearth; then, hopping o'er the floor,
Eyes all the ſmiling family aſkance,
And pecks, and ſtarts, and wonders where he is:
'Till more familiar grown, the table-crumbs
Attract his ſlender feet.

The great beauty of that celebrated poet conſiſts in his elegant and juſt deſcriptions of the œconomy of animals; and the happy uſe* he hath made of natural knowlege, in deſcriptive poetry, ſhines through almoſt every page of his *Seaſons*. The affection this bird has for mankind, is alſo recorded in that antient ballad, † *The babes in the wood*; a compoſition of a moſt beautifull and pathetic ſimplicity. It is the firſt tryal of our humanity: the child that refrains from tears on hearing that read, gives but a bad preſage of the tenderneſs of his future ſenſations.

In the ſpring this bird retires to breed in the thickeſt covers, or the moſt concealed holes of walls and other buildings. The eggs are of a dull white, ſprinkled with reddiſh ſpots. Its ſong is remarkably fine and ſoft; and the more to be valued, as we enjoy it the greateſt part of the winter, and early in the ſpring, and even through great part of

* *Vide our Preface.*

† Reliques of antient *Engliſh* Poetry, Vol. III. p. 170.

the

the summer, but its notes are part of that time drowned in the general warble of the season. Many of the autumnal songsters seem to be the young cock red-breasts of that year.

DESCRIP. The bill is dusky: the forehead, chin, throat and breast are of a deep orange color: the head, hind part of the neck, the back and tail are of a deep ash-color, tinged with green: the wings rather darker; the edges inclining to yellow: the legs and feet dusky.

148. BLACK-CAP.

Atricapilla. *Gesner av.* 371, 384.
Aldr. av. II. 329.
Wil. orn. 226.
La Fauvette a tete noire, Curruca atricapilla. *Brisson av.* III. 380.
Capinera. *Zinan.* 56.
Olina, 9. *Scopoli*, No. 229.
Raii syn. av. 79.

Motacilla atricapilla. *Lin. syst.* 332.
Faun. Suec. sp. 256.
Hav-Skade. *quibusdam* Spikke. *Br.* 228.
Moench mit der Schwartzen Platte (Monk with the black crown) *Frisch*, I. 23.
Schwartz plattl. *Kram.* 377.
Br. Zool. 101. plate S. f. 5.

DESCRIP. THIS bird is among the smallest of this tribe, scarce weighing half an ounce. The crown of the head in the male is black: the hind part of the neck a light ash-color: the back and coverts of the wings are of a greyish green: the quil-feathers and tail dusky, edged with dull green: the breast and upper part of the belly are of a pale ash-color: the vent feathers whitish: the legs of a lead color.

The

The female is diftinguifhed from the male by the fpot on the head, which in that is of a dull ruft-color. The black-cap is a bird of paffage, leaving us before winter. It fings very finely; and on that account is called in *Norfolk* the *mock-nightingale.* It has ufually a full, fweet, deep, loud wild pipe; yet the ftrain is of fhort continuance; and his motions are defultory: but when that bird fits calmly, and in earneft engages in fong, he pours forth very fweet but inward melody; and expreffes great variety of fweet and gentle modulations, fuperior perhaps to thofe of any of our warblers, the nightingale excepted: and while they warble, their throats are wonderfully diftended.

The black-cap frequents orchards and gardens. The laft fpring we difcovered the neft of this bird in a *fpruce fir*, about two feet from the ground; the outfide was compofed of the dried ftalks of the *goofe grafs*, with a little wool and green mofs round the verge; the infide was lined with the fibres of roots, thinly covered with black horfe hair. There were five eggs of a pale reddifh brown, mottled with a deeper color, and fprinkled with a few dark fpots.

Ficedula,

149. PETTY-CHAPS. Ficedula. *Gesner*, 385.
Beccafigo, or Fig eater. *Wil. orn.* 216. *Raii syn. av.* 79.
La Fauvette, curruca. *Brisson av.* III. 372.
Beccafico cinerizio. *Zinan.* 44.
Motacilla Hippolais. M. virescente cinerea subtus flavescens abdomine albido, artubus fuscis, superciliis albidis. *Lin. syst.* 330.
Faun. Suec. sp. 248.
Braune grass-mucke, Kleiner spottvogel. *Kram.* 377.

THIS species is inferior in size to the former. The inside of the mouth is red: the head, neck, back and wings are of an olivaceous ash-color: the quil-feathers darker, edged with olive: the inner coverts of the wings yellow: the breast white, tinged with yellow: the belly of a silvery white: the tail dusky: the legs bluish.

Mr. *Willughby* says, this bird is found in *Yorkshire*, and called the *Beam-bird*, from its nesting under beams in out-buildings, &c.

150. HEDGE. Le petit Mouchet. *Belon av.* 375.
Potamida, *obs.* 12.
Passer sepium *Angl. Aldr. av.* II. 329.
Curruca *Eliotæ* (Zaunschlipfle). *Gesner av.* 371.
Wil. orn. 215.
Raii syn. av. 79.
La Favette de haye, ou la passe buse. Curruca sepiaria. *Brisson av.* III. 394.
Jarnsparf. *Faun. Suec. sp.* 245.
Motacilla, Modularis. *Lin. syst.* 329.
Braunflekkige Grasmücke (Brown spotted Petty-chaps.) *Frisch*, I. 21.
Br. Zool. plate S. 1. f. 3. 4.

DESCRIP. THIS bird weighs twelve drams. Its head is of a deep brown, mixed with ash color, the cheeks

cheeks marked with oblong ſpots of dirty white: the back and coverts of the wings are duſky, edged with reddiſh brown: the quil-feathers and tail duſky: the rump brown, tinged with green: the throat and breaſt are of a dull aſh color: the belly of a dirty white: the ſides, thighs, and vent-feathers are of a pale tawny brown: the legs of a dull fleſh color.

This bird frequents low hedges, eſpecially thoſe of gardens. It makes its neſt in ſome ſmall buſh, and lays four or five eggs of a fine pale blue color: during the breeding ſeaſon has a remarkable flirt with its wings. The male has a ſhort but very ſweet plaintive note, which it begins with the firſt froſty mornings, and continues till a little time in the ſpring. This is the *Motacilla Modularis* of *Linnæus*; the bird which he ſuppoſes to be our hedge ſparrow, and deſcribes under the title of *Motacilla curruca*,* differs in colors of plumage as well as eggs.

* *Faun. Suec. ſp.* 247.

151. YELLOW

Chofti, ou Chanteur. *Belon av.* 344.
Trochilus. *Gesner av.* 726.
Afilus. *Aldr. av.* II. 293.
Little yellowifh Bird. *Wil. orn.* 228.
Raii fyn. av. 80.
Edw. av. 278.
Schnee Rienig (Snow king). *Frifch*, I. 24.
Schmittl. *Kram.* 378.
Le Pouillot, ou chantre. Afilus. *Briffon av.* III. 479.
Motacilla trochilus. *Lin. fyft.* 338. *Scopoli*, No. 238.
Faun. Suec. fp. 264.
Spurre-Konge, Fager-Fiis. *Br.* 286.
Br. Zool. 101. plate S. f. 2. S. 2. f. 1.

THE yellow wren frequents large moift woods, and places where willow trees abound from which it takes one of its names.

DESCRIP.

Its weight is about two drams. The color of the whole upper part of the body is a dufky green: the wings and tail are brown, edged with yellowifh green: above each eye is a yellowifh ftroke; the breaft, belly, and thighs vary in their color in different birds; in fome are of a bright yellow, in others it fades almoft into white.

It builds in hollows in the fides of ditches, making its neft in the form of an egg; with a large hole at the top, as an entrance: the outfide is compofed of mofs and hay, the infide lined with foft feathers. It lays commonly feven white eggs, marked with numerous fmall ruft colored fpots. It has a low plaintive note; and is perpetually creeping up and down the bodies and boughs of trees.

W. WITH

W. WITH forehead and underſide of the body of a fine pale yellow: wings of the ſame color: back and tail of a pale brown. Communicated by Mr. *Latham* of *Dartford*, who thought that it was ſhot in the Highlands of *Scotland*. It was of the ſize of a wren.

152. SCOTCH

153. GOLDEN CRESTED.

La Soulcie. *Belon av.* 345.
Tettigon. *obſ.* 12.
Regulus. *Geſner av.* 727.
Fior rancio. *Olina*, 6.
Aldr. av. II. 290.
Wil. orn. 227.
Raii ſyn. av. 79.
Edw. av. 254.
Cat. Carol. app. 36, 37.
Kratlich. *Scopoli*, No. 240.
Le Poul, ou Souci, ou Roitelet hupé, Calendula. *Briſſon av.* III. 579.
Motacilla regulus. *Lin. ſyſt.* 338.
Kongsfogel. *Faun. Suec. ſp.* 262.
Sommer Zaunkoenig (Summer Wren). *Friſch*, I. 24.
Goldhannel. *Kram.* 378.
Fugle-Konge. *Br.* 285.
Br. Zool. 101. plate S. f. 3.

DESCRIP.

THIS is the leſt of the *Britiſh* birds, weighing only ſeventy-ſix grains. Its length is three inches and a half; the breadth five inches: it may readily be diſtinguiſhed from all other birds, not only by its ſize, but by the beautiful ſcarlet mark on the head, bounded on each ſide by a fine yellow line. The bill is duſky: the feathers of the forehead are green: from the bill to the eyes is a narrow white line: the back and the hind part of the

the neck are of a dull green: the coverts of the wings dusky, edged with green and tipt with white: the quil-feathers and tail dusky, edged with pale green. The throat and lower part of the body white, tinged with green: the legs dull yellow: the claws very long. It frequents woods, and is found principally in oak trees. Though ſo ſmall a bird it indures our winters, for we have frequently ſeen it later than *Chriſtmas.* It is ſeen in autumn as far north as the *Shetland Iſles*, but quits the country before winter; a vaſt flight for ſo minute and delicate a bird.

We have obſerved this bird ſuſpended in the air for a conſiderable time over a buſh in flower, whilſt it ſung very melodiouſly. The note does not much differ from that of the common wren, but is very weak.

354. WREN. Roytelet, Bœuf de Dieu, & Berichot. *Belon av.* 343.
Trilato, *obſ.* 12.
Paſſer troglodytes. *Geſner av.* 651.
Aldr. av. II. 292.
Reatino. *Olina,* 6.
Wil. orn. 229.
Raii ſyn. av. 80.
Streſch; Storſchek. *Scopoli,* No. 239.
Le Roitelet, Regulus. *Briſſon av.* III. 425.
Motacilla troglodytes. *Lin. ſyſt.* 337.
Faun. Suec. ſp. 261.
Nelle-Konge. *Brunnich,* 284.
Schneekoning, Konickerl, Zaunſchlupfrel. *Kram.* 378.
Schneekoenig (Snow king). *Friſch,* I. 24.
Br. Zool. 102.

THE wren may be placed among the fineſt of our ſinging birds. It continues its ſong through-

throughout the winter, excepting during the frosts. It makes its nest in a very curious manner; of an oval shape, very deep, with a small hole in the middle for ingress and egress: the external material is moss, within it is lined with hair and feathers. It lays from ten to eighteen eggs; and as often brings up as many young; which, as Mr. *Ray* observes, may be ranked among those daily miracles that we take no notice of; that it should feed such a number without passing over one, and that too in utter darkness.

DESCRIP.

The head and upper part of the body of the wren are of a deep reddish brown: above each eye is a stroke of white: the back, and coverts of the wings, and tail, are marked with slender transverse black lines: the quil-feathers with bars of black and red. The throat is of a yellowish white. The belly and sides crossed with narrow dusky and pale reddish brown lines. The tail is crossed with dusky bars.

155. SEDGE.

Willow Lark, *Br. Zool.* II. 241.
Sedge Bird. *Br. Zool.* IV. *tab.* X.
Lesser Reed Sparrow. *Wil. orn.* 144.

Passer arundinaceus minor. *Raii syn. av.* 47.
Motacilla salicaria. *Lin. syst.* 330. *Faun. Suec.* No. 249.
La Fauvette babillarde. *Brisson, av.* III. 384.

THIS species is of a slender elegant form: the bill black: the head brown, marked with

C c 3 dusky

duſky ſtreaks: over each eye a line of pure white, over that another of black: cheeks brown: throat, breaſt, and belly white; the two laſt tinged with yellow: hind part of the neck and back of a reddiſh brown: the back ſpotted with black: coverts of the tail tawny: coverts of the wings duſky, edged with pale brown: quil-feathers duſky: tail brown, cuneiform: forming a circle when ſpread: legs duſky.

It is a moſt entertaining polyglot, or mocking bird; ſitting concealed in willows or reeds, in a pleaſing but rather hurrying manner, it imitates the ſwallow, the ſky-lark, the houſe-ſparrow, *&c.* ſings all night, and ſeems to leave us before winter.

156. GRASSHOPPER.

Tit-lark, that ſings like a Graſshopper. *Wil. orn.* 209.
Alauda minima locuſtæ voce. Locuſtella, D. *Johnſon.*
Raii ſyn. av. 70.
Ray's Letters, 108.
Alauda ſepiaria, L'Alouete de Buiſſon. *Briſſon av.* III. 347.
Piep Lerche (Chirping Lark). *Friſch.* I. 16.
Alauda trivalis. *Lin. ſyſt.* 288.
Br. Zool. 95. plate Q. f. 5.

THIS bird was received out of *Shropſhire*: it is the ſame with that Mr. *Ray* deſcribes as having the note of the graſshopper, but louder and ſhriller. It is a moſt artful bird, will ſculk in the middle, and thickeſt part of the hedge, and will keep running along for a hundred yards together, nor can it be forced out but with the greateſt

2 difficulty:

difficulty: it is from this covert that it emits its note, which so much resembles the insect, from which it derives its name, as generally to be mistaken for it. In the height of summer it chirps the whole night: its sibilous note is observed to cease about the latter end of *July*.

The bill is very slender, of a dusky color: the head, and whole upper part of the body is of a greenish brown, spotted with black: the quil-feathers dusky, edged with an olive brown: the tail is very long, composed of twelve sharp pointed feathers; the two middlemost are the longest, the others on each side grow gradually shorter. The under side of the body is of a dull yellowish white, darkest about the breast: the legs are of a dirty white: the hind claw shorter, and more crooked, than is usual in the lark kind.

** With party colored Tails.

157. WHEAT-EAR.

Belon av. 352.
Oenanthe. *Gesner av.* 629.
Aldr. av. II. 332.
Wheat-ear, Fallow-smich, White-tail. *Wil. orn.* 233.
Raii syn. av. 75.
Motacilla oenanthe. *Lin. syst.* 332.
Stensquetta. *Faun. Suec. sp.* 254.
Le Cul blanc, Vitrec, ou Moteux, Vitiflora. *Brisson av.* III. 449.
Culo bianco, Fornarola, Petragnola. *Zinan.* 41.
Norvegis, Steendolp, Steen Squette, Steengylpe. *Brunnich*, 276.
Steinschwaker, Steinschnapperl. *Kram.* 374.
Bella. *Scopoli*, No. 230.
Br. Zool. 102. plate S. 1. f. 5. 6.

THE wheat-ear begins to visit us about the middle of *March*, and continues coming till the beginning of *May*: we have obſerved that the females arrive about a fortnight before the males. They frequent warrens, downs, and the edges of hills, eſpecially thoſe that are fenced with ſtone walls. They breed in the latter, in old rabbet burrows, cliffs, and frequently under old timber: their neſt is large, made of dried graſs, rabbet's down, a few feathers, and horſe hair: and they lay from ſix to eight eggs, of a light blue color.

They grow very fat in autumn, and are eſteemed a delicacy. About *Eaſtbourn* in *Suſſex* they are taken by the ſhepherds in great numbers, in ſnares made of horſe hair, placed under a long turf; being very timid birds, the motion of a cloud, or the appearance of a hawk, will drive them for ſhelter into thoſe traps, and ſo they are taken. The numbers annually enſnared in that diſtrict alone, amount to about 1840 dozen, which ſell uſually at ſix-pence *per* dozen; and what appears very extraordinary, the numbers that return the following year do not appear to be leſſened; as we are aſſured by a very intelligent perſon reſident near that place. The reaſon that ſuch a quantity are taken in the neighbourhood of *Eaſtbourn* is, that it abounds with a certain *fly* which frequents the adjacent hills, for the ſake of the wild thyme they are covered with, which is not only a favorite food of that inſect, but the plant on which it depoſites its eggs?

Wheat-ears are much fatter in a rainy ſeaſon than a dry

a dry one, for they not only feed on insects, but on earth worms, which come out of the ground in greater numbers in wet weather than in dry.

The head and back of the male are of a light grey, tinged with red: over each eye is a white line; beneath that is a broad black stroke, passing across each eye to the hind part of the head: the rump and lower half of the tail are white; the upper half black: the under side of the body is white, tinged with yellow; on the neck it inclines to red: the quil-feathers are black, edged with reddish brown. The colors of the female are more dull: it wants that black stroke across the eyes, and the bar of white on the tail is narrower. These birds disappear in *September*, at least from the northern parts of this kingdom; but in *Hampshire* many of them continue the whole winter.

Le Tarier. *Belon av.* 361.
Rubetra. *Gesner av.* 729.
Le grand Traquet, ou le Tarier. *Brisson av.* III. 432. *tab.* 24. *fig.* 1. The Male.
Wil. orn. 234.
Raii syn. av. 76.

Motacilla rubetra. *Lin. syst.* 332.
Faun. Suec. sp. 255. *Scopoli*, No. 237.
Gestettenschlager. *Kram.* 375.
Grosser Fliegenfuenger (great Fly-catcher). *Frisch*, I. 22.
Br. Zool. 103. plate S. 2. f. 3. 4.

158. WHIN-CHAT.

THIS is in the north of *England*, also a bird of passage: but we are not certain whether it quits this island, but are rather inclined to think

it

it only ſhifts its quarters: in the ſouth it continues the whole year.

DESCRIP. The head and back are of a pale reddiſh brown, regularly ſpotted with black: over each eye is a narrow white ſtroke, beneath that is a broad bed of black, which extends from the bill to the hind part of the head: the breaſt is of a reddiſh yellow: the belly paler: the quil-feathers are brown, edged with a yellowiſh brown: the upper part of the wing is marked with two white ſpots: the lower part of the tail is white, the two middle feathers excepted, which are wholly black: the upper part of the others are of the ſame color.

The colors of the female are far leſs agreeable: in lieu of the white and black marks on the cheeks, is one broad pale brown one: and the white on the wings is in far leſs quantity than that of the male.

159. STONE-CHATTER.

Le Traquet ou Groulard. *Belon av.* 360.
Rubetra. *Aldr. av.* II. 325.
Stone-ſmich, Stone-chatter, or Moortitling. *Wil. orn.* 235.
Raii ſyn. av. 76.
Le Traquet, Rubetra. *Briſſon av.* III. 428. *tab.* 23. *fig.* 1. The Male.
Pontza. *Scopoli*, No. 236.
Occhio di bue. *Zinan.* 52.
Motacilla rubicola. *Lin. ſyſt.* 332.
Criſtoffi. *Kram.* 375.
Br. Zool. 103. plate S. 2. f. 5, 6.

THIS ſpecies is common during ſummer, in gorſy grounds. In the winter they diſperſe into

into marſhes, and other places; but do not quit the iſland. It is a reſtleſs and noiſy bird, and perches frequently on ſome buſh, chattering inceſſantly. The head, neck, and throat are black; but on both ſides the latter is a white bar, ſo that it appears on firſt ſight to be encircled with white: the feathers on the back are black edged with tawny: the lower part of the back juſt above the rump is white: the end and exterior ſide of the two outmoſt feathers of the tail are of a pale ruſt-color, the reſt are black: the breaſt is of a deep reddiſh yellow; the belly of a lighter hue: the quil-feathers are duſky edged with dull red; thoſe next the body are marked with a white ſpot near their bottoms: the coverts of the wings are adorned with another. The head of the female is ferruginous ſpotted with black; and the colors in general leſs vivid. In both ſexes the legs are black; which alſo is the character of the two preceding ſpecies, as well as that next to be deſcribed.

DESCRIP.

Wil. orn. 236.
Raii ſyn. av. 77.
La Meſange cendrée, Parus cinereus. *Briſſon av.* III. 549.

Motacilla ſylvia? *Lin. ſyſt.* 330.
Kogſnetter, Meſar. *Faun. Suec. ſp.* 250.
Br. Zool. 104. plate S. f. 4.

160. WHITE-THROAT.

THIS frequents our gardens in the ſummer time; in the winter it leaves us. It builds in in low bushes near the ground, making its nest externally of the tender stalks of herbs and dry straw; the middle part of fine bents and soft grass, the inside of hair. It lays five eggs of a whitish green color, sprinkled with black spots *. Its note is continually repeated, and often attended with odd gesticulations of the wings: is harsh and displeasing: is a shy and wild bird, avoiding the haunt of man; seems of a pugnatious disposition, singing with an erected crest, and in attitudes of defiance.

DESCRIP. The head of this bird is of a brownish ash color, the throat white: the breast and belly white tinged with red; (in the female wholly white:) the back inclines to red: the lesser coverts of the wings are of a pale brown; the greater dusky, edged with tawny brown; the quil-feathers dusky, edged with reddish brown; the tail the same, except the upper part of the interior side and whole exterior side of the outmost feather, which are white: the legs are of a yellowish brown.

* *Wil. orn.*

W. WITH

DARTFORD WARBLER.

161. DARTFORD.

W. WITH reddiſh *irides:* eye-lids deep crimſon. A ſlender bill a little curved at the point: whole upper part of the head, neck, and back, of a duſky brown tinged with a dull yellow: throat, under ſide of the neck, the breaſt and belly deep ferruginous; the middle of the belly white; quil-feathers duſky edged with white: baſtard wing white: exterior ſide of the interior feather of the tail white, the reſt duſky; and long in proportion to the ſize of the bird: legs yellow.

A pair of theſe were ſhot on a common near *Dartford*, in *April* 1773, and communicated to me by Mr. *Latham*; they fed on flies, which they ſprung on from the furze buſh they ſat on, and then returned to it again.

BILL

XXV. TIT-MOUSE.

BILL ſtrait, ſhort, hard, ſtrong, ſharp-pointed, a little compreſſed.

NOSTRILS round covered with briſtles.

TONGUE as if cut at the end, terminating with two or three briſtles.

162. GREAT.

Nonette ou Meſange. *Belon av.* 376.
Parus major. *Geſner av.* 640.
Aldr. av. II. 319.
Spernuzzola, Paruſſola. *Olina*, 28.
Great Titmouſe, or Ox-eye. *Wil. orn.* 240.
Raii ſyn. av. 73.
Snitza. *Scopoli*, No. 242.
Lin. ſyſt. 341.
Talg-oxe. *Faun. Suec. ſp.* 265.
Le groſſe Meſange ou la Charbonniere. *Briſſon av.* III. 539.
Pl. Enl. 3. f. 1.
Muſvit. *Brunnich*, 287.
Kohlmeiſe. *Kram.* 378. *Friſch*, I. 13.
Br. Zool. 113. plate W. f. 4.

THIS ſpecies ſometimes viſits our gardens; but chiefly inhabits woods, where it builds in hollow trees, laying about ten eggs. This, and the whole tribe feed on inſects, which they find in the bark of trees; in the ſpring they do a great deal of miſchief in the fruit garden, by picking off the tender buds. Like wood-peckers they are perpetually running up and down the bodies of trees in queſt of food. The bird has three chearful notes, which it begins to utter in the month of *February*.

DESCRIP.

The head and throat of this ſpecies are black; the cheeks white; the back green the belly of a yellowiſh

LVII. Nris 162 &c.

1 GREAT, 2 BLUE, 3 COLE, 4 MARSH TITMOUSE.

yellowiſh green, divided in the middle by a bed of black, which extends to the vent; the rump is of a bluiſh grey. The quil-feathers are duſky, edged partly with blue, partly with white: the coverts blue; the greater tipt with white. The exterior ſides of the outmoſt feathers of the tail are white: the exterior ſides of the other bluiſh: their interior ſides duſky: the legs lead color. Toes divided to the origin; and the back toe of the whole genus very large and ſtrong.

163. BLUE.

Belon av. 369.
Parus cœruleus. *Geſner av.* 641.
Aldr. av. II 321.
Blue Titmouſe, or Nun. *Wil. orn.* 242.
Raii ſyn. av. 74.
La Meſange Bleue. *Briſſon av.* III. 544.
Blava ſnitza, Blau mandlitz. *Scopoli*, No. 244.
Pl. Enl. 3. f. 2.
Parozolino, o Fratino. *Zinan.* 76.
Lin. ſyſt. 341.
Blamees. *Faun. Suec. ſp.* 267.
Blaaemeiſe. *Br.* 288.
Blaumeiſe. *Kram.* 379. *Friſch*, I. 14.
Br. Zool. 114. plate W. f. 5.

THIS bird frequents gardens, and does great injury to fruit trees, by bruiſing the young buds in ſearch of the inſects that lurk under them; it breeds in holes of walls, and lays about twelve or fourteen eggs.

DESCRIP.

It is a very beautiful ſpecies, the bill is ſhort and duſky: the crown of the head of a fine blue: from the bill to the eyes is a black line: the forehead

forehead and cheeks are white: the back is of a yellowiſh green: the lower ſide of the body yellow: the wings and tail blue, the former marked tranſverſely with a white bar: the legs of a lead color.

164. COLE. Quatrieſme eſpece de Meſange. *Belon av.* 370.
Parus ater. *Geſner av.* 641.
Aldr. av. II. 321.
Wil. orn. 241.
Raii ſyn. av. 73.
Speermieſe, Creuzmeiſe. *Kram.* 379.
Tannen Meiſe (Pine Titmouſe) *Friſch*, I. 13.
La Meſange a teſte noire, Parus atricapillus. *Briſſon av.* III. 551.
Cat. Carol. app. 37.
P. ater. *Lin. ſyſt.* 341.
Faun. Suec. ſp. 268. *Scopoli*, No. 245.
Br. Zool. 114.

DESCRIP. THE head of the colemouſe is black, marked on the hind part with a white ſpot; the back is of a greeniſh grey; the rump more green; the tail and wings duſky; the exterior feathers edged with green; the coverts of the wings are of a duſky green; the loweſt tipt with white. For a farther account we beg leave to refer to the next deſcription.

Parus

165. MARSH.

Parus paluſtris. *Geſner av.* 641.
Paronzino. *Aldr. av.* II. 32.
Marſh Titmouſe, or Black-cap. *Wil. orn.*
Raii ſyn. av. 73.
Frattino paluſtre. *Zinan.* 77.
La Meſange de Marais ou la Nonette cendrée. *Briſſon av.* III. 555.
Pl. enl. 3. f. 3.
Lin. ſyn. 341.
Entita, Tomlinge. *Faun. Suec. ſp.* 269. *Scopoli,* No. 246.
Aſch Meiſe (Aſh Titmouſe) *Friſch,* I. 13.
Hundſmeiſe. *Kram.* 379.
Norvegis Graae-Meiſe. *Brunnich.* 190.
Br. Zool. 114. plate W. f. 3.

THIS ſpecies is called by *Geſner* the marſh titmouſe; becauſe it frequents wet places. With us they inhabit woods, with the laſt kind; and ſeldom infeſt our gardens: early in *February* it emits two notes, not unlike the whetting of a ſaw.

Mr. *Willughby* obſerves, that this bird differs from the former in theſe particulars, 1ſt, that it is bigger: 2d, that it wants the white ſpot on the head: 3d, it has a larger tail: 4th, its under ſide is white: 5th, it has leſs black under the chin: 6th, it wants the white ſpot on the coverts of the wings. This laſt diſtinction does not hold in general, as the ſubject figured in the *Britiſh Zoology* had thoſe ſpots; yet wanted that on the hind part of the head.

166. LONG TAILED.

Belon av. 368.
Parus caudatus. *Gesner av.* 642.
Monticola. *Aldr. av.* II. 319.
Wil. orn. 242.
Raii syn. av. 74.
Pendolino, Paronzino. *Zinan.* 77.
Gaugartza. *Scopoli,* No. 247.

La Mesange a longue queue, Parus longicaudus. *Brisson av.* III. 570.
Lin. syst. 342.
Alhtita. *Faun. Suec. sp.* 83.
Belzmeise Pfannenstiel. *Kram.* 379.
Langschwaentzige Meise. *Frisch,* I. 14.

DESCRIP.

THE length is five inches and a quarter; the breadth seven inches. The bill is black, very short, thick, and very convex, differing greatly from all others of the titmouse kind: the base is beset with small bristles: the irides are of a hazel color. The top of the head, from the bill to the hind part, is white, mixed with a few dark grey feathers; this bed of white is entirely surrounded with a broad stroke of black, which, rising on each side the upper mandible, passes over each eye, unites at the hind part of the head; and continues along the middle of the back to the rump: the feathers on each side of this black stroke are of a purplish red, as are those immediately incumbent on the tail. The covert feathers of the wings are black: the secondary and quil-feathers are dusky, the largest of the latter are wholly so; the lesser and more remote have their exterior sides edged with white.

The

The tail is the longeſt in proportion to the bulk of any *Britiſh* bird, being in length three inches; the form of it is like that of a magpie, conſiſting of twelve feathers of unequal lengths, the middlemoſt the longeſt, thoſe on each ſide growing gradually ſhorter; the exterior ſides, and the top of the interior ſides of the three outmoſt feathers are white; the reſt of the tail black. The cheeks and throat are white: the breaſt and whole under ſide white, with a caſt of red. The legs, feet, and claws are black.

It forms its neſt with great elegance, of an oval ſhape, and about eight inches deep; near the upper end is a hole for admiſſion: the external materials are moſſes and lichens, curiouſly interwoven with wool; within it is lined very warmly with a thick bed of feathers: it lays from ten to ſeventeen eggs. The young follow the parents the whole winter; and from the ſlimneſs of their bodies, and great length of tail, appear, while flying, like ſo many darts cutting the air. They are often ſeen paſſing through our gardens, going progreſſively from tree to tree, as if in their road to ſome other place, never making any halt.

167. BEARD-ED. Least Butcher Bird. *Edw. av.* 55. Bearded Titmouſe. *Aldr. av.* I. *tab.* 48. *Scopoli*, No. 241. La meſange barbue, ou le mouſtache, Parus barbatus. *Briſſon av.* III. 567. Parus biarmicus. *Lin. ſyſt.* 342. *Br. Zool.* 74. plate C. 2. Least Butcher Bird. *Br. Zool.* Ed. 2d. I. 165.

THIS ſpecies is found in the marſhes near *London:* we have ſeen it near *Glouceſter*; it is alſo frequent among the great tracts of reeds near *Cowbit* in *Lincolnſhire*, where I ſuſpect they breed. It is of the ſame ſhape as the long tailed titmouſe, but rather larger. The bill is ſhort, ſtrong, and very convex; of a box color: *irides* pale yellow: the head is of a fine grey: on each ſide of the bill, beneath the eye, is a long triangular tuft of black feathers: the chin and throat are white: the middle of the breaſt fleſh colored: the ſides and thighs of a pale orange: the hind part of the neck and the back are of an orange bay: the ſecondary feathers of the wings are black edged with orange: the quil-feathers duſky on their exterior, white on their interior ſides: the leſſer quil-feathers tipt with orange. The tail is two inches three quarters long: the two middle feathers of the tail are largeſt, the others gradually ſhorten on each ſide; the outmoſt of which are of a deep orange color. The vent-feathers of the male of a pale black: of the female

DESCRIP.

female of a dull orange. The legs are of a deep ſhining black.

FEMALE.

The *female* wants the black mark on each cheek, and the fine fleſh color on the breaſt: the crown of the head is of a browniſh ruſt color, ſpotted with black: the outmoſt feathers of the tail are black tipt with white.

XXVI
SWALLOW.

Short weak BILLS.
Very wide MOUTHS.
Short weak LEGS.

168. Chimney.

La petite Hirondelle. *Belon av.* 378.
Hirundo domestica. *Gesner av.* 548.
Aldr. av. II. 294.
House or Chimney Swallow. *Wil. orn.* 212.
Raii syn. av. 71.
Rondone. *Zinan.* 47.
L' Hirondelle de Cheminée. *Brisson av.* II. 486.
Hirundo rustica. *Lin. syst.* 343.
Ladu-Swala. *Faun. Suec. sp.* 270.
Forstue-Svale, Mark-Svale. *Brunnich,* 289.
Haus-Schwalbe. *Frisch,* I. 17.
Hauss Schwalbe. *Kram.* 380.
Br. Zool. 96.
Laustaza. *Scopoli,* No. 249.

THIS species appears in *Great Britain* near twenty days before the martin, or any other of the swallow tribe. They leave us the latter end of *September*; and for a few days previous to their departure, they assemble in vast flocks on house tops, churches, and even trees, from whence they take their flight. It is now known that swallows take their winter quarters in *Senegal*, and possibly they may be found along the whole *Morocco* shore. We are indebted to M. *Adanson** for this discovery, who first observed them in the

* Voyage to *Senegal*, p. 121. 163.

month

l. LVIII
No 168
SWALLOW.
SWIFT
No 171

month of *October*, after their migration out of *Europe*, on the ſhores of that kingdom: but whether it was this ſpecies alone, or all the *European* kinds, he is ſilent.

The name of chimny ſwallow may almoſt be confined to *Great Britain*, for in ſeveral other countries they chuſe different places for their neſts. In *Sweden*, they prefer barns, ſo are ſtyled there *Ladu-Swala*, or the barn ſwallow: and in the hotter climates, they make their neſts in porches, gateways, galleries, and open halls.

The houſe ſwallow is diſtinguiſhed from all others by the ſuperior forkineſs of its tail, and by the red ſpot on the forehead, and under the chin.

DESCRIP.

The crown of the head, the whole upper part of the body, and the coverts of the wings are black, gloſſed with a rich purpliſh blue, moſt reſplendent in the male: the breaſt and belly white, that of the male tinged with red: the tail black; the two middle feathers plain: the others marked tranſverſely near their ends with a white ſpot. The exterior feathers of the tail are much longer in the male than in the female.

Its food is the ſame with the others of its kind, viz. inſects; for the taking of which in their ſwifteſt flight, nature hath admirably contrived their ſeveral parts; their mouths are very wide to take in flies, &c. in their quickeſt motion; their wings are long, and adapted for diſtant and continual flight; and their tails are forked, to enable them to turn the readier in purſuit of their prey. This ſpecies, in our country, builds in chimnys, and makes its neſt of clay mixed with ſtraw, leaving the top quite open. It lines the bottom with feathers and graſſes: and uſually lays from four to ſix eggs, white ſpeckled with red; but by taking away one of the eggs daily, it will ſucceſſively lay as far as nineteen, as Doctor *Liſter* has experienced. It breeds earlier than any other ſpecies. The firſt brood are obſerved to quit the neſt the laſt week in *June*, or the firſt in *July*: the laſt brood towards the middle or end of *Auguſt*. The neſt being fixed five or ſix feet deep within the chimny, it is with difficulty that the young can emerge. They even ſometimes fall into the rooms below: but as ſoon as they ſucceed, they perch for a few days on the chimny top, and are there fed by their parents. Their next eſſay is to reach ſome leafleſs bough, where they ſit in rows, and receive their food. Soon after they take to the wing, but ſtill want ſkill to take their own prey. They hover near the place where their parents are in chaſe of flies, attend their motions, meet them, and receive from their mouths the offered ſuſtenance.

It has a ſweet note, which it emits in *Auguſt* and *September*, perching on houſe tops.

Lo

Le Martinet. *Belon av.* 380.
Hirundo ſylveſtris. *Geſner av.* 564. *Friſch,* I. 17.
Aldr. av. II. 311.
Martin, Martlet, or Martinet. *Wil. orn.* 213.
Raii ſyn. av. 71.
Rondone minore, e Graſſolo. *Zinan.* 48.
Huda urnik. *Scopoli,* No. 250.
La petite Hirondelle, ou le Martinet a cul blanc. *Briſſon av.* II. 490.
Hirundo urbica. *Lin. ſyſt.* 344.
Hus-Swala. *Faun. Suec. ſp.* 271.
Speyerl. *Kram.* 380.
Danis Bye v. Tagſkiœg-Svale, *Langelandis,* Rive. *Br.* 290.
Br. Zool. 96. plate Q. f. 2.
Ph. Tr. 1774. p. 196.

169. MARTIN.

DESCRIP.

THE *Martin* is inferior in ſize to the former, and its tail much leſs forked. The head and upper part of the body, except the rump, is black gloſſed with blue: the breaſt, belly and rump are white: the feet are covered with a ſhort white down. This is the ſecond of the ſwallow kind that appears in our country. It builds under the eaves of houſes, with the ſame materials, and in the ſame form as the houſe ſwallow, only its neſt is covered above, having only a ſmall hole for admittance. We have alſo ſeen this ſpecies build againſt the ſides of high cliffs over the ſea. For the time that the young keep the neſt, the old one feeds them, adhering by the claws to the outſide: but as ſoon as they quit it, feeds them flying, by a motion quick and almoſt imperceptible to thoſe who are not uſed to obſerve it.

It is a later breed than the preceding by ſome days:

days: but both will lay twice in the ſeaſon; and the latter brood of this ſpecies have been obſerved to come forth ſo late as the eighteenth of *September*; yet that year (1766) they entirely quitted our ſight by the fifth of *October*; not but they ſometimes continue here much later: the martins and red wing thruſhes having been ſeen flying in view on the ſeventh of *November*. Neſtlings have been remarked in *Hampſhire* as late as the 21ſt. of *October*, 1772.

170. SAND.

L' Hirondelle de rivage. *Belon av.* 379.
Hirundo riparia, ſeu Drepanis. *Geſner av.* 565.
Dardanelli. *Aldr. av.* II. 312.
Sand Martin, or Shore Bird. *Wil. orn.* 213.
Raii ſyn. av 71.
L' Hirondelle de rivage. *Briſſon av.* II. 506.
Cat. Carol. app. 37.
Rondone riparia. *Zinan.* 49.
Hirundo riparia. *Lin. ſyſt.* 344.
Strand-ſwala, Back ſwala. *Faun. Suec. ſp.* 273.
Danis Dig-v. Jord-ſvale, Soil-bakke. *Norveg.* Sand Rænne. *Br.* 291.
Ufer-Schwalbe (Shore Swallow) *Friſch*, I. 18.
Geſtetten-ſchwalbe. *Kram.* 381.
Br. Zool. 97. plate Q. f. 1.

DESCRIP.

THIS is the leſt of the genus that frequents *Great Britain.* The head and whole upper part of the body are mouſe colored: the throat white, encircled with a mouſe colored ring: the belly white: the feet ſmooth and black.

It builds in holes in ſand pits, and in the banks of rivers, penetrating ſome feet deep into the bank, boring

boring through the ſoil in a wonderful manner with its feet, claws, and bill. It makes its neſt of hay, ſtraw, &c. and lines it with feathers: it lays five or ſix white eggs. It is the earlieſt of the ſwallow tribe in bringing out its young.

171. SWIFT.

La grande Hirondelle, Moutardier ou grand Martinet. *Belon av.* 377.
Apus. *Geſner av.* 166.
Aldr. av. II. 312.
Black Martin, or Swift. *Wil. orn.* 214.
Raii ſyn. av. 72.
Rondone. *Zinan.* 47.
Le Martinet. *Briſſon av.* II. 514.
Hirundo apus. *Lin. ſyſt.* 344.
Ring-ſwala. *Faun. Suec. ſp.* 272.
Steen, Kirke-v. Sæe-Svale. *Br.* 292.
Speyer, groſſe thurn ſchwalbe. *Kram.* 380. *Scopoli,* No. 251.
Br. Zool. 97.

THIS ſpecies is the largeſt of our ſwallows; but the weight is moſt diſproportionately ſmall to its extent of wing of any bird; the former being ſcarce one ounce, the latter eighteen inches. The length near eight. The feet of this bird are ſo ſmall, that the action of walking and of riſing from the ground is extremely difficult; ſo that nature hath made it full amends, by furniſhing it with ample means for an eaſy and continual flight. It is more on the wing than any other ſwallows; its flight is more rapid, and that attended with a ſhrill ſcream. It reſts by clinging againſt ſome wall, or other apt body; from whence *Klein* ſtyles this ſpecies

cies *Hirundo muraria*. It breeds under the eaves of houfes, in fteeples, and other lofty buildings; makes its neft of graffes and feathers; and lays only two eggs, of a white color. It is entirely of a gloffy dark footy color, only the chin is marked with a white fpot: but by being fo conftantly expofed to all weathers, the glofs of the plumage is loft before it retires. I cannot trace them to their winter quarters, unlefs in one inftance of a pair found adhering by their claws and in a torpid ftate, in *February* 1766, under the roof of *Longnor Chapel, Shropfhire*: on being brought to a fire, they revived and moved about the room. The feet are of a particular ftructure, all the toes ftanding forward; the left confifts of only one bone; the others of an equal number, viz. two each; in which they differ from thofe of all other birds.

Descrip.

This appears in our country about fourteen days later than the fand martin; but differs greatly in the time of its departure, retiring invariably about the tenth of *Auguft*, being the firft of the genus that leaves us.

The fabulous hiftory of the *Manucodiata*, or bird of *Paradife*, is in the hiftory of this fpecies in great meafure verified. It was believed to have no feet, to live upon the celeftial dew, to float perpetually on the *Indian*, and to perform all its functions in that element.

The Swift actually performs what has been in thefe enlightened times difproved of the former; except

except the ſmall time it takes in ſleeping, and what it devotes to incubation, every other action is done on wing. The materials of its neſt it collects either as they are carried about by the winds, or picks them up from the ſurface in its ſweeping flight. Its food is undeniably the inſects that fill the air. Its drink is taken in tranſient ſips from the water's ſurface. Even its amorous rites are performed on high. Few perſons who have attended to them in a fine ſummer's morning, but muſt have ſeen them make their aerial courſes at a great height, encircling a certain ſpace with an eaſy ſteady motion. On a ſudden they fall into each other's embraces, then drop precipitate with a loud ſhriek for numbers of yards. This is the critical conjuncture, and to be no more wondered at, than that inſects (a familiar inſtance) ſhould diſcharge the ſame duty in the ſame element.

Theſe birds and ſwallows are inveterate enemies to hawks. The moment one appears, they attack him immediately: the ſwifts ſoon deſiſt; but the ſwallows purſue and perſecute thoſe rapacious birds, till they have entirely driven them away.

Swifts delight in ſultry thundry weather, and ſeem thence to receive freſh ſpirits. They fly in thoſe times in ſmall parties with particular violence; and as they paſs near ſteeples, towers, or any edifices where their mates perform the office of incubation, emit a loud ſcream, a ſort of ſerenade, as Mr. *White* ſuppoſes, to their reſpective females.

To

To the curious monographics on the ſwallow tribe, of that worthy correſpondent, I muſt acknowlege myſelf indebted for numbers of the remarks above-mentioned.

OF THE DISAPPEARANCE OF SWALLOWS.

THERE are three opinions among naturaliſts concerning the manner the ſwallow tribes diſpoſe of themſelves after their diſappearance from the countries in which they make their ſummer reſidence. *Herodotus* mentions one ſpecies that reſides in *Egypt* the whole year: *Proſper Alpinus* * aſſerts the ſame; and Mr. *Loten*, late governor of *Ceylon*, aſſured us, that thoſe of *Java* never remove. Theſe excepted, every other known kind obſerve a periodical migration, or retreat. The ſwallows of the cold *Norway* †, and of *North America* ‡, of the diſtant *Kamtſchatka* §, of the temperate parts of *Europe*, of *Aleppo* ||, and of the hot *Jamaica* **, all agree in this one point.

* Hirundines duplicis generis ibi obſervantur; patriæ ſcilicet quæ nunquam ab *Ægypto* diſcedentes, ibi perpetuo morantur, atque peregrinæ, hæ ſunt noſtratibus omnino ſimiles; patriæ vero toto etiam ventre nigricant. *Hiſt. Ægypt.* I. 198.

† *Pontop. hiſt. Norw.* II. 98.

‡ *Cat. Carol.* I. 51. *app.* 8.

§ *Hiſt. Kamtſ.* 162.

|| *Ruſſel Alep.* 70.

** *Phil. Tranſ.* No. 36.

In

In cold countries, a defect of insect food on the approach of winter, is a sufficient reason for these birds to quit them: but since the same cause probably does not subsist in the warm climates, recourse should be had to some other reason for their vanishing.

Of the three opinions, the first has the utmost appearance of probability; which is, that they remove nearer the sun, where they can find a continuance of their natural diet, and a temperature of air suiting their constitutions. That this is the case with some species of *European* swallows, has been proved beyond contradiction (as above cited) by M. *Adanson*. We often observe them collected in flocks innumerable on churches, on rocks, and on trees, previous to their departure hence; and Mr. *Collinson* proves their return here in perhaps equal numbers, by two curious relations of undoubted credit: the one communicated to him by Mr. *Wright*, master of a ship; the other by the late Sir *Charles Wager*; who both described (to the same purpose) what happened to each in their voyages. "Returning home, says Sir *Charles*, in the "spring of the year, as I came into sounding in our "channel, a great flock of swallows came and settled on all my rigging; every rope was covered; "they hung on one another like a swarm of bees; "the decks and carving were filled with them. "They seemed almost famished and spent, and were "only feathers and bones; but being recruited "with

"with a night's reſt, took their flight in the morn-"ing"*. This vaſt fatigue, proves that their journey muſt have been very great, conſidering the amazing ſwiftneſs of theſe birds: in all probability they had croſſed the *Atlantic* ocean, and were returning from the ſhores of *Senegal*, or other parts of *Africa*; ſo that this account from that moſt able and honeſt ſeaman, confirms the later information of M. *Adanſon*.

Mr. *White*, on *Michaelmas* day 1768, had the good fortune to have ocular proof of what may reaſonably be ſuppoſed an actual migration of ſwallows. Travelling that morning very early between his houſe and the coaſt, at the beginning of his journey he was environed with a thick fog, but on a large wild heath the miſt began to break, and diſcovered to him numberleſs ſwallows, cluſtered on the ſtanding buſhes, as if they had rooſted there: as ſoon as the ſun burſt out, they were inſtantly on wing, and with an eaſy and placid flight proceeded towards the ſea. After this he ſaw no more flocks, only now and then a ſtraggler †.

* *Phil. Tranſ.* Vol. LI. Part 2. p. 459.

† In *Kalm*'s Voyage to *America*, is a remarkable inſtance of the diſtant flight of ſwallows; for one lighted on the ſhip he was in, *September* 2d. when he had paſſed only over two thirds of the *Atlantic* ocean. His paſſage was uncommonly quick, being performed from *Deal* to *Philadelphia* in leſs than ſix weeks; and when this accident happened, he was fourteen days ſail from *Cape Hinlopen*.

This

This rendevouz of ſwallows about the ſame time of year is very common on the willows, in the little iſles in the *Thames*. They ſeem to aſſemble for the ſame purpoſe as thoſe in *Hampſhire*, notwithſtanding no one yet has been eye witneſs of their departure. On the 26th of *September* laſt, two Gentlemen who happened to lie at *Maidenhead bridge*, furniſhed at leſt a proof of the multitudes there aſſembled: they went by torch-light to an adjacent iſle, and in leſs than half an hour brought aſhore fifty dozen; for they had nothing more to do than to draw the willow twigs through their hands, the birds never ſtirring till they were taken.

The northern naturaliſts will perhaps ſay, that this aſſembly met for the purpoſe of plunging into their ſubaqueous winter quarters; but was that the caſe, they would never eſcape diſcovery in a river perpetually fiſhed as the *Thames*, ſome of them muſt inevitably be brought up in the nets that haraſs that water.

The ſecond notion has great antiquity on its ſide. *Ariſtotle* * and *Pliny* † give, as their belief, that ſwallows do not remove very far from their ſummer habitation, but winter in the hollows of rocks, and during that time loſe their feathers. The former part of their opinion has been adopted by ſeveral ingenious men; and of late, ſeveral proofs have been brought of ſome ſpecies, at leſt,

* *Hiſt. an.* 935.
† *Lib.* 10. *c.* 24.

having been discovered in a torpid state. Mr. *Collinson** favored us with the evidence of three gentlemen, eye-witnesses to numbers of *sand martins* being drawn out of a cliff on the *Rhine*, in the month of *March* 1762 †. And the Honorable *Daines Barrington* communicated to us the following fact, on the authority of the late Lord *Belhaven*, that numbers of swallows have been found in old dry walls, and in sandhills near his Lordship's seat in *East Lothian*; not once only, but from year to year; and that when they were exposed to the warmth of a fire, they revived. We have also heard of the same annual discoveries near *Morpeth* in *Northumberland*, but cannot speak of them with the same assurance as the two former: neither in the two last instances are we certain of the particular species ‡.

Other witnesses crowd on us to prove the residence of those birds in a torpid state during the severe season.

First, In the chalky cliffs of *Sussex*; as was seen on the fall of a great fragment some years ago.

Secondly, In a decayed hollow tree that was cut down, near *Dolgelli*, in *Merionethshire*.

Thirdly, In a cliff near *Whitby*, *Yorkshire*; where,

* By letter, dated *June* 14, 1764.

† *Phil. Trans.* Vol. LIII. p. 101. art. 24.

‡ *Klein* gives an instance of *swifts* being found in a torpid state, *Hist. av.* 204.

on

on digging out a fox, whole bushels of swallows were found in a torpid condition. And,

Lastly, The Reverend Mr. *Conway*, of *Sychton*, *Flintshire*, was so obliging as to communicate the following fact: A few years ago, on looking down an old lead mine in that county, he observed numbers of swallows clinging to the timbers of the shaft, seemingly asleep; and on flinging some gravel on them, they just moved, but never attempted to fly or change their place; this was between *All Saints* and *Christmas*.

These are doubtless the lurking places of the latter hatches, or of those young birds, who are incapable of distant migrations. There they continue insensible and rigid; but like flies may sometimes be reanimated by an unseasonable hot day in the midst of winter: for very near *Christmas* a few appeared on the moulding of a window of *Merton College*, *Oxford*, in a remarkably warm nook, which prematurely set their blood in motion, having the same effect as laying them before the fire at the same time of year. Others have been known to make this premature appearance; but as soon as the cold natural to the season returns, they withdraw again to their former retreats.

I shall conclude with one argument drawn from the very late hatches of two species.

On the twenty-third of *October* 1767, a *martin* was seen in *Southwark*, flying in and out of its nest: and on the twenty-ninth of the same month, four or five *swallows* were observed hovering round and settling on the county hospital at *Oxford*. As these birds must have been of a late hatch, it is highly improbable that at so late a season of the year, they would attempt from one of our midland counties, a voyage almost as far as the equator to *Senegal* or *Goree:* we are therefore confirmed in our notion, that there is only a partial migration of these birds; and that the feeble late hatches conceal themselves in this country.

The above, are circumstances we cannot but assent to, though seemingly contradictory to the common course of nature in regard to other birds. We must, therefore, divide our belief relating to these two so different opinions, and conclude, that one part of the swallow tribe migrate, and that others have their winter quarters near home. If it should be demanded, why swallows alone are found in a torpid state, and not the other many species of soft billed birds, which likewise disappear about the same time? The following reason may be assigned:

No birds are so much on the wing as swallows, none fly with such swiftness and rapidity, none are obliged to such sudden and various evolutions in their flight, none are at such pains to take their prey, and we may add, none exert their voice more incessantly; all these occasion a vast expence of strength, and of spirits, and may give such a texture to the blood, that other animals cannot experience;

ence; and ſo diſpoſe, or we may ſay, neceſſitate, this tribe of birds, or part of them, at leſt, to a repoſe more laſting than that of any others.

The third notion is, even at firſt ſight, too amazing and unnatural to merit mention, if it was not that ſome of the learned have been credulous enough to deliver, for fact, what has the ſtrongeſt appearance of impoſſibility; we mean the relation of ſwallows paſſing the winter immerſed under ice, at the bottom of lakes, or lodged beneath the water of the ſea at the foot of rocks. The firſt who broached this opinion, was *Olaus Magnus*, Archbiſhop of *Upſal*, who very gravely informs us, that theſe birds are often found in cluſtered maſſes at the bottom of the northern lakes, mouth to mouth, wing to wing, foot to foot; and that they creep down the reeds in autumn, to their ſubaqueous retreats. That when old fiſhermen diſcover ſuch a maſs, they throw it into the water again; but when young inexperienced ones take it, they will, by thawing the birds at a fire, bring them indeed to the uſe of their wings, which will continue but a very ſhort time, being owing to a premature and forced revival *.

That the good Archbiſhop did not want credulity, in other inſtances, appears from this, that after having ſtocked the bottoms of the lakes with birds, he ſtores the clouds with mice, which ſome-

* *Derham's Phyſ. Theol.* note *d.* p. 349. *Pontop. hiſt. Norw.* I. 99.

times

times fall in plentiful ſhowers on *Norway* and the neighboring countries *.

Some of our own countrymen have given credit to the ſubmerſion of ſwallows †; and *Klein* patroniſes the doctrine ſtrongly, giving the following hiſtory of their manner of retiring, which he received from ſome countrymen and others. They aſſerted, that ſometimes the ſwallows aſſembled in numbers on a reed, till it broke and ſunk with them to the bottom; and their immerſion was preluded by a dirge of a quarter of an hour's length. That others would unite in laying hold of a ſtraw with their bills, and ſo plunge down in ſociety. Others again would form a large maſs, by clinging together with their feet, and ſo commit themſelves to the deep ‡.

Such are the relations given by thoſe that are fond of this opinion, and though delivered without exaggeration, muſt provoke a ſmile. They aſſign not the ſmalleſt reaſon to account for theſe birds being able to endure ſo long a ſubmerſion without being ſuffocated, or without decaying, in an element ſo unnatural to ſo delicate a bird;

* *Geſner Icon. An.* 100.

† *Derham's Phyſ. Theol.* 340. 349. *Hildrop's Tracts*, II. 32.

‡ *Klein hiſt. av.* 205, 206. *Ekmarck* migr. av. Amæn. acad. IV. 589.

when

when we know that the otter *, the corvorant, and the grebes, ſoon periſh, if caught under ice, or entangled in nets: and it is well known, that thoſe animals will continue much longer under water than any others to whom nature hath denied that particular ſtructure of heart, neceſſary for a long reſidence beneath that element.

* Though entirely ſatisfied in our own mind of the impoſſibilty of theſe relations; yet, deſirous of ſtrengthening our opinion with ſome better authority, we applied to that able anatomiſt, Mr. *John Hunter*; who was ſo obliging to inform us, that he had diſſected many ſwallows, but found nothing in them different from other birds as to the organs of reſpiration. That all thoſe animals which he had diſſected of the claſs that ſleep during winter, ſuch as lizards, frogs, &c. had a very different conformation as to thoſe organs. That all theſe animals, he believes, do breathe in their torpid ſtate; and, as far as his experience reaches, he knows they do: and that therefore he eſteems it a very wild opinion, that terreſtrial animals can remain any long time under water without drowning.

BILL

XXVII. GOAT-SUCKER.

BILL very ſhort, bent at the end, briſtles round the baſe.

NOSTRILS tubular, very prominent.

TAIL conſiſting of ten feathers, not forked.

172. NOCTURNAL.

L' Effraye ou Freſaye. *Belon av.* 343.
Caprimulgus, Geiſſmelcher. *Geſner av.* 241.
Calcobotto. *Aldr. av.* I. 288.
Fern Owl, Goatſucker, Goat Owl. *Wil. orn.* 107. Alſo,
Churn Owl. *Raii ſyn. av.* 26 *Cat. Carolin.* I. 8.
Dorhawk, accipiter Cantharophagus. *Charlton ex.* 79.
Le Tette Chevre ou Crapaud volant. *Briſſon av.* II. 470. *Tab.* 44.
Covaterra. *Zinanni,* 94. *Scopoli,* No. 254.
Caprimulgus europeus. *Lin. ſyſt.* 346.
Natſkrafa, Natſkarra, Quallknarren. *Faun. Suec. ſp.* 274.
Hirundo cauda æquabili. H. caprimulga. *Klein av.* 81.
Nat-Ravn, Nat-Skade, Aften-bakke. *Brun.* 293.
Mucken ſtecker, Nachtrabb. *Kram.* 381.
Br. Zool. 97. *Tab.* R. R. 1.

KLEIN hath placed this bird in the ſwallow tribe, and ſtyles it a ſwallow with an undivided tail. It has moſt of the characters of that genus; a very ſmall bill, wide mouth, ſmall legs. It is alſo a bird of paſſage; agrees in food with this genus, and the manner of taking it: differs in the time of preying, flying only by night, ſo with ſome juſtice may be called a *nocturnal ſwallow.* It feeds on moths, knats, dorrs or chaffers; from which *Charlton* calls it a *Dorr-hawk,* its food being entirely that ſpecies of beetle during the month of *July,*

M.& F. GOATSUCKERS.

July, the period of that insect's * flight in this country.

MIGRATES.

This bird makes but a short stay with us: appears the latter end of *May*; and disappears in the northern parts of our island the latter end of *August*, but in the southern stays above a month later. It inhabits all parts of *Great Britain*, from *Cornwal* to the county of *Ross*. Mr. *Scopoli* seems to credit the report of their sucking the teats of goats, an error delivered down from the days of *Aristotle*.

NOTES.

Its notes are most singular: the loudest so much resembles that of a large spinning wheel, that the *Welsh* call this bird *aderyn y droell*, or the wheel bird. It begins its song most punctually on the close of day, sitting usually on a bare bough with the head lower than the tail, as expressed in the upper figure in the plate; the lower jaw quivering with the efforts. The noise is so very violent, as to give a sensible vibration to any little building it chances to alight on, and emit this species of note. The other is a sharp squeak, which it repeats often, this seems a note of love, as it is observed to reiterate it when in pursuit of the female among the trees.

EGGS.

It lays its eggs on the bare ground; usually two: they are of a long form, of a whitish hue, prettily marbled with reddish brown.

* *Scarabæus Melolontha.*

The

DESCRIP. The weight of this bird is two ounces and a half: length ten inches and a half: extent twenty-two. Bill very ſhort: the mouth vaſt: irides hazel.

Plumage a beautiful mixture of black, white, aſh-color and ferruginous, diſpoſed in lines, bars and ſpots. The male is diſtinguiſhed from the female by a great oval white ſpot near the end of the three firſt quil-feathers; and another on the outmoſt feathers of the tail: the plumage is alſo more ferruginous.

The legs ſhort, ſcaly and feathered below the knee: the middle toe connected to thoſe on each ſide by a ſmall membrane, as far as the firſt joint: the claw of the middle toe thin, broad, ſerrated.

END OF THE FIRST VOLUME.

BRITISH ZOOLOGY.

CLASS II. BIRDS.

DIV. II. WATER FOWL.

LONDON.
Printed for Benj. White,
MDCCLXXVI.

PLATES

TO

BRITISH ZOOLOGY.

VOL. II. OCTAVO.

Plates.

PLATES.

LXXXVIII.

PLATES.

APPENDIX.

PLATES

APPENDIX.

Plates.

DIVISION II.

WATER-FOWLS.

Pl.LXI. The **COMMON HERON.** No 17.

DIV. II. WATER FOWLS.

SECT. I. WITH CLOVEN FEET.
II. WITH FINNED FEET.
III. WITH WEBBED FEET.

XXVIII. HERON.

BILL long, ſtrong and pointed.
NOSTRILS linear.
TONGUE pointed.
TOES connected as far as the firſt joint by a ſtrong membrane.

173. COMMON.

MALE.

Heron cendrè. *Belon. av.* 182.
Alia ardea. *Geſner av.* 219.
Ardea cinerea major. *Aldr. av.* iii. 157. *Scopoli*, No. 117.
Common Heron, or Heronſhaw. *Wil. orn.* 277.
Ardea cinerea major ſeu pella. *Raii ſyn. av.* 98.
Garza cinerizia groſſa. *Zinan.* 113.
Le Heron hupé. *Briſſon av.* v. 296. *tab.* 35.
Reyger. *Friſch* II. 199.
Blauer Rager. *Kram.* 346.
Ardea major. *Lin. ſyſt.* 236.
Hager. *Faun. Suec. ſp.* 59.
The Heron. *Br. Zool.* 116. *tab.*

FEMALE.

Ardea Pella ſive cinerea. *Geſner av.* 211.
Ardea cinerea tertia. *Aldr. av.* III. 159. *Wil orn.* 279. & *Raii ſyn. av.* 98.
Ardea cinerea. *Lin. ſyſt.* 236.
Danis et *Norvegis* Heyre v. Hegre. *Cimbris* Skid-Heire Skredheire. *Brunnich*, 156.
Le Heron. *Briſſon av.* v. 292. *tab.* 34.
Reyger *Friſch*, II. 198.
Brit. Zool. 116.

THIS bird is remarkably light in proportion to its bulk, ſcarce weighing three pounds and a half: the length is three

F f 2 feet

feet two inches; the breadth five feet four inches. The body is very ſmall, and always lean; and the ſkin ſcarce thicker than what is called gold-beater's ſkin. It muſt be capable of bearing a long abſtinence, as its food, which is fiſh and frogs, cannot be readily got at all times. It commits great devaſtation in our ponds; but being unprovided with webs to ſwim, nature has furniſhed it with very long legs to wade after its prey. It perches and builds in trees, and ſometimes in high cliffs over the ſea, commonly in company with many others, like rooks. At *Creſſi Hall* near *Goſberton* in *Lincolnſhire* I have counted above eighty neſts in one tree. It makes its neſt of ſticks, lines it with wool; and lays five or ſix large eggs of a pale green color. During incubation, the male paſſes much of its time perched by the female. They deſert their neſts during winter, excepting in *February*, when they reſort to repair them. It was formerly in this country a bird of game, heron-hawking being ſo favourite a diverſion of our anceſtors, that laws were enacted for the preſervation of the ſpecies, and the perſon who deſtroyed their eggs was liable to a penalty of twenty ſhillings, for each offence. Not to know the *Hawk* from the *Heronſhaw* was an old proverb*, taken originally from this diverſion; but in courſe of time ſerved to ex-

* In after times this proverb was abſurdly corrupted to, He does not know a *hawk* from a *hand-ſaw*.

preſs

prefs great ignorance in any fcience. This bird was formerly much efteemed as a food; made a favourite difh at great tables, and was valued at the fame rate as a Pheafant. It is faid to be very long lived; by Mr. *Keyfler*'s account it may exceed fixty years *: and by a recent inftance of one that was taken in *Holland* by a hawk belonging to the ftadtholder, its longevity is again confirmed, the bird having a filver plate faftened to one leg, with an infcription, importing it had been before ftruck by the elector of *Cologne*'s hawks in 1735.

The male is a moft elegant bird: the weight about three pounds and a half, the length, three feet three; the breadth, five feet four; the bill fix inches long, very ftrong and pointed: the edges thin and rough; the color dufky above, yellow beneath; noftrils linear; the irides of a deep yellow; orbits and fpace between them and the bill covered with a bare greenifh fkin.

The forehead and crown white, the hind part of the head adorned with a loofe pendent creft of long black feathers waving with the wind; the upper part of the neck is of a pure white, and the coverts of the wings of a light grey; the back clad only with down, covered with the fcapulars; the fore part of the neck white fpotted with a double row of black: the feathers are white, long, narrow,

* *Keyfler's Travels*, I. 70.

unwebbed, falling loose over the breast; the scapulars of the same texture, grey streaked with white.

The ridge of the wing white, primaries and bastard wing black; along the sides beneath the wings is a bed of black feathers, very long, soft and elegant; in old times used as egrets for the hair, or ornaments to the caps of Knights of the garter; the breast, belly, and thighs white: the last dashed with yellow. The tail consists of twelve short cinereous feathers: the legs are of a dirty green: the toes long, the claws short, the inner edge of the middle claw finely serrated.

FEMALE

The head of the female is grey: it wants the long crest, having only a short plume of dusky feathers: the feathers above the breast short; the scapulars grey and webbed: the sides grey. This has hitherto been supposed to be a distinct species from the former; but later observations prove them to be the same.

174. BITTERN.

Le Butor. *Belon av.* 192.
Brrind, Rordump. *Gesner av.* 215.
The Myredromble. *Turner.*
Trombone, Terrabuso. *Aldr. av.* III. 164.
Bittour, Bittern, or Miredrum. *Wil. orn.* 282.
Raii syn. av. 100.
Botaurus, le Butor. *Brisson av.* V. 444. *tab.* 37.
Garza bionda, o di color d'oro. *Zinan.* 112. *Scopoli,* No. 125.
Rohrtrummel, Mosskuh. *Kram.* 348.
Rohrdommel. *Frisch,* II. 205.
Ardea stellaris. *Lin. syst.* 239.
Rordrum. *Faun. Suec. sp.* 164.
Danis Rordrum. *Brunnich,* 155.
Br. Zool. 117. *tab.* A. 1.

THE bittern is a very retired bird, concealing itself in the midst of reeds and rushes in marshy

marſhy places. It is with great difficulty provoked to flight, and when on wing has ſo dull and flagging a pace, as to acquire among the *Greeks* the title of οκνος * or the lazy. It has two kinds of notes; the one croaking, when it is diſturbed: the other bellowing, which it commences in the ſpring and ends in autumn. Mr. *Willughby* ſays, that in the latter ſeaſon it ſoars into the air with a ſpiral aſcent to a great height, making at the ſame time a ſingular noiſe. From the firſt obſervation, we believe this to be the ſpecies of heron that *Virgil* alludes to among the birds that forbode a tempeſt,

> In ſicco ludunt fulicæ, notaſque paludes
> Deſerit, atque altam ſupra volat *Ardea* nubem †.

For the antients mention three kinds ‡; the *Leucon*, or white heron; the *Pellos*, ſuppoſed to be the common ſort; and the *Aſterias*, or bittern; which ſeems to have acquired that name from this circumſtance of its aſpiring flight, as it were attempting, at certain ſeaſons, the very ſtars; though at other times its motion was ſo dull, as to merit the epithet of *lazy*.

Some commentators have ſuppoſed this to have been the *Taurus* of *Pliny*; but as he has expreſsly declared that to be a ſmall bird, remarkable for

* *Ariſt. hiſt. an.* 1056.

† *Georg.* I. 363.

‡ *Ariſt. hiſt. an.* 1006. *Plin.* lib.. x. c. 60.

imitating the lowing of oxen, we muſt deny the explanation; and wait for the diſcovery of the *Roman* naturaliſt's animal from ſome of the *literati* of *Arles*, in which neighbourhood *Pliny* ſays the bird was found *. In ſize it is inferior to the heron: the bill is weaker, and only four inches long: the upper mandible a little arched; the edges of the lower jagged: the rictus or gape is ſo wide, that the eyes ſeem placed in the bill: the irides are next the pupil yellow; above the yellow incline to hazel: the ears are large and open. The crown of the head is black; the feathers on the hind part form a ſort of ſhort pendent creſt: at each corner of the mouth is a black ſpot: the plumage of this bird is of very pale dull yellow, ſpotted, barred, or ſtriped with black: the baſtard wing, the greater coverts of the wings, and the quil-feathers are of a bright ferruginous color, regularly marked with black bars: the lower belly is of a whitiſh yellow: the tail is very ſhort, and conſiſts of only ten feathers. The feathers on the breaſt are very long, and hang looſe: the legs are of a pale green. All the claws are long and ſlender: the inner ſide of the middle claw finely ſerrated to hold its prey the better; its hind claw is remarkably long, and being a ſuppoſed preſervative for the teeth, is ſometimes ſet in ſilver and uſed as a tooth-pick. Beſides this common ſpecies, Mr. *Edwards* mentions a

DESCRIP.

* Lib. x. c. 42.

ſmall

Pl. LXII. The WHITE HERON. No. 17

ſmall one of the ſize of a lapwing, ſhot near *Shrewſbury*. He adds no more than that the crown of the head was black: as this anſwers the deſcription of a kind frequent in *Switzerland* and *Auſtria* *, we imagine it to be a ſtrayed bird from thoſe parts.

It builds its neſt with the leaves of water plants on ſome dry clump among the reeds, and lays five or ſix eggs, of a cinereous green color. This bird and the heron are very apt to ſtrike at the fowler's eyes, when only maimed. The food of the bittern is chiefly frogs; not that it rejects fiſh, for ſmall trouts have been met with in their ſtomachs. In the reign of *Henry* VIII. it was held in much eſteem at our tables; and valued at one ſhilling. Its fleſh has much the flavour of a hare; and nothing of the fiſhineſs of that of the heron.

175. WHITE.

Le Heron blanc. *Belon av.* 191.
Ardea alba. *Geſner av.* 213. *Turner.*
Wil. orn. 279.
Raii ſyn. av. 99.

Ardea candida, le Heron blanc. *Briſſon av.* V. 428.
Groſſer weiſſer Rager. *Kram.* 346. *Scopoli*, No. 126.
Ardea alba. *Lin. ſyſt.* 239.
Faun. Suec. ſp. 166.
Br. Zool. 117.

THIS bird has not fallen within our obſervation; therefore we muſt give Mr. *Willughby*'s

* *Kramer Elench. anim. Auſtriæ*, 348.

account

account of it. The length to the end of the feet is fifty-three inches and a half, to that of the tail only forty; the breadth fixty inches; the weight forty ounces.

The bill is yellowifh; the naked fkin between that and the eyes green; the edges of the eye-lids, and the irides, are of a pale yellow; the legs are black; the inner edge of the middle claw ferrated: the whole plumage is of a fnowy whitenefs. This bird is very common in many parts of *Europe*; *Turner* fays, that in his time this fpecies bred (though rarely) in the fame places with the common fort: but we believe it to be feldom found with us at prefent, any more than the fmall fpecies of crefted white heron mentioned by *Leland*, under the name of *Egritte*, in one of the bills of fare in the magnificent feafts of our anceftors *.

* *Leland's collectanea*, Vol. 6. L' Aigrette. *Briffon av.* V. 431.

BILL

Pl. LXIII. No 176.

CURLEW.

BILL long, ſlender, incurvated.
NOSTRILS linear, placed near the baſe.
TONGUE ſhort, ſharp pointed.
TOES connected as far as the firſt joint by a ſtrong membrane.

XXIX. CURLEW.

Le Corlieu. *Belon av.* 204.
Arquata, ſive numenius. *Geſner av.* 221.
Arcaſe Torquato. *Aldr. av.* III. 169.
Wil. orn. 294.
Raii ſyn. av. 103.
Le Courly. *Briſſon av.* V. 311.
Goiſſer, Brach-ſcknepf. *Kram.* 350. *Friſch*, II. 229.
Scolopax arquata. *Lin. ſyſt.* 242.
Faun. Suec. ſp. 168.
Danis Heel-ſpove. Regn. Spaaer. Regn. Spove. *Brunnich*, 158.
Br. Zool. 118.

176. CURLEW.

THESE birds frequent our ſea coaſts and marſhes in the winter time in large flocks, walking on the open ſands; feeding on ſhells, frogs, crabs, and other marine inſects: in ſummer they retire to the mountanous and unfrequented parts of the country, where they pair and breed. Their eggs are of a pale olive color, marked with irregular but diſtinct ſpots of pale brown. Their fleſh is very rank and fiſhy, notwithſtanding an old *Engliſh* proverb in its favour.

DESCRIP.

Curlews differ much in weight and ſize; ſome weighing thirty-ſeven ounces, others not twenty-two: the length of the largeſt to the tip of the tail

tail twenty-five inches; the breadth three feet five inches; the bill is ſeven inches long: the head, neck, and coverts of the wings are of a pale brown; the middle of each feather black; the breaſt and belly white, marked with narrow oblong black lines: the back is white, ſpotted with a few black ſtrokes: the quil-feathers are black, but the inner webs ſpotted with white: the tail white, tinged with red and beautifully barred with black; the legs are long, ſtrong, and of a bluiſh grey color: the bottoms of the toes flat and broad, to enable it to walk on the ſoft mud, in ſearch of food.

177. WHIMBREL.

Phæopus altera, vel arquata minor. *Geſner av.* 499.
Tarangolo, Girardello. *Aldr. av.* III. 180.
Wil. orn. 294.
Raii ſyn. av. 103.
Edw. av. 307.
Scolopax Phæopus. *Lin. ſyſt.* 243. *Scopoli*, No. 132.
Windſpole, Spof. *Faun. Suec. ſp.* 169.
Kleiner Goiſſer. *Kram.* 350.
Kleine Art Brachvogel or Regenvogel. *Friſch*, II. 225.
Le petit Courly, ou le Courlieu. Numenius minor. *Briſſon av.* V. 317. *tab.* 27.
Danis Mellum-Spove. *Norveg.* Smaae Spue. *Br.* 159.
Br. Zool. 119.

THE whimbrel is much leſs frequent on our ſhores than the curlew; but its haunts, food, and general appearance are much the ſame. It is obſerved to viſit the neighbourhood of *Spalding* (where it is called the *Curlew knot)* in vaſt flocks

WHIMBREL.

flocks in *April*, but continues there no longer than *May*; nor is it ſeen there any other time of year: it ſeems at that ſeaſon to be on its paſſage to its breeding place, which I ſuſpect to be among the Highlands of *Scotland*.

The ſpecific difference is the ſize; this never exceeding the weight of twelve ounces. The bill is two inches three quarters long; duſky above, red below: the feathers on the head and neck are brown tinged with red, marked in the middle with an oblong black ſpot: the cheeks of a paler color: the upper part of the back, the coverts of the wings, the ſcapulars, and the fartheſt quil-feathers, are of the ſame color with the neck, but the black ſpots ſpread out tranſverſely on each web: the quil-feathers duſky; their ſhafts white; and their exterior webs marked with large ſemicircular white ſpots. The breaſt, belly, and lower part of the back are white: the coverts of the tail, and the tail itſelf, are of a very pale whitiſh brown, croſſed with black bars. The legs and feet are of a dull green, and formed like thoſe of the curlew.

Descrip.

I received one from *Invercauld*, ſhot on the *Grampian Hills*, whoſe length was ſixteen inches; the bill two: the head round, black on the top, divided length-ways by a white line: chin white: cheeks, neck, breaſt, and upper part of the belly whitiſh brown, marked with ſtreaks of black pointing down, with narrow ſtreaks on the neck; broad on the belly: lower belly and vent white: back

and

and coverts of the wings dusky: the sides of each feather spotted with reddish white: lower part of the back white: rump white barred with black: tail barred with dusky and white: quil-feathers black, with large white spots on the inner webs; the secondaries on both webs: legs black.

BILL

REDSHANK.

No

WOODCOCK.

No

XXX. SNIPE

BILL long, slender, weak and strait.
NOSTRILS linear, lodged in a furrow.
TONGUE pointed, slender.
TOES divided, or very slightly connected, back toe very small.

178. WOODCOCK.

La Beccasse. *Belon av.* 272.
Rusticola, seu Perdix rustica major (Grosser schnepff). *Gesner av.* 501.
Aldr. av. III. 182.
Wil. orn. 289.
Raii syn. av. 104.
La Beccasse. *Brisson av.* V. 292.
Beccaccia, Acceggia. *Zinan.* 101.
Schniffa. *Scopoli*, No. 134.
Wald schnepf. *Kram.* 351. *Frisch*, II. 226. foem. 227.
Scolopax rusticola. *Lin. syst.* 243.
Morkulla. *Faun. Suec. sp.* 170.
Norvegis Blom-Rokke, Rutte, *quibusdam* Krog-quist. *Danis* Holt Sneppe. *Brunnich*, 164.
Br. Zool. 119.
Fauna Scotica. No. 142.

THESE birds during summer are inhabitants of the *Alps* *, of *Norway*, *Sweden*, *Polish Prussia*, the march of *Brandenburg* †, and the northern parts of *Europe:* they all retire from those countries the beginning of winter, as soon as the frosts commence; which force them into milder climates, where the ground is open, and adapted to their manner of feeding. The time of their appear-

* *Wil. orn.* 290.
† *Frisch*, II. 226.

appearance and disappearance in *Sweden*; coincides most exactly with that of their arrival in, and their retreat from *Great Britain**. They live on worms and insects, which they search for with their long bills in soft ground and moist woods. Woodcocks generally arrive here in flocks, taking advantage of the night, or a mist: they soon separate; but before they return to their native haunts, pair. They feed and fly by night; beginning their flight in the evening, and return the same way, or through the same glades to their day retreat. They leave *England* the latter end of *February*, or beginning of *March*; not but they have been known to continue here accidentally. In *Case-wood*, about two miles from *Tunbridge*, a few breed almost annually: the young having been shot there the beginning of *August*, and were as healthy and vigorous as they are with us in the winter, but not so well tasted: a female with egg was shot in that neighbourhood in *April*; the egg

* M. *de Geer*'s and Dr. *Wallerius*'s letters to myself. M. *de Geer* expresses himself thus; *La Becasse (Scolopax rusticola) part d'ici vers l'automne, Je ne scais pas au juste dans quel mois. On la trouve ici assez en abondance dans l'eté. Elle a coutume au soleil couchant de faire sa volèe en cercle ou toujours en rond en l'air revenant toujours dans le meme endroit a plusieurs reprises, et c'est alors qu'on peut la tirer a coup de fusil. En hiver on ne voit aucune, elle partent alors toutes.*

M. *Wallerius* gave me this account of them. *Scolopaces rusticolæ penes nos nidificant. Sed autumnali tempore abeunt, ac vernali redeunt.*

was

was the ſize of that of a pigeon. They are remarkably tame during incubation; a perſon who diſcovered one on its neſt, has often ſtood over, and even ſtroaked it: notwithſtanding which it hatched the young; and in due time diſappeared with them.

Theſe birds appear in *Scotland* firſt on the eaſtern coaſts, and make their progreſs from *Eaſt to Weſt*. They do not arrive in *Breadalbane*, a central part of the kingdom till the beginning or middle of *November*: and the coaſts of *Nether Lorn*, or of *Roſſhire*, till *December* or *January*: are very rare in the more remote *Hebrides*, or in the *Orknies*. A few ſtragglers now and then arrive there. They are equally ſcarce in *Cathneſs*. I do not recollect that any have been diſcovered to have bred in North *Britain*.

Their autumnal and vernal appearances on the coaſt of *Suffolk* have been moſt accurately marked by Sir *John Cullum*, Bar[t]. who favoured me with the following curious account.

From ſome old and experienced ſportſmen, who live on the coaſt, I collected the following particulars. They come over ſparingly in the firſt week of *October*, the greater numbers not arriving till the months of *November* and *December*, and always after ſun-ſet. It is the wind and not the moon that determines the time of their arrival: and it is probable that this ſhould be the caſe, as they come hither in queſt of food, which fails then in the places they leave. If the wind has favoured their flight, their ſtay on the coaſt, where they drop, is very ſhort, if any: but if they have been forced to ſtruggle with an adverſe gale (ſuch as a ſhip can hardly make way with) they take a day's reſt, to recover their fatigue: and ſo greatly has their ſtrength been exhauſted, that they have been taken by hand in *Southwald* ſtreets. They arrive not gregarious, but ſeparate and diſperſed. When the *Red wing* appears on the coaſt in autumn, it is certain the *Woodcocks* are at hand; when they *Royſton Crow*, they are come. Between the twelfth and twenty-fifth of *March* they flock towards the coaſt to be ready for their departure: the firſt law of nature bringing them to us, in autumn; the ſecond carrying them from us in ſpring. If the wind be propitious, they are gone immediately; but if contrary, they are detained in the neighboring woods, or among the ling and furze on the coaſt. It is in this criſis that the ſportſman finds extraordinary diverſion: the whole country around echoes with the diſcharge of guns; even ſeventeen brace have been killed by one perſon in a day: but if they are kept any time on the dry heaths, they become ſo lean, that they are a prey hardly worth purſuing, at leſt eating. The inſtant a fair wind ſprings up, they ſeize the opportunity, and where the ſportſman has ſeen hundreds one day, he will not find a ſingle bird the next. As this extraordinary diverſion depends on the winds, it

muſt

muſt neceſſarily be precarious; and it accordingly ſometimes happens, that the ſportſmen on the coaſt, for ſome years together know not preciſely the time of the *Woodcocks* departure. They have the ſame harbingers (the *Red wings*) in ſpring, as in autumn.

MIGRATIONS.

In the ſame manner we know they quit *France*, *Germany* and *Italy*; making the northern and cold ſituations their general ſummer rendezvous. They viſit *Burgundy* the latter end of *October*, but continue there only four or five weeks; it being a dry country they are forced away for want of ſuſtenance by the firſt froſts. In the winter they are found in vaſt plenty as far ſouth as *Smyrna* and *Aleppo* *, and in the ſame ſeaſon in *Barbary* †, where the *Africans* call them, the *aſs* of the partridge: and we have been told, that ſome have appeared as far ſouth as *Ægypt*, which are the remoteſt migrations we can trace them to on that ſide the eaſtern world; on the other ſide, they are found very common in *Japan* ‡. The birds that reſort into the countries of the *Levant*, probably come from the deſarts of *Siberia* or *Tartary* §, or the cold mountains of *Armenia*.

Our ſpecies of woodcock is unknown in *North*

* *Ruſſel's hiſt. Aleppo.* 64.

† *Shaw's travels,* 253.

‡ *Kæmpfer's hiſt. Japan.* I. 129.

§ *Bell's travels,* I. 198.

America; but a kind is found there that has the general appearance of it; but is ſcarce half the ſize, and wants the bars on the breaſt and belly.

Descrip. The weight of the woodcock is uſually about twelve ounces: the length near fourteen inches: the breadth twenty-ſix: the bill is three inches long, duſky towards the end, reddiſh at the baſe: tongue ſlender, long, ſharp, and hard at the point: the eyes large, and placed near the top of the head, that they may not be injured when the bird thruſts its bill into the ground: from the bill to the eyes is a black line: the forehead is a reddiſh aſh-color: the crown of the head, the hind part of the neck, the back, the coverts of the wings, and the ſcapulars are prettily barred with a ferruginous red, black and grey; but on the head the black predominates: the quil-feathers are duſky, indented with red marks.

The chin is of a pale yellow: the whole underſide of the body is of a dirty white, marked with numerous tranſverſe lines of a duſky color. The tail conſiſts of twelve feathers, duſky, or black on the one web, and marked with red on the other: the tips above are aſh-colored, below white; which, when ſhooting on the ground was in vogue, was the ſign the fowler diſcovered the birds by. The legs and toes are livid; the latter divided almoſt to their very origin, having only a very ſmall web between the middle and interior toes; as are thoſe of the two ſpecies of ſnipes found in *England*.

Godwit

Godwit, Yarwhelp, or Yarwip. *Wil. orn.* 290.
Raii syn. av. 105.
Scolopax ægocephala. *Lin. syst.* 246.

Limosa grisea major. La grande.
Barge grise. *Brisson av.* V. 272. *Tab.* 24. *fig.* 2.
Br. Zool. 120. *Tab.*

179. GODWIT.

DESCRIP.

THIS species weighs twelve ounces and a half; the length is sixteen inches; the breadth twenty-seven; the bill is four inches long, turns up a little, black at the end, the rest a pale purple: from the bill to the eye is a broad white stroke: the feathers of the head, neck, and back, are of a light reddish brown, marked in the middle with a dusky spot: the belly and vent feathers white: the tail regularly barred with black and white.

The six first quil-feathers are black; their interior edges of a reddish brown: the legs in some are dusky, in others of a greyish blue; which perhaps may be owing to different ages: the exterior toe is connected as far as the first joint of the middle toe, with a strong serrated membrane. The male is distinguished from the female by some black lines on the breast and throat; which in the female are wanting.

These birds are taken in the fens, in the same season, and in the same manner with the ruffs and reeves, and when fattened are esteemed a great delicacy, and sell for half a crown, or five shillings a piece. A ſtale of the ſame ſpecies is placed in the net. They appear in ſmall flocks on our coaſt in *September*, and continue with us the whole winter; they walk on the open ſands like the curlew; and feed on inſects.

M. *Briſſon* has figured this bird very accurately, but has given it the ſynonym of our *greenſhanks*. *Turner* ſuſpects this bird to have been the *attagen* or *attagas* of the antients. *Ariſtophanes* names it in an addreſs to the birds that inhabit the fens; therefore ſome commentators conclude it to be a water-fowl; though in a line or two after he ſpeaks of thoſe that frequent the beautiful meadows of *Marathon*. He then deſcribes the bird in very ſtriking terms, under the title of the *attagas*, *the bird with painted wings*; and in another place he ſtyles it the *ſpotted attagas**. This alone would be inſufficient to prove what ſpecies the poet intended; we muſt therefore have recourſe to *Athenæus*, who is particular in his deſcription of the *attagas*, and evinces it to be of the partridge tribe.

He ſays it is leſs than that bird; that the back is ſpotted with different colors, ſome of a pot color, but more red; that by reaſon of the ſhortneſs of the wings and heavineſs of the body, it is taken

* Ὄρνις τε πτεροποίκιλος
ἀτταγᾶς.
Ἀτταγας ἐτος παρ' ἡμιν ποικίλος κεκλήσεται.
Av. 249. 762.

easily

eaſily by the fowlers. That it rolls in the duſt, brings many young, and feeds on ſeeds.

We are ſorry to own our ſmall acquaintance with the zoology of *Attica*, conſidering the various opportunities our countrymen have had of informing themſelves of it. We therefore cannot pronounce, that the *attagas* ſtill exiſts on the plains of *Marathon*; but we diſcover it in *Samos*, an iſland of *Ionia*, a country celebrated by the antients for producing the fineſt kinds:

> Inter ſapores fertur alitum primus
> *Ionicarum* guſtus *attagenarum*,

Is the opinion of *Martial* *; and *Horace* †, and *Pliny* ‡, both ſpeak of it with applauſe. *Tournefort* § has given us the figure of the bird itſelf, which he found in the *marſhes* of *Samos*, whoſe painted and ſpotted plumage exactly anſwers the deſcriptions of *Ariſtophanes* and *Athenæus*. It is of the partridge genus, and known to the *Italians* by the name of *Francolino*. Thoſe who wiſh to ſee it in its proper colors, and to be ſatisfied how well they agree with the deſcriptions of the antients, need only conſult the 246th plate of the works of our ingenious friend the late Mr. *Edwards*.

* *Epig. Lib.* XIII. *Ep.* 61.
† *Epod.* II.
‡ *Lib.* X. *c.* 48.
§ *Voy.* Vol. I. 311. 4*to. ed.*

180. CINEREOUS.

THIS ſpecies was ſhot near *Spalding*, and the deſcription communicated to me by the Rev. Doctor *Buckworth*.

The bill was two inches and a half long. The head, neck, and back variegated with aſh-color and white: the tail ſlightly barred with cinereous. The throat and breaſt white: the laſt marked with a few aſh-colored ſpots. The legs long, ſlender, and aſh-colored.

This was about the ſize of my *Green-ſhanks:* approaches it nearly in colors: but the bill was ſo much thicker, as to form a ſpecific diſtinction.

181. RED.

Scolopax Lapponica. *Lin. ſyſt.* 246.

Faun. Suec. ſp. 174.
Br. Zool. add. plates.

DESCRIP.

THE red godwit is ſuperior in ſize to the common kind: the bill is three inches three-quarters long; not quite ſtrait, but a little reflected upwards; the lower half black, the upper yellow: the head, neck, breaſt, ſides, ſcapulars, and upper part of the back, are of a bright ferruginous color: the head marked with oblong duſky lines: the neck is plain: the breaſt, ſides, ſcapulars, and back varied with tranſverſe black bars, and

CENEREOUS GODWIT.

RED GODWIT.

and the edges of the feathers with a pale cinereous brown: the middle of the belly is white, marked ſparingly with ſimilar ſpots.

The leſſer coverts of the wings are of a light brown: the greater tipt with white: the ſhafts and lower interior webs of the greater quil-feathers are white: the exterior webs and upper part of the interior black: the upper half of the ſecondary feathers are of the ſame color; the lower half white: the coverts, and the lower part of the feathers of the tail are white; the upper part black; the white gradually leſſening from the outmoſt feathers on each ſide: the legs are black, and four inches long: and the thighs above the knees are naked for the ſpace of an inch and three-quarters.

Theſe birds vary in their colors, ſome that we have ſeen being very ſlightly marked with red, or only marbled with it on the breaſt: but the reflected form of the bill is ever ſufficient to determine the ſpecies. This is not a very common ſpecies in *England*; we have known it to have been ſhot near *Hull*; and have once met with it in a poulterer's ſhop in *London*. Mr. *Edwards* has figured a bird from *Hudſon's Bay*, that ſeems related to this; but the difference in the colors of the tail, forbids our placing it among the ſynonyms. And *Linnæus* omitting a deſcription of that part, in his *Fauna Suecica*, obliges us to queſtion whether it be the ſame with the above.

La

182. LESSER. La Barge. *Belon av.* 205.
The second sort of Godwit, the *Totanus* of *Aldrovand*; called at *Venice*, *Vetola*. *Wil. orn.* 293.
Fedoa nostra secunda, the Stone Plover *Raii syn. av.* 105.
Limosa, la Barge. *Brisson av.* V. 262.
Br. Zool. 120.

DESCRIP. MR. *Ray* (for we are not acquainted with this species) describes it thus. Its weight is nine ounces; the length to the tail seventeen inches; to the toes twenty-one; its breadth twenty-eight: the bill like that of the former: the chin white, tinged with red: the neck ash-colored; the head of a deep ash-color, whitish about the eye; the back of a uniform brownness, not spotted like that of the preceding: the rump encompassed with a white ring: the two middle feathers of the tail black: the outmost, especially on the outside web, white almost to the tips; in the rest the white part grew less and less to the middlemost.

Besides these, Mr. *Willughby* mentions a third species, called in *Cornwal* the *Stone Curlew*; but describes it no farther than saying it has a shorter and slenderer bill than the preceding.

Limosa;

Limosa, et glottis. *Gesner av.* 519, 520.
Piviero. *Aldr. av.* III. 207.
Greater Plover of *Aldrovand. Wil. orn.* 298.
Raii syn. av. 106.

Scolopax glottis. *Lin. syst.* 245.
Glut. *Faun. Suec. sp.* 171.
Pivier Maggiore. *Zinan.* 102.
Norvegis Hoest-Fugl. 167. *Brunnich.*
Br. Zool. 121.
Tschoket. *Scopoli,* No. 137.

183. GREEN SHANK.

DESCRIP.

THESE birds are not so common as the former: appearing on our coasts and wet grounds in the winter time in small flocks. The length to the end of the tail is fourteen inches, to that of the toes twenty; its breadth twenty-five. The bill is two inches and a half long: the upper mandible black, strait, and very slender; the lower reflects a little upwards: the head and upper part of the neck are ash-colored, marked with small dusky lines pointing down: over each eye passes a white line: the coverts of the wings, the scapulars, and upper part of the back are of a brownish ash-color: the quil-feathers dusky, but the inner webs speckled with white: the breast, belly, thighs, and lower part of the back are white: the tail white, marked with undulated dusky bars: the inner coverts of the wings finely crossed with double and treble rows of a dusky color.

It is a bird of an elegant shape, and small weight in proportion to its dimensions, weighing only six ounces.

The

The legs are very long and ſlender, bare above two inches higher than the knees. The exterior toe is united to the middle toe, as far as the ſecond joint, by a ſtrong membrane which borders their ſides to the very end.

Theſe birds are the *Chevaliers aux pieds verds* of the *French*; as the ſpotted redſhanks are the *Chevaliers aux pieds rouges.*

184. RED SHANK.

Gallinula erythropus. *Geſner av.* 504.
Totanus *Aldr. av.* III. 171.
Redſhank, or Pool-ſnipe. *Wil. orn.* 299.
Raii ſyn. av. 107.
Totanus, le Chevalier. *Briſſon av.* V. 188. *Tab.* 17. *fig.* 1.
Scolopax Calidris. *Lin. ſyſt.* 245.
Sc. Totanus. *Faun. Suec. ſp.* 167.
Rothfuſsler *Kram.* 353.
Kleiner grau-und-weiſbunter Sandlœuffer? *Friſch*, II. 240.
Hœmantopus, magnitudine inter Vanellum et Gallinaginem minorem media. *Ray's itin.* 247.
Br. Zool. 124.

THIS ſpecies is found on moſt of our ſhores: in the winter time it conceals itſelf in the gutters; and is generally found ſingle, or at moſt in pair.

DESCRIP. It weighs five ounces and a half: the length is twelve inches: the breadth twenty-one: the bill near two inches long, red at the baſe, black towards the point. The head, hind part of the neck, and ſcapulars, are of a duſky aſh-color, obſcurely ſpotted with black: the back is white, ſprinkled with

with black ſpots: the tail elegantly barred with black and white: the cheeks, under ſide of the neck, and upper part of the breaſt are white, ſtreaked downward with duſky lines: the belly white: the exterior webs of the quil-feathers are duſky: the legs long, and of a fine bright orange color: the outmoſt toe connected to the middle toe by a ſmall membrane; the inmoſt by another ſtill ſmaller.

It breeds in the fens, and marſhes; and flies round its neſt when diſturbed, making a noiſe like a *lapwing*. It lays four eggs, whitiſh tinged with olive, marked with irregular ſpots of black chiefly on the thicker end.

185. CAMBRIDGE.

I DISCOVERED this in the collection of the Rev. Mr. *Green*, ſhot near *Cambridge*.

It is larger than the common redſhank. The head, upper part of the neck, and the back are of a cinereous brown: the leſſer coverts of the wings brown edged with dull white, and barred with black: the primaries duſky, whitiſh on their inner ſides: ſecondaries barred with duſky and white: under ſide of neck and breaſt of a dirty white: belly and vent white: tail barred with cinereous and black: legs of an orange red.

Le

186. SPOTTED REDSHANK.

Le chevalier rouge. *Belon av.* 207.
Aldr. av. III. 171.
The other Totano. *Wil. orn.* 299.
Le Chevalier rouge. *Brisson av.* V. 192.

DESCRIP.

THIS ſpecies we found in the collection of *Taylor White*, Eſq. In ſize it is equal to the greenſhank: the head is of a pale aſh-color, marked with oblong ſtreaks of black: the back duſky, varied with triangular ſpots of white: the coverts of the wings aſh-colored, ſpotted in the ſame manmanner: the quil-feathers duſky; breaſt, belly, and and thighs white, the firſt thinly ſpotted with black: the middle feathers of the tail are aſh-colored; the ſide feathers are whitiſh, barred with black: the legs very long, and of a bright red.

187. COMMON SN.

La Becaſſine ou Becaſſeau. *Belon av.* 215.
Gallinago, ſeu ruſticola minor. *Geſner av.* 503.
Aldr. av. III. 184.
The Snipe, or Snite. *Wil. orn.* 290.
Raii ſyn. av. 105.
La Beccaſſine. *Briſſon av.* V. 298. *Tab.* 26. *fig.* 1.
Pizzarda, Pizzardella. *Zinan.* 101.
Mooſs ſchnepf. *Kram.* 352.
Friſch, II. 229.
Scolopax gallinago. *Lin. ſyſt.* 244.
Horſgjok. *Faun. Suec. ſp.* 173.
Capella cœleſtis. *Klein av.* 100.
Iſlandis Myr Snippe. *Norvegis* Trold Ruke. *Cimbris quibuſd.* Hoſſegioeg. *Danis* Dobbelt Sneppe, Steen Sneppe. *Br.* 160.
Br. Zool. 121.
Koſitza. *Scopoli*, No. 138.

IN the winter time ſnipes are very frequent in all our marſhy and wet grounds, where they lie concealed

LXVIII.
Nº 189
JACK SNIPE.
SNIPE.
Nº 187

concealed in the rushes, &c. In the summer they disperse to different parts, and are found in the midst of our highest mountains, as well as our low moors: their nest is made of dried grass; they lay four eggs of a dirty olive color, marked with dusky spots; their young are so often found in *England*, that we doubt whether they ever entirely leave this island. When they are disturbed much, particularly in the breeding season, they soar to a vast height, making a singular bleating noise; and when they descend, dart down with vast rapidity: it is also amusing to observe the cock (while his mate sits on her eggs) poise himself on his wings, making sometimes a whistling and sometimes a drumming noise. Their food is the same with that of the woodcock; their flight very irregular and swift, and attended with a shrill scream. They are most universal birds, found in every quarter of the globe, and in all climates.

DESCRIP.

This species weighs four ounces; the length, to the end of the tail, is near twelve inches: the breadth about fourteen: the bill is three inches long, of a dusky color, flat at the end, and often rough like shagrin above and below. The head is divided lengthways with two black lines, and three of red, one of the last passing over the middle of the head, and one above each eye: between the bill and the eyes is a dusky line: the chin is white: the neck is varied with brown and red.

The scapulars are beautifully striped lengthways with

with black and yellow: the quil-feathers are duſky, but the edge of the firſt is white, as are the tips of the ſecondary feathers: the quil-feathers next the back are barred with black and pale red: the breaſt and belly are white: the coverts of the tail are long, and almoſt cover it: they are of a reddiſh brown color. The tail conſiſts of fourteen feathers; black on their lower part, then croſſed with a broad bar of deep orange, another narrow one of black; and the ends white, or pale orange. The vent feathers a dull yellow: the legs pale green: the toes divided to their origin.

188. GREAT SNIPE.

THIS ſpecies is rarely found in *England*. A fine ſpecimen, ſhot in *Lancaſhire*, is preſerved in the *Muſeum* of *Aſhton Lever*, Eſq.

The weight eight ounces. The head divided lengthways by a teſtaceous line, bounded on each ſide by another of black: above and beneath each eye is another: neck and breaſt of a yellowiſh white, finely marked with ſemicircular lines of black: belly, with cordated ſpots: ſides undulated with black.

Back, coverts of wings, and ſcapulars teſtaceous, ſpotted with black and edged with white. Primaries duſky. Tail ruſt-colored, barred with black. Legs black?

Gid,

189. JACK SNIPE.

Gid, Jackſnipe, and Judcock. *Wil. orn.* 291.
Raii ſyn. av. 105.
La petite Beccaſſine. *Briſſon av.* V. 303. *tab.* 26. *fig.* 2.
Pokerl. *Scopoli,* No. 139.
Pizzardina. *Zinan.* 101.

Scolopax gallinula. *Lin. ſyſt.* 244.
Danis Roer-Sneppe. *Brunnich,* 163.
Haar-Schnepfe, Pudel-Schnepfe, Kleinſte Schnepfe. *Friſch,* II. 231.
Br. Zool. 121

THE haunts and food of this ſpecies are the ſame with thoſe of the former; it alſo feeds on ſmall ſnails: it is much leſs frequent among us, and very difficult to be found, lying ſo cloſe as to hazard being trod on before it will riſe: the flight is never diſtant, and its motion is more ſluggiſh than that of the larger kind.

DESCRIP.

Its weight is leſs than two ounces, inferior by half to that of the ſnipe; for which reaſon the *French* call them *deux pour un,* we the *half ſnipe.* The dimenſions bear not the ſame proportion; the length of the ſnipe being twelve inches; this eight and a half: the bill an inch and a half long: crown of the head black, tinged with ruſt color: over each eye is a yellow ſtroke; the neck varied with white, brown, and pale red. The ſcapular feathers narrow, very long, brown, bordered with yellow. The rump a gloſſy bluiſh purple: the belly and vent white; the greater quil-feathers dusky: the tail brown, edged with tawny; consisting of twelve pointed feathers: the legs are of a cinereous green.

BILL

XXXI. SAND-PIPER*.

BILL ſtraight, ſlender, not an inch and half long.
NOSTRILS ſmall.
TONGUE ſlender.
TOES divided; generally the two outmoſt connected at the bottom by a ſmall membrane.

190. LAPWING.

Le Vanneau, Dixhuit, Papechieu. *Belon av.* 209.
Zweiel. *Geſner av.* 765.
Pavonzino. *Aldr. av.* III. 202.
Pavoncella. *Olina*, 21.
Lapwing, baſtard Plover, or Pewit. *Wil. orn.* 307.
Vanellus, le Vanneau. *Briſſon av.* V. 94. *tab.* 8. *fig.* 1.
Raii ſyn. av. 110.
Kiwik. *Kram.* 353. *Friſch*, II. 213.
Tringa vanellus. *Lin. ſyſt.* 248.
Wipa, Kowipa, Blæcka. *Faun. Suec. ſp.* 176.
Danis Vibe, Kivit. *Brunnich*, 170.
Br. Zool. 122. *Scopoli*, No. 141.

THIS elegant ſpecies inhabits moſt of the heaths and marſhy grounds of this iſland. It lays four eggs, making a ſlight neſt with a few bents. The eggs have an olive caſt, and are ſpotted with black. It is worthy of notice, that among water fowl, congenerous birds lay the ſame number of eggs; for example, all of this tribe, alſo of the plo-

* This genus, the *Tringa* of *Linnæus*, wanting an *Engliſh* name, we have given it that of the *Sandpipers*; moſt of the ſpecies being converſant about ſhores; and their note whiſtling or piping.

vers, lay four a-piece; the puffin genus only one; and the duck tribe, in general, are numerous layers, producing from eight to twenty.

The young as ſoon as hatched, run like chickens: the parents ſhew remarkable ſolicitude for them, flying with great anxiety and clamour near them, ſtriking at either men or dogs that approach, and often flutter along the ground like a wounded bird, to a conſiderable diſtance from their neſt, to elude their purſuers; and to aid the deceit, become more clamorous when moſt remote from it: the eggs are held in great eſteem for their delicacy; and are ſold by the *London* poulterers for three ſhillings the dozen. In winter, lapwings join in vaſt flocks; but at that ſeaſon are very wild: their fleſh is very good, their food being inſects and worms. During *October* and *November*, they are taken in the fens in nets, in the ſame manner that *Ruffs* are, but are not preſerved for fattening, being killed as ſoon as caught.

DESCRIP. Their weight is about eight ounces: the length thirteen inches and a half: the breadth two feet and a half. The bill is black, and little more than an inch long: the crown of the head of a ſhining blackneſs: the creſt of the ſame color, conſiſting of about twenty ſlender unwebbed feathers of unequal lengths, the longeſt are four inches: the cheeks and ſides of the neck are white; but beneath each eye is a black line: the throat and fore part of the neck are black: the plumage on the hind part mixed

mixed with white, ash-color and red: the back and scapulars are of a most elegant glossy green; and the latter finely varied with purple: the lesser covert feathers of the wings are of a resplendent black blue and green: the greater quil-feathers black, but the ends of the four first are marked with a white spot: the upper half of the lesser quil-feathers are black, the lower white: those next the body of the same colors with the scapulars: the breast and belly are white: the vent-feathers and the coverts of the tail orange color: the tail consists of twelve feathers; the outmost on each side is white, marked on the upper end of the inner web with a dusky spot; the upper half of all the others are black, tipt with a dirty white; their lower half of a pure white: the legs are red: the irides hazel.

The female is rather less than the male.

Merret, in his *Pinax*, p. 182. says, that there is in *Cornwal* a bird related to this; but less than a thrush, having blue feathers, and a long crest.

191. GREY.

Le pluvier gris. *Belon av.* 262.
Pivier montano. *Aldr. av.* III. 207.
Wil. orn. 309.
Raii ſyn. av. 111.
Tringa ſquatarola. *Lin. ſyſt.* 252.
Faun. Suec. ſp. 186.
Vanellus griſeus, le Vanneau gris. *Briſſon av.* v. 100. *tab.* 9 *fig.* 1.
Piviero montano. *Zinan.* 102.
Bornholmis Floyte-Tyten, Dolken, *Brunnich*, 176.
Br Zool. 122. *Scopoli*, No. 145.

DESCRIP.

IT weighs ſeven ounces: the length to the tip of the tail is twelve inches: the breadth twenty-four: the bill black, about an inch long, ſtrong and thick: the head, back, and coverts of the wings black, edged with greeniſh aſh-color, and ſome white: cheeks and throat white, marked with oblong duſky ſpots: the belly and thighs white: the exterior webs of the quil-feathers black: the lower part of the interior webs of the four firſt white: the rump white: the tail marked with tranſverſe bars of black and white: the legs of a dirty green: the back toe very ſmall.

Theſe appear in ſmall flocks in the winter time, but are not very common: their fleſh is very delicate.

Avis

Pl. LXIX.

REEVE.

RUFF

No. 19

192. RUFF.

Avis pugnax. *Aldr. av.* III. 167.
Wil. orn. 302.
Raii ſyn. av. 107.
Kroſsler. *Kram.* 352.
Tringa pugnax. *Lin. ſyſt.* 247.
Bruſhane. *Faun. Suec. ſp.* 175.
Le Combattant, ou Paon de mer. *Briſſon av.* v. 240. *tab.* 22.
Danis Bruuſhane. *Brunnich,* 168.
Streitſchnepfe, Rȧmpf hæhnlein. *Friſch,* II. 232, 235.
Br. Zool. 123. *Scopoli,* No. 140.

DESCRIP.

THE males, or *Ruffs,* aſſume ſuch variety of colors in ſeveral parts of their plumage, that it is ſcarce poſſible to ſee two alike; but the great length of the feathers on the neck, that gives name to them, at once diſtinguiſhes theſe from all other birds. On the back of their necks is a ſingular tuft of feathers ſpreading wide on both ſides. Theſe, and the former, in ſome are black; in others white, yellow, or ferruginous; but this tuft and the ruffs frequently differ in colors in the ſame bird. The feathers that bear an uniformity of coloring through each individual of this ſex, are the coverts of the wings, which are brown inclining to aſh-color: the feathers on the breaſt, which are often black or duſky: the four exterior feathers of the tail, which are of a cinereous brown; and the four middle, which are barred with black and brown: the bill is black towards the end; red at the baſe. The legs in all, are yellow. In moulting they loſe the character of the long neck-feathers, nor do they recover it till after their return to the *fens* the ſpring following. It is then they regain that ornament, and at the ſame time a ſet of ſmall pear ſhaped yellow pimples break out in great numbers on their face above the bill.

The *Stags* or male birds of the firſt year want theſe marks, and have ſometimes been miſtaken for a new ſpecies of *Tringa*; but they may be eaſily known by the colors of the coverts of the wings, and the middle feathers of the tail.

The older the birds are, the more numerous the pimples, and the fuller and longer the ruffs.

The length of the male to the tip of the tail is one foot, the breadth two; of the *Reeve* ten inches, the breadth nineteen: the weight of the former when juſt taken is ſeven ounces and a half; of the latter only four.

The *Reeves* never change their colors, which are pale brown: the back ſpotted with black, ſlightly edged with white: the tail brown; the middle feathers ſpotted with black: the breaſt and belly white: the legs of a pale dull yellow.

Theſe birds appear in the fens in the earlieſt ſpring, and diſappear about *Michaelmas*. The *Reeves* lay four eggs in a tuft of graſs, the firſt week in *May*, and ſit about a month. The eggs are white, marked with large ruſty ſpots. Fowlers avoid in general the taking of the females, not only becauſe they are ſmaller than the males; but that they may be left to breed.

Soon

Soon after their arrival, the males begin to *bill*, that is to collect on fome dry bank near a fplafh of water, in expectation of the females, who refort to them.

Each male keeps poffeffion of a fmall piece of ground, which it runs round till the grafs is worn quite away, and nothing but a naked circle is left. When a female lights, the ruffs immediately fall to fighting. I find a vulgar error, that ruffs muft be fed in the dark leaft they fhould deftroy each other by fighting on admiffion of light. The truth is, every bird takes its ftand in the room as it would in the open fen. If another invades its circle, an attack is made, and a battle enfues. They make ufe of the fame action in fighting as a cock, place their bills to the ground and fpread their ruffs. I have fet a whole room full a fighting by making them move their ftations; and after quitting the place, by peeping through a crevice, feen them refume their circles and grow pacific.

When a fowler difcovers one of thofe *bills*, he places his net over night, which is of the fame kind as thofe that are called *clap* or *day nets*, only it is generally fingle, and is about fourteen yards long and four broad.

The fowler reforts to his ftand at day break, at the diftance of one, two, three, or four hundred yards from the nets, according to the time of the feafon; for the later it is, the fhyer the birds grow. He then makes his firft pull, taking fuch birds

birds that he finds within reach: after that he places his ſtuft birds or ſtales to entice thoſe that are continually traverſing the fen. An old fowler told me, he once caught forty-four birds at the firſt hawl, and in all ſix dozen that morning. When the ſtales are ſet, ſeldom more than two or three are taken at a time. A fowler will take forty or fifty dozens in a ſeaſon.

Theſe birds are found in *Lincolnſhire*, the *Iſle of Ely*, and in the eaſt riding of *Yorkſhire* *; where they are taken in nets, and fattened for the table, with bread and milk, hempſeed, and ſometimes boiled wheat; but if expedition is required, ſugar is added, which will make them in a fortnight's time a lump of fat: they then ſell for two ſhillings or half a crown a piece. Judgement is required in taking the proper time for killing them, when they are at the higheſt pitch of fatneſs, for if that is neglected, the birds are apt to fall away. The method of killing them is by cutting off their head with a pair of ſciſſars: the quantity of blood that iſſues is very great, conſidering the ſize of the bird. They are dreſſed like the woodcock, with their inteſtines; and, when killed at the critical time, ſay the *Epicures*, are reckoned the moſt delicious of all morſels.

* They viſit a place called *Martin-Mere* in *Lancaſhire*, the latter end of *March* or beginning of *April*, but do not continue there above three weeks.

Wil.

Wil. orn. 302.
Raii syn. av. 108.
Edw. av. 276.
Le Canut. *Brisson av.* V. 258.
Tringa canutus. *Lin. syst.* 251.
Faun. Suec. sp. 183.

Islandis Sidlingar-Kall. *Norvegis* FiærePist. Fiær-Kurv, Fiær-Muus. *Bornholmis* Rytteren.
Brunnich, Tringa maritima. 182.
Br. Zool. 123.

193. KNOT.

DESCRIP.

THE specimens that we had opportunity of examining, differ a little in colors, both from Mr. *Willughby*'s description, and from Mr. *Edwards*'s figure: the forehead, chin, and lower part of the neck in ours were brown, inclining to ash color: the back and scapulars deep brown, edged with ash color: the coverts of the wings with white, the edges of the lower order deeply so, forming a white bar: the breast, sides, and belly white; the two first streaked with brown: the coverts of the tail marked with white and dusky spots alternately: the tail ash colored, the outmost feather on each side white: the legs were of a bluish grey; and the toes, as a special mark, divided to the very bottom: the weight four ounces and a half.

These birds, when fattened, are preferred by some to the ruffs themselves. They are taken in great numbers on the coasts of *Lincolnshire*, in nets such as employed in taking ruffs; with two or three

three dozens of ſtales of wood painted like the birds, placed within: fourteen dozens have been taken at once. Their ſeaſon is from the beginning of *Auguſt* to that of *November*. They diſappear with the firſt froſts. *Camden** ſays they derive their name from king *Canute*, *Knute*, or *Knout*, as he is ſometimes called; probably becauſe they were a favorite diſh with that monarch. We know that he kept the feaſt of the purification of the *Virgin Mary* with great pomp and magnificence at *Ely*, and this being one of the fen birds, it is not unlikely but he met with it there †. *Shakeſpear* in his *Othello*, ſpeaking of *Roderigo* (if Mr. *Theobald*'s reading is juſt) makes the *Knot* an emblem of a dupe:

> " I have rubb'd this young *Knot* almoſt to the ſenſe;
> " And he grows angry." *Othello.*

194. ASH COLORED.

Tringa cinerea. *Brunnich, ornith.* 53.
Br. Zool. 124.

Braun und Weisſbunter Sandlœuffer? *Friſch*, II. 237.

DESCRIP.

THIS ſpecies weighs five ounces: the length is ten inches: the breadth nineteen: the head is of a browniſh aſh color, ſpotted with black: the

* *Camden Brit.* 971.
† *Dugdale on embanking*, 185.

whole

whole neck ash color, marked with dusky oblong streaks: the back and coverts of the wings elegantly varied with concentric semicircles of ash color, black and white: the coverts of the tail barred with black and white: the tail ash colored, edged with white: the breast and belly of a pure white: the legs of a greenish black: the toes bordered with a narrow membrane, finely scolloped.

These birds appear on the shores of *Flintshire*, in the winter time, in large flocks.

195. BROWN.

THIS species is in the collection of Mr. *Tunstal*, is of the size of a jack-snipe. The bill is black: the head, upper part of the neck, and back, are of a pale brown, spotted with black: coverts of the wings dusky, edged with dirty white: under side of the neck white, streaked with black: the belly white: tail cinereous: legs black.

Bought in the *London* market.

196. SPOTTED.

Spotted Tringa. *Edw. av.* 277.
Turdus aquaticus, la Grive d'Eau. *Brisson av.* V. 255.
Tringa macularia. *Lin. syst.* 249.
Br. Zool. 124.

THIS bird is common to *Europe* and *America*; according to Mr. *Edwards*'s figure, it is less than the preceding.

The

DESCRIP. The bill is of the ſame colors with that of the red ſhank: the head, upper part of the neck, the back and coverts of the wings, are brown, inclining to olive, and marked with triangular black ſpots: above each eye is a white line: the greater quil-feathers are wholly black, the leſſer tipt with white: the middle feathers of the tail are brown: the ſide feathers white, marked with duſky lines: the whole under ſide, from neck to tail, is white, marked with dusky ſpots: the female has none of theſe ſpots, except on the throat: the legs of a dusky fleſh color. Mr. *Edwards* imagines theſe to be birds of paſſage; the bird he toke his deſcription from was ſhot in *Eſſex*.

197. BLACK. MR. *Bolton* favored us with a deſcription of this ſpecies ſhot in *Lincolnſhire*.

DESCRIP. It was the ſize of a thruſh: the beak ſhort, blunt at the point and dusky: the noſtrils black: the irides yellow: the head ſmall and flatted at top: the color white, moſt elegantly ſpotted with grey: the neck, ſhoulders, and back mottled in the ſame manner, but darker, being tinged with brown; in ſome lights theſe parts appeared of a perfect black and gloſſy: the wings were long: the quil-feathers black, croſſed near their baſe with a white line: the throat, breaſt, and belly white, with faint brown and black ſpots of a longiſh

Pl. LXX.

GAMBET.

longiſh form, irregularly diſperſed; but on the belly become larger and more round; the tail ſhort, entirely white, except the two middle feathers, which are black:. the legs long and ſlender, and of a reddiſh brown color.

198. GAMBET.

Tringa Gambetta. *Lin. ſyſt.* 248. *Faun. Suec.* No. 177.
Gambetta. *Wil. orn.* 300. *Raii ſyn. av.* 117. *Aldr. av.*
Totanus ruber. *Briſſon,* V. 192. *Scopoli,* No. 142.
Tringa varieguta. *Brunnich,* No. 181.

THIS ſpecies is of the ſize of the *Green-ſhank*: the head, back, and breaſt cinereous brown, ſpotted with dull yellow: the coverts of the wings, ſcapulars, cinereous, edged with yellow: the primaries dusky: the ſhaft of the firſt feather white: belly white: tail dusky, bordered with yellow: legs yellow.

This ſpecies has been ſhot on the coaſt of *Lincolnſhire.*

199. TURNSTONE.

Turnſtone, or Sea Dottrel. *Wil. orn.* 311.
Cat. Carol. I. 72.
Morinellus Marinus. *Raii ſyn. av.* 112.
La Coulon-chaud, Arenaria. *Briſſon av.* V. 132.
Tringa Morinellus. *Lin. ſyſt.* 249.
Br. Zool. 125.

THIS ſpecies is about the ſize of a thruſh: the bill is an inch in length, a little prominent on on the top; is very ſtrong; black at the tip, and at the baſe whitiſh: the forehead and throat are aſh colored: the head, whole neck and coverts of the wings are of a deep brown, edged with a pale reddiſh brown: the ſcapular feathers are of the ſame color, very long, and cover the back: that and the rump are white; the laſt marked with a large triangular black ſpot: the tail conſiſts of twelve feathers, their lower half is white, the upper black, and the tips white: the quil-feathers are duſky, but from the third or fourth the bottoms are white, which continually increaſes, till from about the nineteenth the feathers are entirely of that color: the legs are ſhort and of an orange color.

Theſe birds take their name from their method of ſearching for food, by turning up ſmall ſtones with their ſtrong bills to get at the inſects that lurk under them. The bird we toke our deſcription from was ſhot in *Shropſhire.* Mr. *Ray* obſerved them flying three or four in company on the coaſts of *Cornwal* and *Merionethſhire:* and Sir *Thomas Brown* of *Norwich* diſcovered them on the coaſt of *Norfolk*; communicating the picture of one to Mr. *Ray*, with the name of *Morinellus marinus*, or ſea dottrel.

Tringa

Tringa interpres. *Lin. syst.* 248. *Faun. Suec.* No. 178.
Turnstone from *Hudson's Bay*. *Edw.* 141.
Arenaria, Le Coulon-chaud. *Brisson*, V. 132.

200. HEBRIDAL.

THIS species is often shot in the north of *Scotland*, and its islands; also in *North America*.

Is of the size of a thrush: forehead, throat, and belly white: breast black: neck surrounded with a black collar; from thence another bounds the sides of the neck, and passes over the forehead: head and lower part of the neck behind white; the first streaked with dusky lines: back ferruginous, mixed with black: coverts of the tail white, crossed with a black bar: tail black, tipt with white: coverts of the wings cinereous brown; the lower order edged with white: primaries and secondaries black; the ends of the last white: tertials ferruginous and black: legs rather short, and of a full orange.

201. GREEN. Cinclus. *Belon av.* 216.
Gallinæ aquaticæ secunda species de nov. adject. *Gesner av.* 511.
Giarolo, Gearoncello. *Aldr. av.* III. 185.
The *Tringa* of *Aldrovand. Wil. orn.* 300.
Raii syn. av. 108.
Tringa ochropus. *Lin. syst.* 251.
Weisſpunotirto Sandlæuffer. *Frisch*, II. 239.
Faun. Suec. sp. 180.
Le Beccasseau ou Cul-blanc, Tringa. *Brisson av.* V. 177. *tab.* 16. *fig.* 1.
Danis Horse-Gioeg. *Islandis* Hrossagaukr. *Norvegis* Skodde Foll, Skod-de-Fugl. Jordgeed. Makkre-Gouk, Rœs Jouke. *Brunnich*, 183.
Br. Zool. 125.

DESCRIP. THIS beautiful species is not very common in these kingdoms. The head and hind part of the neck are of a brownish ash color, streaked with white; the under part mottled with brown and white: the back, scapulars, and coverts of the wings are of a dusky green, glossy and resplendent as silk, and elegantly marked with small white spots: the lesser quil-feathers of the same colors: the under sides of the wings are black, marked with numerous white lines, pointing obliquely from the edges of the feather to the shaft, representing the letter V: the rump is white; the tail of the same color: the first feather plain, the second marked near the end with one black spot, the third and fourth with two, the fifth with three, and the sixth with four.

Except in pairing time, it is a solitary bird: it is never found near the sea; but frequents rivers, lakes, and other fresh waters. In *France* it is highly

highly efteemed for its delicate tafte; and is taken with limed twigs placed near its haunts.

Mr. *Fleifcher* favored us with a bird from *Denmark*, which, in all refpects, refembled this, except that the fpots were of a pale ruft color: *Linnæus* defcribes it under the title of *Tringa littorea, Faun. Suec. fp.* 185. but we believe it does not differ fpecifically from that above defcribed.

Tringa Icelandica. *Lin. fyft. inter addenda.*
Tringa ferruginea *Iflandis*

Randbriflanger. *Brunnich,* No. 180.

202. RED.

BIRDS of this fpecies have appeared in great flocks on the coaft of *Effex*, on the eftate of Col. *Schutz*.

Crown of the head fpotted with black and ferruginous. The lower fide of the neck, the breaft, and belly of a full ferruginous color: back marked with black and ruft color: coverts of the wings afh color: legs black: bill ftrong, an inch and a half long: the whole length of the bird ten inches.

La Maubeche tachetée. *Briffon* V. 229?

203. ABERDEEN.

THIS was communicated by the late Doctor *David Skene* of *Aberdeen*.

I i 2 Bill

Bill ſlender and black: head, back, leſſer coverts of the wings, and the ſcapulars, of a dull ferruginous color, ſpotted with black: the greater coverts tipt with white: quil-feathers duſky, edged on the exterior ſide with white: breaſt reddiſh brown, mixed with duſky: belly and vent white: tail cinereous; two middle feathers longer than the reſt: legs black; ſize of the former.

204. COMMON.

Gallinula hypoleucos (Fyſterlin). *Geſner av.* 509.
Aldr. av. III. 182.
Wil. orn. 301.
Raii ſyn. av. 108.
Sandlaufferl. *Kram.* 353.
Tringa hypoleucos. *Lin. ſyſt.* 250.
Snappa, Strandſittare. *Faun. Suec. ſp.* 182.

Guinetta, la Guignette. *Briſſon av.* V. 183. *tab.* 16. *fig.* 1.
Norvegis der lille Myrſtikkel. *Bornholmis* Virlen. *Brunnich*, 174.
Br. Zool. 125.
Martin's *Scopoli*, No. 143.

THIS ſpecies agrees with the former in its manners and haunts; but is more common: its note is louder and more piping than others of this genus. Its weight is about two ounces: the head is brown, ſtreaked with downward black lines; the neck an obſcure aſh color: the back and coverts of the wings brown, mixed with a gloſſy green, elegantly marked with tranſverſe duſky lines: over each eye is a white ſtroke: the breaſt and belly are of a pure white: the quil-feathers are brown, the firſt entirely ſo, the nine next marked on the inner

DESCRIP.

web

PURR *No. 206*

SANDPIPER *No. 204*

web with a white ſpot: the middle feathers of the tail brown; edges ſpotted with black and pale red: the exterior tipt and barred with white: the legs of a dull pale green.

Wil. orn. 205.
Raii ſyn. av. 109.
Tringa alpina. *Lin. ſyſt.* 249.
Faun. Suec. ſp. 181.
La Beccaſſine d'Angleterre. *Briſſon av.* V. 309.

Danis Domſneppe, Ryle. *Brunnich*, 167, & 173.
Kleinſte Schnepfe, or Kleinſte Sandlœuffer. *Friſch*, II. 241.
Br. Zool. 126. *tab. fig.* 2.

205. DUNLIN.

DESCRIP.

THIS ſpecies is at once diſtinguiſhed from the others by the ſingularity of its colors. The back, head, and upper part of the neck are ferruginous, marked with large black ſpots: the lower part of the neck white, marked with ſhort duſky ſtreaks: the coverts of the wings aſh color: the belly white, marked with large black ſpots, or with a black creſcent pointing towards the thighs: the tail aſh colored, the two middle feathers the darkeſt: legs black: toes divided to their origin. In ſize it is ſuperior to that of a lark. Theſe birds are found on our ſea coaſts; but may be reckoned among the more rare kinds. They lay four eggs of a dirty white color, blotched with brown round the thicker end, and marked with a few ſmall ſpots of the ſame color on the ſmaller end. I received the eggs from *Denmark*; but as I have ſhot theſe birds in *May*, and again in *Auguſt*, on the ſhores of

Flintshire, suppose they breed with us; but I never discovered their nest. They are common on the *Yorkshire* coasts, and esteemed a great delicacy.

206. PURRE. L'Allouette de Mer. *Belon av.* 213.
Cinclus sive Motacilla Maritima, Lysklicker. *Gesner av.* 616.
Giarolo. *Aldr. av.* III. 188.
The Stint. *Wil. orn.* 305.
Stint, in *Sussex* the Ox-eye. *Raii syn. av.* 110.
N. Com. Petr. IV. 428.
L'Allouette de Mer, Cinclus. *Brisson av.* V. 211. *tab.* 19. *fig.* 1.
Tringa cinclus. *Lin. syst.* 251.
Br. Zool. 126.

DESCRIP. THIS bird weighs about an ounce and a half: length seven inches and a half; extent fourteen inches: the head and hind part of the neck are ash colored, marked with dusky lines: a white stroke divides the bill and eyes: the chin white: underside of the neck mottled with brown: the back is of a brownish ash color: the breast and belly white: the coverts of the wings and tail a dark brown, edged with light ash color or white: the greater coverts dusky, tipt with white: the upper part of the quil-feathers dusky, the lower white: the two middle feathers of the tail dusky, the rest of a pale ash color, edged with white: the legs of a dusky green; the toes divided to their origin. The bill an inch and a half long, slender and black; irides dusky.

These birds come in prodigious flocks on our sea

ſea coaſts during the winter: in their flight they perform their evolutions with great regularity; appearing like a white, or a dusky cloud, as they turn their backs or their breaſts towards you. They leave our ſhores in ſpring, and retire to ſome unknown place to breed.

They were formerly a well known diſh at our tables; known by the name of *Stints.*

207. LITTLE.

THIS is the leſt of the genus, ſcarcely equalling a hedge ſparrow in ſize. The head, upper ſide of the neck, the back, and coverts of the wings brown, edged with black and pale ruſty brown. Breaſt and belly white.

The greater coverts dusky, tipt with white: the primaries and ſecondaries of the ſame colors. The tail dusky. Legs black.

This ſpecimen was communicated to me by the Rev. Mr. *Green*, of *Trinity College*, *Cambridge*; and was ſhot near that place in *September*. It is common to *North America* and *Europe*.

XXXII. PLOVER.

BILL ſtrait, no longer than the head.
NOSTRILS linear.
TONGUE
TOES, wants the hind toe.

208. GOLDEN.

Le Pluvier Guillemot. *Belon av.* 260.
Pluvialis. *Geſner av.* 714.
Pivier. *Aldr. av.* III. 206.
Wil. orn. 308.
Raii ſyn. av. 111.
Brachhennl. *Kram.* 354.
Rechter Brachvogel. *Friſch*, II. 217.
Charadrius Pluvialis. *Lin. ſyſt.* 254.
Dalekarlis Akerhona, *Lappis* Hutti. *Faun. Suec. ſp.* 190.
Pluvialis aurea, le Pluvier doré. *Briſſon av.* V. 43. *Tab.* 4. *fig.* 1.
Piviero verde. *Zinan.* 102.
Norvegis Akerloe, *Cimbris* Brok-Fugl. *Brunnich*, 187.
Br. Zool. 128.

DESCRIP.

THIS elegant ſpecies is often found on our moors and heaths, in the winter time, in ſmall flocks. Its weight is nine ounces: its length eleven inches: its breadth twenty-four: the bill is ſhort and black: the feathers on the head, back, and coverts of the wings are black, beautifully ſpotted on each ſide with light yellowiſh green: the breaſt brown, marked with greeniſh oblong ſtrokes: the belly white: the middle feathers of the tail barred with black and yellowiſh green: the reſt with black and brown: the legs black. We have obſerved ſome variety in theſe birds, but cannot determine whether it is owing to age or ſex: we

RED SAND-PIPER.

GOLDEN PLOVER.

we have ſeen ſome with black bellies, others with a mixture of black and white; others with bluiſh legs, and ſome with a ſmall claw in the place of the hind toe.

They lay four eggs, ſharply pointed at the leſſer end, of a dirty white color, and irregularly marked, eſpecially at the thicker end, with black blotches and ſpots. It breeds on ſeveral of our unfrequented mountains; and is very common on thoſe of the iſle of *Rum*, and others of the loftier *Hebrides*. They make a ſhrill whiſtling noiſe: and may be inticed within ſhot by a ſkilful imitator of the note.

This ſpecies, on account of its ſpots, has been ſuppoſed to have been the *Pardalis* of *Ariſtotle:* but his account of the bird makes no mention of that diſtinction: perhaps he thought that the name implied it. The *Romans* ſeem to have been unacquainted with the plover: for the name never once occurs in any of their writings. We derive it from the *French* PLUVIER, *pource qu'on le prend mieux en temps pluvieux qu'en nulle autre ſaiſon**.

* *Belon Oyſeaux.* 260.

Le

209. LONG LEGGED.

Le grand Chevalier d'Italie. *Belon Portr. d'Oyseaux*, 53.
Aldr. av. III. 176.
Gesner av. 546.
Himantopus. *Wil. orn.* 297.
Raii syn. av. 106.
Sibb. Scot. 19. *Tab.* 11. 13.
L'Echasse. *Brisson av.* V. 33. *Tab.* 3. *fig.* 1.
Charadrius himantopus. *Lin. syst.* 255. *Scopoli*, No. 148.
Br. Zool. 128. *add.* plates.

THIS is the most singular of the *British* birds. The legs are of a length, and weakness greatly disproportioned to the body, which is inferior in size to that of the green plover: this, added to the defect of the back toe, must render its paces aukward and infirm. The naked part of the thigh is three inches and a half long; the legs four and a half: these, and the feet are of a blood red: the bill is black, above two inches long. The length from its tip to the end of the tail is thirteen inches: the breadth from tip to tip of the wing twenty-nine inches: the forehead, and whole under side of the body are white: the crown of the head, back, and wings black: on the hind part of the neck are a few black spots: the tail is of a greyish white: the wings when closed extend far beyond it. These birds are extremely rare in these islands: Sir *Robert Sibbald* records a brace that were shot in *Scotland:* another was shot a few years ago on *Stanton-Harcourt* common near *Oxford*, and we have seen them

DESCRIP.

SANDERLING.

DOTTEREL. No. 210

them often in the cabinets of the curious at *Paris*, taken on the *French* coasts.

Morinellus avis anglica. *Gesner av.* 615.
Wil. orn. 309.
Raii syn. av. 111.
Camden. Brit. I. 570.
Pluvialis minor, sive morinellus, le petit Pluvier, ou le Guignard. *Brisson av.* V. 54. *Tab.* 4. *fig.* 2.
Charadrius morinellus. *Lin. syst.* 254.
Lappis Lahul. *Faun. Suec. sp.* 188.
Caii opusc. 96.
Cimbris Pomerants Fugl. *Norvegis* Bold Ticet. Mindre Akerloe. *Brunnich*, 185.
Br. Zool. 129.

210. DOTTREL.

THE female dottrel, according to Mr. *Willughby*, weighs more than four ounces; the male above half an ounce less. The length of the female ten inches; the breadth nineteen and a half: the male not so large. The bill black, slender, depressed in the middle, and not an inch long: the forehead, top and back of the head black, the former spotted with white; a broad white stroke that presses over the eyes, surrounds the whole: the cheeks and throat are white: the neck of a cinereous olive color: the middle of the feathers of the back, and coverts of the wings and tail olive; but their edges of a dull deep yellow: the quill-feathers are brown, with brown shafts; but the exterior side and the shaft of the first feather is white. The tail consists of twelve feathers of a brown olive color, barred near their ends with black,

DESCRIP.

black, and tipped with white. The breaſt and ſides are of a dull orange color; but immediately above that is a line of white, bounded above with a very narrow one of black. The belly (in the male) is black: thighs and vent-feathers white: legs yellowiſh green: toes duſky.

FEMALE. The colors of the female in general are duller: the white over the eye is leſs; and the crown of the head is mottled with brown and white. The white line croſs the breaſt is wanting. The belly is mixed with black and white.

PLACE. Theſe birds are found in *Cambridgeſhire*, *Lincolnſhire*, and *Derbyſhire*: on *Lincoln-heath*, and on the moors of *Derbyſhire* they are migratory, appearing there in ſmall flocks of eight or ten only in the latter end of *April*, and ſtay there all *May* and part of *June*, during which time they are very fat, and much eſteemed for their delicate flavor. In the months of *April* and *September* they are taken on the *Wiltſhire* and *Berkſhire* downs: they are alſo found in the beginning of the former month on the ſea ſide at *Meales* in *Lancaſhire*, and continue there about three weeks, attending the barly fallows: from thence they remove northward to a place called *Leyton Haws*, and ſtay there about a fortnight; but where they breed, or where they reſide during winter, we have not been able to diſcover. They are reckoned very fooliſh birds, ſo that a dull fellow is proverbially called a *Dottrel*. They were alſo believed to mimick the action of the

the fowler; to ſtretch out a wing when he ſtretched out an arm, &c. continuing their imitation, regardleſs of the net that was ſpreading for them.

To this method of taking them, *Michael Drayton* alludes in his panegyrical verſes on *Coryate's Crudities:*

> Moſt worthy man with thee it is even thus,
> As men take *Dottrels*, ſo haſt thou ta'en us;
> Which as a man his arme or leg doth ſet,
> So this fond bird will likewiſe counterfeit.

At preſent, ſportſmen watch the arrival of the *Dottrels*, and ſhoot them; the other method having been long diſuſed.

211. RINGED.

Charadrius ſive hiaticula. *Aldr. av.* III. 207.
Wil. orn. 310.
Raii ſyn. av. 112.
Grieſshennl. *Kram.* 354.
Charadrius hiaticula. *Lin. ſyſt.* 253. *Scopoli*, No. 147.
Strandpipare, Grylle, Trulls, *Lappis* Pago. *Faun. Suec. ſp.* 187.
Pluvialis torquata minor, le petit Pluvier a collier. *Briſſon av.* V. 63. *Tab.* 5. *fig.* 2.
Bornholmis Prœſte-Krave, Sand-Vriſter. *Brunnich*, 184. *Friſch*, II. 214.
Sea Lark. *Br. Zool.* II. 383.

DESCRIP.

IT weighs near two ounces. The length is ſeven inches and a half; the breadth ſixteen: the bill is half an inch long; the upper half orange color; the lower black; from it to the eyes is a black

black line; the cheeks are of the ſame color; the forehead white, bounded by a black band that paſſes over from eye to eye; the crown of the head is of a fine light brown; the upper part of the neck is incircled with a white collar; the lower part with a black one; the back and coverts of the wings of a light brown; the breaſt and belly white; the tail brown, tipt with a darker ſhade; the legs yellow.

Theſe birds frequent our ſhores in the ſummer, but are not numerous. They lay four eggs of a dull whitiſh color, ſparingly ſprinkled with black: at approach of winter they diſappear.

212. SANDERLING.

Sanderling, or Curwillet. *Wil. orn.* 303.
Raii ſyn. av. 109.
Towillee. *Borlaſe hiſt. Cornwal.* 247.
Calidris griſea minor, la petite Maubeche griſe. *Briſſon av.* V. 236. *Tab.* 20. *fig.* 2.
Charadrius Caladris. *Lin. ſyſt.* 255.
Br. Zool. 129. *add.* plates.

WE have received this ſpecies out of *Lancaſhire*; but it is found in greater plenty on the *Corniſh* ſhores, where they fly in flocks. The ſanderling weighs little more than one ounce three quarters. Its length is eight inches; extent fifteen. Its body is of a more ſlender form than others of the genus. The bill is an inch long, weak and black. The head, and hind part of the neck are aſh-

DESCRIP.

aſh-colored, marked with oblong black ſtreaks; the back and ſcapulars are of a browniſh grey, edged with dirty white; the coverts of the wings, and upper parts of the quil-feathers duſky: the whole under ſide of the body is white; in ſome ſlightly clouded with brown. The tail conſiſts of twelve ſharp pointed feathers of a deep aſh color; the legs are black.

BILL

XXXIII. OYSTER CATCHER.

BILL long, compreſſed, the end cuneated.
NOSTRILS linear.
TONGUE, a third the length of the bill.
TOES, only three.

213. PIED.

La Pie, Becaſſe de mer. *Belon av.* 203.
Hæmatopus. *Geſner av.* 548.
Aldr. av. III. 176.
Wil. orn. 297.
Raii ſyn. av. 105.
L'Hutrier, Pie de mer. *Briſſon av.* V. 38. *tab.* 3. *fig.* 1.
The Oyſter Catcher. *Cat. Carol.* I. 85. Hœmatopus oſtralegus. *Lin. ſyſt.* 257.
Marſpitt, Strandſkjura, *Faun. Suec. ſp.* 192.
Pica marina. *Caii opuſc.* 62. *N. Com. Petr.* IV. 425.
Tirma, or Trilichan. *Martin's voy.* St. *Kilda.* 35.
Iſlandis mas Tialldur, fœmina Tilldra. *Feroenſibus* Kielder. *Norvegis* Tield v. Kield, Glib, Strand-Skiure. *Danis* Strand-Skade. *Brunnich*, 189.
Br. Zool. 127.

SEA Pies are very common on moſt of our coaſts; feeding on marine inſects, oyſters, limpets, &c. Their bills, which are compreſſed ſideways, and end obtuſely, are very fit inſtruments to inſinuate between the limpet and the rock thoſe ſhells adhere to; which they do with great dexterity to get at the fiſh. On the coaſt of *France*, where the tides recede ſo far as to leave the beds of oyſters bare, theſe birds feed on them; forcing the ſhells open with their bills. They keep in ſummer time in pairs, laying their eggs on the bare ground: they

OYSTER-CATCHER.

they lay four of a whitiſh brown hue, thinly ſpotted and ſtriped with black: when any one approaches their young, they make a loud and ſhrill noiſe. In winter they aſſemble in vaſt flocks, and are very wild.

DESCRIP.

Weight ſixteen ounces; length ſeventeen inches. Bill three inches, compreſſed, obtuſe at the end, of a rich orange color: *irides* crimſon: edges of the eye-lids orange; beneath the lower a white ſpot. Head, neck, ſcapulars, and coverts of the wings a fine black; in ſome the neck marked with white: wings duſky, with a broad tranſverſe band of white: the back, breaſt, belly, and thighs white: tail ſhort, conſiſts of twelve feathers; the lower half white; the end black: legs thick and ſtrong; of a dirty fleſh color: middle toe connected to the exterior toe as far as the firſt joint by a ſtrong membrane: the claws duſky, ſhort and flat.

XXXIV. RAIL.

BILL ſlender, a little compreſſed, and ſlightly incurvated.

NOSTRILS ſmall.

TONGUE rough at the end.

TAIL very ſhort.

214. WATER.

Le Raſle noir. *Belon av.* 112.
Gallina cinerea (aſhhunlin). *Geſner av.* 515.
Ralla aquatica. *Aldr. av.* III. 179.
Water-rail, Bilcock, or Brook Ouzel. *Wil. orn.* 314.
Raii ſyn. av. 113.
Waſſer hennl *Kram.* 348.
Rallus aquaticus. *Lin. ſyſt.* 262.
Faun. Suec. ſp. 195.
Rallus aquaticus, le Raſle d'Eau. *Briſſon av.* 151. *tab.* 12. *fig.* 2. *Scopoli*, No 155.
Norvegis Vand-Rixe. *Feroenſibus* Jord-Koene. *Brunnich*, 193.
Br. Zool. 130.

THE water rail is a bird of a long ſlender body, with ſhort concave wings. It delights leſs in flying than running; which it does very ſwiftly along the edges of brooks covered with buſhes: as it runs, every now and then flirts up its tail; and in flying hangs down its legs: actions it has in common with the water hen.

DESCRIP.

Its weight is four ounces and a half. The length to the end of the tail twelve inches: the breadth ſixteen. The bill is ſlender, ſlightly incurvated, one inch three quarters long: the upper mandible black, edged with red; the lower orange colored: the

Nº 214.

WATER-RAIL.

CRAKE GALLINULE.

Nº 216.

the irides red: the head, hind part of the neck, the back, and coverts of the wings and tail are black, edged with an olive brown; the baſe of the wing is white; the quil-feathers and ſecondaries duſky: the throat, breaſt, and upper part of the belly are aſh-colored: the ſides under the wings as far as the rump finely varied with black and white bars. The tail is very ſhort, conſiſts of twelve black feathers; the ends of the two middle tipt with ruſt-color; the feathers immediately beneath the tail white. The legs are placed far behind, and are of a duſky fleſh-color. The toes very long, and divided to their very origin; though the feet are not webbed, it takes the water; will ſwim on it with much eaſe; but oftener is obſerved to run along the ſurface.

This bird is properly *ſui generis*, agreeing with no other, ſo forms a ſeparate tribe. M. *Briſſon* and *Linnæus* place it with the land Rail, and Mr. *Ray* with the water hens, which have their peculiar characters, ſo very diſtinct from the Rail, as to conſtitute another genus, as may be obſerved in the generical table preceding this claſs.

XXXV. GALLINULE.

BILL thick at the baſe ſloping to the point, the upper mandible reaching far up the forehead, callous.

WINGS ſhort and concave.

BODY compreſſed.

TOES long, divided to the origin.

215. SPOTTED.

Gallinula ochra (Wynkernell). *Geſner. av.* 513.
Porcellana, Porzana, Grugnetto. *Aldr. av.* III. 181.
Grinetta. *Wil. orn. ſp.* 8. p. 315.
Raii ſyn. av. 115. *ſp.* 7.
Rallus aquat. minor, ſive Maruetta, le petit Raſle d'Eau, ou la Marouette.
Briſſon av. V. 155. *tab.* 13. *fig.* 1.
Couchouan ou Marouette. *Argenv. Lithcl.* 533. *tab.* 25.
Kleines geſprenkeltes Waſſerhuhn. *Friſch*, II. 211.
Rallus porzana. *Lin. ſyſt.* 262.
Br. Zool. 130.

THIS ſpecies is not very frequent in *Great Britain*, and is ſaid to be migratory. Inhabits the ſides of ſmall ſtreams, concealing itſelf among the buſhes. Its length is nine inches; its breadth fifteen: its weight four ounces five drachms. The head is brown, ſpotted with black; the neck a deep olive, ſpotted with white; from the bill beyond the eyes is a broad grey bar: the feathers of the back are black next their ſhafts, then olive colored, and edged with white: the ſcapulars are olive, finely

DESCRIP.

finely marked with two ſmall white ſpots on each web: the legs of a yellowiſh green.

216. CRAKE.

Le Raſle rouge ou de Genet. *Belon av.* 212.
Ortygometra, Crex. *Geſner av.* 361, 362.
Aldr. av. III. 179.
Rail, or Daker Hen. *Wil. orn.* 170. *Phil. Tranſ.* II. 853.
Raii ſyn. av. 58.
Corn-crek. *Sib. Scot.* 16.
Corn-craker. *Martin's Weſt. Iſles*, 71.
Rallus geniſtarum, le Raſle de Genet, ou Roi des Cailles. *Briſſon av.* V. 159. *Tab.* 13. *fig.* 2.
Wachtel-konig. *Kram.* 349.
Rallus Crex. *Lin. ſyſt.* 261.
Angſnarpa, Korknarr, Seydreifwer. *Faun. Suec. ſp.* 194.
Danis & Norv. Vagtel-Konge. Aker-Rixe. Skov-Snarre, *Norvegis quibuſdam* Agerhoene. *Brunnich*, 192.
Br. Zool. 131.
Roſtz. *Scopoli*, No. 154.

THIS ſpecies has been ſuppoſed by ſome to be the ſame with the water rail, and that it differs only by a change of color at a certain ſeaſon of the year: this error is owing to inattention to their characters and nature, both which differ entirely. The bill of this ſpecies is ſhort, ſtrong, and thick; formed exactly like that of the water hen, and makes a generical diſtinction. It never frequents watery places, but is always found among corn, graſs, broom, or furze. It quits this kingdom before winter; but the water rail endures our ſharpeſt ſeaſons. They agree in their averſion to flight; and the legs, which are remarkably long for the ſize of the bird, hang down whilſt

Kk 3 they

they are on the wing; they truſt their ſafety to their ſwiftneſs of foot, and ſeldom are ſprung a ſecond time but with great difficulty. The land rail lays from twelve to twenty eggs, of a dull white color, marked with a few yellow ſpots; notwithſtanding this, they are not very numerous in this kingdom. Their note is ſingular, reſembling the word *Crex* often repeated They are in greateſt plenty in *Angleſea*, where they appear about the twentieth of *April*, ſuppoſed to paſs over from *Ireland*, where they abound: at their firſt arrival it is common to ſhoot ſeven or eight in a morning. They are found in moſt of the *Hebrides*, and the *Orknies*. On their arrival they are very lean, weighing only ſix ounces; but before they leave this iſland, grow ſo fat as to weigh above eight.

Descrip. The feathers on the crown of the head, hind part of the neck, and the back, are black, edged with bay color: the coverts of the wings of the ſame color; but not ſpotted: the tail is ſhort, and of a deep bay: the belly white: the legs aſh-colored.

La

La Poulette d'eau. *Belon av.* 211.
Ein wasserhen. *Gesner av.* 501.
Chloropus major nostra. *Aldr. av.* III. 177.
Common Water-hen, or Moor-hen. *Wil. orn.* 312.
Raii syn. av. 112.
Gallinula, la Poule d'eau. *Brisson av.* VI. 3. *Tab.* 1.
Gallinella aquatica, Porzanone. *Zinan.* 109.
Wasserhennl. *Kram.* 358.
Rothblæssige Kleine Wasserhuhn. *Frisch*, II. 209.
Fulica chloropus. *Lin. syst.* 258.
Brunnich, 191. *Scopoli*, No. 153.
Br. Zool. 131.

217. COMMON.

DESCRIP.

THE male of this species weighs about fifteen ounces. Its length to the end of the tail fourteen inches: the breadth twenty-two. The crown of the head, hind part of the neck, the back, and coverts of the wings are of a fine, but very deep olive green. Under side of the body cinereous: the chin and belly mottled with white: quil-feathers and tail dusky: exterior side of the first primary feather, and the ridge of the wings white: vent black: feathers just beneath the tail white: legs dusky green. The colors of the plumage in the female, are much less brilliant than that of the male: in size it is also inferior. Mr. *Willughby* in his description takes no notice of the beautiful olive gloss of the plumage of these birds; nor that the bill assumes a fuller and brighter red in the courting season.

It gets its food on grassy banks, and borders near

Kk 4 fresh

fresh waters, and in the very waters, if they be weedy. It builds upon low trees and shrubs by the water side; breeding twice or thrice in the summer; and when the young are grown up, drives them away to shift for themselves. They lay seven eggs of a dirty white color, thinly spotted with rust color. It strikes with its bill like a hen; and in the spring has a shrill call. In flying it hangs down its legs: in running often flirts up its tail, and shews the white feathers. We may observe, that the bottoms of its toes are so very flat and broad (to enable it to swim) that it seems the bird that connects the cloven-footed aquatics with the next tribe; the fin toed.

Eggs.

RED AND GREY SCOLLOP TOED SAND-PIPER.

SECT. II. FIN-FOOTED BIRDS.

XXXVI. PHALAROPE.

BILL ſtrait and ſlender.
NOSTRILS minute.
BODY and LEGS like the Sandpiper.
TOES furniſhed with ſcalloped membranes.

218. GREY.

Grey Coot footed Tringa. *Edw. av.* 308.
Phil. Tranſ. Vol. 50.
Le Phalarope. *Briſſon av.* VI. 12.

Tringa Lobata. *Lin. ſyſt.* 249.
Faun. Suec. ſp. 179.
Brunnich, 171.
Br. Zool. 126.

DESCRIP.

THIS is about the ſize of the common *Purre,* weighing one ounce. The bill black, not quite an inch long, flatted on the top, and channeled on each ſide; and the noſtrils are placed in the channels: the eyes are placed remarkably high in the head: the forehead white: the crown of the head covered with a patch of a duſky hue, ſpotted with white and a pale reddiſh brown; the reſt of the head, and whole under part of the neck and body are white: the upper part of the neck of a light grey: the back and rump a deep dove color, marked with duſky ſpots: the edges of the ſca-

pulars

pulars are dull yellow: the coverts dusky; the lower or larger tipt and edged with white: the eight first quil-feathers dusky; the shafts white; the lower part of the interior side white: the smaller quil-feathers are tipt with white: the wings closed, reach beyond the tail: the feathers on the back are either wholly grey or black, edged on each side with a pale red: the tail dusky, edged with ash-color: the legs are of a lead color: the toes extremely singular, being edged with scolloped membranes like the coot: four scollops on the exterior toe, two on the middle, and the same on the interior; each finely serrated on their edges.

This bird was shot in *Yorkshire*, and communicated to us by Mr. *Edwards*.

219. RED.

Mr. *Johnson*'s small cloven footed Gull. *Wil. orn.* 355.
Ray's collection of *English* words, &c. p. 92.

Larus fidipes alter nostras. *Raii syn. av.* 132.
Edw. av. 143.
Tringa hyperborea. *Lin. syst.* 249.

THIS species was shot on the banks of a fresh water pool on the isle of *Stronsa*, *May* 1769. It is of the size of the Purre. The bill is an inch long, black, very slender, and strait almost to the end which bends downwards: the crown of the head, the hind part of the neck and the coverts of the wings are of a deep lead color; the back and scapulars

ſcapulars the ſame, ſtriped with dirty yellow: the quil-feathers dusky; the ſhafts white: croſs the greater coverts is a ſtripe of white: the chin and throat white: the under part and ſides of the neck bright ferruginous: the breaſt dark, cinereous: belly white: coverts of the tail barred with black and white; tail ſhort, cinereous: legs and feet black.

Mr. *Ray* ſaw this ſpecies at *Brignal* in *Yorkſhire*: Mr. *Edwards* received the ſame kind from *North America*, being common to the *North* of *Europe* and *America*.

Short

XXXVII. COOT.

Short thick BILL, with a callus extending up the forehead.

NOSTRILS narrow and pervious.

TOES furniſhed with broad ſcalloped membranes.

220. COMMON.

La Poulle d'eau. *Belon av.* 181.
Fulica recentiorum. *Geſner av.* 390.
Follega, Follata, Fulca. *Aldr. av.* III. 39, 42.
Wil. orn. 319.
Raii ſyn. av. 116.
La Foulque, ou Morrelle. *Briſſon av.* VI. 23. *tab.* 2. *fig.* 1.
Folaga, o Polon. *Zinan.* 108.
Rohr-hennl, Blasſl. *Kram.* 357.
Weiſblæſſige groſſe Waſſer-huhn. *Friſch*, II. 208.
Fulica atra. *Lin. ſyſt.* 257.
Blas-klacka. *Faun. Suec. ſp.* 193.
Danis Vand-Hoene, Bles-Hoene. *Brunnich*, 190.
Br. Zool. 132.
Liſka. Scopoli, No. 149.

DESCRIP.

THESE birds weigh from twenty-four to twenty-eight ounces. Their belly is aſh-colored; and on the ridge of each wing is a line of white: every part beſides is of a deep black: the legs are of a yellowiſh green: above the knee is a yellow ſpot.

Coots frequent lakes and ſtill rivers: they make their neſt among the ruſhes, with graſs, reeds, &c. floating on the water, ſo as to riſe and fall with it. They lay five or ſix large eggs, of a dirty whitiſh hue, ſprinkled over with minute deep ruſt color ſpots; and we have been credibly informed that they

LXXVII. No 217.

COMMON GALLINULE.

COOT. No 220.

they will ſometimes lay fourteen and more. The young when juſt hatched are very deformed, and the head mixed with a red coarſe down. In winter they often repair to the ſea: we have ſeen the channel near *Southampton* covered with them: they are often brought to that market, where they are expoſed to ſale, without their feathers, and ſcalded like pigs. We once ſaw at *Spalding*, in *Lincolnſhire*, a coot ſhot near that place that was white, except a few of the feathers in the wings, and about the head.

221. GREAT.

Fulica aterrima. *Lin.* 258. *Scopoli*, No. 150.
Greater Coot. *Wil. orn.* 320. *Belon* 182.
La grand foulque ou la Macroule. *Briſſon av.* VI. 28.

THIS ſpecies differs from the preceding only in its ſuperior ſize; and the exquiſite blackneſs of the plumage.

Diſcovered in *Lancaſhire* and in *Scotland*.

BILL

XXXVIII. GREBE *.

BILL ſtrong, ſtrait, ſharp pointed.

TAIL, none.

LEGS flat, thin, and ſerrated behind with a double row of notches.

222. TIPPET. Colymbus major. *Geſner av.* 138.
Aldr. av. III. 104.
Greater Loon, or Arsfoot. *Wil. orn.* 339.
Greater Dobchick. *Edw. av.* 360. *fig.* 2.
Raii ſyn. av. 125.
Colymbus, la Grebe. *Briſſon av.* VI. 34. *tab.* 3. *fig.* 1.
Colymbus urinator *Lin. ſyſt.* 223 *Scopoli*, No. 102.
Br. Zool. 133.

THIS differs from the *great creſted Grebe* in being rather leſs, and wanting the creſt and ruff. The ſides of the neck are ſtriped downwards from the head with narrow lines of black and white: in other reſpects the colors and marks agree with that bird.

This ſpecies has been ſhot on *Roſterne Mere* in *Cheſhire*; is rather ſcarce in *England*, but is common in the winter time on the lake of *Geneva*. They appear there in flocks of ten or twelve: and are killed for the ſake of their beautiful ſkins.

* The *Grebes* and *Divers* are placed in the ſame genus, *i. e.* of *Colymbi*, by Mr. *Ray* and *Linnæus*; but the difference of the feet, forbade our judicious friend, M. *Briſſon*, from continuing them together; whoſe example we have followed.

The

№ 225.
DUSKY GREBE.

TIPPET GREBE.
№ 222.

The under side of them being drest with the feathers on, are made into muffs and tippets; each bird sells for about fourteen shillings.

223. GREAT CRESTED.

Grand Plongeon de riviere. *Belon av.* 178.
Ducchel. *Gesner av.* 138.
Aldr. av. III 104.
Avis pugnax 8va. *Aldr.* 169.
Greater crested and horned Doucker *Wil orn.* 340.
Ash-colored Loon of Dr. *Brown, ibid. Raii syn. av.* 124.
Plott's hist. Staff. 229. *tab.* 22.
The Cargoose. *Charleton ex.* 107.
Pet. Gaz. I. *tab.* 43. *fig.* 12.
Colymbus cristatus. *Lin. syst.* 222. *Scopoli*, No. 99.
Faun. Suec. sp. 151.
La Grebe hupée. *Brisson av.* VI. 38. *tab.* 4. et Colymbus cornutus. 45. *tab.* 5. *fig.* 1.
Smergo, Fisolo marino. *Zinan.* 107.
Danis Topped og Halskraved Dykker, Topped Hav Skiœre. *Brunnich*, 135.
Gehoernter Scehahn, Noerike. *Frisch*, II. 183.
Br. Zool. 132.

DESCRIP.

THIS species weighs two pounds and a half. Its length is twenty-one inches: the breadth thirty: the bill is two inches one-fourth long; red at the base; black at the point: between the bill and the eyes is a stripe of black naked skin: the irides are of a fine pale red: the tongue is a third-part shorter than the bill, slender, hard at the end, and a little divided: on the head is a large dusky crest, separated in the middle. The cheeks and throat are surrounded with a long pendent ruff, of a bright tawny color, edged with black: the chin is white: from the bill to the eye is a black line,

line, and above that a white one: the hind part of the neck, and the back are of a footy hue: the rump, for it wants a tail, is covered with long foft down.

The covert feathers on the fecond and third joints of the wing, and the under coverts are white: all the other wing feathers, except the fecondaries, are dufky, thofe being white: the breaft and belly are of a moft beautiful filvery white, gloffy as fattin, and equal in elegance to thofe of the *Grebe* of *Geneva*; and are applied to the fame ufes: the plumage under the wings is dufky, blended with tawny: the outfide of the legs, and the bottom of the feet are dufky: the infide of the legs, and the toes of a pale green.

Thefe birds frequent the *Meres* of *Shropfhire* and *Chefhire*, where they breed; and in the great *Eaft Fen* in *Lincolnfhire*, where they are called *Gaunts*. Their fkins are made into tippets, which are fold at as high a price as thofe that come from *Geneva*.

This fpecies lays four eggs, white, and of the fize of thofe of a pigeon; the neft is formed of the roots of bugbane, ftalks of water lilly, pond weed and water violet, floating independent among the reeds and flags; the water penetrates it, and the bird fits and hatches the eggs in that wet condition; the neft is fometimes blown from among the flags into the middle of the water: in thefe circumftances, the fable of the *Halcyon*'s neft, its

fluctivaga

fluctivaga domus, as *Statius* expreſſes it, may in ſome meaſure be vindicated.

> Fluctivagam ſic ſæpe domum, madidoſque penates
> *Halcyone* deſerta gemit; cum pignora ſævus
> Auſter, et algentes rapuit *Thetis* invida nidos.
>
> *Thebaid.* lib. ix. 360.

It is a careful nurſe of its young, being obſerved to feed them moſt aſſiduouſly, commonly with ſmall ells; and when the infant brood are tired, will carry them either on its back or under its wings. This bird preys on fiſh, and is almoſt perpetually diving: it does not ſhew much more than the head above water, and is very difficult to be ſhot, as it darts down on the appearance of the leſt danger. It is never ſeen on land; and though diſturbed ever ſo often, will not fly farther than the end of the lake. Its ſkin is out of ſeaſon about *February*, loſing then its bright color: and in the breeding time its breaſt is almoſt bare. The fleſh of this bird is exceſſively rank: but the fat is of great virtue in rheumatic pains, cramps and paralytic contractions.

224. EARED. Eared dobchick. *Edw. av.* 96. *fig.* 2.
La Grebe a Oreilles. *Briſſon av.* VI. 54.
Colymbus auritus. *Lin. ſyſt.* 223. *Scopoli*, No. 100.
Norvegis Sav-Orre, Soe-Orre. *Bornholmis* Soe-Hoene. *Iſlandis* Flaueſkitt. *Brunnich*, 136.
Br. Zool. 133.

DESCRIP. THE length of this ſpecies to the rump is one foot; the extent twenty-two inches: the bill black, ſlender and very ſlightly recurvated: the *irides* crimſon: the head and neck are black; the throat ſpotted with white: the whole upper ſide of a blackiſh brown, except the ridge of the wing about the firſt joint, and the ſecondary feathers, which are white: the breaſt, belly, and inner coverts of the wings are white: the ſubaxillary feathers, and ſome on the ſide of the rump, furruginous: behind the eyes, on each ſide, is a tuft of long looſe ruſt colored feathers, hanging backwards: the legs of a duſky green.

Theſe birds inhabit the fens near *Spalding*, where they breed. I have ſeen both male and female, but could not obſerve any external difference. They make their neſt not unlike that of the creſted grebe; and lay four or five ſmall white eggs.

The

BLACK CHIN GREBE.

EARED GREBE. No. 224.

The black and white Dobchick. *Edw. av.* 96. *fig* 1.
Colymbus minor, la petite Grebe. *Brisson av.* VI. 56.
Br. Zool. 133.
Colymbus nigricans? *Scopoli,* No. 101.

225. DUSKY.

DESCRIP.

THE length from the bill to the rump eleven inches: the extent of wings twenty: the bill was little more than an inch long. The crown of the head, and whole upper side of the body dusky: the inner coverts, the ridge of the wing, and the middle quil-feathers were white; the rest of the wing dusky: a bare skin of a fine red color joined the bill to the eye: the whole underside from the breast to the rump was a silvery white: on the thighs were a few black spots. In some birds the whole neck was ash colored: so probably they might have been young birds, or different in sex. Inhabits the Fens of *Lincolnshire.*

Le Castagneux, ou Zoucet. *Belon av.* 177.
Mergulus fluviatilis (Ducchelin, Arsfuss). *Gesner av.* 141.
Trapazorola arzauolo, Piombin. *Aldr. av.* III. 105.
Didapper, Dipper, Dobchick, small Doucker, Loon, or Arsfoot. *Wil orn.* 340.
Raii syn. av. 125.
Colymbus fluviatilis, la Grebe de Riviere, ou le Castagneux. *Brisson av.* VI. 59.
Colymbus auritus. *Lin. syst.* 223.
Kleiner Seehahn, or Noerike. *Frisch,* II. 184.
Faun. Suec. sp. 152.
Br. Zool. 134.

226. LITTLE.

DESCRIP.

THE weight of this species is from six to seven ounces. The length to the rump ten inches:

L l 2 to

to the end of the toes thirteen: the breadth ſix-teen. The head is thick ſet with feathers, thoſe on the cheeks, in old birds, are of a bright bay: the top of the head, and whole upper ſide of the body, the neck and breaſt, are of a deep brown, tinged with red: the greater quil-feathers duſky: the interior webs of the leſſer white: the belly is aſh colored, mixed with a ſilvery white, and ſome red: the legs of a dirty green.

The wings of this ſpecies, as of all the other, are ſmall, and the legs placed far behind: ſo that they walk with great difficulty, and very ſeldom fly. They truſt their ſafety to diving; which they do with great ſwiftneſs, and continue long under water. Their food is fiſh, and water plants. This bird is found in rivers, and other freſh waters. It forms its neſt near their banks, in the water; but without any faſtening, ſo that it riſes and falls as that does. To make its neſt it collects an amazing quantity of graſs, water-plants, &c. It lays five or ſix white eggs; and always covers them when it quits the neſt. It ſhould ſeem wonderful how they are hatched, as the water riſes through the neſt, and keeps them wet; but the natural warmth of the bird bringing on a fermentation in the vege-tables, which are full a foot thick, makes a hot bed fit for the purpoſe.

NEST.

GR.

GR. with a black chin. Fore part of the neck ferruginous: hind part mixed with duſky. Belly cinereous and ſilver intermixed. Rather larger than the laſt.

227. BLACK CHIN.

Inhabits *Tiree*, one of the *Hebrides*.

SECTION III. WEB-FOOTED BIRDS.

XXXIX. AVOSET.

BILL long, ſlender, very thin, depreſſed, bending upwards.

NOSTRILS narrow, pervious. TONGUE ſhort.

LEGS very long. FEET palmated. Back toe very ſmall.

228. SCOOPING.

Recurviroſtra. *Geſner av.* 231.
Avoſetta, Beccoſtorto, Beccoroella, Spinzago d'acqua. *Aldr. av.* III. 114.
Wil. orn. 321.
Raii ſyn. av. 117.
The Scooper. *Charlton ex.* 102.
The crooked Bill. *Dale's hiſt. Harwich*, 402.
Plott's hiſt. Staff. 231.
Avoſetta, L'Avocette. *Briſſon av.* VI. 538. *Tab.* 47. *fig.* 2.
Krumbſchnabl. *Kram.* 348.
Recurviroſtra Avoſetta. *Lin. ſyſt.* 256. *Scopoli*, No. 129.
Skarflacka, Alfit. *Faun. Suec. ſp.* 191.
Danis Klyde, Loufugl, Forkeert Regnſpove. *Br.* 188.
Br. Zool. 134.

AN *Avoſet* that we ſhot weighed thirteen ounces. Its length to the end of the tail was eighteen inches, to that of the toes twenty-two: the breadth thirty. This bird may at once be diſtinguiſhed from all others, by the ſingular form of its bill; which is three inches and a half long, ſlender, compreſſed very thin, flexible, and of a ſubſtance like whalebone; and contrary to the bills of

LXXX.
N.º 228.
AVOSET

of other birds, is turned up for near half its length. The noſtrils are narrow and pervious: the tongue ſhort: the head very round: that, and half the hind part of the neck black; but above and beneath each eye is a ſmall white ſpot: the cheeks, and whole under ſide of the body from chin to tail is of a pure white: the back, exterior ſcapular feathers, the coverts on the ridge of the wings, and ſome of the leſſer quil-feathers, are of the ſame color; the other coverts, and the exterior ſides and ends of the greater quil-feathers, are black: the tail conſiſts of twelve white feathers: the legs are very long, of a fine pale blue color, and naked far above the knees: the webs duſky, and deeply indented: the back toe extremely ſmall.

Theſe birds are frequent in the winter on the ſhores of this kingdom: in *Glouceſterſhire*, at the *Severn's Mouth*; and ſometimes on the lakes of *Shropſhire*. We have ſeen them in conſiderable numbers in the breeding ſeaſon near *Foſſdike Waſh* in *Lincolnſhire*. Like the lapwing when diſturbed they flew over our heads, carrying their necks and long legs quite extended, and made a ſhrill noiſe *(Twit)* twice repeated, during the whole time. The country people, for this reaſon, call them *Yelpers*; and ſometimes diſtinguiſh them by the name of *Picarini*. They feed on worms and inſects that they ſcoop with their bills out of the ſand; their ſearch after food is frequently to be diſcerned on our ſhores by alternate ſemicircular marks in the ſand, which ſhew their progreſs. They lay two eggs about the ſize of thoſe of a pigeon, white tinged with green, and marked with large black ſpots.

BILL

Pl. LXXXI.

No. 229.

GREAT AUK.

XL. AUK.

BILL strong, thick, compressed.

NOSTRILS linear; placed near the edge of the mandible.

TONGUE almost as long as the bill.

TOES, no back toe.

229. GREAT.

Goirfugel. *Clusii exot.* 367.
Penguin. *Wormii*, 300.
Wil. orn. 323.
Raii syn. av. 119.
Edw. av. 147.
Martin's voy. St. Kilda. 27.
Avis, *Gare* dicta. *Sib. Scot.* III. 22.
Alca major, le grand Pingoin. *Brisson av.* VI. 85. *Tab.* 7.
Esorokitsok *. *Crantz's Greenl.* I. 82.
Alca impennis. *Lin. syst.* 210.
Faun. Suec. sp. 140.
Islandis Gyr-v Geyrfugl. *Norvegis* Fiært, Anglemaage, Penguin, Brillefugl. *Brunnich*, 105.
Br. Zool. 136.

ACCORDING to Mr. *Martin*, this bird breeds on the isle of St. *Kilda*; appearing there the beginning of *May*, and retiring the middle of *June*. It lays one egg, which is six inches long, of a white color; some are irregularly marked with purplish lines crossing each other, others blotched with black and ferruginous about the thicker end: if the egg is taken away, it will not lay another

* Or little wing.

that

that ſeaſon. A late writer * informs us, that it does not viſit that iſland annually, but ſometimes keeps away for ſeveral years together; and adds, that it lays its egg cloſe to the ſea-mark; being incapable, by reaſon of the ſhortneſs of its wings, to mount higher.

The length of this bird, to the end of its toes, is three feet; the bill, to the corner of the mouth, four inches and a quarter: part of the upper mandible is covered with ſhort, black, velvet like feathers; it is very ſtrong, compreſſed and marked with ſeveral furrows that tally both above and below: between the eyes and the bill on each ſide is a large white ſpot: the reſt of the head, the neck, back, tail and wings, are of a gloſſy black: the tips of the leſſer quil-feathers white: the whole under ſide of the body white: the legs black. The wings of this bird are ſo ſmall, as to be uſeleſs for flight: the length, from the tip of the longeſt quil-feathers to the firſt joint, being only four inches and a quarter.

This bird is obſerved by ſeamen never to wander beyond *ſoundings*; and according to its appearance they direct their meaſures, being then aſſured that land is not very remote. Thus the modern ſailors pay reſpect to *auguries*, in the ſame manner

* *Macaulay's hiſt. St. Kilda.* p. 156.

as

LITTLE AUK.

RAZOR BILL. Nº 23

as *Ariſtophanes* tells us thoſe of *Greece* did above two thouſand years ago.

> Προερεῖ τις ἀεὶ τῶν ορνὶθων μαντευομένω ϖερι τȢ πλȢ,
> Νυνὶ μὴ πλεῖ, χειμων ἔϛαι, νυνὶ πλεῖ, κερδος επεϛαι.
>
> *Aves.* 597.

> From birds, in ſailing men inſtructions take,
> Now lye in port; now ſail and profit make.

230. RAZOR-BILL.

Razor-bill, Auk, Murre. *Wil. orn.* 325.
Raii ſyn. av. 119.
The Falk. *Martin's voy. St. Kilda.* 33.
The Marrot. *Sib. hiſt. Fife,* 48.
Edw. av. 358. *fig.* 2.
Alca, le Pingoin. *Briſſon av.* VI. 89. *Tab.* 8. *fig.* 1.
Alca torda. *Lin. ſyſt.* 210.
Tord, Tordmule. *Faun. Suec. ſp.* 139.
Norvegis Klub-Alke, Klympe. *Iſlandis* Aulka, Klumbr, Klumburnevia. *Groenlandis* Awarſuk. *Danis* Alke. *Brunnich,* 100.
Br. Zool. 136. *Scopoli,* No. 94.

DESCRIP.

THESE ſpecies weigh twenty-two ounces and a half. The length about eighteen inches: the breadth twenty-ſeven. The bill is two inches long, arched, very ſtrong and ſharp at the edges; the color black: the upper mandible is marked with four tranſverſe grooves; the lower with three; the wideſt of which is white, and croſſes each mandible. The inſide of the mouth is of a fine pale yellow: from the eye to the bill is a line of white: the head, throat, and whole upper ſide of the body are black; the wings of the ſame color, except

cept the tips of the leſſer quil-feathers, which are white: the tail conſiſts of twelve black feathers, and is ſharp pointed: the whole under ſide of the body is white: the legs black.

PLACE. Theſe birds, in company with the *Guillemot*, appear in our ſeas the beginning of *February*; but do not ſettle on their breeding places till they begin to lay, about the beginning of *May*. They inhabit the ledges of the higheſt rocks that impend over the ſea, where they form a groteſque appearance; ſitting cloſe together, and in rows one above the other. They properly lay but one egg a piece, of an extraordinary ſize for the bulk of the bird, being three inches long: it is either white, or of a pale ſea green, irregularly ſpotted with black: if this egg is deſtroyed, both the auk and guillemot will lay another; if that is taken, then a third: they make no neſt, depoſiting their egg on the bare rock: and though ſuch multitudes lay contiguous, by a wonderful inſtinct each diſtinguiſhes its own. What is alſo matter of great amazement, they fix their egg on the ſmooth rock, with ſo exact a balance, as to ſecure it from rolling off; yet ſhould it be removed, and then attempted to be replaced by the human hand, it is extremely difficult, if not impoſſible to find its former equilibrium.

The eggs are food to the inhabitants of the coaſts they frequent; which they get with great hazard; being lowered from above by ropes, truſting to the

the ſtrength of their companions, whoſe footing is often ſo unſtable that they are forced down the precipice, and periſh together.

Alca minor, le petit pingoin. *Briſſon av.* VI. 92. *Tab.* 8. *fig.* 2.
Alca Pica. *Lin. ſyſt.* 210.
Alca uniſulcata. *Brunnich,* 102.
Br. Zool. 137.

231. BLACK BILLED.

DESCRIP.

THIS weighs only eighteen ounces: the length fifteen inches and a half: the breadth twenty-five inches. The bill is of the ſame form with the Auk's, but is entirely black. The cheeks, chin, and throat are white; in all other reſpects it agrees with the former ſpecies: we can only obſerve, that this was ſhot in the winter, when the common ſort have quitted the coaſts.

When this bird was killed, it was obſerved to have about the neck abundance of lice, reſembling thoſe that infeſt the human kind, only they were ſpotted with yellow.

The *Alca Balthica* of *Brunnich*, No. 115, a variety in all reſpects like the common kind, only the under ſide of the neck white, is ſometimes found on our coaſts.

232. PUFFIN. Puphinus anglicus. *Gesner av.* 725.
Pica marina. *Aldr. av.* III. 92.
Puffin, Coulterneb, &c. *Wil. orn.* 325.
Raii syn. av. 120.
Edw. av. 358. *fig.* 1.
The Bowger. *Martin's voy. St. Kilda.* 34.
Fratercula, le Macareux. *Brisson av.* VI. 81. *Tab.* 6. *fig.* 1.
Caii opusc. 97.
Anas arctica. *Clusii Exot.* 104.
Alca arctica. *Lin. syst.* 211.
Faun. Suec. sp. 141.
Islandis & Norveg. Lunde, *hujus pulli* Lund Toller. *Danis* Islandsk Papegoye. *Brunnich*, 103.
See-Papagey, or See-Taucher. *Frisch*, II. 192.
Br. Zool. 135.

DESCRIP. THIS bird weighs about twelve ounces: its length is twelve inches: the breadth from tip to tip of the wings extended, twenty-one inches: BILL. the bill is short, broad at the base, compressed on the sides, and running up to a ridge, triangular and ending in a sharp point: the base of the upper mandible is strengthened with a white narrow prominent rim full of very minute holes: the bill is of two colors, the part next the head of a bluish grey, the lower part red: in the former is one transverse groove or furrow, in the latter three: the size of the bills of these birds vary: those of *Priestholm Isle* are one inch and three quarters long; and the base of the upper mandible one inch broad: but in the birds from the *Isle of Man* these proportions are much less.

NOSTRILS. The nostrils are very long and narrow; commence

mence at the above-mentioned rim, terminate at the firſt groove, and run parallel with the lower edge of the bill.

EYES.

The *irides* are grey, and the edges of the eye-lids of a fine crimſon: on the upper eye-lid is a ſingular callous ſubſtance, grey, and of a triangular form: on the lower is another of an oblong form: the crown of the head, whole upper part of the body, tail, and covert feathers of the wings are black; but in ſome the feathers of the back are tinged with brown: the quil-feathers are of a duſky hue.

HEAD.

The cheeks are white, and ſo full of feathers as to make the head appear very large and almoſt round: the chin of the ſame color; bounded on each ſide by a broad bed of grey: from the corner of each eye is a ſmall ſeparation of the feathers terminating at the back of the head. The neck is encircled with a broad collar of black: but the whole lower part of the body as far as is under water is white, which is a circumſtance in common with moſt of this genus.

Tail black, compoſed of ſixteen feathers: legs ſmall, of an orange color, and placed ſo far behind as to diſqualify it from ſtanding, except quite erect: reſting not only on the foot, but the whole length of the leg: this circumſtance attends every one of the genus, but not remarked by any naturaliſt, except *Wormius*, who has figured the *Penguin*, a bird of this genus, with great propriety: this makes the riſe

rise of the Puffin from the ground very difficult, and it meets with many falls before it gets on wing; but when that is effected, few birds fly longer or stronger.

PLACE.

These birds frequent the coasts of several parts of *Great Britain* and *Ireland*; but no place in greater numbers than *Priestholm Isle**, where their flocks may be compared to swarms of bees for multitude. These are birds of passage; resort there annually about the fifth or tenth of *April*, quit the place (almost to a bird) and return twice or thrice before they settle to burrow and prepare for ovation and incubation. They begin to burrow the first week in *May*; but some few save themselves that trouble, and dislodge the rabbets from their holes, and take possession of them till their return from the isle. Those which form their own burrows, are at that time so intent on the work as to suffer themselves to be taken by the hand. This task falls chiefly to the share of the males, for on dissection ten out of twelve proved of that sex. The males also assist in incubation; for on dissection several males were found sitting.

The first young are hatched the beginning of *July*, the old ones shew vast affection towards them; and seem totally insensible of danger on the breeding season. If a parent is taken at that time, and suspended by the wings, it will in a sort of despair

* Off the coast of *Anglesea*.

treat

treat itſelf moſt cruelly by biting every part it can reach; and the moment it is looſed, will never offer to eſcape, but inſtantly reſort to its unfledged young: but this affection ceaſes at the ſtated time of migration, which is moſt punctually about the eleventh of *Auguſt*, when they leave ſuch young as cannot fly, to the mercy of the *Peregrine Falcon*, who watches the mouths of the houſe for the appearance of the little deſerted puffins which forced by hunger are compelled to leave their burrows. The Rev[d]. Mr. *Hugh Davies*, of *Beaumaris*, to whom I am indebted for much of this account, informed me that on the twenty-third of *Auguſt*, ſo entire was the migration, that neither Puffin, Razor-Bill, Guillemot, or Tern was to be ſeen there.

I muſt add, that they lay only one egg, which differ much in form; ſome have one end very acute; others have both extremely obtuſe; all are white.

Their fleſh is exceſſive rank, as they feed on ſea weeds and fiſh, eſpecially *Sprats*: but when pickled and preſerved with ſpices, are admired by thoſe who love high eating. Dr. *Caius* tells us, that in his days the church allowed them in lent, inſtead of fiſh: he alſo acquaints us, that they were taken by means of ferrets, as we do rabbits: at preſent they are either dug out, or drawn from their burrows by a hooked ſtick: they bite extremely hard, and keep ſuch faſt hold on whatſoever they faſten, as not to be eaſily diſengaged. Their noiſe, when taken, is very disagreeable; being like the efforts of a dumb person to speak.

NOTE OF SEA FOWL.

The notes of all the sea birds are extremely harsh or inharmonious: we have often rested under the rocks attentive to the various sounds above our heads, which, mixed with the solemn roar of the waves swelling into and retiring from the vast caverns beneath, have produced a fine effect. The sharp voice of the sea gulls, the frequent chatter of the guillemots, the loud note of the auks, the scream of the herons, together with the hoarse, deep, periodical croak of the corvorants, which serves as a base to the rest; has often furnished us with a concert, which, joined with the wild scenery that surrounded us, afforded, in a high degree, that species of pleasure which arises from the novelty, and we may say gloomy grandeur of the entertainment.

The winter residence of this genus, and that of the guillemot, is but imperfectly known: it is probable they live at sea, in some more temperate climate, remote from land; forming those multitudes of birds that navigators observe in many parts of the ocean: they are always found there at certain seasons, retiring only at breeding time: repairing to the northern latitudes; and during that period are found as near the *Pole* as navigators have penetrated.

During winter *Razor-bills* and *Puffins* frequent the coast of *Andalusia*, but do not breed there.

Rotges

Rotges *Martin's* Spitzberg. 85.
Little black and white Diver. *Wil. orn.* 343.
Mergulus Melanoleucos roſtro acuto brevi. *Raii ſyn. av.* 125.
Edw. av. 91.
Uria minor, le petit Guillemot. *Briſſon av.* VI. 73.
Alca alle. *Lin. ſyſt.* 211.
Faun. Suec. ſp. 142.
Iſlandis Halkioen, Havdirdell. *Norvegis* Soe Konge, Soeren Jakob, Perdrikker, Perſuper, Boefiær, Borrefiær, Hys Thomas. *Feroenſibus* Fulkop. *Groenlandis* Akpaliarſok. *Brunnich*, 106.
Gunner tab. 6.
Br. Zool. 137.

233. LITTLE.

DESCRIP.

THE bird our deſcription was made from was taken in *Lancaſhire*; its bulk was not ſuperior to that of a blackbird. The bill convex, ſhort, thick, and ſtrong; its color black. That of the crown of the head, the hind part of the neck, the back, and the tail black; the wings the ſame color; but the tips of the leſſer quil-feathers white: the inner coverts of the wings grey: the cheeks, throat, and whole under ſide of the body white: the ſcapular feathers black and white: the legs and feet covered with dirty greeniſh white ſcales; the webs black.

Mr. *Edwards* has figured a bird that varies very little from this: and has added another, which he imagines differs only in ſex: in that, the head and neck are wholly black; and the inner coverts of the wings barred with a dirty white. We met with the laſt in the cabinet of Doctor *David Skene* at *Aberdeen*; it was ſhot on the coaſt north of *Slains* in the ſpring of the year,

BILL

BILL ſlender, ſtrong, pointed. The upper mandible ſlightly bending towards the end. Baſe covered with ſoft ſhort feathers.

XLI. GUILLEMOT.

NOSTRILS lodged in a hollow near the baſe.

TONGUE ſlender, almoſt the length of the bill.

TOES, no back toe.

234. FOOLISH

Guillem, Guillemot, Skout, Kiddaw, Sea-hen. *Wil. orn.* 324.
Raii ſyn. av. 120.
The Lavy. *Martin's voy. St. Kilda*, 32.
Edw. av. 359. *fig.* 1.
Uria, le Guillemot. *Briſſon av.* VI. 70. *Tab.* 6. *fig.* 1.
Lommia. *N. Com. Petr.* IV. 414.
Colymbus Troile. *Lin. ſyſt.* 220.
Faun. Suec. ſp. 149.
Iſlandis & Norvegis Lomvie, Langivie, Lomrifvie, Storfugl. *Brunnich*, 108.
Sea-Taube, or Groenlandiſcher Taucher. *Friſch*, II. 185.
Br. Zool. 138.

DESCRIP.

THIS ſpecies weighs twenty-four ounces: the length ſeventeen inches: the breadth twenty-ſeven and a half: the bill is three inches long; black, ſtrait, and ſharp pointed: near the end of the lower mandible is a ſmall proceſs; the inſide of the mouth yellow: the feathers on the upper part of the bill are ſhort, and ſoft like velvet: from the eye to the hind part of the head is a ſmall diviſion of the feathers. The head, neck, back, wings, and tail are of a deep mouſe color;

the tips of the leſſer quil-feathers white: the whole under part of the body is of a pure white: the ſides under the wings marked with duſky lines. Immediately above the thighs are ſome long feathers that curl over them. The legs duſky.

These birds are found in amazing numbers on the high cliffs on ſeveral of our coaſts, and appear at the ſame time as the auk. They are very ſimple birds; for notwithſtanding they are ſhot at, and ſee their companions killed by them, they will not quit the rock. Like the auk, they lay only one egg, which is very large; ſome are of a fine pale blue, others white, ſpotted, or moſt elegantly ſtreaked with lines croſſing each other in all directions. The Rev. Mr. *Low* of *Birſa* aſſures me, that they continue about the *Orknies* the whole winter.

EGG.

235. LESSER. Uria Svarbag. *Iſlandis* Stutnefur, Svartbakur. *Br. Zool.* 138.

Ringuia. *Brunnich*, No. 110. *Scopoli*, No. 103.

DESCRIP.

THE weight is nineteen ounces: the length ſixteen inches: the breadth twenty-ſix. The bill two inches and a half long, ſhaped like the Guillemot's, but weaker. The top of the head, the whole upper part of the body, wings and tail are of a darker color than the former: the cheeks, throat,

Nº 235.

LESSER GUILLEMOT.

SPOTTED GUILLEMOT. Nº 236.

throat, and all the lower ſide of the body are white: from the corner of the eye is a duſky ſtroke, pointing to the hind part of the head: the tips of the ſecondary feathers white: the legs are black: the tail very ſhort, and conſiſts of twelve feathers.

Theſe birds frequent the *Welch* coaſts in the winter time; but that very rarely: where they breed is unknown to us; having never obſerved them on the rocks among the congenerous birds. Theſe and the black-billed *Auks* haunt the *Firth* of *Forth* during winter in flocks innumerable, in purſuit of ſprats. They are called there *Morrots*: they all retire before ſpring.

236. BLACK.

Greenland-dove, or Sea-turtle. *Wil. orn.* 326.
Raii ſyn. av. 121.
Ray's itin. 183, 192.
Feiſte. *Gunner. tab.* 4.
Turtur maritimus inſulæ Baſs. *Sib. hiſt. Fife*, 46.
The Scraber. *Martin's voy. St. Kilda.* 32.
Cajour, Pynan. *N. Com. Petr.* IV. 418.
Uria minor nigra, le petit Guillemot noir. *Briſſon av.* VI. 76.
Colymbus Grylle. *Lin. ſyſt.* 220.
Faun. Suec. ſp. 148.
Iſlandis Teiſta. *Norvegis* Teiſte. *Groenlandis* Sarpak. *Brunnich*, 113.
Groenlandiſche Taube. *Friſch*, II. 185.
Br. Zool. 138.

DESCRIP.

THE length of this ſpecies is fourteen inches: the breadth twenty-two: the bill is an inch and a half long; ſtrait, ſlender, and black: the inſide of the mouth red: on each wing is a

Mm 4 large

large bed of white, which in young birds is ſpotted: the tips of the leſſer quil-feathers, and the inner coverts of the wings, are white: except theſe, the whole plumage is black. In winter it is ſaid to change to white: and a variety ſpotted with black and white* is not uncommon in *Scotland*. The tail conſiſts of twelve feathers: the legs are red.

Theſe birds are found on the *Baſs iſle* in *Scotland*; in the *iſle of St. Kilda*; and, as Mr. *Ray* imagines, in the *Farn iſlands* off the coaſt of *Northumberland*; we have alſo ſeen it on the rocks of *Llandidno* in *Caernarvonſhire*. Except at breeding time, it keeps always at ſea; and is very difficult to be ſhot, diving at the flaſh of the pan. The *Welch* call this bird *Caſgan Longwr*, or the failor's hatred, from a notion that its appearance forebodes a ſtorm. It viſits *St. Kilda*'s in *March*: makes its neſt far under ground; and lays a grey egg; or, as *Steller* ſays, whitiſh ſpotted with ruſt, and ſpeckled with aſh color.

* The ſpotted *Greenland* Dove of Mr. *Edwards*, plate 50.

BILL.

No. 23

IMBER.

NORTHERN DIVER.

No. 25

XLII. DIVER.

BILL strong, strait, pointed. Upper mandible longest; edges of each bending in.
NOSTRILS linear.
TONGUE pointed, long, serrated near the base.
LEGS thin and flat.
TOES, exterior the longest: back toe joined to the interior by a small membrane.
TAIL short, consisting of twenty feathers.

237. NORTHERN.

Clusius's. *Wil. orn.* 342.
Raii syn. av. 125.
Mergus maximus Farrensis, sive Arcticus. *Clusii exot.* 102.
Colymbus maximus stellatus nostras. *Sib. hist. Scot.* 20. *Tab.* 15.
Le grand Plongeon tacheté. *Brisson av.* VI. 120. *Tab.* 11 *fig.* 1.
Colymbus glacialis. *Lin. syst.* 221.
Norvegis Brusen. *Groenlandis* Tiulik. *Brunnich, orn.* 134.
Grosse Halb-Ente, Meer-Nœring. *Frisch,* II. 185. *A.*
Br. Zool. 139.

DESCRIP.

THE length of this species is three feet five inches: its breadth four feet eight: the bill to the corners of the mouth four inches long; black and strongly made. The head and neck are of a deep black: the hind part of the latter is marked with a large semilunar white band: immediately under the throat is another; both marked with black oblong strokes pointing down: the lower part of the neck is of a deep black, glossed with

with a rich purple: the whole under ſide of the body is white: the ſides of the breaſt marked with black lines: the back, coverts of the wings, and ſcapulars, are black, marked with white ſpots: thoſe on the ſcapulars are very large, and of a ſquare ſhape; two at the end of each feather.

The tail is very ſhort, and almoſt concealed by the coverts, which are duſky ſpotted with white: the legs are black. Theſe birds inhabit the northern parts of this iſland, live chiefly at ſea, and feed on fiſh: we do not know whether they breed with us, as they do in *Norway*; which has many birds in common with *Scotland*. In the laſt it is called *Mur-buachaill*, or the *Herdſman* of the ſea, from its being ſo much in that element.

238. IMBER.

Colymbus immer. *Lin. ſyſt.* 222.
Geſner's greater Doucker. *Wil. orn.* 342. *Raii ſyn. av.* 126. No. 8. Fluder. *Geſner av.* 140.
Immer. *Brunnich*, No. 129.
Ember Gooſe. *Sibbald Scot.* 21. *Wallace Orkney*, 16. *Debes Feroe Iſles*, 138. *Pontoppidan*, II. 80.
Le grand Plongeon. *Briſſon*, VI. 105. *Tab.* X.

THIS ſpecies inhabits the ſeas about the *Orknies*; but in ſevere winters viſits the ſouthern parts of *Great Britain*. It lives as much at ſea as the former; ſo that credulity believed that it never quitted the water, and that it hatched its young in a hole a hole formed by nature under the wing for that end.

It is ſuperior in ſize to a gooſe. The head duſky: the back, coverts of the wings, and tail clouded with lighter and darker ſhades of the ſame. Primaries and tail black: under ſide of the neck ſpotted with duſky: the breaſt and belly ſilvery: legs black.

The ſkins of the birds of this genus are uncommonly tough; and in the northern countries have been uſed as leather.

239. SPECKLED.

Greateſt ſpeckled Diver, or Loon. *Wil. orn.* 341.
Raii ſyn. av. 125.
Colymbus caudatus ſtellatus. *N. Com. Petr.* IV. 424.
Le petit Plongeon. *Briſſon av.* VI. 108. *Tab.* 10. *fig.* 2.
Mergus Stellatus, *Danis* Soe-Hane. *Brunnich*, 130.
Br. Zool. 139.

DESCRIP.

THIS ſpecies weighs two pounds and a half: its length twenty-ſeven inches: its breadth three feet nine. The bill three inches long, and turns a little upwards; the mandibles, when cloſed at the points, do not touch at the ſides. The head is of a duſky grey, marked with numerous white ſpots: the hind part of the neck an uniform grey: the whole upper part of the body, and greater coverts of the wings duſky, ſpeckled with white: the leſſer coverts duſky, and plain. The tail conſiſts of about twenty black feathers; in ſome tipt with white.

white. The cheeks and whole under ſide of the body of a fine gloſſy white: and the feathers, as in all this genus, which reſides almoſt perpetually on the water, are exceſſively thick, and cloſe ſet: the legs are duſky.

Theſe birds frequent our ſeas, lakes and rivers in the winter. On the *Thames* they are called *ſprat loons*, for they attend that fiſh during its continuance in the river. They are ſubject to vary in the diſpoſition and form of their ſpots and colors: ſome having their necks ſurrounded with a ſpeckled ring: in ſome the ſpots are round, in others oblong.

240. RED THROATED.

Edw. av. 97.
Gunner. Tab. 2. f. 2.
Colymbus ſeptentrionalis. *Lin. ſyſt.* 220.
Le Plongeon a gorge rouge. *Briſſon av.* VI. 111. *Tab.* II. *fig.* 1.
Iſlandis & Norvegis Loom v. Lumme, *Danis* Lomm. *Brunnich,* 132.
Br. Zool. 140.

THIS ſpecies breeds in the northern parts of *Scotland*, on the borders of the lakes: but migrates ſouthward during winter. It lays two eggs. The ſexes do not differ in colors; and are a diſtinct kind from the black throated, the *Lumme* of the *Norwegians*. Its ſhape is more elegant than that of the others. The weight is three pounds:

DESCRIP.

the length, to the tail end, two feet; to that of the toes, two feet four inches: the breadth three feet

RED THROATED DIVER.

BLACK THROATED DIVER. №. 241.

feet five inches. The head fmall and taper: the bill ftrait, and lefs ftrong: the fize about a fourth lefs than the preceding. The head and chin are of a fine uniform grey: the hind part of the neck marked with dufky and white lines, pointing downwards: the throat is of a dull red: the whole upper part of the body, tail and wings of a deep grey almoft dufky; but the coverts of the wings, and the back, are marked with a few white fpots: the under fide of the body white: the legs dufky.

241. BLACK THROATED.

Lumme. *Worm. Muf. Brunnich*, No. 133.
Northern Doucker. *Wil. orn.* 343. *Raii fyn. av.* 125.
Colymbus arcticus. *Lin. fyft.* 221. *Faun. Suec.* No.
Speckled Diver. *Edw.* 146.

A SPECIES fomewhat larger than the laft. Bill black: front black: hind part of the head and neck cinereous: fides of the neck marked with black and white lines pointing downwards: fore part of a gloffy variable black, purple and green.

Back, fcapulars, and coverts of wings black, marked (the two firft with fquare) the laft with round fpots of white: quil feathers dufky: breaft and belly white. Tail fhort and black: legs partly dufky, partly reddifh.

BILL

XLIII. GULL.

BILL ſtrong, ſtrait, bending near the end; an angular prominency on the lower mandible.

NOSTRILS linear.

TONGUE a little cloven.

BODY light, wings large.

LEG and back toe ſmall, naked above knee.

242. BLACK BACKED.

Wil. orn. 344.
Raii ſyn. av. 127.
Le Goiland noir. *Briſſon av.* VI. 158.
Larus marinus. *Lin. ſyſt.* 225.
Faun. Suec. ſp. 155.
Danis Blaae maage, *Norvegis* Svartbag, Havmaaſe. *Brunnich*, 145.
Br. Zool. 140.

DESCRIP.

THE weight of this ſpecies is near five pounds: the length twenty-nine inches: the breadth five feet nine. The bill is very ſtrong and thick, and almoſt four inches long; the color a pale yellow; but the lower mandible is marked with a red ſpot, with a black one in the middle. The irides yellow: the edges of the eye-lids orange color: the head, neck, whole under ſide, tail and lower part of the back, are white: the upper part of the back, and wings, are black: the quil-feathers tipt with white: the legs of a pale fleſh color.

This kind inhabits our coaſts in ſmall numbers; and breeds in the higheſt cliffs. It feeds not only on fiſh: but like the Raven, very greedily devours carrion;

carrion. Its egg is very blunt at each end; of a duſky olive color, quite black at the greater end; and the reſt of it thinly marked with duſky ſpots.

I have ſeen on the coaſt of *Angleſea*, a bird that agrees in all reſpects with this except in ſize, in wanting the black ſpot on the bill, and in the color of the legs, which in this are of a bright yellow: the extent of wings is only four feet five: the length only twenty-two inches: the weight one pound and a half. This ſpecies, or perhaps variety (for I dare not aſſert which) rambles far from the ſea, and has been ſhot at *Bulſtrode*, in *Middleſex*.

243. SKUA.

Our *Cataracta*, I ſuppoſe the *Corniſh* Gannet. *Wil. orn.* 348.
Raii ſyn. av. 128.
Cataractes. *Sibb. Scot. tab.* 14.
Sea Eagle. *Sibb. hiſt. Fife.* 46.
Le Stercoraire rayè. *Briſſon av.* VI. 152.
Pontopp. Norw. II. 96.
Skua Hoirei. *Cluſii Exot.* 368, 369.
Larus Cataractes. *Lin. ſyſt.* 226.
Skua. *Brunnich, ornith.* 33.
Feroenſibus Skue. *Iſlandis* Skumr. *Norvegis* Kav-Oern. *Brunnich*, 125.
Brown and ferruginous Gull. *Br. Zool.* 140.

DESCRIP.

THE length of this ſingular *Gull* is two feet: the extent four feet and a half: the weight three pounds: the bill two inches one fourth long, very much hooked at the end, and very ſharp: the upper mandible covered more than half way with

with a black cere or ſkin as in the hawk kind: the noſtrils placed near the bend, and are pervious.

The feathers on the head, neck, back, ſcapulars and coverts of the wings are of a deep brown, marked with rüſt color, (brighteſt in the male). The ſhafts of the primaries are white: the end and exterior ſide of the firſt is deep brown; the ends only of the reſt brown: the lower parts on both ſides being white; the ſecondaries marked in like manner; forming a great bar of white. The breaſt, belly and vent ferruginous, tinged with aſh color. The tail when ſpread is circular, of a deep brown, white at the root; and with ſhafts of the ſame color.

The legs are covered with great black ſcales: the talons black, ſtrong and crooked; the interior remarkably ſo.

History. This bird inhabits *Norway*, the *Ferroe* iſles, *Shetland*, and the noted rock *Foula*, a little weſt of them. It is alſo a native of the *South ſea*. It is the moſt formidable Gull, its prey being not only fiſh, but what is wonderful in a web-footed bird, all the leſſer ſort of water fowl, ſuch as teal, &c. Mr. *Schroter*, a Surgeon in the *Ferroe* iſles, relates that it likewiſe preys on ducks, poultry, and even young lambs *. It has all the fierceneſs of the eagle in defending its young; when the inhabitants of thoſe iſlands viſit the neſt, it attacks them with

* *Hoier in Clus. exot.* 369. *Brunnich,* 35.

great

great force, so that they hold a knife erect over their heads, on which the *Skua* will transfix itself in its fall on the invaders.

The Rev. Mr. *Low*, minister of *Birsa*, in *Orkney*, from whom an accurate history of those islands, and of *Shetland* may be expected, confirmed to me part of the above. On approaching the quarters of these birds, they attacked him and his company with most violent blows; and intimidated a bold dog of Mr. *Low's* in such a manner, as to drive him for protection to his master. The natives are often very rudely treated by them, while they are attending their sheep on the hills; and are obliged to guard their heads by holding up their sticks, on which the birds often kill themselves. In *Foula* it is a priveleged bird, because it defends the flocks from the eagle, which it beats and pursues with great fury; so that even that rapacious bird seldom ventures near its quarters. The natives of *Foula* on this account lay a fine on any person who destroys one: they deny that it ever injures their flocks or poultry, but imagine it preys on the dung of the *Arctic*, and other larger gulls, which it persecutes till they mute for fear.

Mr. *Ray* and Mr. *Smith** suppose this to be the *Cornish Gannet*; but in our account of that bird we shall shew that it is a different species. Mr. *Macauly*† mentions a gull that makes great ha-

* *Hist. Kerry.*

† *Hist. St. Kilda.* p. 158.

voke among the eggs and ſea fowl of *St. Kilda*; it is there called *Tuliac:* his deſcription ſuits that of the *herring Gull*; but we ſuſpect he confounds theſe two kinds, and has transferred the manners of this ſpecies to the latter.

Linnæus involves two ſpecies in the article *Larus Cataracta*; this, and the *arctic* bird of Mr. *Edwards*, birds of very different characters. M. *Briſſon* does not ſeem perfectly acquainted with this bird; for the ſynonym of the *Skua*, given by him to his fifth gull (our brown and white gull) belongs to this ſpecies; and his print of the *Stercoraire rayé*, p. 152. *tab* 13. *tom.* VI. to which he has given the ſynonym of Mr. *Edwards*'s arctic bird, ſeems to be the very ſame which we have here deſcribed.

244. BLACK TOED.

Cepphus. *Aldr. av.* III. 38.
Wil. orn. 351.
Raii ſyn. av. 129.

The Cepphus. *Phil. Tranſact.* Vol. 52. 135.
Catharacta Cepphus, Strandhoeg. *Brunnich, ornith.* 126.

DESCRIP.

THIS ſpecies weighs eleven ounces: its length is fifteen inches: its breadth thirty-nine: the bill is one inch and a half long, the upper part covered with a brown cere: the noſtrils like thoſe of the former; the end black and crooked. The feathers of the forehead come pretty low on the bill: the head and neck are of a dirty white: the hind

.LXXXVI. No. 248.

WINTER GULL.

PL. LXXXVII.

No. 245

ARCTIC GULLS.

hind part of the latter plain, the reſt marked with oblong duſky ſpots.

The breaſt and belly are white, croſſed with numerous duſky and yellowiſh lines: the feathers on the ſides and the vent, are barred tranſverſely with black and white: the back, ſcapulars, coverts of the wings and tail, are black, beautifully edged with white or pale ruſt color: the ſhafts and tips of the quil-feathers are white: the exterior web, and upper half of the interior web black, but the lower part of the latter white: the tail conſiſts of twelve black feathers tipt with white; the two middle of which, are near an inch longer than the others: the ſhafts are white; and the exterior webs of the outmoſt feather is ſpotted with ruſt color. The legs are of a bluiſh lead color: the lower part of the toes and webs black.

A bird of this kind was taken near *Oxford*, and communicated to the Royal Society by Dr. *Lyſons* of *Glouceſter*.

245. ARCTIC.

The Struntjagger, or Dunghunter. *Marten's Spitzberg* 87.
The Arctic Bird. *Edw. av.* 148. 149.

Larus Paraſiticus. *Lin. ſyſt.* 226.
Swartlaſſe, Labben, Elof. *Faun Suec. ſp.* 156.
Brunnich, 127.

THESE birds are very common in the *Hebrides.* I ſaw numbers in *Jura*, *Ilay* and *Rum*,

 where

where they breed in the heath; if diſturbed they fly about like the lapwing, but ſoon alight. They are alſo found in the *Orknies*, where they appear in *May*, and retire in *Auguſt*. It is alſo found on the coaſt of *Yorkſhire*, where it is known by the name of *Feaſer*. All writers that mention it agree, that it has the property of purſuing the leſſer gulls ſo long, that they mute for fear, and that it catches up and devours their excrement before they drop into the water; from which the name. *Linnæus* wittily calls it the *Paraſite*, alluding to its ſordid life.

Descrip. The length of this ſpecies is twenty-one inches: the bill is duſky, about an inch and a half long, pretty much hooked at the end, but the ſtrait part is covered with a ſort of cere. The noſtrils are narrow, and placed near the end, like the former.

Male. In the *male*, the crown of the head is black: the back, wings, and tail duſky; but the lower part of the inner webs of the quil-feathers white: the hind part of the neck, and whole underſide of the body white: the tail conſiſts of twelve feathers, the two middlemoſt near four inches longer than the others: the legs black, ſmall, and ſcaly.

Female. The *female* is entirely brown; but of a much paler color below than above: the feathers in the middle of the tail only two inches longer than the others. The ſpecimen from which Mr. *Edwards* toke the figure of his female *Arctic* bird, had loſt thoſe

Pl. LXXXVIII.

those long feathers, so he has omitted them in the print.

Linnæus has separated this from its mate, his *Larus parasiticus*, and made it a synonym to his *L. Cataractes*, a bird as different from this as any other of the whole genus.

246. HERRING.

Burgermeister *Martin's Spitzberg*. 84.
Herring Gull. *Wil. orn.* 345.
Larus cinereus maximus. *Raii syn av.* 127.
Le Goiland gris. *Brisson av.* VI. 162.
Larus fuscus. *Lin. syst.* 125. *Faun. Suec. sp.* 154.
Danis Silde-Maage. *Islandis* Veydebjalla. *Brunnich*, 142.
Grosse Staff Moeur. *Frisch*, II. 218.
Br. Zool. 141.

DESCRIP.

THIS gull weighs upwards of thirty ounces: the length twenty-three inches; its breadth fifty-two. The bill yellow, and the lower mandible marked with an orange colored spot: the irides *straw color:* the edges of the eye-lids red: the head, neck, and tail white: the back, and coverts of the wings ash colored: the upper part of the five first quil-feathers are black, marked with a white spot near their end: the legs of a pale flesh color. These birds breed on the ledges of rocks that hang over the sea: they make a large nest of dead grass, and lay three eggs of a dirty white, spotted with black. The young are ash colored, spotted with brown; they do not come to their proper color the first year: this is common to other gulls; which has greatly multiplied the ſpecies among authors, who are inattentive to theſe particulars. This gull is a great devourer of fiſh, eſpecially of that from which it takes its name: it is a conſtant attendent on the nets, and ſo bold as to ſeize its prey before the fiſhermens faces.

(A.)
247. WAGEL.

Great grey Gull, the Corniſh Wagel. *Wil. orn.* 349. *Raii ſyn. av.* 130.
Le Goiland variè, ou le Griſard. *Briſſon av.* VI. 167. *tab.* 15.

Larus Nævius. *Lin. ſyſt.* 225.
Danis Graae-Maage. *Iſlandis* Kablabrinkar. *Brunnich,* 150.
Brown and White Gull. *Br. Zool.* II. 422.

DESCRIP.

THESE birds vary much in their ſize; one we examined weighed three pounds ſeven ounces: the length was two feet two inches: the breadth five feet ſix: others again did not weigh two pounds and a half: the irides are duſky: the bill black, and near three inches long. The whole plumage of the head and body, above and below, is a mixture of white, aſh color, and brown: the laſt color occupies the middle of each feather; and in ſome birds is pale, in others dark: the quil-feathers black: the lower part of the tail is mottled with black and white; towards the end is a brown black bar, and the tips are white: the legs are of a dirty white.

Some have ſuppoſed this to be the young of the preceding

preceding ſpecies, which (as well as the reſt of the gull tribe) ſcarce ever attains its true colors till after the firſt year: but it muſt be obſerved, that the firſt colors of the irides, of the quil-feathers, and of the tail, are in all birds permanent; theſe, as we have remarked, differ in each of theſe gulls ſo greatly, as ever to preſerve unerring notes of diſtinction.

This ſpecies is likewiſe called by ſome the *Dung Hunter*; for the ſame reaſon as the laſt is ſtyled ſo.

Winter Mew, or Coddy Moddy. *Wil. orn.* 350.
Raii ſyn. av. 130.

Gavia Hyberna, le Mouette d'hiver. *Briſſon av.* VI. 189.
Br. Zool. 142.

248. WINTER.

DESCRIP.

THIS weighs from fourteen to ſeventeen ounces: the length eighteen inches; the breadth three feet nine. The irides are hazel: the bill two inches long, but the ſlendereſt of any gull: it is black at the tip, whitiſh towards the baſe. The crown of the head, and hind part, and ſides of the neck, are white, marked with oblong duſky ſpots; the forehead, throat, middle of the breaſt, belly, and rump, are white; the back and ſcapulars are of a pale grey; the laſt ſpotted with brown; the coverts of the wings are of a pale brown, edged with white; the firſt quil-feather is black; the ſucceeding are tipt with white: the tail is white, croſſed near the end with a black bar; the legs of a dirty bluiſh white.

This kind frequents, during winter, the moiſt meadows in the inland parts of *England*, remote from the ſea. The gelatinous ſubſtance, known by the name of *Star Shot*, or *Star Gelly*, owes its origin to this bird, or ſome of the kind; being nothing but the half digeſted remains of earth-worms, on which theſe birds feed, and often diſcharge from their ſtomachs*.

Linnæus, p. 224. makes this ſpecies ſynonymous with the *Larus tridactylus* or *Tarrock*; but as we have had opportunity of examining ſeveral of each ſpecies, and find in all thoſe ſtrong diſtinctions remarked in our deſcriptions, we muſt decline aſſenting to the opinion of that eminent naturaliſt.

249. COMMON.

Galedor, Crocala, Galetra. *Aldr. av.* III. 34.
Common Sea Mall. *Wil. orn.* 345.
Common Sea Mall, or Mew. *Raii ſyn. av.* 127.

La Mouette cendrée. *Briſſon av.* VI. 175. *tab.* 16. *fig.* 1.
Gabbiano minore. *Zinan.* 115.
Larus canus. *Lin. ſyſt.* 224.
Br. Zool. 142. *Scopoli*, No. 104.

THIS is the moſt numerous of the genus. It breeds on the ledges of the cliffs that im-

* Vide *Morton's Nat. Hiſt. Northampt.* p. 353.

pend

LXXXIX.
Nº 250
KITTIWAKE
COMMON GULL.
Nº 249.

pend over the ſea: in winter they are found in vaſt flocks on all our ſhores. They differ a little in ſize; one we examined weighed twelve ounces and a half: its length was ſeventeen inches: its breadth thirty-ſix: the bill yellow: the head, neck, tail, and whole under ſide of the body, a pure white: the back, and coverts of the wings, a pale grey: near the end of the greater quil-feathers was a black ſpot: the legs a dull white, tinged with green.

DESCRIP.

250. KITTIWAKE.

Larus Riſſa. *Lin. ſyſt.* 224.
Ritſa *Iſlandis*, incolis *Chriſtianſoe*, Lille Solvet, Rotteren. *Brunnich*, No. 140.
Kittiwake. *Sibbald's hiſt. Scotl.* 20.

THE length of this ſpecies is fourteen inches: the extent three feet two. When arrived at full age, the head, neck, belly, and tail are of a ſnowy whiteneſs; behind each ear is ſometimes a duſky ſpot: the back and wings grey: the exterior edge of the firſt quil-feather, and tips of the four or five next, are black: the bill yellow, tinged with green; inſide of the mouth orange: legs duſky, with only a knob inſtead of the back toe.

It inhabits the romantic cliffs of *Flamborough-head* (where it is called *Petrel)* the *Baſs Iſle*, the vaſt rocks near the Caſtle of *Slains*, in the county of *Aberdeen*, and *Prieſtholm Iſle*.

The young of theſe birds are a favorite diſh in

in *North Britain*, being ſerved up roaſted, a little before dinner, in order to provoke the appetite; but, from their rank taſte and ſmell, ſeem much more likely to produce a contrary effect.

251. TARROCK.

La Mouette cendrèe. *Briſſon av.* 169.
Gavia cinerea alia. *Aldr. av.* III. 35.
Wil. orn. 346.
Raii ſyn. av. 128.
Larus tridactylus. *Lin. ſyſt.* 224.
Faun. Suec. 157. *ſp.*
La Mouette cendrèe tachetèe. *Briſſon av.* VI. 185. *tab.* 17. *fig.* 2.
Br. Zool. 142.

DESCRIP.

THE length is fourteen inches; the breadth three feet: the weight only ſeven ounces. The bill is black, ſhort, thick, and ſtrong; the head large: the color of that, the throat, neck, and whole under ſide are white: near each ear, and under the throat, is a black ſpot: on the hind part of the neck is a black creſcent, the horns pointing to the throat.

The back and ſcapulars are of a bluiſh grey: the leſſer coverts of the wings duſky, edged with grey; the larger next to them of the ſame color; the reſt grey: the exterior ſides, and ends of the four firſt quil-feathers are black: the tips of the two next black; all the reſt wholly white: the ten middle feathers of the tail white, tipt with black; the two outmoſt quite white: the legs of a duſky aſh color.

In

In lieu of the back toe, it has only a ſmall protuberance.

This ſpecies breeds on *Prieſtholme Iſle*, alſo among the former in *Scotland*. I muſt retract my opinion of its being the young of that ſpecies.

252. BLACK HEAD.

Cepphus *Turneri*. *Geſner av.* 249.
Larus cinereus tertius. *Aldr. av.* III. 35.
Pewit, or Black Cap, Sea Crow, Mire Crow. *Wil. orn.* 347.
Raii ſyn. av. 128. *itin.* 217.
Pewit. *Plott's hiſt. Staff.* 231.
Puit. *Fuller's Brit. Worthies.* 318.
La Mouette rieuſe a pattes rouges. *Briſſon av.* VI. 196.
Gabbiano cinerizio col roſtro, e col li piedi roſſi. *Zinan.* 115.
Larus ridibundus. *Lin. ſyſt.* 225.
Br. Zool. 143.

THESE birds breed in vaſt numbers in the iſlands of certain pools in the county of *Stafford*; and, as Dr. *Fuller* tells us, in another on the *Eſſex* ſhores; alſo in the Fens of *Lincolnſhire*. They are birds of paſſage; reſort there in the ſpring; and after the breeding ſeaſon diſperſe to the ſea coaſts: they make their neſt on the ground, with ruſhes, dead graſs, and the like; and lay three eggs of a dirty olive color, marked with black. The young were formerly highly eſteemed, and numbers were annually taken and fattened for the table. *Plott* gives a marvellous account of their attachment to the lord of the ſoil they inhabit; inſomuch, that on on his death, they never fail to ſhift their quarters for a certain time.

Whitelock, in his annals, mentions a piece of ground near *Portſmouth*, which produced to the owner forty pounds a year by the ſale of *Pewits*, or this ſpecies of gull. Theſe are the *See-gulles* that in old times were admitted to the noblemens tables*.

DESCRIP. The notes of theſe gulls diſtinguiſh them from any others; being like a hoarſe laugh. Their weight is about ten ounces: their length fifteen inches; their breadth thirty-ſeven: their irides are of a bright hazel: the edges of the eye-lids of a fine ſcarlet; and on each, above and below, is a ſpot of white feathers. Their bills and legs are of a ſanguine red: the heads and throats black or duſky: the neck, and all the under ſide of the body, and the tail, a pure white: back and wings aſh colored: tip, and exterior edge of the firſt quil-feather black; the reſt of that feather white; the next to that tipt with black, and marked with the ſame on the inner web.

A VARIETY. La Grande Mouette blanche. *Belon.* 170. *Wil. orn.* 348. *Raii ſyn. av.* Larus canus. *Scopoli*, No. 104.

THIS was taken in a trap near my houſe, *January* 25th, 1772, and ſeemed only a variety

* *Vide* Appendix.

ty of the former. It differed in having the edges of the eye-lids covered with white ſoft feathers. The forepart of the head white; the ſpace round the eyes duſky: from the corner of each eye is a broad duſky bar, ſurrounding the hind part of the head; behind that is another reaching from ear to ear: the ends, interior and exterior edges of the three firſt quil-feathers black; the ends and interior ſides only of the two next black, but the ſhafts and middle part white; the tips of the two next white; beneath a black bar: the reſt, as well as the ſecondaries, aſh color.

In all other reſpects it reſembled the common pewit gull. The fat was of a deep orange color.

253. BROWN.

The brown Tern. *Wil. orn.* 352.
Sterna fuſca. *Raii ſyn. av.* 131.
Sterna nigra. *Lin. ſyſt.* 227.
Faun. Suec. ſp. 159.
Br. Zool. 143.

MR. *Ray* has left us the following obſcure account of this bird; communicated to him by Mr. *Johnſon*, a *Yorkſhire* gentleman. "The whole "under ſide is white; the upper brown: the "wings partly brown, partly aſh color: the head "black: the tail not forked: theſe birds fly in "companies."

DESCRIP.

From

From the defcription, we fufpect this bird to be the young of the *greater Tern*, that had not yet attained its proper colors, nor the long feathers of the tail, which it does not acquire till mature age.

BILL

Pl. XC.

No 255.

GREAT & LESSER TERNS.

No 254.

XLIV.
TERN*.

BILL ſtrait, ſlender, pointed.
NOSTRILS linear.
TONGUE ſlender and ſharp.
WINGS very long.
TAIL forked.
TOES, a ſmall back toe.

254. GREAT.

Sterna (Stirn, Spyrer, Schnirring) *Geſner av.* 586.
Aldr. av. III. 35.
The Sea Swallow. *Wil. orn.* 352.
Raii ſyn. av. 131.
Sterna major, la grande Hirondelle de mer. *Briſſon av.* VI. 203. *tab.* 19. *fig.* 1.
Sterna hirundo. *Lin. ſyſt.* 227.
Tarna. *Faun. Suec. ſp.* 159.
The Kirmew. *Marten's Spitzberg.* 92.
Iſlandis Kria. *Norvegis* Tenne, Tende, Tendelobe, Sand-Tolle, Sand-Tærrne. *Danis* Tærne. *Bornholmis* Kirre, Krop-Kirre. *Brunnich*, 151.
Grauer fiſcher. *Kram.* 345.
Schwartzplattige Schwalben Moewe. *Friſch*, II. 219.
Br. Zool. 144.
Makauka. *Scopoli*, No. 3.

DESCRIP.

THIS kind weighs four ounces, one-quarter: the length is fourteen inches; the breadth thirty: the bill and feet are of a fine crimſon; the former tipt with black, ſtrait, ſlender, and ſharp pointed: the crown, and hind part of the head, black: the throat, and whole under ſide of the

* A name theſe birds are known by in the *North of England*; and which we ſubſtitute inſtead of the old compound one of *Sea Swallow*; which was given them on account of their forked tails.

body,

body, white: the upper part, and the coverts of the wings, a fine pale grey: the tail confifts of twelve feathers; the exterior edges of the three outmoft are grey, the reft white: the exterior, on each fide, is two inches longer than the others: in flying, the bird frequently clofes them together, fo as to make them appear one flender feather.

Thefe birds frequent the fea fhores, banks of lakes and rivers: they feed on fmall fifh, and water infects; hovering over the water, and fuddenly darting into it, catch up their prey. They breed among fmall tufts of rufhes; and lay three or four eggs, of a dull olive color, fpotted with black. All the birds of this genus are very clamorous.

255. Lesser.

Larus pifcator (Fifcherlin, Fel.) *Gefner av.* 587. *fig.* 588.
Aldr. av. III. 35.
Leffer Sea Swallow. *Wil. orn.* 353.
Raii fyn. av. 131.
La petite Hirondelle de mer. *Briffon av.* VI. 206. *tab.* 19. *fig.* 2.
Larus Minuta. *Lin. fyft.* 228.
Hætting Tærne. *Brunnich,* 152.
Br. Zool. 144.

Descrip.

THE weight is only two ounces five grains: the length eight inches and a half; the breadth nineteen and a half. The bill is yellow, tipt with black: the forehead and cheeks white: from the eyes to the bill is a black line: the top of the head, and hind part black: the breaft, and under fide of the body cloathed with feathers fo clofely fet together,

ther, and of ſuch an exquiſite rich gloſs, and ſo fine a white, that no ſatin can be compared to it: the back and wings of a pale grey: the tail ſhort, leſs forked than that of the former, and white: the legs yellow: the irides duſky.

Theſe two ſpecies are very delicate, and ſeem unable to bear the inclemency of the weather on our ſhores * during winter: for we obſerve they quit their breeding places at the approach of it; and do not return till ſpring.

The manners, haunts, and food of this are the ſame with thoſe of the former; but theſe are far leſs numerous.

256. BLACK.

Larus niger (Meyvogelin) *Geſner av.* 588. *fig.* 589.
Aldr. av. III. 35.
The Scare Crow. *Wil. orn.* 353.
Our black cloven-footed Gull. *Idem.* 354.
Raii ſyn. av. 131. *Idem.* 132. No. 6.
L'Epouvantail. *Briſſon av.* VI. 211. *tab.* 20. *fig.* 1.
Sterna fiſſipes. *Lin. ſyſt.* 228.
Sicælandis Glitter. *Brunnich,* 153.
Kleinote Moewe. *Friſch,* II. 220.
Br. Zool. 145.

DESCRIP.

THIS is of a middle ſize, between the firſt and ſecond ſpecies. The uſual length is ten inches; the breadth twenty-four; the weight two ounces and a half. The head, neck, breaſt, and

* *North Wales.*

belly, as far as the vent, are black; beyond is white: the male has a white ſpot under its chin: the back and wings are of a deep aſh color: the tail is ſhort and forked; the exterior feather on each ſide is white; the others aſh colored: the legs and feet of a duſky red. Mr. *Ray* calls this a cloven-footed gull; as the webs are depreſſed in the middle, and form a creſcent. Theſe birds frequent freſh waters; breed on their banks, and lay three ſmall eggs of a deep olive color, much ſpotted with black.

They are found during ſpring and ſummer in vaſt numbers in the Fens of *Lincolnſhire*; make an inceſſant noiſe, and feed as well on flies as water inſects and ſmall fiſh.

Birds of this ſpecies are ſeen very remote from land. *Kalm* ſaw flocks of hundreds in the *Atlantic* ocean, midway between *England* and *America*; and a later voyager aſſured me he ſaw one 240 leagues from the *Lizard*, in the ſame ocean.

BILL

STORMY PETREL.

FULMAR. №. 257.

XLV.
PETREL.

BILL ſtrait, hooked at the end.
NOSTRILS cylindric, tubular.
LEGS naked above the knees.
BACK TOE none: inſtead, a ſharp SPUR pointing downwards.

257. FULMAR.

Wil. orn. 395.
Fulmar. *Martin's voy. St. Kilda.* 30. *Deſcr. weſt. Iſles.* 283.
Fulmer. *Macauly's hiſt. St. Kilda.* 145.
Haffheſt. *Cluſii exot.* 368.
Procellaria cinerea, le Petrel cendrè. *Briſſon av.* VI. 143. *tab.* 12. *fig.* 2.
Pl. enl. 59.
Lin. ſyſt. 213.
The Mallemucke. *Martin's Spitzberg.* 93.
Hav-Heſt. *Gunner, tab.* 1.
Procellaria glacialis. *Brunnich ornith.* 118.
Norvegis Hav-Heſt, Mallemoke V. Mallemuke. *Brunnich*, 118.
Br. Zool. 145.

THIS ſpecies inhabits the iſle of *St. Kilda*; makes its appearance there in *November*, and continues the whole year, except *September* and *October*; it lays a large, white, and very brittle egg; and the young are hatched the middle of *June*. No bird is of ſuch uſe to the iſlanders as this: the *Fulmar* ſupplies them with oil for their lamps, down for their beds, a delicacy for their tables, a balm for their wounds, and a medicine for their diſtempers. The *Fulmar* is alſo a certain prognoſticator of the change of the wind; if it comes to land, no weſt wind is expected for ſome time; and the contrary when it returns and keeps the ſea.

The whole genus of *Petrels* have a peculiar faculty of ſpouting from their bills, to a conſiderable diſtance, a large quantity of pure oil; which they do by way of defence, into the face of any that attempts to take them: ſo that they are, for the ſake of this *panacæa*, ſeized by ſurprize; as this oil is ſubſervient to the above-mentioned medical uſes. *Martin* tells us, it has been uſed in *London* and *Edinburgh* with ſucceſs, in *Rheumatic* caſes.

DESCRIP. The ſize of this bird is rather ſuperior to that of the common gull: the bill very ſtrong, much hooked at the end, and of a yellow color. The noſtrils are compoſed of two large tubes, lodged in one ſheath: the head, neck, whole under ſide of the body, and tail, are white; the back, and coverts of the wings aſh colored: the quil-feathers duſky: the legs yellowiſh. In lieu of a back toe, it has only a ſort of ſpur, or ſharp ſtrait nail. Theſe birds feed on the blubber or fat of whales, &c. which, being ſoon convertible into oil, ſupplies them conſtantly with means of defence, as well as proviſion for their young, which they caſt up into their mouths. They are likewiſe ſaid to feed on *ſorrel*, which they uſe to qualify the unctious diet they live on.

Frederick Martens, who had opportunity of ſeeing vaſt numbers of theſe birds at *Spitzbergen*, obſerves, that they are very bold, and reſort after the whale fiſhers

fishers in great flocks, and that when a whale is taken, will, in spite of all endeavours, light on it and pick out large lumps of fat, even when the animal is alive. That the whales are often discovered at sea by the multitudes of *Mallemuckes* flying; and that when one of the former are wounded, prodigious multitudes immediately follow its bloody track. He adds, that it is a most gluttonous bird, eating till it is forced to disgorge its food.

258. SHEAR-WATER.

Avis Diomedea, Artenna. *Aldr. av.* III. 36.
Manks Puffin. *Wil. orn.* 333.
Raii syn. av. 134.
Shear water. *Idem.* 133.
Wil. orn. 334.
Patines de oviedo. *Raii syn. av.* 191.
Edw. av. 359.
Procellaria Puffinus. *Lin. syst.* 213.
Puffinus, le Puffin. *Brisson av.* VI. 131. *tab.* 12. *fig.* 1. is a variety of it.
Feroensibus Skrabe. *Norvegis* Skraap, Pullus. *Feroensibus* Liere. *Brunnich,* 119.
Manks Petrel. *Br. Zool.* 146.

DESCRIP.

THE length of this species is fifteen inches; the breadth thirty-one: the weight seventeen ounces: the bill is an inch and three-quarters long; nostrils tubular, but not very prominent: the head, and whole upper side of the body, wings, tail, and thighs, are of a sooty blackness; the under side from chin to tail, and inner coverts of the wings, white: the legs weak, and compressed sideways; dusky behind, whitish before.

These birds are found in the *Calf* of *Man:* and

as Mr. *Ray* ſuppoſes in the *Scilly-iſles:* they reſort to the former in *February*; take a ſhort poſſeſſion of the rabbet burrows, and then diſappear till *April:* they lay one egg, white and blunt at each end; and the young are fit to be taken the beginning of *Auguſt*; when great numbers are killed by the perſon who farms the iſle: they are ſalted and barelled; and when they are boiled, are eaten with potatoes. During the day they keep at ſea, fiſhing; and towards evening return to their young; whom they feed, by diſcharging the contents of their ſtomachs into their mouths; which by that time is turned into oil: by reaſon of the backward ſituation of their legs they ſit quite erect. They quit the iſle the latter end of *Auguſt*, or beginning of *September*; and, from accounts lately received from navigators, we have reaſon to imagine, that like the *ſtorm-finch*, they are diſperſed over the whole *Atlantic* ocean.

This ſpecies inhabits alſo the *Orkney* iſles, where it makes its neſt in holes on the earth near the ſhelves of the rocks and headlands; it is called there the *Lyre*; and is much valued there, both on account of its being a food, and for its feathers. The inhabitants take and ſalt them in *Auguſt* for winter proviſions, when they boil them with cabbage. They alſo take the old ones in *March*; but they are then poor, and not ſo well taſted as the young: they appear firſt in thoſe iſlands in *February*.

The

259. STORMY

The Storm-finck. *Clusii exot.* 368.
Wil. orn. 395.
Small Petrel. *Edw. av.* 90.
Borlase's Cornwal. 247. *tab.* 29.
The Gourder. *Smith's hist. Kerry.* 186.
Assilag. *Martin's voy. St. Kilda.* 34.
Sib. hist. Fife. 48.
Procellaria, le Petrel. *Brisson av.* VI. 140. *tab.* 13. *fig.* 1.
Procellaria pelagica. *Lin. syst.* 212. *Scopoli,* No. 95.
Stromwaders vogel. *Faun. Suec. sp.* 143.
Norvegis Soren Peder. St. Peders Fugl, Vesten-vinds Are Sonden-vinds Fugl, Uveyrs Fugl: *nonnullis,* Hare. *Feroensibus* Strunkvit. *Brun.* 117.
Little Petrel. *Br. Zool.* 146.

DESCRIP.

THIS bird is about the bulk of the house swallow: the length six inches; the extent of wings thirteen. The whole bird is black, except the coverts of the tail and vent-feathers, which are white: the bill is hooked at the end: the nostrils tubular: the legs slender, and long. It has the same faculty of spouting oil from its bill as the other species: and Mr. *Brunnich* tells us, that the inhabitants of the *Ferroe isles* make this bird serve the purposes of a candle, by drawing a wick through the mouth and rump, which being lighted, the flame is fed by the fat and oil of the body. Except in breeding time it is always at sea; and is seen all over the vast *Atlantic* ocean, at the greatest distance from land; often following the vessels in great flocks, to pick up any thing that falls from on board: for trial sake chopped straw has been flung over, which they would ſtand on with expanded wings; but were never obſerved to ſettle on, or ſwim in the water: it preſages bad weather, and cautions the ſeamen of the approach of a tempeſt, by collecting under the ſtern of the ſhips: it braves the utmoſt fury of the ſtorm, ſometimes ſkimming with incredible velocity along the hollows of the waves, ſometimes on the ſummits: *Cluſius* makes it the *Camilla* of the ſea.

Vel mare per medium fluctu ſuſpenſa tumenti
Ferret iter, celeres nec tingeret æquore plantas. VIRGIL.

She ſwept the ſeas, and as ſhe ſkim'd along,
Her flying feet unbath'd on billows hung. DRYDEN.

Theſe birds are the *Cypſelli* of *Pliny*, which he places among the *Apodes* of *Ariſtotle*; not becauſe they wanted feet, but were Κακόποδα *, or had bad, or uſeleſs ones; an attribute he gives to theſe ſpecies, on a ſuppoſition they were almoſt always on the wing. *Hardouin*, a critick quite unſkilled in natural hiſtory, imagines them to be martins, the *Cypſelli* of *Ariſtotle* †: but a little attention to the text of each of thoſe antient naturaliſts, is ſufficient to evince that they are very different birds; the latter very accurately deſcribes the characters of that ſpecies of ſwallow: while *Pliny* expreſſes the very manner of life of our *Petrel.*

* *Ariſt.* 17.
† P. 1067.

" Nidificant

" Nidificant in ſcopulis, hæ ſunt quæ *toto mari cernuntur:* nec unquam tam longo naves, tamque continuo curſu recedunt a terra, ut non circumvolitent eas Apodes." *Lib.* x. *c.* 39.

In *Auguſt* 1772, I found them on the rocks called *Macdonald's Table*, off the north end of the *Iſle of Skie*; ſo conjecture they breed there. They lurked under the looſe ſtones, but betrayed themſelves by their twittering noiſe.

BILL

XLVI. MERGANSER.

BILL ſlender, furniſhed at the end with a crooked nail. Edges of each mandible ſharply ſerrated.

NOSTRILS near the middle of the mandible. Small, ſub-ovated.

TONGUE ſlender.

FEET, exterior toe longer than the middle.

260. GOOSANDER.

Mergus cirrhatus *(fæm.) Geſner av.* 134. Merganſer (Merrach) 135.
Aldr. av. III. 113.
Gooſander. *Wil. orn.* 335.
Dun diver, or Sparling-fowl. *ibid.*
Raii ſyn. av. 134.
Merganſer, l'Harle. *Briſſon av.* VI. 231. *Tab.* 22.
Meer-rache. *Kram.* 343.
See-Rache. *Friſch,* II. 190, 191.
Mergus merganſer. *Lin. ſyſt.* 208.
Wrakfogel, Kjorkfogel, Ard, Skraka. *Faun. Suec. ſp.* 135.
Pekſok. *Crantz's Greenl.* I. 80.
Iſlandis Skior-And. *Danis* Skalleſluger. *Brunnich,* 92, & 93.
Br. Zool. 147.

THESE birds frequent our rivers, and other freſh waters, eſpecially in hard winters; they are great divers, and live on fiſh. They are never ſeen in the ſouthern parts of *Great Britain* during ſummer; when they retire far north to breed; for in that ſeaſon they have been ſhot in the *Hebrides.* They are uncommonly rank, and ſcarcely eatable.

DESCRIP. The male weighs four pounds: its length is two feet four inches; the breadth three feet two. The

M. & F. GOOSANDER.

MALE.

The bill is three inches long, narrow, and finely toothed, or ſerrated: the color of that, and the irides, is red.

The head is large, and the feathers on the hind part long and looſe: the color black, finely gloſſed with green: the upper part of the neck the ſame: the lower part, and under ſide of the body of a fine pale yellow: the upper part of the back, and inner ſcapulars are black: the lower part of the back, and the tail are aſh colored: the tail conſiſts of eighteen feathers: the greater quil-feathers are black, the leſſer white, ſome of which are edged with black: the coverts at the ſetting on of the wing are black; the reſt white: the legs of a deep orange color.

DUN DIVER.

The *dun Diver*, or female, is leſs than the male: the head, and upper part of the neck are ferruginous; the throat white: the feathers on the hind part are long, and form a pendent creſt: the back, the coverts of the wings, and the tail are of a deep aſh color: the greater quil feathers are black, the leſſer white: the breaſt, and middle of the belly are white, tinged with yellow.

We believe that *Belon* * deſcribes this ſex under the title of *Bieure oyſeau*, and aſſerts, that it builds its neſt on rocks and in trees like the Corvorant.

* *Belon av.* 163.

Anas

261. RED BREASTED.

Anas Longiroftra. *Gefner av.* 133. *Aldr. av.* III. 113.
The Serula. *Wil. orn.* 336.
Raii fyn. av. 135.
Leffer toothed Diver. *Morton's Northampt.* 429.
L' Harle hupé. *Briffon av.* VI. 237.

Braun kopfiger Tilger, Taucher. *Kram.* 343.
Mergus ferrator. *Lin. fyft.* 208.
Pracka. *Faun. Suec. fp.* 136.
Danis Fifk-And. *Brunnich,* 96.
Br. Zool. 147.

DESCRIP.

THIS fpecies weighs two pounds: the length is one foot nine inches; the breadth two feet feven: the bill is three inches long; the lower mandible red; the upper dufky: the irides a purplifh red: head and throat a fine changeable black and green: on the firft a long pendent creft of the fame color: upper part of the neck, of the breaft, and the whole belly white: lower part of the breaft ferruginous, fpotted with black: upper part of the back black: near the fetting on of the wings fome white feathers, edged and tipt with black: the exterior fcapulars black; the interior white: lower part of the back, the coverts of the tail, and feathers on the fides under the wings and over the thighs grey, elegantly marked with ziczag lines of black: coverts on the ridge of the wings dufky; then fucceeds a broad bar of white: the greater coverts half black, half white: the fecondaries next the quil feathers marked in the fame manner; the reft white, edged on one fide with black: the quil feathers

M. & F. RED-BREASTED GOOSANDER.

feathers duſky. Tail ſhort and brown: legs orange colored.

The head and upper part of the neck of the *female* of a deep ruſt color: the creſt ſhort: throat white: fore part of the neck and breaſt marbled with deep aſh color: belly white: great quil-feathers duſky: lower half of the neareſt ſecondaries black; the upper white; the reſt duſky: back, ſcapulars, and tail aſh colored. The upper half of the firſt ſecondary feathers white; the lower half black: the others duſky.

Theſe birds breed in the northern parts of *Great Britain*; we have ſeen them and their young on *Loch Mari* in the county of *Roſs*, and in the iſle of *Ilay*.

262. SMEW.

La Piette. *Belon av.* 171.
Mergus rhenanus. *Geſner av.* 131.
Aldr av. III. 111.
White Nun. *Wil. orn.* 337.
Lough Diver. 338.
Raii ſyn. av. 135.
Mergus albellus. *Lin. ſyſt.* 209.
Faun. Suec. ſp. 137.
Le petit harle hupè ou le Piette. *Briſſon av.* VI. 243. *Tab.* 24. *fig.* 1. & 2.
Kram. 344.
Kreutz-Ente (Croſs-Duck) *Friſch*, II. 172.
Cimbris Hviid Side. *Brunnich*, 97.
Br. Zool. 148. *Scopoli*, No. 89.

DESCRIP.

ITS weight is thirty-four ounces: the length eighteen inches; the breadth twenty-ſix. The bill is near two inches long, and of a lead color: the

the head is adorned with a long creſt, white above, black beneath: from a little beyond the eye to the bill, is a large oval black ſpot, gloſſed with green; the head, neck, and whole under ſide of the body are of a pure white; on the lower part of the neck are two ſemilunar black lines pointing forward: the inner ſcapulars, the back, the coverts on the ridge of the wing, and the greater quil-feathers are black; the middle rows of coverts are white; the next black, tipt with white; the leſſer quil feathers the ſame; the ſcapulars next the wings white: the tail deep aſh color: the legs a bluiſh grey.

The female, or *lough diver*, is leſs than the male. The marks in the wings are the ſame in both ſexes: the back, the ſcapulars, and the tail are duſky: the head, and hind part of the neck ferruginous: chin, and fore part of the neck white: the breaſt clouded with grey: the belly white: the legs duſky.

263. Red headed.

The Weeſel Coot. *Ald. av.* I. *p.* 84. *Tab.* 88.
Mergus minutus. *Lin. ſyſt.* 209.
Faun. Suec. ſp. 138.
L' Harle etoilé. *Briſſon av.* VI. 252.
Br. Zool. 148.

Descrip.

THIS bird weighs fifteen ounces: the length is one foot four inches; the breadth one foot eleven inches: the bill is of a lead color: the head is is ſlightly creſted, and of a ruſt color: from beyond the eyes to the bill is an oval black ſpot: the cheeks and throat are white: the hind part of the neck is of a deep grey; the fore part clouded with a lighter: the belly white: the back and tail are of a duſky aſh color: the legs of a pale aſh color: the wings have exactly the ſame marks and colors with the ſmew; and as the ſpaces between the eyes and bill are marked with a ſimilar ſpot in both, if authors did not agree to make the *lough diver* the female of that bird, we ſhould ſuppoſe this to be it.

BILL.

XLVII. DUCK.

BILL ſtrong, flat, or depreſſed, and commonly furniſhed at the end with a nail. Edges divided into ſharp *lamellæ*.

NOSTRILS ſmall and oval.

TONGUE broad, edges near the baſe fringed.

FEET; middle toe the longeſt.

264. WILD SWAN.

Geſner av. 373.
Wild Swan, Elk, or Hooper. *Wil. orn.* 356.
Raii ſyn. av. 136.
Edw. av. 150.
Le Cygne ſauvage. *Briſſon av.* VI. 292. *Tab.* 28.
Labod. *Scopoli*, No. 66.
Schwane. *Kram.* 338.
Anas Cygnus ferus. *Lin. ſyſt.* 194.
Swan. *Faun. Suec. ſp.* 107.
Danis Vild Svane. *Cimbris* Snabel-Svane. *Brunnich*, 94.
Br. Zool. 149. *add. plates.*

THE wild ſwan frequents our coaſts in hard winters in large flocks, but as far as we can inform ourſelves does not breed in *Great Britain*. *Martin* * acquaints us, that ſwans come in *October* in great numbers to *Lingey*, one of the *Weſtern Iſles*; and continue there till *March*, when they retire more northward to breed. A few continue in *Mainland*, one of the *Orknies*, and breed in the little iſles of the freſh water lochs; but the multitude retires at approach of ſpring. On that account, ſwans are there the country man's almanack: on

* *Deſcr. Weſt. Iſles*, 71.

their

their quitting the iſland, they preſage good weather; on their arrival, they announce bad. Theſe, as well as moſt other water fowl, prefer for the purpoſe of incubation thoſe places that are leſt frequented by mankind: accordingly we find that the lakes and foreſts of the diſtant *Lapland* are filled during ſummer with myriads of water fowl, and there ſwans, geeſe, the duck tribe, gooſanders, divers, &c. paſs that ſeaſon; but in autumn return to us, and to other more hoſpitable ſhores *.

DESCRIP.

This ſpecies is leſs than the tame ſwan: length five feet to the end of the feet; to that of the tail four feet ten inches: extent of wing ſeven feet three inches: weight from thirteen to ſixteen pounds. The lower part of the bill is black; the baſe of it, and the ſpace between that and the eyes, is covered with a naked yellow ſkin; the eyelids are bare and yellow: the whole plumage in old birds is of a pure white; the down is very ſoft and thick: the legs black. The cry of this kind is very loud, and may be heard at a great diſtance, from which it is ſometimes called the Hooper.

* *Flora Lapponica*, 273. *Oeuvres de M. de Maupertuis. Tom.* III. p. 141, 175. According to the obſervation of that illuſtrious writer, the *Lapland* lakes are filled with the *larvæ* of the Knat (culex pipiens. *Lin. ſyſt.* 602.) or ſome other inſect, that depoſites its eggs in the water; which being an agreeable food to water fowl, is another cauſe of their reſort to thoſe deſerts.

265. TAME SWAN.

Le Cygne. *Belon av.* 151.
Gesner av. 371.
Cygno, Cisano. *Aldr. av.* III. 1.
Wil. orn. 355.
Raii syn. av. 136.
Edw. av. 150.
Plott's hist. Staff. 228.
Le Cygne. *Brisson av.* VI. 288.
Anas Cygnus mansuetus. *Lin. syst.* 194.
Swan. *Faun. Suec. sp.* 107.
Schwan. *Frisch,* II. 152.
Danis Tam Svane. *Brunnich,* 44.
Br. Zool. 149. *add. plates.*

DESCRIP.

THIS is the largest of the *British* birds. It is distinguished externally from the wild swan; first, by its size, being much larger: secondly, by the bill, which in this is red, and the tip and sides black, and the skin between the eyes and bill is of the same color. Over the base of the upper mandible projects a black callous knob: the whole plumage in old birds is white; in young ones ash colored till the second year: the legs dusky: but Dr. *Plott* mentions a variety found on the *Trent* near *Rugely*, with red legs. The swan lays seven or eight eggs, and is near two months in hatching: it feeds on water plants, insects and shells. No bird perhaps makes so inelegant a figure out of the water, or has the command of such beautiful attitudes in that element as the swan: almost every poet has taken notice of it, but none with that justice of description, and in so picturesque a manner, as our *Milton*.

The

> The ſwan with arched neck
> Between her white wings mantling, proudly rows
> Her ſtate with oary feet. *Par. Loſt*, B. VII.

But we cannot help thinking that he had here an eye to that beautiful paſſage in *Silius Italicus* on the ſame ſubject, though the *Engliſh* poet has greatly improved on it.

> Haud ſecus *Eridani* ſtagnis, ripâve *Cayſtri*
> Innatat albus olor, pronoque immobile corpus
> Dat fluvio, et pedibus tacitas eremigat undas. *Lib.* XIV.

In former times it was ſerved up at every great feaſt, when the elegance of the table was meaſured by the ſize and quantity of the good cheer. Cygnets are to this day fattened at *Norwich* about *Chriſtmas*, and are ſold for a guinea a piece.

Swans were formerly held in ſuch great eſteem in *England*, that by an act of *Edward* IV. *c.* 6. " no one that poſſeſſed a freehold of leſs clear yearly value than five marks, was permitted to keep any, *other than the ſon of our ſovereign lord the* king." And by the eleventh of *Henry* VII. *c.* 17. the puniſhment for taking their eggs was impriſonment for a year and a day, and a fine at the king's will. Though at preſent they are not ſo highly valued as a delicacy, yet great numbers are preſerved for their beauty; we ſee multitudes on the *Thames* and *Trent*, but no where greater numbers than on the ſalt water inlet of the ſea, near *Abbotſbury* in *Dorſetſhire*.

These birds were by the ancients consecrated to *Apollo* and the *Muses*;

—— ενθα κυκνος μελωδος
Μουσας θεραπευει. *Eurip. Iphig. in Taur. lin.* 1104.

And *Callimachus*, in his hymn upon the island of *Delos*, is still more particular:

—— Κυκνοι δε θεου μελποντες αοιδοι
Μηονιον πακτωλον εκυκλωσαντο λιποντες
Εβδομακις περι Δηλον. επηεισαν δε λοχειη
Μουσαων ορνιθες, αοιδοτατοι πετεηνων
Ενθεν ο παις τοσσασδε λυρη ενεδησατο χορδας
Υστερον, οσσακι κυκνοι επ ωδινεσσιν αεισαν:
Ογδοον εκ ετ αεισαν, ο δ' εκθορεν.

—— When from *Pactolus'* golden banks
Apollo's tuneful songsters, snowy swans
Steering their flight, seven times their circling course
Wheel round the island, caroling mean time
Soft melody, the favourites of the Nine,
Thus ushering to birth with dulcet sounds
The God of harmony, and hence sev'n strings
Hereafter to his golden lyre he gave,
For ere the eighth soft concert was begun
He sprung to birth. *Dod's Callimachus, p.* 115.

Upon this idea of their being peculiarly consecrated to *Apollo* and the *Muses*, (the deities of harmony) seems to have been ingrafted, the notion the antients had of swans being endowed with a musical voice. Tho' this might be one reason for the fable; yet, to us there appears another still stronger, which

which arose from the *Pythagorean* doctrine of the transmigration of the soul into the bodies of animals; from the belief, that the body of the swan was allotted for the mansion of departed poets. Thus *Plato* makes his prophet say, ιδειν μεν γαρ ψυχην εφη την ποτε Ορφεως γενομενην κυκνε βιον αιρεμενην *. "I saw the soul of *Orpheus* prefer the life of a swan."

After the antients had thus furnished these birds with such agreeable inmates, it is not to be doubted but they would attribute to them the same powers of harmony, that poets possessed, previous to their transmigration: but the vulgar not distinguishing between the sweetness of numbers, and that of voice, ignorantly believed that to be real, which philosophers and poets only meant metaphorically.

In time a swan became a common trope for a Bard; *Horace* calls *Pindar Dircæum Cygnum*, and in one ode even supposes himself changed into a swan; *Virgil* speaks of his poetical brethren in the same manner,

> *Vare*, tuum nomen
> Cantantes sublime ferent ad sydera cygni. *Eclog.* IX.

when he speaks of them figuratively, he ascribes to them melody, or the power of musick; but when he talks of them as birds, he lays aside fiction, and like a true naturalist gives them their real note,

> Dant sonitum *rauci* per stagna loquacia cygni. *Æneid.* Lib. X. I.

* *De Republ. Lib.* X. *sub fine.*

Thus he, as well as *Pliny* *, in fact, gave no credit to the musick of swans. *Aristotle* speaks of it only by hearsay †, but, when once an error is started, it is not surprizing that it is adopted, especially by poets, geniuses of all others of the most unbounded imaginations. For this reason poets were said to animate swans, from the notion that they flew higher than any other birds, and *Hesiod* distinguishes them by the epithet of κυκνοι αεροιποται ‡, "the lofty flying swans"; Thus *Horace*, whilst he humbly compares himself to a bee, contenting itself with the creeping thyme, sends his *Dircæum Cygnum* into the clouds

> Multa *Dircæum* levat aura *cygnum*,
> Tendit, *Antoni*, quoties in altos
> Nubium tractus. *Ode*. II. Lib. 4.

but when he finds himself struck with a true poetical spirit, he at once assumes the form of this favourite bird,

> Non usitata nec tenui feror
> Penna, biformis per liquidum æthera
> Vates:
> —— et album mutor in alitem. *Ode*. XX. Lib. 2.

And doubtless he was on the wing in his first ode,

> Sublimi feriam sydera vertice.

* Lib. X. c. 33.
† *Hist. an.* 1045.
‡ *Scut. Herc.* l. 316.

Besides

Befides thefe opinions, the antients held another ftill more fingular, imagining that the fwan foretold its own end: to explain this we muft confider the twofold character of the poet, *Vates* and *Poeta*, which the fable of the tranfmigration continue to the bird, or they might be fuppofed to derive that faculty from *Apollo* * their patron deity, the god of prophecy and divination.

As to their being fuppofed to fing more fweetly at the approach of death, the caufe is beautifully explained by *Plato*, who attributes that unufual melody, to the fame fort of *Ecftafy* that good men are fometimes faid to enjoy at that awful hour, forefeeing the joys that are preparing for them on putting off mortality, Μαντικοι τε εισι, και προειδοτες τα εν Αδȣ αγαθα, αδȣσι τε, και τερπονται εκεινην την ημεραν διαφεροντως η, εν τω προσθεν χρονω †. "They become prophetic, and forefeeing the happinefs which they fhall enjoy in another ftate, are in greater ecftafy than they have before experienced".

This notion, tho' accounted for by *Plato*, feems to have been a popular one long before his time, for *Æfchylus* alludes to it in his *Agamemnon*; *Clytemneftra* fpeaking of *Caffandra*, fays,

—— η δε τοι, κυκνȣ δικην,
Τον υςατον μελψασα θανασιμον γοον,
Κειται.

—— She like the fwan
Expiring, dies in melody.

* *Platonis Phædo*. Ed. Cantab. 1683. p. 124.
† *Ibid*.

266. GREY LAG.

Grey Lag, the Fen-Goose of *Lister. Ph. Tranf. abr.* II. 852.

Raii fyn. av. 136.

Gofs (the tame). *Scopoli*, No. 69.

DESCRIP. THIS is our largeft fpecies; the heavieft weigh ten pounds: the length is two feet nine; the extent five feet.

The bill is large and elevated; of a flefh color, tinged with yellow: the nail white: the head and neck cinereous, mixed with ochraceous yellow: the the hind part of the neck very pale; and at the bafe of a yellowifh brown.

Breaft and belly whitifh, clouded with grey or afh color: back grey: leffer coverts of the wings almoft white; the middle row, deep cinereous flightly edged with white: the primaries grey, tipt with black, and edged with white: fecondaries entirely black; grey only at their bafe: the fcapulars of a deep afh color, edged with white.

The coverts of the tail, and the vent feathers of a pure white: the middle feathers of the tail dufky, tipt with white; the exterior feathers almoft wholly white. The legs of a flefh color.

HISTORY. This fpecies refides in the fens the whole year: breeds there, and hatches about eight or nine young which are often taken, eafily made tame, and efteemed moft excellent meat, fuperior to the do-

domeſtic gooſe. The old geeſe which are ſhot, are plucked and ſold in the market as fine tame ones; and readily bought, the purchaſer being deceived by the ſize, but their fleſh is coarſe. Towards winter they collect in great flocks, but in all ſeaſons live and feed in the fens.

The *Grey Lag* is the origin of the *domeſtic* gooſe; it is the only ſpecies that the *Britons* could take young, and familiarize: the other two never breed here, and migrate during ſummer. The mallard comes within the ſame deſcription, and is the ſpecies to which we owe our tame breed of ducks: both preſerve ſome of the marks of their wild ſtate; the gooſe the whiteneſs of the coverts of the tail and vent-feathers; the drake its curled feathers. The gooſe in other colors ſports leſs in the tame kind than the other.

TAME GOOSE

Tame geeſe are of vaſt longevity. Mr. *Willughby* gives an example of one that attained eighty years.

Tame geeſe are keep in vaſt multitudes in the fens of *Lincolnſhire*; a ſingle perſon will keep a thouſand old geeſe, each of which will rear ſeven; ſo that towards the end of the ſeaſon he will become maſter of eight thouſand. I beg leave to repeat here part of the hiſtory of their œconomy from my tour in *Scotland*, in order to complete my account.

During the breeding ſeaſon theſe birds are lodged in the ſame houſes with the inhabitants, and

even

even in their very bed-chambers: in every apartment are three rows of coarſe wicker pens, placed one above another; each bird has its ſeparate lodge divided from the other, which it keeps poſſeſſion of during the time of ſitting. A perſon, called a *Gozzard*, i. e. *Gooſe-herd*, attends the flock, and twice a day drives the whole to water; then brings them back to their habitations, helping thoſe that live in the upper ſtories to their neſts, without ever miſplacing a ſingle bird.

FEATHERS. The geeſe are plucked five times in the year: the firſt plucking is at *Lady-Day*, for feathers and quils, and the ſame is renewed, for feathers only, four times more between that and *Michaelmas*. The old geeſe ſubmit quietly to the operation, but the young ones are very noiſy and unruly. I once ſaw this performed, and obſerved, that goſlins of ſix weeks old were not ſpared; for their tails were plucked, as I was told, to habituate them early to what they were to come to. If the ſeaſon proves cold, numbers of the geeſe die by this barbarous cuſtom. At the time, about ten pluckers are employed, each with a coarſe apron up to his chin.

Vaſt numbers of geeſe are driven annually to *London* to ſupply the markets, among them all the ſuperannuated geeſe and ganders (called here *Cagmags)* which, by a long courſe of plucking, prove uncommonly tough and dry.

The feathers are a conſiderable article of commerce;

merce; thofe from *Somerfetfhire* are efteemed the beft; and thofe from *Ireland* the worft.

It will not here be foreign to the fubject to give fome account of the feathers that other birds and other countries fupply our *Ifland* with, which was communicated to us by an intelligent perfon in the feather trade.

Eider down is imported from *Denmark*, the ducks that fupply it being inhabitants of *Hudfon's-Bay*, *Greenland*, *Iceland* and *Norway*; our own iflands weft of *Scotland* breed numbers of thefe birds, and might turn out a profitable branch of trade to the poor inhabitants. *Hudfon's-Bay* alfo furnifhes a very fine feather, fuppofed to be of the goofe kind.

The down of the fwan is brought from *Dantzick*. The fame place alfo fends us great quantity of the feathers of the cock and hen. The *London* poulterers fell a great quantity of the feathers of thofe birds, and of ducks and turkies; thofe of ducks being a weaker feather, are inferior to thofe of the goofe; turkey's feathers are the worft of any.

The beft method of curing feathers is to lay them in a room in an expofure to the fun, and when dried to put them in bags, and beat them well with poles to get the dirt off.

We have often been furprized that no experiments had been made on the feathers of the *Auk* tribe, as fuch numbers refort to our rocks annually,

ally, and promiſe, from the appearance of their plumage, to furniſh a warm and ſoft feather; but we have lately been informed, that ſome unſucceſsful trials have been made at *Glaſgow:* a gentleman who had made a voyage to the weſtern iſles, and brought ſome of the feathers home with a laudable deſign of promoting the trade of our own country, attempted to render them fit for uſe, firſt by baking, then by boiling them; but their ſtench was ſo offenſive, that the *Glaſgow* people could not be prevaled on to leave off their correſpondence with *Dantzick.* The diſagreeable ſmell of theſe feathers muſt be owing to the quantity of oil that all water fowls uſe from the glandules of their rump to preſerve and ſmooth their feathers; and as ſea birds muſt expend more of this unction than other water fowl, being almoſt perpetually on that element, and as their food is entirely fiſh, that oil muſt receive a great rankneſs, and communicate it to the plumage, ſo as to render it abſolutely unfit for uſe.

L'Oye

Pl. XCIV. No. 267

BEAN GOOSE.

1

1. WHITE FRONTED WILD GOOSE. No. 268.

L'Oye privèe, L'Oye Sauvage. *Belon av* 156. 158.
Gesner av. 142. 158.
Aldr. av. III 42. 67. *Phil. Tr*. II. 852.
Tame Goose, common wild Goose. *Wil. orn*. 358. *sp*. 1, 2.
Raii syn. av. 136. *sp*. 3, 4.
L'Oye domestique, L'Oye Sauvage. *Brisson av*. VI. 262, 265.
Oca domestica, Salvatica, Baletta. *Zinan*. 104.
Gus dikaya. *Russ. N. Com*. Petr. IV. 418.
Wild gansf, Einheimische gansf. *Kram*. 338. *Frisch*, II. 155, 157.
Anas anser mansuetus—ferus. *Lin. syst*. 197.
Gas—will gas. *Faun. Suec*. *sp*. 114.
Crantz's Greenl. I. 80.
Danis Tam Gaas. *Brunnich*, 55.
Br. Zool. 150.

267. BEAN-GOOSE.

DESCRIP.

THE length of this species is two feet seven inches: the extent four feet eleven: the weight six pounds and a half. The bill which is the chief specific distinction between this and the former is small, much compressed near the end, whitish and sometimes pale red in the middle; and black at the base and nail: head and neck are cinereous brown, tinged with ferruginous: breast and belly dirty white, clouded with cinereous: sides and scapulars dark ash color, edged with white: the back of a plain ash color: coverts of the tail white: lesser coverts of the wings light grey, nearly white; the middle deeper tipt with white: primaries and secondaries grey, tipt with black: feet and legs saffron color: claws black.

HISTORY.

This species arrives in *Lincolnshire* in autumn, and

and is called there the *bean goose*, from the likeness of the nail of the bill to a horse bean. They always light on corn fields, and feed much on the green wheat.

They never breed in the fens; but all disappear in *May*. They retreat to the sequestred wilds of the north of *Europe*: in their migration they fly a great height, cackling as they go. They preserve a great regularity in their motions, sometimes forming a strait line, at others assume the shape of a wedge, which facilitates their progress; for they cut the air the readier in that form than if they flew pell-mell.

268. White Fronted.

The laughing Goose. *Edw. av.* 153.
Anas erythropus (*fœm.*). *Lin. syst.* 197.
Fiællgas. *Faun. Suec. sp.* 116.
L'Oye sauvage du nord. *Brisson av.* VI. 269.
Polnische Gansz. *Kram.* 339.
Danis Vild Gaas. *Brunnich,* 53.
Br. Zool. 150.

Descrip.

THE weight of this kind is about five pounds and a half: the length two feet four: the extent four feet six: the bill elevated, of a pale yellow color, with a white nail. The forehead white: head and neck of the same color with those of the former: the coverts of the wing; the primaries and secondaries darker: in the tail the ash color predominates: it is like the two preceding, surrounded

rounded with a white ring. The breaſt and belly of a dirty white, marked with great ſpots of black: the legs yellow: the nails whitiſh.

HISTORY.

Theſe viſit the fens and other parts of *England* during winter, in ſmall flocks: they keep always in marſhy places, and never frequent the corn lands. They diſappear in the earlieſt ſpring, and none are ſeen after the middle of *March*. *Linnæus* makes this gooſe the female of the *Bernacle*; but we think his opinion not well founded.

Doctor *Liſter* adds two other ſpecies to the liſt of *Engliſh* geeſe; one he calls the *great Black Gooſe* or *Whilk*; the other the *ſmall Spaniſh Gooſe*, which he ſays is of the ſame color with the common gooſe; but is no larger than the *Brent*; but each ſpecies has hitherto eluded our moſt diligent enquiry.

I muſt conclude this ſubject with obſerving that the gooſe was one of the forbidden foods of the *Britons* in the time of *Cæſar*.

269. BERNACLE.

L'Oye nonnette ou Cravant. *Belon av.* 158.
Brenta, vel Bernicla. *Geſner av.* 109. 110.
Aldr. av. III. 73. *Phil. Tr.* II. 853.
Bernacle, or Clakis. *Wil. orn.*
Raii ſyn. av. 137.
Sibb. hiſt. Scot. 21.
Gerard's Herbal. 1587.
La Bernache. *Briſſon av.* VI. 300.
Anas Erythropus *(mas)*, *Lin. ſyſt.* 197.
Fiællgas. *Faun. Suec. ſp.* 116.
Schottische Gans. *Friſch*, II. 189.
Anſer brendinus. *Caii opuſc.* 87.
Crantz's Greenl. I. 80.
Br. Zool. 150.

DESCRIP.

THIS bird weighs about five pounds; the length is two feet one inch; the breadth four feet

feet five inches; the bill is black, and only one inch three-eights long; the head is ſmall; the forehead and cheeks white; from the bill to the eyes is a black line; the hind part of the head, the whole neck, and upper part of the breaſt and back are of a deep black; the whole underſide of the body, and coverts of the tail are white; the back, ſcapulars and coverts of the wings, are beautifully barred with grey, black, and white; the tail is black, the legs of the ſame color, and ſmall.

Theſe birds appear in vaſt flocks during winter, on the north weſt coaſts of this kingdom: are very ſhy and wild; but on being taken, grow as familiar as our tame geeſe in a few days; in *February* they quit our ſhores, and retire as far as *Lapland*, *Greenland* and even *Spitzbergen* to breed *.

They live to a great age: the Rev. Doctor *Buckworth* of *Spalding* had one which was kept in the family above two and thirty years; but was blind during the two laſt: what its age was when firſt taken was unknown.

Theſe are the birds that about two hundred years ago were believed to be generated out of wood, or rather a ſpecies of ſhell that is often found ſticking to the bottoms of ſhips, or fragments of them; and were called *Tree-geeſe* †. Theſe were alſo

* *Amœn Acad.* VI. 585. *Barent's voy.* 19.

† The ſhell here meant is the lepas anatifera. *Lin. ſyſt.* 668. *Argenville Conch.* tab. 7. the animal that inhabits it is furniſhed with a feathered beard; which, in a credulous age, was believed to be part of the young bird.

thought

thought by ſome writers to have been the *Chenalopeces* of *Pliny*: they ſhould have ſaid *Chenerotes*; for thoſe were the birds that naturaliſt ſaid were found in *Britain*; but as he has ſcarce left us any deſcription of them; it is difficult to ſay which ſpecies he intended. I ſhould imagine it to be the following; the *Brent-gooſe*, which is far inferior in ſize to the wild gooſe, and very delicate food *: in both reſpects ſuiting his account of the *Cheneros*.

270. BRENT.

Les Canes de Mer. *Belon av.* 166.
Aldr. av. III. 73.
Wil. orn. 360.
Raii ſyn. av. 137.
Bernacle. *Nat. hiſt. Ireland.* 192.
Brenta, le Cravant. *Briſſon av.* VI. 304. *tab.* 31.
Anas Bernicla. *Lin. ſyſt.* 198.
Belgis Rotgans, *Calmarienſibus* Prutgas. *Faun. Suec. ſp.* 115.
Cimbris Ray-v Rad-Gaas. *Norvegis* Raat-v. Raatne-Gaas. *item* Goul-v. Gagl. *Brnnnich*, 52.
Baum-Gans. *Friſch*, II. 156.
Br. Zool. 151.
Branta Bernicla. *Scopoli*, No. 84.

DESCRIP.

THIS is inferior in ſize to the former: the bill is one inch and an half long; the color of that, the head, neck, and upper part of the breaſt is black; on each ſide the ſlendereſt part of the neck is a white ſpot; the lower part of the breaſt, the ſcapulars, and coverts of the wings are aſh colored, clouded with a deeper ſhade; the feathers

* Anſerini generis ſunt *Chenalopeces*: et quibus lautiores epulas non novit *Britannia Chenerotes*, fere anſere minores. *Lib.* x. c. 22.

above and below the tail are white; the tail and quil-feathers black; the legs black.

These birds frequent our coasts in the winter: in *Ireland* they are called *Bernacles*, and appear in great quantities in *August*, and leave it in *March*. They feed on a sort of long grass growing in the water; preferring the root and some part above it, which they dive for, bite off and leave the upper part to drive on shore. They abound near *Londonderry*, *Belfast*, and *Wexford*; and are taken in flight time in nets placed a-cross the rivers; and are much esteemed for their delicacy. The *Rat* or *Roadgoose*, of Mr. *Willughby* *, agrees in so many respects with this kind, that we suspect it only to be a young bird not come to full feathers: the only difference consisting in the feathers next the bill, and on the throat and breast being brown. We have the greater reason to imagine it to be so as Mr. *Brunnich* informs us that the *Danish* and *Norvegian* names for this bird are *Radgaas* and *Raatgaas*, which agree with those given it by Mr. *Willughby*. Mr. *Willughby*, Mr. *Ray*, and M. *Brisson* very properly describe the *Bernacle* and *Brent* as different species, but *Linnæus* makes these synonymous, and describes the true *Bernacle* as the female of the white fronted wild goose. Vide *Faun. Suec.* 116.

* Page 361.

Wormius's

EIDER DRAKE & DUCK.

271. EIDER.

Wormius's Eider, or soft feathered Duck, the Cuthbert Duck. *Wil. orn.* 362.
Raii syn. av. 141.
Great black and white Duck. *Edw. av.* 98.
Eider anas. *Sib. Scot.* 21.
The Colk. *Martin's description of the western isles.* 25.
Anser lanuginosus, l'Oye a duvet. *Brisson av.* VI. 294. *tab.* 29. *et* 30.

Anas mollissima. *Lin. syst.* 198.
Ada, Eider, Gudunge, Æra. *Faun. Suec. sp.* 117.
Pontop. hist. Norway. II. 70.
Hor. hist. Icel. 65. *Debes Feroe* 137.
Egede's hist. Greenland. 92.
Mittek. *Crantz's Greenl.* I. 81.
Edder. *Brunnich,* 57. 66. *Monogr. tab.* 1. 2.
Duntur Goose. *Sib. Scot.* 21.

THIS useful species is found in the *western isles* of *Scotland*, particularly on *Oransa*, *Barra*, *Rona*, and *Heisker*, and on the *Farn isles*; but in greater numbers in *Norway*, *Iceland*, and *Greenland*: from whence a vast quantity of the down, known by the name of *Eider* or *Edder*, which these birds furnish, is annually imported: its remarkably light, elastic, and warm qualities, make it highly esteemed as a stuffing for coverlets, by such whom age or infirmities render unable to support the weight of common blankets. This down is produced from the breast of the bird in the breeding season. It lays its eggs among the stones or plants, near the shore: and prepares a soft bed for them, by plucking the down from its own breast; the natives watch the opportunity, and take away both eggs and nest: the duck lays again, and repeats the plucking of its breast; if she is robbed after that, ſhe will ſtill lay; but the drakes muſt ſupply the down, as her ſtock is now exhauſted; but if her eggs are taken a third time, ſhe wholly deſerts the place.

When I viſited the *Farn iſles**, I found the ducks ſitting, and toke ſome of the neſts, the baſe of which were formed of ſea plants, and covered with the down. After ſeparating it carefully from the plants, it weighed only three quarters of an ounce, yet was ſo elaſtic as to fill a larger ſpace than the crown of the greateſt hat. Theſe birds are not numerous on the iſles; and it was obſerved that the drakes kept on thoſe moſt remote from the ſitting places. The ducks continue on their neſts till you come almoſt cloſe to them, and when they riſe are very ſlow fliers. The number of eggs in each neſt were from three to five, warmly bedded in the down; of a pale olive color, and very large, gloſſy and ſmooth.

DESCRIP. This kind is double the ſize of the common duck: its bill is black; the feathers of the forehead and cheeks advance far into the baſe, ſo as to form two very ſharp angles: the forehead is of a full velvet black: from the bill to the hind part of the head is a broad black bar, paſſing acroſs the eyes on each ſide: on the hind part of the neck, juſt beneath the ends of theſe bars, is a broad pea-green mark, that looks like a ſtain:

* *July* 15th, 1769.

the

M. & F. VELVET DUCK.

the crown of the head, the cheeks, the neck, back, ſcapulars and coverts of the wings are white; the lower part of the breaſt, the belly, tail, and quil feathers are black; the legs are green.

FEMALE.

The female is of a reddiſh brown, barred tranſverſely with black; but the head and upper part of the neck are marked with duſky ſtreaks pointing downward; the primary feathers are black; the greater or laſt row of coverts of the wings, and the leſſer row of quil feathers tipt with white: the tail is duſky; the belly of a deep brown, marked obſcurely with black. One I weighed was three pounds and a half.

272. VELVET.

Anas nigra, roſtro nigro rubro et luteo. *Aldr. av.* III. 97.
The black Duck. *Wil. orn.* 363.
Raii ſyn. av. 141.
Dale's hiſt. Harwich, 405.
Turpan. *N. Com. Petr.* IV. 420.
La grande Macreuſe. *Briſſon av.* VI. 423.
Anas fuſca. *Lin. ſyſt.* 190.
Faun. Suec. ſp. 109.
Gunner. Tab. V.
Incolis Chriſtianſoe Svœrte. *Norvegis* Soe-Orre, Hav-Orre v. Sav-Orre, quibuſdam Sorte. *Brunnich*, 48.
Nordiſche ſchwartze Ente. *Friſch*, II. 165. Supl.
Br. Zool. 152. *Scopoli*, No. 68.

DESCRIP.

THE male of this ſpecies is larger than the tame duck. The bill is broad and ſhort, yellow on the ſides, black in the middle, and the hook red: the head, and part of the neck is black tinged with green: behind each ear is a white ſpot; and in each wing is a white feather; all the reſt of the plumage is of a fine black, and of the ſoft and delicate appearance of velvet: the legs and feet are red; the webs black. The female is entirely of a deep brown color; the marks behind each ear and on the wings excepted: the bill is of the ſame colors with that of the male; but wants the protuberance at the baſe of it, which *Linnæus* gives the male *.

273. Scoter. Black Diver, or Scoter. *Wil. orn.* 366.
Raii ſyn. av. 141.
La Macreuſe. *Ray's Letters*, 161.
Dale's hiſt. Harwich, 405.
La Macreuſe. *Briſſon av.* VI. 420. *Tab.* 38. *fig.* 2.
Anas nigra. *Lin. ſyſt.* 196.
Faun Suec. ſp. 110.
Br. Zool. 153.

Descrip. THIS ſpecies weighs two pounds nine ounces: the length is twenty-two inches; the breadth thirty-four: the middle of the bill is of a fine yellow, the reſt is black: both male and female want the hook at the end; but on the baſe of the bill of the former is a large knob, divided by a fiſſure in the middle. The tail conſiſts of ſixteen ſharp pointed feathers, of which the middle are the longeſt. The color of the whole plumage is black, that of the head and neck gloſſed over with purple: the legs are black.

* *Faun. Suec. laſt edit.* 39.

This

This bird is allowed in the *Romish church* to be eaten in *Lent*, and is the *macreuse* of the *French*. It is a great diver, said to live almost constantly at sea, and to be taken in nets placed under water.

274. TUFTED

Un petit Plongeon espece de Canard. *Belon av.* 175.
Strauss endt. *Gesner av.* 107.
Querquedula Cristata. *Aldr. av.* III. 91.
Wil. orn. 365.
Raii syn. av. 142.
Le petit Morillon. *Brisson av.* VI. 411. *Tab.* 27. *fig.* 1.
Kram. 341.
Anas fuligula. *Lin. syst.* 207.
Wigge. *Faun. Suec. sp.* 132.
Norvegis Trol-And. *Brunnich*, 90.
Reiger-Ente, Strauss-Ente. *Frisch*, II. 171.
Br. Zool. 153. *Scopoli*, No. 78.

DESCRIP.

THIS scarcely weighs two pounds: the length is fifteen inches and a half: the bill is broad, of a bluish grey, the hook black: the *irides* of a fine yellow. The head is adorned with a thick, but short pendent crest. The belly, and under coverts of the wings are of a pure white: the quil feathers dusky on their exterior sides and ends; part of their interior webs white; the secondaries white tipt with black. The rest of the plumage is black, varied about the head with purple: the tail is very short, and consists of fourteen feathers: the legs of a bluish grey; the webs black. The female wants the crest.

When young, this sex is of a deep brown; and the sides of the head next the bill of a pale yellow: but it preſerves the other marks of the old duck. In this ſtate it has been deſcribed in the *Ornith. boreal.* 91, under the title of *anas latiroſtra.*

275. SCAUP.

Bollenten. *Geſner av.* 120.
Scaup Duck. *Wil. orn.* 365.
Raii ſyn. av. 142.
Anas marila *Lin. ſyſt.* 196.
Faun. Suec. ſp. 111.
Le petit Morillon rayè. *Briſſon av.* VI. 416.
Danis Polſk Edelmand. *Brunnich*, 50, 51.
Schwartze wilde Ente. *Friſch*, II. 193.
Br. Zool. 153. *add. plates.*

DESCRIP.

THIS we deſcribed from ſome ſtuft ſkins very well preſerved *. It ſeemed leſs than the common duck. The bill was broad, flat, and of a greyiſh blue color: the head and neck black gloſſed with green: the breaſt black: the back, the coverts of the wings, and the ſcapulars finely marked with numerous narrow tranſverſe bars of black and grey: the greater quil feathers are duſky: the leſſer white, tipt with black: the belly is white: the tail and feathers, both above and below, are black; the thighs barred with duſky and white ſtrokes: the legs duſky.

Mr. *Willughby* acquaints us, that theſe birds take their name from feeding on *ſcaup*, or broken ſhell fiſh: they differ infinitely in colors; ſo

* When this happens, we have recourſe to Mr. *Willughby* for the weight and meaſurements, whenever he hath noted them.

that

that in a flock of forty or fifty there are not two alike.

Clangula. *Gesner av.* 119.
Aldr. av. III. 94.
Wil. orn. 368.
Raii syn. av. 142.
Le Garrot. *Brisson av.* VI. 416. *Tab.* 37. *fig.* 2.
Schwartzkopfige Enten-Taucher. *Frisch*, II. 183, 184.
Eifs Ente. *Kram.* 341.
Anas clangula. *Lin. syst.* 201.
Knippa, Dopping. *Faun. Suec. sp.* 122.
Norvegis Ring-Oye, Hviin-And v. Quiin-And, Lund-And. *Incolis Christiansoe*, Bruus-Kop v. Blanke-Kniv. *Br.* 70, 71.
Br. Zool. 154. *add. plates.*
Scopoli, No. 71.

276. GOLDEN EYE.

DESCRIP. MALE.

THIS species weighs two pounds: the length is nineteen inches; the breadth thirty-one. The bill is black, short, and broad at the base: the head is large, of a deep black glossed with green: at each corner of the mouth is a large white spot; for which reason the *Italians* call it *Quatt'occhii*, or four eyes: the *irides* are of a bright yellow: the upper part of the neck is of the same color with that of the head: the breast and whole under side of the body are white.

The scapulars black and white: the back, tail, and the coverts on the ridge of the wings, black: the fourteen first quil feathers, and the four last are black; the seven middlemost white, as are the coverts immediately above them: the legs of an orange color.

The

FEMALE. The head of the female * is of a deep brown, tinged with red: the neck grey: breaſt and belly white: coverts and ſcapulars duſky and aſh colored: middle quil feathers white; the others, together with the tail, black: the legs duſky. Theſe birds frequent freſh water, as well as the ſea; being found on the *Shropſhire* meres during winter.

277. MORILLON. Le Morillon. *Belon*, 165. *Wil. orn.* 368. *Raii ſyn. av.* 144. Anas glaucion? *Lin. ſyſt.* 201. *Scopoli*, No. 72. Grey headed Duck. *Br. Zool.* Ed. 2d. II. 471.

THIS ſpecies is rather leſs than the laſt. The bill of a yellowiſh brown: the *irides* gold color: the head of a duſky ruſt color: round the upper part of the neck is a collar of white; beneath that a broader of grey. The back and coverts duſky, with a few white lines: the greater coverts duſky, with a few great ſpots of white: the primaries black: the ſecondaries white. Breaſt and belly white: tail duſky: the ſides above the thighs black: the legs yellow.

This was bought in the *London* market. I am doubtfull of the ſex. Conſult *Briſſon*, VI. 406. *tab.* XXXVI.

* The ſmaller red headed Duck. *Wil. orn.* 369. *Raii ſyn. av.* 143.

Mr.

Mr. *Cockfield*, of *Stratford* in *Essex*, favored me with an account of two birds of this species, shot near the same time. Both agreed in colors; but one weighed twenty-six ounces, the other only nineteen.

278. SHIELDRAKE.

La Tadorne. *Belon av.* 172.
Anas maritima. *Gesner av.* 803, 804.
Vulpanser Tadorne. *Aldr. av.* III. 71, 97.
Shieldrake, or burrough Duck. *Wil. orn.* 363.
Raii syn. av. 140.
Anas tadorna. *Lin. syst.* 195.
Jugas. *Faun. Suec. sp.* 113.
La Tadorne. *Brisson av.* VI. 344. *tab.* 33. *fig.* 2.
Pl. enl. 53.
Bergander *Turneri*. Chenalopex *Plinii*.
Danis Brand-Gaas, Grav-Gaas. *Norvegis* Ring-Gaas, Fager-Gaas, Ur Gaas, Rodbelte. *Feroensibus* Hav-Simmer. *Islandis* Avekong. *Br.* 47.
Kracht-Ente. *Frisch*, II. 166.
Br. Zool. 154.

DESCRIP.

THE male of this elegant species weighs two pounds ten ounces: the length is two feet; the breadth three and a half. The bill is of a bright red, and at the base swells into a knob, which is most conspicuous in the spring: the head and upper part of the neck is of a fine blackish green; the lower part of the neck white: the breast, and upper part of the back is surrounded with a broad band of bright orange bay: the coverts of the wings, and the middle of the back are white; the nearest scapulars black, the others white; the greater quil feathers are black; the exterior webs

webs of the next are a fine green, and thoſe of the three ſucceeding orange; the coverts of the tail are white; the tail itſelf of the ſame color, and except the two outmoſt feathers tipt with black; the belly white, divided lengthways by a black line; the legs of a pale fleſh color.

Theſe birds inhabit the ſea coaſts, and breed in rabbet holes. When a perſon attempts to take their young, the old birds ſhew great addreſs in diverting his attention from the brood; they will fly along the ground as if wounded, till the former are got into a place of ſecurity, and then return and collect them together. From this inſtinctive cunning, *Turner*, with good reaſon, imagines them to be the *chenalopex* *, or *fox-gooſe* of the antients: the natives of the *Orknies* to this day call them the *ſlygooſe*, from an attribute of that quadruped. They lay fifteen or ſixteen eggs, white, and of a roundiſh ſhape. In winter they collect in great flocks. Their fleſh is very rank and bad.

* *Plinii, Lib.* X. *c.* 22.

Les

Pl. XCVII. WILD DUCKS. No 279

279. MALLARD.

Les Canards et les Canes. *Belon av.* 160.
Anas fera torquata minor. Anas domeftica. *Gefner av.* 113, 96.
Aldr. av. III. 83, 85.
Common wild Duck and Mallard. Common tame Duck. *Wil. orn.* 371, 380.
Raii fyn. av. 145, 150.
Le Canard domeftique, le Canard fauvage. *Briffon av.* VI. 308, 318.
Einheimifche ent. Stock ent. *Kram.* 341.
Anitra, Anitra falvatica, Cifone. *Zinan.* 105, 106.
Anas bofchas. Anas domeftica. *Lin. fyft.* 205.
Gras-and, Blanacke. *Faun. Suec. fp.* 131.
Fera, *Norvegis* Blaaehals v. Græs-And, aliis Stok-And. *Danis* Vild-And. *Brunnich*, 87.
Domeftica, *Danis* Tam-And. *ibid.* 88.
Wilde Ente. *Frifch*, II. 158. fæmina. 159.
Br. Zool. 155.
Ratza. *Scopoli*, No. 77.

DESCRIP.

THE mallard ufually weighs two pounds and an half: the length is twenty-three inches; the breadth thirty-five: the bill is of a yellowifh green: the head and neck are of a deep and fhining green: more than half round the lower part of the neck is an incomplete circle of white: the upper part of the breaft is of a purplifh red; and the beginning of the back of the fame color: the breaft and belly of a pale grey, marked with tranfverfe fpeckled lines of a dufky hue.

The fcapulars white, elegantly barred with brown: the fpot on the wing is of a rich purple: the tail confifts of twenty-four feathers. What diftinguifhes the male of this fpecies from all others are the four middle feathers, which are black and ftrongly ſtrongly curled upwards; but the females want this mark. Their plumage is of a pale reddiſh brown, ſpotted with black. The legs are of a ſaffron color.

The common tame ſpecies of ducks take their origin from theſe, and may be traced to it by unerring characters. The drakes, howſoever they vary in colors, always retain the curled feathers of the tail: and both ſexes the form of the bill of the wild kind. Nature ſports in the colors of all domeſtic animals; and for a wiſe and uſeful end; that mankind may the more readily diſtinguiſh and clame their reſpective property. Wild ducks pair in the ſpring, and breed in all marſhy grounds, and lay from ten to ſixteen eggs. They abound in *Lincolnſhire*, the great magazine of wild fowl in this kingdom; where prodigious numbers are taken annually in the decoys.

DECOYS.

A decoy is generally made where there is a large pond ſurrounded with wood, and beyond that a marſhy and uncultivated country: if the piece of water is not thus ſurrounded, it will be attended with the noiſe and other accidents, which may be expected to fright the wild fowl from a quiet haunt, where they mean to ſleep (during the daytime) in ſecurity.

If theſe noiſes or diſturbances are wilful, it hath been held, that an action will lye againſt the diſturber.

As ſoon as the evening ſets in, the decoy *riſes* (as

(as they term it) and the wild fowl feed during the night. If the evening is ftill, the noife of their wings, during their flight, is heard at a very great diftance, and is a pleafing, though rather melancholy found. This *rifing* of the decoy in the evening, is in *Somerfetfhire* called *rodding*.

The decoy ducks are fed with hempfeed, which is flung over the fkreens in fmall quantities, to bring them forwards into the pipes, and to allure the wild fowl to follow, as this feed is fo light as to float.

There are feveral *pipes* (as they are called) which lead up a narrow ditch, that clofes at laft with a funnel net. Over thefe pipes (which grow narrower from the firft entrance) is a continued arch of netting, fufpended on hoops. It is neceffary to have a pipe or ditch for almoft every wind that can blow, as upon this circumftance it depends which pipe the wild fowl will take to; and the decoy-man always keeps on the leeward fide of the ducks, to prevent his effluvia reaching their fagacious noftrils. All along each pipe, at certain intervals, are placed fkreens made of reeds, which are fo fituated, that it is impoffible the wild fowl fhould fee the decoy-man, before they have paffed on towards the end of the pipe, where the purfe-net is placed. The inducement to the wild fowl to go up one of thefe pipes is, becaufe the decoy-ducks, trained to this, lead the way, either after hearing the whiftle of the decoy-man, or enticed

by

by the hempſeed; the latter will dive under water, whilſt the wild fowl fly on, and are taken in the purſe.

It often happens, however, that the wild fowl are in ſuch a ſtate of ſleepineſs and dozing, that they will not follow the decoy-ducks. Uſe is then generally made of a dog, who is taught his leſſon: he paſſes backwards and forwards between the reed ſkreens (in which are little holes, both for the decoy-man to ſee, and for the little dog to paſs through) this attracts the eye of the wild fowl, who not chuſing to be interrupted, advance towards this ſmall and contemptible animal, that they may drive him away. The dog, all this time, by direction of the decoy-man, plays among the ſkreens of reeds, nearer and nearer to the purſe-net; till at laſt, perhaps, the decoy-man appears behind a ſkreen, and the wild fowl not daring to paſs by him in return, nor being able to eſcape upwards on account of the net-covering, ruſh on into the purſe-net. Sometimes the dog will not attract their attention, if a red handkerchief, or ſomething very ſingular, is not put about him.

The general ſeaſon for catching fowl in decoys, is from the latter end of *October* till *February*; the taking of them earlier is prohibited by an act 10. *George* II. *c.* 32. which forbids it from *June* 1, to *October* 1, under the penalty of five ſhillings for each bird deſtroyed within that ſpace.

The *Lincolnſhire* decoys are commonly ſet at a certain

certain annual rent, from five pounds to twenty pounds a year: and we have heard of one in *Somerſetſhire* that pays thirty. The former contribute principally to ſupply the markets of *London*. Amazing numbers of ducks, wigeons, and teal are taken: by an account ſent us of the number caught, a few winters paſt, in one ſeaſon, and in only ten decoys, in the neighborhood of *Wainfleet*, it appeared to amount to thirty-one thouſand two hundred, in which is included ſeveral other ſpecies of ducks; it is alſo to be obſerved, that in the above particular, wigeon and teal are reckoned but as one, and conſequently ſell but at half the price of the ducks. This quantity makes them ſo cheap on the ſpot, that we have been aſſured ſeveral decoy-men would be glad to contract for years to deliver their ducks at *Boſton* for ten-pence the couple. The account of the numbers here mentioned, relates only to thoſe that were ſent to the *Capital*.

It was cuſtomary formerly to have in the fens an annual *driving* of the young ducks before they took wing. Numbers of people aſſembled, who beat a vaſt tract, and forced the birds into a net placed at the ſpot where the ſport was to terminate. A hundred and fifty dozens have been taken at once: but this practice being ſuppoſed to be detrimental, has been aboliſhed by act of parlement.

280. SHOVELER.

Anas latiroſtra (ein Breitſchnabel.) *Geſner av.* 120.
Aldr. av. III. 94.
Wil. orn. 370.
Raii ſyn. av. 143.
Phaſianus marinus. *Charlton ex.* 105.
Blue-wing Shoveler *(fæm.) Cat. Carol.* I. 96.
Le Souchet. *Briſſon av.* VI. 329. *Tab.* 32. *fig.* 1.
Schaufll-ente, Loffl-ente. *Kram.* 342.
Anas clypeata. *Lin. ſyſt.* 200.
Faun. Suec. ſp. 119.
Kertlutock *. *Krantz's Greenl.* I. 80.
Danis Krop-And, *Norvegis* Stok-And. *Cimbris* Leffel-And. *Brunnich*, 67. 68.
Schield-Ente, Loeffel-Ente. *Friſch*, II. 161, 162. fæm. 163.
Br. Zool. 155. *Scopoli*, No. 70.

DESCRIP.

THIS weighs twenty-two ounces: its length twenty-one inches. The bill is black, three inches long, ſpreads near the end to a great breadth, is furniſhed with a ſmall hook, and the edges of each mandible are pectinated, or ſupplied with thin laminæ, that lock into each other when the mouth is cloſed. The irides are of a bright yellow: the head and upper part of the neck of a blackiſh green: the lower part of the neck, the breaſt, and the ſcapulars are white: the back brown: the coverts of the wings of a fine ſky blue; thoſe next the quil feathers tipt with white: the greater quil feathers are duſky; the exterior webs of thoſe in the middle, are of a gloſſy green. The tail conſiſts of fourteen feathers; the outmoſt are white;

* *i. e.* Broad bill.

thoſe

thofe in the middle black, edged with white: the belly is of a bay color: the vent feathers black: the legs red. The female has the fame marks in the wings as the male, but the colors are lefs bright: the reft of the plumage refembles that of the common wild duck.

281. RED BREASTED SHOVELER.

WE are indebted to Mr. *Bolton* for the defcription of this bird, who informed us that it was fometimes taken in the decoys in *Lincolnfhire*.

DESCRIP.

It is the fize of a common duck. The bill large, broad, ferrated at the fides, and entirely of a brownifh yellow color: the head large: eyes fmall: irides yellow: the breaft and throat of a reddifh brown, the latter paler, but both quite free from any fpots. The back is brown, growing paler towards the fides. The tips and pinions of the wings grey: the quil-feathers brown; the reft of a greyifh brown: the *fpeculum* or fpot purple, edged with white: in the female, the fpot is blue, and all the other colors are fainter. The tail is fhort and white: the vent feathers of a bright brown, fpotted with darker: the legs fhort and flender: the feet fmall, of a reddifh brown color.

282. PIN-TAIL.

Anas caudacuta (ein ſpitz-ſchwantz) *Geſner av.* 121.
Aldr. av. III. 97.
Sea Pheaſant, or Cracker. *Wil. orn.* 376.
Le Canard a longue queue. *Briſſon av.* VI. 369. *tab.* 34.
Schwalbenſcheif. *Kram.* 340.
Raii ſyn. av. 147.
Anas acuta. *Lin. ſyſt.* 202.
Aler, Ahlvogel. *Faun. Suec. ſp.* 126.
Faſan-Ente. *Friſch*, II. 160.
Brunnich in append.
Aglek. *Crantz's Greenl.* I. 80.
Br. Zool. 156. *Scopoli*, No. 73.

DESCRIP.

THE form of this ſpecies is ſlender, and the neck long: its weight twenty-four ounces: its length twenty-eight inches; its breadth one yard two inches. The bill is black in the middle, blue on the ſides: the head is ferruginous, tinged behind the ears with purple; from beneath the ears commences a white line, which runs ſome way down the neck; this line is bounded by black: the hind part of the neck, the back, and ſides are elegantly marked with white and duſky waved lines: the fore part of the neck, and belly are white.

The ſcapulars ſtriped with black and white: the coverts of the wings aſh colored; the loweſt tipt with dull orange: the middle quil-feathers barred on their outmoſt webs with green, black and white: the exterior feathers of the tail are aſh colored: the two middle black, and three inches longer than the others: the feet of a lead color. The female is of a light brown color, ſpotted with black. Mr. *Hartlib*, in the appendix to his *Lega-*

cy,

LONG TAILED DUCK.

WHITE THROATED DUCK.

cy, tells us that theſe birds are found in great abundance in *Connaught* in *Ireland*, in the month of *February* only; and that they are much eſteemed for their delicacy.

283. LONG TAILED.

Wil. orn. 364.
Raii ſyn. av. 145.
Long tailed Duck. *Edw. av.* 280.
Le Canard a longue queue d'Iſlande. *Briſſon av.* VI. 379.
Anas glacialis. *Lin. ſyſt.* 203.
Norvegis Ungle, Angeltaſke v. Troefoerer. *Feroenſibus* Oedel. *Iſlandis* Ha-Ella v. Ha-Old. *Incolis Chriſtianſoe* Gadiſſen, Klaeſhahn Dykker. *Brunnich*, 75, 76.
Br. Zool. 156. *Scopoli*, No. 74.

DESCRIP.

THIS is inferior in ſize to the former. The bill is ſhort, black at the tip and baſe, orange colored in the middle; the cheeks are of a pale brown: the hind part of the head, and the neck both before and behind are white; the ſides of the upper part of the neck are marked with a large duſky bar, pointing downwards; the breaſt and back are of a deep chocolate color; the ſcapulars are white, long, narrow, and ſharp pointed. The coverts of the wings, and greater quil feathers duſky; the leſſer of a reddiſh brown: the belly white: the four middle feathers of the tail are black; and two of them near four inches longer than the others, which are white: the legs duſky. Theſe birds breed in the moſt northern parts of the world, and only viſit our coaſts in the ſevereſt winters.

284. POCHARD. La Cane a teste rousse. *Belon av.* 173.
Anas fera fusca, vel media (ein wilte grauwe ente, Rotent.) *Gesner av.* 116.
Aldr. av. III. 93.
Poker, Pochard, or red headed Wigeon. *Wil. orn.* 367.
Raii syn. av. 143.
Anas ferina. *Lin. syst.* 203.
Faun. Suec. sp. 127.
Penelope, le Millouin. *Brisson av.* VI. 384. *tab.* 35. *fig.* 1.
Danis Brun-Nakke. *Norvegis* Rod-Nakke. *Brunnich,* 80.
Br. Zool. 156.

DESCRIP. ITS weight is about one pound twelve ounces: its length nineteen inches; its breadth two feet and a half. The bill is of a deep lead color: the head and neck are of a bright bay color: the breast and part of the back where it joins the neck, are black: the coverts of the wings, the scapulars, back and sides under the wings are of a pale grey, elegantly marked with narrow lines of black: the quil feathers dusky: the belly ash colored and brown: the tail consists of twelve short feathers, of a deep grey color: the legs lead colored: the irides of a bright yellow, tinged with red.

FEMALE. The head of the female is of a pale reddish brown: the breast is rather of a deeper color: the coverts of the wings a plain ash color: the back marked like that of the male: the belly ash colored. These birds frequent fresh water as well as the sea; and being very delicate eating, are much sought for in the *London* markets, where they are known by the name of *Dun birds.*

Anas

No. 285. FERRUGINOUS DUCK.

LONG TAILED DUCK, a Variety. No. 28.

Anas rufa roſtro pedibuſque cinereis. *Faun. Suec. ſp.* 47.

285. FERRUGINOUS.

DESCRIP.

THE deſcription of this ſpecies was ſent to us by Mr. *Bolton*. The weight was twenty ounces: the bill is long and flatted, rounded a little at the baſe, ſerrated along the edges of each mandible, and furniſhed with a nail at the end of the upper. The color a pale blue. The head, neck, and whole upper part of the bird is of an agreeable reddiſh brown: the throat, breaſt and belly of the ſame color, but paler: the legs of a pale blue; but the webs of the feet black.

This ſpecies, he informed us, was killed in *Lincolnſhire*. We do not find it mentioned by any writer, except *Linnæus*, who toke his deſcription from *Rudbeck*'s paintings; and adds, that it is found, though rarely, in the *Swediſh* rivers.

Anas fiſtularis (ein Pfeifente) *Geſner av.* 121.
Penelope. *Aldr. av.* III. 92.
Wigeon, or Whewer. *Wil. orn.* 375.
Raii ſyn. av. 146.
Anas penelops. *Lin. ſyſt.* 202.
Wriand. *Faun. Suec. ſp.* 124.
Anas fiſtularis, le Canard ſiffleur. *Briſſon av.* VI. 391. *tab.* 35. *fig.* 2.
Eiſſent mit weiſſer platten. *Kram.* 342.
Danis Bles-And. *Brunnich*, 72.
Br. Zool. 157. *add. plates.*

286. WIGEON

DESCRIP.

THE wigeon weighs near twenty-three ounces: the length is twenty inches; the breadth two feet three. The bill is lead colored; the end of it black; the head, and upper part of the neck is of a bright light bay; the forehead paler, in fome almoft white: the plumage of the back, and fides under the wings are elegantly marked with narrow, black and white undulated lines: the breaft is of a purplifh hue, which fometimes though rarely is marked with round black fpots: the belly white: the vent feathers black. In fome birds the coverts of the wings are almoft wholly white; in others of a pale brown, edged with white: the greater quil feathers are dufky; the outmoft webs of the middle feathers of a fine green, the tips black; the laft are elegantly ftriped with black and white. The two middle feathers of the tail are longer than the others, black and fharp pointed; the reft afh colored: the legs dufky. The head of the female is of a rufty brown, fpotted with black; the back is of a deep brown, edged with a paler: the tips of the leffer quil feathers white: the belly white.

FEMALE.

287. BIMACULATED.

THE length is twenty inches; extent twenty-five and a half. Bill a deep lead color: nail black.

Crown, brown changeable with green, ending in a ftreak of brown at the hind part of the head, with a fmall creft. Between the bill and the eye, and

Pl. C.

№ 275

SCAUP DUCK.

BIMACULATED DUCK.

№ 287.

and behind each ear, a ferruginous ſpot. The firſt round: the laſt oblong and large. Throat of a fine deep purple. The reſt of the head of a bright green, continued in ſtreaks down the neck. Breaſt a light ferruginous brown, ſpotted with black: hind part of the neck, and back, dark brown waved with black.

Coverts of the wings aſh colored: lower coverts ſtreaked with ruſt color: ſcapulars cinereous: quil feathers browniſh cinereous. Secondaries of a fine green, ending in a ſhade of black, and edged with white.

Coverts of the tail a deep changeable green. Twelve feathers in the tail: two middlemoſt black; the others brown edged with white. Belly duſky, finely granulated. Legs ſmall, and yellow. Webs duſky.

Taken in a decoy near in 1771. Communicated to me by *Poore*, Eſq.

288. GADWALL.

Anas ſtrepera (ein Leiner). *Geſner av.* 121.
Aldr. av. III. 97.
Gadwall, or Gray. *Wil. orn.* 374.
Raii ſyn. av. 145.
Le Chipeau. *Briſſon av.* VI. 339. *tab.* 33. *fig.* 1.

Anas ſtrepera. *Lin. ſyſt.* 200.
Faun. Suec. ſp. 121.
Cimbris Knarre-Gaas. *Brunnich*, 91.
Br. Zool. 157.
Grave mittel-ente. *Friſch*, II. 168.

DESCRIP.

THIS ſpecies is rather inferior in ſize to the wigeon. The bill is two inches long, black, and

and flat; the head, and upper part of the neck, are of a reddiſh brown, ſpotted with black; the lower part, the breaſt, the upper part of the back, and the ſcapulars, are beautifully marked with black and white lines; the belly is of a dirty white; the rump above and below is black; the tail aſh colored, edged with white; the coverts on the ridge of the wing are of a pale reddiſh brown; thoſe beneath are of purpliſh red, the loweſt of a deep black: the greater quil-feathers are duſky: the inner web of three of the leſſer quil-feathers are white; which forms a conſpicuous ſpot; the legs are orange colored. The breaſt of the female is of a reddiſh brown, ſpotted with black: the back of the ſame color; and though it has the ſame marks on the wings, they are far inferior in brightneſs to thoſe of the male.

289. GARGANEY.

La Sarcelle. *Belon av.* 175.
Querquedula varia. *Geſner av.* 107.
Scavolo, Cercevolo, Garganello. *Aldr. av.* III. 89, 90.
Wil. orn. 377.
Querquedula prima Aldr. *Raii ſyn. av.* 148*.
La Sarcelle. *Briſſon av.* VI. 427. *tab.* 39.
Krickantl. *Kram.* 343.
Anas Querquedula. *Lin. ſyſt.* 203.
Faun. Suec. ſp. 128.
Kriech-Ente. *Friſch,* II. 176.
Norvegis Krek-And. *Quibuſd.* Saur-And. *Brunnich,* 81.
Br. Zool. 158. *Scopoli,* No. 75.

DESCRIP.

THE length of this ſpecies is ſeventeen inches; the extent twenty-eight. The bill is of

* Mr. *Ray,* in his *ſyn. av.* 147. deſcribes a duck under the name of *Phaſeas*; in *Yorkſhire* it is called the widgeon: he ſays,

GARGANEY.

FEMALE GARGANEY.

of a deep lead color; the crown of the head is duſky, marked with oblong ſtreaks; on the chin is a large black ſpot; from the corner of each eye is a long white line, that points to the back of the neck: the cheeks, the upper part of the neck, are of a pale purple, marked with minute oblong lines of white, pointing downwards; the breaſt is of a light brown, marked with ſemicircular bars of black: the belly is white; the lower part and vent varied with ſpecks, the bars of a duſky hue; the coverts of the wings are grey; but the loweſt are tipt with white; the firſt quil-feathers are aſh colored; the exterior webs of thoſe in the middle green; the ſcapulars are long and narrow, and elegantly ſtriped with white, aſh color, and black; the tail duſky: the legs lead color.

FEMALE.

The female has an obſcure white mark over the eye; the reſt of the plumage is of a browniſh aſh color, not unlike the hen teal, but the wing wants the green ſpot, which ſufficiently diſtinguiſhes theſe birds.

In many places theſe birds are called the *Summer Teal.*

ſays, the head and neck are brown, ſpotted with triangular black marks: the body, wings, and tail duſky, edged with a paler color: in the wings is a double line of white: belly white: bill and legs blue. We ſuſpect it to be a young bird of this ſpecies, but wait for further information before we can determine it,

Querquedula.

290. TEAL. Querquedula. *Gesner av.* 106.
Garganei. *Aldr. av.* III. 90.
Wil. orn. 377.
Raii syn. av. 147.
La petite Sarcelle. *Brisson av.* VI. 436. *tab.* 40. *fig.* 1.
Rothantl, Pfeiffantl. *Kram.* 343.
Spiegel-Entlein. *Frisch,* II. 174.
Anas Crecca. *Lin. syst.* 204.
Arta, Kræcka. *Faun. Suec. sp.* 129.
Cimbris Atteling-And. *Norvegis* Heftelort-And. *Danis* Communiter Krik-And. *Brunnich,* 82, 83.
Br. Zool. 158. *add. plates.*

DESCRIP. THE Teal weighs about twelve ounces: the length is fourteen inches; the breadth twenty-three: the weight of a drake twelve ounces; of a duck nine: the bill black: the head, and upper part of the neck are of a deep bay: from the bill to the hind part of the head is a broad bar of gloffy changeable green, bounded on the lower fide by a narrow white line: the lower part of the neck, the beginning of the back, and the fides under the wings, are elegantly marked with waved lines of black and white.

The breaft and belly are of a dirty white; the firft beautifully fpotted with black: the vent black: the tail fharp pointed, and dufky: the coverts of the wings brown: the greater quil-feathers dufky; the exterior webs of the leffer marked with a gloffy green fpot; above that another of black, and the tips white: the irides whitifh; the legs dufky. The female is of a brownifh afh color, fpotted with

with black; and has a green ſpot on the wing like the male.

SUMMER TEAL.

By the deſcription Mr. *Willughby* has left of the *Summer Teal*, p. 378. we ſuſpect that it differs not in the ſpecies from the common kind, only in ſex. *Linnæus* hath placed it among the birds of his country *; but leaves a blank in the place of its reſidence; and hath evidently copied Mr. *Willughby*'s imperfect deſcription of it: and to confirm our ſuſpicion that he has followed the error of our countryman; we obſerved that a bird ſent us from the *Baltic* ſea, under the title of *anas circia*, the Summer Teal of *Linnæus*, was no other than the female of our teal.

* *Fauna Suecica*, *ſp.* 130.

BILL

XLVIII. CORVORANT*.

BILL ſtrong, ſtrait; end either hooked or ſloping.

NOSTRILS, either totally wanting, or ſmall, and placed in a longitudinal furrow.

FACE naked.

GULLET naked, capable of great diſtenſion.

TOES, all four webbed.

291. CORVORANT.

Mergus *Plinii* lib. x. c. 33.
Le Cormorant. *Belon av.* 161.
Corvus aquaticus, Carbo aquaticus. 136.
Phalacrocorax. *Geſner av.* 683. 350.
Aldr. av. III. 108.
The Cormorant. *Wil. orn.* 329.
Raii ſyn. av. 122.
Pelecanus Carbo. *Lin. ſyſt.* 216.
N. Com. Petr. IV. 423.
Le Cormoran. *Briſſon av.* VI. 511. *tab.* 45. *The Male.*
Norvegis Skarv, Strand-Ravn. *Danis* Aalekrage. *Iſlandis* Skarfur. *Brunnich,* 120, 121.
Scharb, or See-Rabe. *Friſch,* II. 187.
Br. Zool. 159. *Scopoli,* No. 98.

DESCRIP.

I HAVE weighed a bird of this ſpecies that exceeded ſeven pounds: the length three feet four: the extent four feet two: the bill duſky, five inches long, deſtitute of noſtrils; the baſe of the lower mandible is covered with a naked yellowiſh ſkin, that extends under the chin, and forms a ſort of pouch: a looſe ſkin of the ſame color

* The learned Dr. *Kay*, or *Caius*, derives the word *Corvorant*, from *Corvus vorans*, from whence corruptly our word *Cormorant*. *Caii opuſc.* 99.

reaches

reaches from the upper mandible round the eyes, and angles of the mouth: the head and neck are of a footy blacknefs; but under the chin of the male the feathers are white: and the head in that fex is adorned with a fhort loofe pendent creft; in fome the creft and hind part of the head are ftreaked with white. The coverts of the wings, the fcapulars, and the back, are of a deep green, edged with black, and gloffed with blue: the quil-feathers and tail dufky: the laft confifts of fourteen feathers: the breaft and belly black: in the midft of the laft is often a bed of white: on the thighs of the male is a tuft of white feathers: the legs are fhort, ftrong, and black; the middle claw ferrated on the infide: the irides are of a light afh color.

NEST.

EGGS.

Thefe birds occupy the higheft parts of the cliffs that impend over the fea: they make their nefts of fticks, fea tang, grafs, &c. and lay fix or feven white eggs of an oblong form. In winter they difperfe along the fhores, and vifit the frefh waters, where they make great havoke among the fifh. They are remarkably voracious, having a moft fudden digeftion, promoted by the infinite quantity of fmall worms that fill their inteftines. The corvorant has the rankeft and moft difagreeable fmell of any bird, even when alive. Its form is difagreeable; its voice hoarfe and croaking, and its qualities bafe. No wonder then that *Milton* fhould make *Satan* perfonate this bird, to *furvey*

undelighted

undelighted the beauties of Paradiſe: and *ſit deviſing death* on the tree of life *.

Theſe birds have been trained to fiſh like falcons to fowl. *Whitelock* tells us, that he had a caſt of them *manned* like hawks, and which would come to hand. He took much pleaſure in them, and relates, that the beſt he had was one preſented him by Mr. WOOD, *Maſter of the Corvorants to* CHARLES I. It is well known that the *Chineſe* make great uſe of theſe birds, or a congenerous ſort, in fiſhing; and that not for amuſement, but profit †.

292. SHAG.

Corvus aquaticus minor. *Aldr. av.* III. 109.
The Shag, called in the North of *England* the Crane. *Wil. orn.* 330.
Corvus aquaticus minor. Graculus palmipes dictus. *Raii ſyn. av.* 123.
Le petit Cormoran. *Briſſon av.* VI. 516.
Pelecanus graculus. *Lin. ſyſt.* 217.
Phalacrocorax criſtatus. *Norvegis* Top Skarv. *Brunnich ornith.* 123.
Br. Zool. 159.

DESCRIP.

THE ſhag is much inferior in ſize to the corvorant: the length is twenty-ſeven inches; the breadth three feet ſix: the weight three pounds three quarters. The bill is four inches long, and more ſlender than that of the preceding: the head is adorned with a creſt two inches long, pointing

* *Paradiſe Loſt, Book* IV. l. 194, &c.
† *Duhalde* I. 316.

back-

Pl. CII.

No 292

SHAG.

backward: the whole plumage of the upper part of this bird is of a fine and very ſhining green, the edges of the feathers a purpliſh black; but the lower part of the back, the head, and neck, wholly green: the belly is duſky: the tail conſiſts of only twelve feathers, of a duſky hue, tinged with green; the legs are black, and like thoſe of the corvorant. During my voyage among the *Hebrides*, I ſaw ſeveral birds of this ſpecies ſhot: they agreed in all reſpects, but in being deſtitute of a creſt; whether they were females, a variety, or diſtinct ſpecies, muſt be left to future naturaliſts to determine.

Both theſe kinds agree in their manners, and breed in the ſame places: and, what is very ſtrange in webbed footed birds, will perch and build in trees: both ſwim with their head quite erect, and are very difficult to be ſhot; for, like the *Grebes* and *Divers*, as ſoon as they ſee the flaſh of the gun, pop under water, and never riſe but at a conſiderable diſtance.

We are indebted for this bird to the late Mr. *William Morris* of *Holyhead*, with whom we had a conſtant correſpondence for ſeveral years, receiving from that worthy man and intelligent naturaliſt, regular and faithful accounts of the various animals frequenting that vaſt promontory.

293. GANNET.

Anſer *Baſſanus* ſive *Scoticus.* *Geſner av.* 163.
Aldr. av. 68.
Sula. *Hoieri Cluſ. ex.* 367.
Hector Boeth. 6.
Soland Gooſe. *Wil. orn.* 328.
Raii ſyn. av. 122.
Itin. 191. 269. 279.
Sibb. hiſt. Scot. 20. *tab.* 9.
Sibb. hiſt. Fife. 45. 47.
Jaen van Gent. *Martin's Spitzberg.* 97.
Solan Gooſe. *Martin's voy. St. Kilda.* 27.
Deſcript. Weſt. Iſles. 281.
Macauly's hiſt. St. Kilda. 133.
Sula Baſſana, le Fou de Baſſan. *Briſſon av.* VI. 503. *tab.* 44.
Pelecanus Baſſanus. *Lin. ſyſt.* 217.
Norvegis Sule, Hav-Sul. *Brunnich,* 124.
Br. Zool. 160.

DESCRIP.

THIS ſpecies weighs ſeven pounds: the length is three feet one inch; the breadth ſix feet two inches. The bill is ſix inches long, ſtrait almoſt to the point, where it inclines down; and the ſides are irregularly jagged, that it may hold its prey with more ſecurity: about an inch from the baſe of the upper mandible is a ſharp proceſs pointing forward; it has no noſtrils; but in their place a long furrow, that reaches almoſt to the end of the bill: the whole is of a dirty white, tinged with aſh color. The tongue is very ſmall, and placed low in the mouth: a naked ſkin of a fine blue ſurrounds the eyes, which are of a pale yellow, and are full of vivacity: this bird is remarkable for the quickneſs of its ſight: *Martin* tells us that *Solan* is derived from an *Iriſh* word expreſſive of that quality.

From

GANNET

Edwards Pinx.t

R. Murray Sc.t

From the corner of the mouth is a narrow ſlip of black bare ſkin, that extends to the hind part of the head: beneath the chin is another, that like the pouch of the *Pelecan*, is dilatable, and of ſize ſufficient to contain five or ſix entire herrings; which, in the breeding ſeaſon, it carries at once to its mate or young.

The neck is very long: the body flat, and very full of feathers: the crown of the head, and a ſmall ſpace on the hind part of the neck is buff colored: the reſt of the plumage is white: the baſtard wing and greater quil-feather excepted, which are black; the legs and toes are black; but the fore part of both are marked with a ſtripe of fine pea green. The tail conſiſts of twelve ſharp pointed feathers, the middle of which is the longeſt.

YOUNG.

The young birds, during the firſt year, differ greatly in color from the old ones; being of a duſky hue, ſpeckled with numerous triangular white ſpots; and at that time reſemble in colors the *ſpeckled Diver*. Each bird, if left undiſturbed, would only lay one egg in the year; but if that be taken away, they will lay another; if that is alſo taken, then a third; but never more that ſeaſon. A wiſe proviſion of nature, to prevent the extinction of the ſpecies by accidents, and to ſupply food for the inhabitants of the places where they breed; their egg is white, and rather leſs than that of the common gooſe: the neſt is large, and formed of any thing the bird finds floating on the

EGG.

NEST.

Sf2 water,

water, ſuch as graſs, ſea plants, ſhavings, &c. Theſe birds frequent the *Iſle of Ailſa*, in the *Firth of Clyde*; the rocks adjacent to *St. Kilda*, the *Stack* of *Souliſkery*, near the *Orkneys*; the *Skelig Iſles*, off the coaſts of *Kerry*, *Ireland* *, and the *Baſs Iſle*, in the *Firth of Edinburgh:* the multitudes that inhabit theſe places are prodigious. Dr. *Harvey's* elegant account of the latter, will ſerve to give ſome idea of the numbers of theſe, and of the other birds that annually migrate to that little ſpot.

" *There is a ſmall iſland, called by the* Scotch, " Baſs Iſland, *not more than a mile in circumfe-* " *rence; the ſurface is almoſt wholly covered du-* " *ring the months of* May *and* June *with neſts, eggs,* " *and young birds; ſo that it is ſcarcely poſſible to* " *walk without treading on them: and the flocks of* " *birds in flight are ſo prodigious, as to darken the* " *air like clouds; and their noiſe is ſuch, that you can-* " *not, without difficulty, hear your next neighbour's* " *voice. If you look down upon the ſea, from the* " *top of the precipice, you will ſee it on every ſide* " *covered with infinite numbers of birds of different* " *kinds, ſwimming and hunting for their prey: if in* " *ſailing round the iſland you ſurvey the hanging cliffs,* " *you may ſee in every cragg or fiſſure of the broken*

* This information we owe to that worthy prelate, the late Dr. *Pocock*, Biſhop of *Meath*; who had viſited the *Skeligs*. Mr. *Smith*, in his hiſtories of *Cork* and *Kerry*, confounds this bird with the Gull deſcribed by Mr. *Willughby*; from whom he has evidently borrowed the whole deſcription.

" *rocks,*

" *rocks, innumerable birds of various ſorts and ſizes,*
" *more than the ſtars of heaven when viewed in a*
" *ſerene night: if from afar you ſee the diſtant*
" *flocks, either flying to or from the iſland, you would*
" *imagine them to be a vaſt ſwarm of bees* *."

Nor do the rocks of *St. Kilda* ſeem to be leſs frequented by theſe birds; for *Martin* aſſures us, that the inhabitants of that ſmall iſland conſume annually no leſs than 22,600 young birds of this ſpecies, beſides an amazing quantity of their eggs; theſe being their principal ſupport throughout the year; they preſerve both eggs and fowls in ſmall pyramidal ſtone buildings, covering them with turf aſhes, to preſerve them from moiſture. This is a dear bought food, earned at the hazard of their lives, either by climbing the moſt difficult and

* *Eſt inſula parva, Scoti Baſſe nominant, haud amplius mille paſſuum circuitu amplitudo ejus clauditur. Hujus inſulæ ſuperficies, menſibus Maio & Junio nidis ovis pulliſque propemodum tota inſtrata eſt, adeo ut vix, præ eorum copia pedem liberè ponere liceat: tantaque ſuper-volantium turba, ut nubium inſtar, ſolem cœlumque auferant: tantuſque vociferantium clangor & ſtrepitus, ut propè alloquentes vix audias. Si ſubjectum mare inde, tanquam ex edita turri & altiſſimo præcipitio deſpexeris, idem quoquo verſûm, infinitis diverſorum generum avibus natantibus prædæque inhiantibus, opertum videas. Si circumnavigando imminentem clivum ſuſpicere libuerit; videas in ſingulis prærupti loci crepidinibus & receſſibus, avium cujuſlibet generis & magnitudinis, ordines innumerabiles, plures ſanè quam nocte, ſereno cœlo, ſtellæ conſpiciuntur. Si advolantes avolanteſque eminùs adſpexeris, apum profectò ingens examen credas.* De generat. Animal. Exercit. 11.

narrow paths, where (to appearance) they can barely cling, and that too, at an amazing height over the raging ſea: or elſe being lowered down from above, they collect their annual proviſion, thus hanging in midway air; placing their whole dependance on the uncertain footing of one perſon who holds the rope, by which they are ſuſpended at the top of the precipice. The young birds are a favorite diſh with the *North Britons* in general: during the ſeaſon they are conſtantly brought from the *Baſs Iſle* to *Edinburgh*, ſold at 20d. a piece, are roaſted, and ſerved up a little before dinner as a whet.

The *Gannets* are birds of paſſage. Their firſt appearance in thoſe iſlands is in *March*; their continuance there till *Auguſt* or *September*, according as the inhabitants take or leave their firſt egg; but in general, the time of breeding, and that of their departure, ſeems to coincide with the arrival of the herring, and the migration of that fiſh (which is their principal food) out of thoſe ſeas. It is probable that theſe birds attend the herring and pilchard during their whole circuit round the *Britiſh* iſlands; the appearance of the former being always eſteemed by the fiſhermen as a ſure preſage of the approach of the latter. It migrates in queſt of food as far ſouth as the mouth of the *Tagus*, being frequently ſeen off *Liſbon* during the month of *December*, plunging for *Sardinæ*, fiſh reſembling, if not the ſame with our *Pilchard*.

I have

I have in the month of *August* obſerved in *Cathneſs* their northern migrations: I have ſeen them paſſing the whole day in flocks, from five to fifteen in each: in calm weather they fly high; in ſtorms they fly low and near the ſhore; but never croſs over the land, even when a bay with promontories intervenes, but follow, at an equal diſtance, the courſe of the bay, and regularly double every cape. I have ſeen many of the parties make a ſort of halt for the ſake of fiſhing: they ſoared to a vaſt height, then darting headlong into the ſea, made the water foam and ſpring up with the violence of their deſcent; after which they purſued their route. I enquired whether they ever were obſerved to return ſouthward in the ſpring, but was anſwered in the negative; ſo it appears that they annually encircle the whole iſland.

Name.

They are well known on moſt of our coaſts but not by the name of the *Soland-Gooſe*. In *Cornwal* and in *Ireland* they are called *Gannets*; by the *Welſh Gan*. The excellent Mr. *Ray* ſuppoſed the *Corniſh Gannet* to be a ſpecies of large Gull; a very excuſeable miſtake, for during his ſix months reſidence in *Cornwal*, he never had an opportunity of ſeeing that bird, except flying; and in the air it has the appearance of a gull. On that ſuppoſition he gave our *Skua*, p. 417. the title of *Cataracta*, a name borrowed from *Ariſtotle* *, and which admirably expreſſes the rapid deſcent of this bird on

* Page 1045.

S f 4 its

its prey. Mr. *Moyle* firſt detected this miſtake*; and the Rev. Doctor *William Borlaſe*, by preſenting us with a fine ſpecimen of this bird, confirms the opinion of Mr. *Moyle*; at the ſame time he favored us with ſo accurate an account of ſome part of the natural hiſtory of this bird, that we ſhall uſe the liberty he indulged us with, of adding it to this deſcription.

" The *Gannet* comes on the coaſts of *Cornwal* " in the latter end of ſummer, or beginning of au- " tumn; hovering over the ſhoals of pilchards that " come down to us through *St. George's Channel* " from the northern ſeas. The *Gannet* ſeldom " comes near the land, but is conſtant to its prey, " a ſure ſign to the fiſhermen that the pilchards are " on the coaſts; and when the pilchards retire, ge- " nerally about the end of *November*, the *Gannets* " are ſeen no more. The bird now ſent was killed " at *Chandour*, near *Mountſbay*, *Sept.* 30, 1762, af- " ter a long ſtruggle with a water ſpaniel, aſſiſted " by the boatmen; for it was ſtrong and pugna- " cious. The perſon who took it obſerved that it " had a tranſparent membrane under the eye-lid, " with which it covered at pleaſure the whole eye, " without obſcuring the ſight or ſhutting the eye- " lid; a gracious proviſion for the ſecurity of the " eyes of ſo weighty a creature, whoſe method of " taking its prey is by darting headlong on it

* *Moyle's Works*, I. 424.

" from

" from a height of a hundred and fifty feet or more
" into the water. About four years ago, one of
" these birds flying over *Penzance*, (a thing that
" rarely happens) and seeing some pilchards lying
" on a fir-plank, in a cellar used for curing fish,
" darted itself down with such violence, that it
" struck its bill quite through the board (about
" an inch and a quarter thick) and broke its neck."

These birds are sometimes taken at sea by a deception of the like kind. The fishermen fasten a pilchard to a board, and leave it floating; which inviting bait decoys the unwary *Gannet* to its own destruction.

In the *Cataracta* of *Juba** may be found many characters of this bird: he says, that the bill is toothed: that its eyes are fiery; and that its color is white: and in the very name is expressed its furious descent on its prey. The rest of his accounts savors of fable.

We are uncertain whether the *Gannet* breeds in any other parts of *Europe* besides our own islands; except (as Mr. *Ray* suspects, the *Sula*, described in *Clusius's Exotics*, which breeds in the *Ferroe Isles)* be the same bird. In *America* there are two species of birds of this genus, that bear a great resemblance to it in their general form and their manner of preying. Mr. *Catesby* has given the figure of the head of one, which he calls the *Greater*

* *Plinii*, lib. x. c. 44.

Booby;

Booby; his defcription fuits that of the young *Gannet*; but the angle on the lower mandible made us formerly fufpect that it was not the fame bird; but from fome late informations we have been favored with, we find it is common to both countries, and during fummer frequents *North America*. Like the *Penguin*, it informs navigators of the approach of *foundings*, who on fight of it drop the plummet. *Linnæus* claffes our bird with the *Pelecan*; in the tenth edition of his fyftem, he confounds it with the bird defcribed by Sir *Hans Sloane*, hift. *Jam.* vol. I. p. 31. *preface*, whofe colors differ from the *Gannet* in each ftage of life: but in his laft edition he very properly feparates them. We continue it in the fame clafs, under the generical name of *Corvorants*, as more familiar to the *Englifh* ear than that of *Pelecan*.

APPENDIX.

APPENDIX.

ROUGH LEG'D FALCON.

APPENDIX.

Birds now extinct in GREAT BRITAIN, or ſuch as wander here accidentally.

LAND BIRDS.

I. ROUGH LEG'D FALCON.

THIS ſpecies is a native of *Denmark*, but was ſhot in and is preſerved in the *Leverian Muſeum.*

Its length is two feet two inches: that of the wing, when cloſed, eighteen inches: the bill duſky; the cere yellow: the head, neck, and breaſt of a yellowiſh white, marked in ſome parts with oblong brown ſtrokes: the belly of a deep brown: thighs and legs of a pale yellow, marked with brown: the ſcapulars blotched with brown and yellowiſh white: coverts of the wings brown, edged with ruſt: ends of the primaries deep brown; the lower parts white: the extreme half of the tail brown, tipt with dirty white: that next to the body white. Legs covered with feathers as low as the feet: the feet yellow.

II. ROLLER.

II. ROLLER.

Roller. *Wil. orn.* 131.
Garrulus *argentoratensis. Raii syn. av.* 41.
Galgulus, le Rollier. *Brisson av.* II. 64. *tab.* 5.
Coracias Garrula. *Lin. syst.* 159.
Spransk Kraka, Blakraka, Allekraka. *Faun. Suec. sp.* 94.
Edw. 109.
The Shagarag. *Shaw's Travels.* 252.
Ellekrage. *Brunnich*, 35.
Birk-Heker; Blaue-Racke. *Frisch*, I. 57.

OF these birds we have heard of only two being seen at large in our island; one was shot near *Helston-bridge, Cornwal*, and an account of it transmitted to us by the Reverend Doctor *William Borlase*. They are frequent in most parts of *Europe*, and we have received them from *Denmark*.

FEMALE.

In size it is equal to a jay. The bill is black, strait, and hooked at the point; the base beset with bristles: the space about the eyes is bare and naked: behind each ear is also another bare spot, or protuberance: the head, neck, breast, and belly are of a light bluish green: the back, and feathers of the wings next to it, are of a reddish brown: the coverts on the ridge of the wings are of a rich blue; beneath them of a pale green: the upper part and tips of the quil-feathers are dusky; the lower parts of a fine deep blue; the rump is of the same color: the tail consists of twelve feathers, of which the outmost on each side are considerably longer

The ROLLER. App.x II.

The NUTCRACKER. App.x.

longer than the reſt; are of a light blue, and tipt with black, beneath that a ſpot of deep blue; as is the caſe with ſuch part of the quil-feathers that are black above: the other feathers of the tail are of a dull green: the legs ſhort, and of a dirty yellow.

It is remarkable for making a chattering noiſe, from which it is by ſome called *Garrulus*.

III. NUTCRACKER.

Caryocatactes. *Wil. orn.* 132. *Raii ſyn. av.* 42.
Nucifraga, le Caſſe-noix. *Briſſon av.* II. 59. *tab.* 5.
Corvus Caryocatactes. *Lin. ſyſt.* 157.
Notwecka, Notkraka. *Faun. Suec. ſp.* 19.
Tannen-Heher (Pine-Jay) *Friſch*, I. 56.
Edw. 240.
Danis Noddekrige. *Norvegis* Not-kraake. *Brunnich*, 34.

THE ſpecimen we toke our deſcription from, is the only one we ever heard was ſhot in theſe kingdoms; is was killed near *Moſtyn*, *Flintſhire*, *October* 5, 1753.

DESCRIP. It was ſomewhat leſs than the jackdaw: the bill ſtrait, ſtrong, and black: the color of the whole head and neck, breaſt and body, was a ruſty brown: the crown of the head and the rump were plain: the other parts marked with triangular white ſpots: the wings black: the coverts ſpotted in in the ſame manner as the body: the tail rounded at the end, black tipt with white: the vent-feathers white: the legs duſky.

This bird is alſo found in moſt parts of *Europe*. We received a ſpecimen from *Denmark*, by means of Mr. *Brunnich*, author of the *Ornithologia Borealis*, a gentleman to whoſe friendſhip we owe a numerous collection of the curioſities of his country.

It feeds on nuts, from whence the name.

IV. The ORIOLE.

Oriolus Galbula. *Lin. ſyſt.* 160. *Faun. Suec.* No. 95.
Scopoli, No. 45. *Kramer*, 360.
Oriolus. *Geſner av.* 713. *Aldr. av.* I. 418.
The Witwal. *Wil. orn.* 198. *Raii ſyn. av.* 68.
Le Loriot. *Briſſon* II. 320.
Golden Thruſh. *Edw.* 185.

HISTORY. THIS beautiful bird is common in ſeveral parts of *Europe*; where it inhabits the woods, and hangs its neſt very artificially between the ſlender branches on the ſummits of antient oaks. Its note is loud, and reſembles its name. I have heard of only one being ſhot in *Great Britain*, and that in *South Wales*.

DESCRIP. It is of the ſize of a thruſh: the head and whole body of the male is of a rich yellow: the bill red; from

THE ORIOLE M. & F.

The ROSE COLORED OUZEL. *App:*

from that to the eye a black line: the wings black, marked with a bar of yellow: the ends of the feathers of the same color: the two middle feathers of the tail black; the rest black, with the ends of a fine yellow: the legs dusky.

The body of the female is of a dull green: the wings dusky: the tail of a dirty green: the ends of the exterior feathers whitish.

V. The ROSE COLORED OUSEL.

Merula rosea. *Raii syn. av.* 67. *Aldr. av.* II. 283.
Wil. orn. 194.
Le Merle Couleur de Rose. *Brisson av.* II. 250.
Turdus roseus. *Lin. syst.* 294.
Faun. Suec. sp. 219.
Edw. 20.

MR. *Edwards* discovered this beautiful bird twice in our island, near *London*, at *Norwood*, and another time in *Norfolk*. The figure of this and the preceding, were copied, by permission, from his beautiful and accurate designs, which we gratefully acknowledge, as well as every other assistance from our worthy friend; whose pencil has done as much honor to our country, as the integrity of his heart, and communicative disposition, has procured him esteem from a numerous and respectable acquaintance.

The size of this bird appears by the print to be equal

DESCRIP.

equal to that of a ſtare. The bill at the point is black, at the baſe a dirty fleſh color: the head is adorned with a creſt hanging backwards. The head, creſt, neck, wings, and tail are black, gloſſed with a changeable blue, purple and green: the breaſt, belly, back, and leſſer coverts of the wings, are of a roſe color, mixed with a few ſpots of black: the legs of a dirty orange color.

This bird is found in *Lapland*, *Italy*, and *Syria.* About *Aleppo* it is called the *locuſt bird*, poſſibly from its food; and appears there only in ſummer *. In *Italy* it is ſtyled the ſea-ſtare; and as *Aldrovandus* ſays, frequents heaps of dung †. And Mr. *Ekmarck* ‡ informs us, that it reſides in *Lapland*, never paſſing beyond the limits of that frozen region. We have mentioned very oppoſite climes, but believe it to be a ſcarce bird in all, at leſt in *Europe.*

* *Ruſſel's hiſt. Alep.* 70. *Tavernier*, 146.

† *Aldr. av.* II. 283.

‡ *Migr. av. Amæn. acad.* IV. 594.

WATER

The CRANE. App.

WATER FOWL.

VI. The CRANE.

Le Grue. *Belon av.* 187.
Grus. *Gesner av.* 528.
A Crane. *Turner.*
Gru, Grua. *Aldr. av.* III. 132.
Wil. orn. 274.
Raii syn. av. 95.
La Grue. *Brisson av.* V. 374. *tab.* 33.
Kranich. *Kram.* 345.
Kranich. *Frisch*, II. 194.
Ardea Grus. *Lin. syst.* 234.
Trana. *Faun. Suec. sp.* 161.
Danis Trane. *Brunnich.*
Br. Zool. 118.

THIS species was placed, in the folio edition of the *Zoology*, among the *British* birds, on the authority of Mr. *Ray*; who informs us, that in his time, they were found during the winter in large flocks in *Lincolnshire* and *Cambridgeshire*: but on the strictest enquiry we learn, that at present the inhabitants of those counties are scarcely acquainted with them; we therefore conclude, that these birds have forsaken our island. A single bird was killed near *Cambridge* about three years ago, and is the only instance I ever knew of the crane being seen in this island in our time. They were formerly in high esteem at our tables, for the delicacy of their flesh; for they feed only on grain, herbs, or insects; so have nothing of the rankness of the piscivorous birds of this genus.

DESCRIP. Its weight is about ten pounds; the length ſix feet; the bill of a darkiſh green, four inches long; and a little depreſſed on the top of the upper mandible: the top of the head covered with black briſtles; the back of the head bald and red, beneath which is an aſh colored ſpot: from the eyes, of each ſide, is a broad white line the whole length of the neck: the fore part as far as the breaſt is black: the quil-feathers are black: the tail aſh colored, tipt with black: all the reſt of the plumage is aſh colored. The legs are black.

No author, except *Geſner*, takes notice of a large tuft of feathers that ſpring out of one pinion on each wing: they are unwebbed, and finely curled at the ends, which the birds have power to erect or depreſs; when depreſſed they hang over and cover the tail. *Geſner* tells us, that theſe feathers uſed in his time to be ſet in gold, and worn as ornaments in caps. Though this ſpecies ſeems to have forſaken theſe iſlands at preſent, yet it was formerly a native, as we find in *Willughby*, p. 52. that there was a penalty of twenty-pence for deſtroying an egg of this bird; and *Turner* relates, that he has very often ſeen their young in our marſhes. *Marſigli* * ſays, that the crane lays two eggs like thoſe of a gooſe, but of a bluiſh color.

* *Hiſt. Danub.* V. p. 8.

VII. The

The EGRET.

VII. The EGRET.

Leſſer White Heron. *Wil. orn.* 280.
Ardea Garzetta. *Lin. ſyſt.* 237.
Ardea Alba minor. *Raii ſyn. av.* 99.
Dwarf Heron. *Barbot*, 29.
L'Aigrette. *Briſſon av.* V. 431.
Kleiner Weiſſer Rager. *Kram.* 345.

WE once received out of *Angleſea*, the feathers of a bird ſhot there, which we ſuſpect to be the *Egret*; this is the only inſtance perhaps of its being found in our country. That formerly this bird was very frequent here, appears by ſome of the old bills of fare: in the famous feaſt of Archbiſhop *Nevill*, we find no leſs than a thouſand *Aſterides* *, *Egrets* or *Egrittes*, as it is differently ſpelt. Perhaps the eſteem they were in as a delicacy during thoſe days, occaſioned their extirpation in our iſlands; abroad they are ſtill common, eſpecially in the ſouthern parts of *Europe*, where they appear in flocks.

DESCRIP.

The *Egret* is a moſt elegant bird; it weighs about one pound; the length is twenty-four inches, to the end of the legs thirty-two: the bill is ſlender and black: the ſpace about the eyes naked and

* *God-win de Præſul. Angl. com. Leland's* Collect.

green: the irides of a pale yellow: the head adorned with a beautiful creſt, compoſed of ſome ſhort, and of two long feathers, hanging backward; theſe are upwards of four inches in length: the whole plumage is of a reſplendent whiteneſs: the feathers on the breaſt, and the ſcapulars, are very delicate, long, ſlender, and unwebbed, hanging in the lighteſt and looſeſt manner: the legs are of a dark green almoſt black: the ſcapulars and the creſt were formerly much eſteemed as ornaments for caps and head-pieces; ſo that *aigrette* and *egret* came to ſignify any ornament to a cap, though originally the word was derived from *aigre, a cauſe de l' aigreur de ſa voix* *.

We never met with this bird or the crane in *England*, but formed our deſcriptions from ſpecimens in the elegant cabinet of Doctor *Mauduit* in *Paris*.

* *Belon av.* 195.

VIII. The

The LITTLE BITTERN. *App.x VI*

VIII. The LITTLE BITTERN.

Ardeola (le Blongios) *Brisson av.* V. 497. *tab.* 40. *fig.* 1.
Ardea vertice dorsoque nigris, collo antice et alarum tectricibus lutescentibus. (Stauden Ragerl, Kleine Mooss-kuh.) *Kram.* 348.
Boonk or long Neck. *Shaw's Travels*, 255.
Ardea Minuta. *Lin. syst.* 240.
Kleiner Rohrdommel. *Frisch*, II. 206. 207.
Edw. av. 275.

THIS species was shot as it perched on one of the trees in the *Quarry* or public walks in *Shrewsbury*, on the banks of the *Severn*; it is frequent in many other parts of *Europe*, but the only one we ever heard of in *England*.

DESCRIP.

The length to the tip of the tail was fifteen inches, to the end of the toe twenty. The bill to the corners of the mouth two inches and a half long, dusky at the point; the sides yellow; the edge jagged: the bulk of the body not larger than that of a *fieldfare*.

The top of the head, the back, and tail were black, glossed with an obscure green: the neck is very long, the forepart of which, the breast and thighs, were of a buff color: the belly and vent-feathers white: the hind-part of the neck bare of feathers, but covered with those growing on the side of it: on the ſetting on of the wing is a large cheſnut ſpot: the leſſer coverts of a yellowiſh buff; the larger coverts whitiſh: the web of that next the back half buff and half black: the quil-feathers black: the legs and toes duſky; and what is ſingular in a bird of this genus, the feathers grow down to the knees: the inſide of the middle claw is ſerrated.

For this deſcription, and the drawing, we are indebted to Mr. *Plymley*.

IX. The SPOON-BILL.

Pelecanus ſeu Platea. *Geſner av.* 666.
Albardeola. *Aldr. av.* III. 160.
Spoon-bill. *Wil. orn.* 288.
Raii ſyn. av. 102.
Platalea Leucorodia. *Lin. ſyſt.* 231. *Faun. Suec.* No. 160.
Briſſon V. 352.
Loffel-gans. *Scopoli*, No. 115.

A FLOCK of theſe birds migrated into the marſhes near *Yarmouth*, in *Norfolk*, in *April*, 1774. Theſe birds inhabit the continent of *Europe*. In Mr. *Ray*'s time, they bred annually in a wood at *Sevenhuys*, not remote from *Leyden*: but the wood is now deſtroyed; and theſe birds, with ſeveral others that formerly frequented the country, are at preſent become very rare.

Mr.

SPOONBILL.

Mr. *Joseph Sparshall* of *Yarmouth* favored me with the following very accurate description:

The length from the end of the beak to the extremity of the middle toe forty inches: breadth of the wings, extended, fifty-two inches: bill, length of the upper mandible seven inches; of the lower six three-fourths ditto: breadth of the spoon, near the point, two inches; ditto of the nether mandible one inch seven-eighths: breadth of both, in the narrowest part, near the middle, three-fourths of an inch: a bright orange colored spot, about the breadth of a sixpence, just above the point of the upper mandible, which is a little hooked, or bent downward at its extremity. At the angles of the bill, on each cheek, a spot of a bright orange color: the skin between the sides of the lower mandible, and extending about three inches downward on the throat or neck, covered with very fine down, almost imperceptible, which, with the skin on that part, are of a very bright orange color: irides of the eyes a bright flame color, very lively and vivid: the whole bill (except the above spot) of a fine shining black: its upper surface elegantly waved with dotted protuberances: a depressed line extending from the nostrils (which are three-eights of an inch long, and situate half an inch below the upper part of the bill) is continued round it about one eighth of an inch from its edge: its substance has something of the appearance of whale bone, thin, light, and elastic. Inside of of the mouth a dark aſh color, almoſt black: the tongue (remarkably ſingular) being very ſhort, heart ſhaped, and when drawn back, ſerving as a valve to cloſe the entrance of the throat, which it ſeems to do effectually; when pulled forward has the appearance of a triangular button: the ears, or auditory apertures large, and placed an inch behind the angles of the mouth. Plumage of the whole body, wings, and tail white: on the back-part of the head a beautiful creſt of white feathers, hanging pendent behind the neck; their length about five inches; which, in the living ſubject, gives it a very beautiful appearance.

Weight of the fowl, three days after killed, was three pounds and a half.

The legs black, their length ſix inches, and thighs the ſame; the latter naked about half their length; toes connected by a ſmall web, extending to the firſt joint on each.

No.

No. I.

ADDITIONS TO THE HISTORY OF THE HORSE.

HORSE.

THE reprefentative of this fpecies is a native of *Yemine*, in *Arabia Fælix*; the property of Lord *Grofvenour*, taken from a picture in poffeffion of his Lordfhip, painted by Mr. *Stubbs*, an artift not lefs happy in reprefenting animals in their ftiller moments, than when agitated by their furious paffions; his matchlefs paintings of horfes will be lafting monuments of the one, and that of the lion and panther of the other.

This horfe, by its long refidence among us, may be faid to be naturalized, therefore we hope to be excufed for introducing it here, notwithftanding its foreign defcent. From its great beauty it may be prefumed that it derives its lineage from *Monaki Shaduki, of the pure race of horfes, purer than milk* *.

* Vide the *Arabian* certificate, in a following note, for the meaning of this phrafe.

Arabia

Arabia produces theſe noble animals in the higheſt perfection; firſt, becauſe they take their origin from the wild unmixed breeds that formerly were found in the deſerts *, which had as little degenerated from their primæval form and powers as the lion, tiger, or any other creature which ſtill remains in a ſtate of nature unchanged by the diſcipline of man, or harveſted proviſion.

The *Arabs* place their chief delight in this animal; it is to them † as dear as their family, and is indeed part of it: men, women, children, mares, and foals all lie in one common tent, and they lodge promiſcuouſly without fear of injury.

This

* *Leo Africanus*, who wrote in the time of *Leo* X. ſays, that in his days great numbers of wild horſes were found in the *Numidian* and *Arabian* Deſerts, which were broke for uſe. He adds, that the trial of their ſwiftneſs was made againſt the *Lant*, or the *Oſtrich*; and if they could overtake either of thoſe animals, were valued at a hundred camels. *Hiſt. Africa*, 339.

† As a proof of this, receive the following lamentation of an *Arab*, obliged, thro' poverty, to part with his mare: *My eyes*, ſays he, to the animal, *my ſoul, muſt I be ſo unfortunate as to have ſold thee to ſo many maſters, and not to keep thee myſelf? I am poor, my* ANTELOPE. *You know well enough, my honey, I have brought thee up as my child; I never beat nor chid thee; I made as much of thee as ever I could for my life. God preſerve thee my deareſt; thou art pretty; thou art lovely; God defend thee from the looks of the envious.* To underſtand the firſt part of this ſpeech, it muſt be obſerved, that it is uſual for many

This conſtant intercourſe produces a familiarity that could not otherwiſe be effected; and creates a tractability in the horſes that could ariſe only from a regular good uſage; little acts of kindneſs, and a ſoothing language, which they are accuſtomed to from their maſters: they are quite unacquainted with the ſpur; the leſt touch with the ſtirrup ſets theſe airy courſers in motion; they ſet off with a fleetneſs that ſurpaſſes that of the Oſtrich *, yet they are ſo well trained as to ſtop in their moſt rapid ſpeed by the ſligheſt check of the rider: there are ſometimes inſtances of their being mounted without either bridle or ſaddle, when they ſhew ſuch compliance to their rider's will, as to be directed in their courſe by the meer motion of a ſwitch †.

Paret in obſequium lentæ moderamine virgæ,
Verbera ſunt præcepta fugæ, ſunt verbera fræna ‡.

Several things concur to maintain this perfection in the horſes of *Arabia*, ſuch as the great care the *Arabs* take in preſerving the breed genuine, by permitting none but ſtallions of the firſt form to have

many *Arabs*, of the poorer rank, to join in the purchaſe of a horſe, the original owner generally retaining one ſhare. This, as well as moſt of the other particulars relating to the *Arabian* horſe, are taken from *M. D'Arvieux*'s curious account of *Arabia*, p. 167, *London*, 1732.

* For an account of its ſpeed, *vide Adanſon's voy.* 85.

† *Tavernier's Travels*, I. 63.

‡ *Nemeſion Cyneg.* 267.

acceſs

accefs to the mares: this is never done but in the prefence of a witnefs, the fecretary of the *Emir*, or fome public officer; he afferts the fact, records the name of the horfe, mare, and whole pedigree of each, and thefe atteftations* are carefully preferved, for on thefe depend the future price of the foal.

The

* The reader is here prefented with an original atteftation, fome of which *M. D'Arvieux* fays have been preferved for above 500 years in the public records.

Taken before ABDORRAMAN, KADI of ACCA.

The Occafion of this prefent Writing or Inftrument is that at Acca in the Houfe of *Bati* legal eftablifh'd Judge, appear'd in Court *Thomas Usgate* the Englifh Conful and with him *Sheikh Morad Ebn al Hajj Abdollah*, *Sheikh* of the County of *Safad*, and the faid Conful defir'd from the aforefaid *Sheikh* proof of the Race of the Grey Horfe which he bought of him, and He affirm'd to be *Monaki Shadubi**, but he was not fatisfied with this but defir'd the Teftimony of the *Arabs*, who bred the Horfe and knew how he came to *Sheikh Morad*; whereupon there appear'd certain *Arabs* of Repute whofe names are undermention'd, who teftified and declar'd that the Grey Horfe which the Conful formerly bought of *Sheikh Morad*, is *Monaki Shaduki* of the pure Race of Horfes, purer than Milk †, and that the Beginning of the Affair

* *Thefe are the Names of the two Breeds of Arab Horfes, which are reckoned pure and true, and thofe which are of both thefe Breeds by Father and Mother, are the moft noble and free from Baftardy.*

† *A Proverbial Expreffion.*

was,

The *Arabs*, whofe riches are their horfes, take all imaginable care of them; they have it not in their power to give them grafs in their hot climate, except in the fpring; their conftant food is barley, and that given only in the night, being never fuffered to eat during the day.

In the day-time they are kept faddled at the door of the tent, ready for any excurfion their mafters may make; the *Arabs* being fond of the chace, and live by the plundering of travellers. The horfes are never hurt by any fervile employ, never injured by heavy burthens, or by long journies, enjoy a pure dry air, due exercife, great temperance, and great care.

was, that *Sheikh Saleh*, *Sheikh* of *Alsabal*, bought him of the *Arabs* of the Tribe of *al Mohammadat*, and *Sheikh Saleh* fold him to *Sheikh Morad Ebn al Hajj Abdollah*, *Sheikh* of *Safad*, and *Sheikh Morad* fold him to the Conful aforefaid, when thefe Matters appear'd to us, and the Contents were known, the faid Gentleman defir'd a Certificate thereof, and Teftimony of the Witneffes, whereupon we wrote him this Certificate, for him to keep as a Proof thereof. Dated Friday 28 of the latter *Rabi* in the Year 1135.

i. e. 29 *January*, 1722.

Witneffes,

Sheikh Jumat al Falibau of the *Arabs* of *al Mohammadat*.
Ali Ebn Taleb al Kaabi.
Ibrahim his Brother.
Mohammed al Adhra Skeikh Alfarifat.
Khamis al Kaabi.

Every

Every horſe in *Arabia* (except thoſe which by way of contempt are called *Guidich*, or pack horſes) has a degree of good qualities ſuperior to thoſe of any other places; but it is not to be ſuppoſed, but that there are certain parts of that country, which have attained a higher perfection in the art of management than the others.

Thus we find by ſome late information *, that *Yemine* in *Arabia Fælix*, is at preſent in great repute for its breed; for the jockies of that part have acquired ſuch a ſuperior name, as to be able to ſell their three year old horſes for two or three hundred guineas a-piece, and when they can be prevailed on to part with a favorite *ſtallion*, they will not take leſs for it than fifteen hundred guineas. It is from this country that the great men in *India* are ſupplied with horſes, for *India* itſelf is poſſeſſed of a very bad kind; theſe noble animals being much neglected there, from the conſtant uſe of the *Buffalo*, not only in tillage, but even in riding.

It may be allowed here to give ſome account of the horſes of other countries, which derive their origin, or at leſt receive their improvement from the *Arabian* kind, for whereſoever the *Saracens* ſpread their victorious arms, they, at the ſame time, introduced their generous race of horſes.

Thoſe of *Perſia* are light, ſwift, and very like thoſe of *Arabia*, but formed very narrow before:

* *Wall* on horſes, 74.

they

they are fed with chopped ſtraw, mixed with barley, and inſtead of ſoiling, are fed with new eared or green barley for about fourteen or twenty days *.

Æthiopia has with ſome writers the credit of having originally furniſhed *Arabia* with its fine race of horſes; but we believe the reverſe, and that they were introduced into that empire by the *Arabian* princes, whoſe lineage to this day fills that throne. The horſes of that country are ſpirited and ſtrong, and generally of a black color: they are never uſed in long journies, but only in battle or in the race, for all ſervile work is done by mules: the *Æthiopians* never ſhoe them, for which reaſon, on paſſing through ſtony places, they diſmount, and ride on mules, and lead their horſes †; ſo from this we may collect, that this nation is not leſs attached to theſe animals than the *Arabs*.

Ægypt has two breeds of horſes, one its own, the other *Arabian*; the laſt are moſt eſteemed, and are bought up at a great price, in order to be ſent to *Conſtantinople*; but ſuch is the diſcouragement, ariſing from the tyranny of the government, that the owners often wilfully lame a promiſing horſe ‡, leſt the *Beys* ſhould like it and force it from them.

Barbary owes its fine horſes to the ſame ſtock, but in general they are far inferior in point of value; and for the ſame reaſon as is given in the laſt arti-

* *Tavernier's Travels*, I. 145.

† *Ludolph. hiſt. Æthiop.* 53.

‡ *Univ. modern hiſt.* quotted from *Maillet* and *Pocock*.

cle, the great insecurity of property under the *Turkish* government. The breed was once very famous: M. *D'Arvieux* * says, that when he was there in 1668, he met with a mare that he thought worthy of the stud of his *grand Monarque*, when in the height of his glory; but Doctor *Shaw* informs us, that at present the case is entirely altered †.

Notwithstanding *Spain* has been celebrated of old for the swiftness of its horses, yet it must have received great improvement from those brought over by their conquerors, the *Saracens*. According to *Oppian* ‡, the *Spanish* breed had no other merit than that of fleetnese, but at present we know that they have several other fine qualities.

To sum up the account of this generous animal, we may observe, that every country that boasts of a fine race of horses, is indebted to *Arabia*, their primæval seat. No wonder then, that the poetic genius of the author of the book of *Job*, who not only lived on the very spot, but even at time when the animal creation still enjoyed much of its original perfection, should be able to compose that sublime description which has always been the admiration of every person of genuine taste §.

* *D'Arvieux*, 173.

† *Shaw's Travels*, 238.

‡ *Cyneg. lib.* I. V. 284.

§ *Job*. ch. XXXIX. v. 19. to 25.

No.

No. II.

Of the TAKING of WOLVES, &c.

Ex Autographo penes Dec. et Capit. Exon.

From Bp. Lyttelton's Collections.

JOHAN. comes *Moreton* omnibus hominibus et amicis suis *Francis* et *Anglicis* presentibus et futuris salutem sciatis nos concess.se reddidisse et hac cartamea confirmasse comit. baron militibus et omnibus libere tenentibus clericis et laicis in *Devenescire* libertates suas foreste quas habuerunt tempore *Henrici* Reg. proavi mei tenendas et habendas illis et heredibus suis de me et heredibus meis et nominatim quod habeant arcus et pharetras, et sagittas in terris suis deferendas extra regardum foreste mee, et quod canes sui vel hominum suorum, non sint espaltati extra regardum foreste, et quod habeant canes suos et alias libertates, sicut melius et liberius illas habuerunt tempore ejusd. *Henrici* Regis et Reisellos suos, et quod capiant Capreolum, Vulpem, Cattum, Lupum, Leporem, Lutram, ubicunque illam inveniunt ex-

U u 2 tra

tra regardum forefte mee. Et ideo vobis firmiter precipio, quod nullus eis, de hiis vel aliis libertatibus fuis moleftiam inferat vel gravamen. Hiis teftibus *Will. Marefcall. Will.* comite *Sarefbur. Will.* com. de *Vernon. Steph. Ridell* cancellario meo, *Will. de Wenn. Hamone de Valoin*, *Rogero de Novoburgo*, *Ingelram de Pincoll. Rob. de Moritomari*, *Waltero Maltravers. Rad. Morin. Walt. de Cantelu. Gilberti Morin* et multis aliis.

Seal appendant, an armed man on horfeback, and on the reverfe, a fmall impreffion from an antique head — the legend broken.

No.

No. III.

OF THE CHOICE OF HIS MAJESTY'S HAWKS.

TO all thoſe to whome this preſent Writinge ſhall come I Sr. *Anthony Pell* Knight Maiſter Faulkner Surveyor and Keeper of his Majeſties Hawkes ſend greetinge, Whereas I am credibly informed that divers perſons who doe uſuallie bringe Haukes to ſell doe commonlye convey them from ſhipbord and cuſtome howſe before ſuch tyme as I or my ſervants or deputies have any ſight or choiſe of them for his Majeſties uſe whereby his Highneſs is not nor hath not lately beene furniſhed with the number of Hawkes as is moſt meete, Wherefore theis are in his Majeſties name to will charge and commaund you and every of you that ſhall at any tyme hereafter bringe any Hawkes to ſell, That neither you nor any of you nor any others for you or by your appointment doe remove or convey awaye any of your Hawkes whatſoever from ſhipbord or the cuſtome houſe untill ſuch tyme as the bearer hereof my welbeloved friend *William Spence* Gent. have his firſt choiſe

Uu3 for

for his Majesties service, And that you and every one of you do quietly permitt and suffer the said *Wm. Spence* the bearer hereof to take his choise and make tryal of such of your Hawkes as he shall thinke meete with a gorge or two of meat before such tyme as his Majesties price be paide beeinge as hereafter followeth, viz for a Faulcon twenty six shillings and eight pence, for a Tassell gentle thirteene shillings and four pence, for a Lanner twenty six shillings and eight pence, for a Lannarett thirteene shillings and foure pence, for a Goshawke twentie shillings, for a Tassell of a Goshawke thirteene shillings and foure pence, for a Gerfaulkon thirtie shillings, for a Jerkin thirteen shillings and fourepence, hereof fayle you not as you will answere the contrary at your perills. Dated the six and twentieth day *Januarie* Anno Domini 1621.

This warrant to endure untill the
first daye of *August* next comeinge,

No.

No. IV.

Of the SMALL BIRDS of FLIGHT,

By the Honble. Daines Barrington.

IN the fuburbs of *London* (and particularly about *Shoreditch*) are feveral weavers and other tradefmen, who, during the months of *October* and *March*, get their livelihood by an ingenious, and we may fay, a fcientific method of bird-catching, which is totally unknown in other parts of *Great Britain*.

The reafon of this trade being confined to fo fmall a compafs, arifes from there being no confiderable fale for finging birds except in the *metropolis*: as the apparatus for this purpofe is alfo heavy, and at the fame time muft be carried on a man's back, it prevents the bird-catchers going to above three or four miles diftance.

This method of bird-catching muft have been long practifed, as it is brought to a moft fyftematical perfection, and is attended with a very confiderable expence.

The nets are a moft ingenious piece of mechanifm, are generally twelve yards and a half long, and two yards and a half wide; and no one on bare inſpection would imagine that a bird (who is ſo ſo very quick in all its motions) could be catched by the nets flapping over each other, till he becomes eye witneſs of the pullers ſeldom failing *.

The wild birds *fly* (as the bird-catchers term it) chiefly during the month of *October*, and part of *September* and *November*; as the flight in *March* is much leſs conſiderable than that of *Michaelmaſs*. It is to be noted alſo, that the ſeveral ſpecies of *birds of flight* do not make their appearance preciſely at the ſame time, during the months of *September*, *October* and *November*. The Pippet †, for example, begins to *fly* about *Michaelmaſs*, and then the Woodlark, Linnet, Goldfinch, Chaffinch, Greenfinch, and other birds of flight ſucceed; all of which are not eaſily to be caught, or in any numbers, at any other time, and more particularly the Pippet and the Woodlark.

Theſe birds, during the *Michaelmaſs* and *March* flights, are chiefly on the wing from day break to noon, though there is afterwards a ſmall *flight* from two till night; but this however is ſo incon-

* Theſe nets are known in moſt parts of *England* by the name of *day-nets* or *clap-nets*; but all we have ſeen are far inferior in their mechaniſm to thoſe uſed near *London*.

† A ſmall ſpecies of Lark, but which is inferior to other birds of that *Genus* in point of ſong.

ſiderable,

ſiderable, that the bird-catchers always take up their nets at noon.

It may well deſerve the attention of the naturaliſt whence theſe periodical flights of certain birds can ariſe. As the ground however is ploughed during the months of *October* and *March* for ſowing the winter and lent corn, it ſhould ſeem that they are thus ſupplied with a great profuſion both of ſeeds and inſects, which they cannot ſo eaſily procure at any other ſeaſon.

It may not be improper to mention another circumſtance, to be obſerved during their flitting, *viz.* that they fly always againſt the wind; hence, there is great contention amongſt the bird-catchers who ſhall gain that point, if (for example) it is weſterly, the bird catcher who lays his nets moſt to the eaſt, is ſure almoſt of catching every thing, provided his call-birds are good: a gentle wind to the ſouth-weſt generally produces the beſt ſport.

The bird-catcher, who is a ſubſtantial man, and hath a proper apparatus for this purpoſe, generally carries with him five or ſix *linnets* (of which more are caught than any ſinging bird) two *goldfinches*, two *greenfinches*, one *woodlark*, one *redpoll*, a *yellowhammer*, *titlark*, and *aberdavine*, and perhaps a *bullfinch*; theſe are placed at ſmall diſtances from the nets in little cages. He hath, beſides, what are called *flur-birds*, which are placed within in the nets, are raiſed upon the *flur* *, and gently let down at the time the wild bird approaches them. Theſe generally conſiſt of the *linnet*, the *goldfinch*, and the *greenfinch*, which are ſecured to the *flur* by what is called a *brace* †; a contrivance that ſecures the birds without doing any injury to their plumage.

It having been found that there is a ſuperiority between bird and bird, from the one being more *in ſong* than the other; the bird-catchers contrive that their *call birds* ſhould moult before the uſual time. They, therefore, in *June* or *July*, put them into a cloſe box, under two or three folds of blankets, and leave their dung in the cage to raiſe a greater heat; in which ſtate they continue, being perhaps examined but once a week to have freſh water. As for food, the air is ſo putrid, that they eat little during the whole ſtate of confinement, which laſts about a month. The birds frequently die under the operation ‡; and hence the value of a *ſtopped bird* riſes greatly.

When

* A moveable perch to which the bird is tied, and which the bird-catcher can raiſe at pleaſure, by means of a long ſtring faſtened to it.

† A ſort of bandage, formed of a ſlender ſilken ſtring that is faſtened round the bird's body, and under the wings, in ſo artful a manner as to hinder the bird from being hurt, let flutter ever ſo much in the raiſing.

‡ We have been lately informed by an experienced bird-catcher,

When the bird hath thus prematurally moulted, he is *in ſong*, whilſt the wild birds are *out of ſong*, and his note is louder and more piercing than that of a wild one; but it is not only in his note he receives an alteration, the plumage is equally improved. The black and yellow in the wings of the *goldfinch*, for example, become deeper and more vivid, together with a moſt beautiful gloſs, which is not to be ſeen in the wild bird. The bill, which in the latter is likewiſe black at the end, in the *ſtopped bird* becomes white and more taper, as do its legs: in ſhort, there is as much difference between a wild and a *ſtopped bird*, as there is between a horſe which is kept in body cloaths, or at graſs.

When the bird-catcher hath laid his nets, he diſpoſes of his *callbirds* at proper intervals. It muſt be owned, that there is a moſt malicious joy in theſe *call-birds* to bring the wild ones into the ſame ſtate of captivity; which may likewiſe be obſerved with regard to the decoy ducks.

Their ſight and hearing infinitely excels that of the bird-catcher. The inſtant that the * wild birds are perceived, notice is given by one to the reſt of

catcher, that he purſues a cooler regimen in *ſtopping* his birds, and that he therefore ſeldom loſes one: but we ſuſpect that there is not the ſame certainty of making them moult.

* It may be alſo obſerved, that the moment they ſee a hawk, they communicate the alarm to each other by a plaintive note; nor will they then *jerk* or call though the wild birds are near,

the

the *call-birds*, (as it is by the firſt hound that hits on the ſcent, to the reſt of the pack) after which, follows the ſame ſort of tumultuous ecſtacy and joy. The *call-birds*, while the bird is at a diſtance, do not ſing as a bird does in a chamber; they invite the wild ones by what the bird-catchers call *ſhort jerks*, which when the birds are good, may be heard at a great diſtance. The aſcendency by this call or invitation is ſo great, that the wild bird is ſtopped in its courſe of flight, and if not already acquainted with the nets*, lights boldly within twenty yards of perhaps three or four bird-catchers, on a ſpot which otherwiſe it would not have taken the leſt notice of. Nay, it frequently happens, that if half a flock only are caught, the remaining half will immediately afterwards light in the nets, and ſhare the ſame fate; and ſhould only one bird eſcape, that bird will ſuffer itſelf to be pulled at till it is caught, ſuch a faſcinating power have the *call-birds*.

While we are on this ſubject of the *jerking* of birds, we cannot omit mentioning, that the bird-catchers frequently lay conſiderable wagers whoſe *call-bird* can *jerk* the longeſt, as that determines the ſuperiority. They place them oppoſite to each other, by an inch of candle, and the bird who

* A bird, acquainted with the nets, is by the bird-catchers termed a *ſharper*, which they endeavour to drive away, as they can have no ſport whilſt it continues near them.

jerks

jerks the ofteneſt, before the candle is burnt out, wins the wager. We have been informed, that there have been inſtances of a bird's giving a hundred and ſeventy *jerks* in a quarter of an hour; and we have known a linnet, in ſuch a trial, perſevere in its emulation till it ſwooned from the perch: thus, as *Pliny* ſays of the nightingale, *victa morte finit ſæpe vitam, ſpiritu prius deficiente quàm cantu* *.

It may be here obſerved, that birds when near each other, and in ſight, ſeldom *jerk* or ſing. They either fight, or uſe ſhort and wheedling calls; the *jerking* of theſe *call-birds*, therefore, face to face, is a moſt extraordinary inſtance of contention for ſuperiority in ſong.

It may be alſo worthy of obſervation, that the female of no ſpecies of birds ever ſings: with birds, it is the reverſe of what occurs in human kind: among the feathered tribe, all the cares of life fall to the lot of the tender ſex: theirs is the fatigue of incubation; and the principal ſhare in nurſing the helpleſs brood: to alleviate theſe fatigues, and to ſupport her under them, nature hath given to the male the ſong, with all the little blandiſhments and ſoothing arts; theſe he fondly exerts (even after courtſhip) on ſome ſpray contiguous to the neſt, during the time his mate is performing her parental duties. But that ſhe ſhould be ſilent, is alſo a-

* *Lib.* x. c. 29.

nother

nother wife provifion of nature, for her fong would difcover her neft; as would a gaudinefs of plumage, which, for the fame reafon, feems to have been denied her.

To thefe we may add a few particulars that fell within our notice during our enquiries among the bird-catchers, fuch as, that they immediately kill the hens of every fpecies of birds they take, being incapable of finging, as alfo being inferior in plumage; the *pippets* likewife are indifcriminately deftroyed, as the cock does not fing well: they fell the dead birds for three-pence or four-pence a dozen.

Thefe fmall birds are fo good, that we are furprized the luxury of the age neglects fo delicate an acquifition to the table. The modern *Italians* are fond of fmall birds, which they eat under the common name of *Beccaficos*: and the dear rate a *Roman Tragedian* paid for one difh of finging birds* is well known.

Another particular we learned, in converfation with a *London* bird-catcher, was the vaft price that is fometimes given for a fingle fong bird, which

* *Maximè tamen infignis eft in hac memoria*, Clodii Æfopi *tragici hiftrionis patina fexcentis H. S. taxata*; *in quo pofuit aves cantu aliquo, aut humano fermone, vocales*. Plin. lib. x. c. 51. The price of this expenfive difh was about 6843 *l.* 10 *s.* according to *Arbuthnot*'s Tables. This feems to have been a wanton caprice, rather than a tribute to epicurifm. T. P.

had

had not learned to whiſtle tunes. The greateſt ſum we heard of, was five guineas for a *chaffinch*, that had a particular and uncommon note, under which it was intended to train others: and we alſo heard of five pounds ten ſhillings being given for a *call-bird linnet*.

A third ſingular circumſtance, which confirms an obſervation of *Linnæus*, is, that the male *chaffinches* fly by themſelves, and in the *flight* precede the females; but this is not peculiar to the *chaffinches*. When the *titlarks* are caught in the beginning of the ſeaſon, it frequently happens, that forty are taken and not one female among them: and probably the ſame would be obſerved with regard to other birds (as has been done with relation to the *wheat-ear)* if they were attended to.

An experienced and intelligent bird-catcher informed us, that ſuch birds as breed twice a year, generally have in their firſt brood a majority of males, and in their ſecond, of females, which may in part account for the above obſervation.

We muſt not omit mention of the *bulfinch*, though it does not properly come under the title of a ſinging bird, or a bird of *flight*, as it does not often move farther than from hedge to hedge; yet, as the bird ſells well on account of its learning to whiſtle tunes, and ſometimes flies over the fields where the nets are laid; the bird-catchers have often a *call-bird* to enſnare it, though moſt of them can imitate the call with their mouths. It is remarkable

markable with regard to this bird, that the female anſwers the purpoſe of a *call-bird* as well as the male, which is not experienced in any other bird taken by the *London* bird-catchers.

It may perhaps ſurprize, that under this article of *ſinging birds*, we have not mentioned the *nightingale*, which is not a bird of *flight*, in the ſenſe the bird-catchers uſe this term. The *nightingale*, like the *robin*, *wren*, and many other ſinging birds, only moves from hedge to hedge, and does not take the periodical *flights* in *October* and *March*. The perſons who catch theſe birds, make uſe of ſmall trap-nets, without call-birds, and are conſidered as inferior in dignity to other bird-catchers, who will not rank with them.

The nightingale being the firſt of ſinging birds, we ſhall here inſert a few particulars relating to it, that were tranſmitted to us ſince the deſcription of that bird was printed.

Its arrival is expected, by the trappers in the neighborhood of *London*, the firſt week in *April*; at the beginning none but cocks are taken, but in a few days the hens make their appearance, generally by themſelves, though ſometimes a few males come along with them.

The latter are diſtinguiſhed from the females not only by their ſuperior ſize, but by a great ſwelling of their vent, which commences on the firſt arrival of the hens.

They

They do not build till the middle of *May*, and generally chuſe a quickſet to make their neſt in.

If the nightingale is kept in a cage, it often begins to ſing about the latter end of *November*, and continues its ſong more or leſs till *June*.

A young *Canary bird*, *linnet*, *ſkylark*, or *robin* (who have never heard any other bird) are ſaid beſt to learn the note of a *nightingale*.

They are caught in a net-trap; the bottom of which is ſurrounded with an iron ring; the net itſelf is rather larger than a cabbage net.

When the trappers hear or ſee them, they ſtrew ſome freſh mould under the place, and bait the trap with a meal-worm from the baker's ſhop.

Ten or a dozen nightingales have been thus caught in a day.

No. V.

EXPERIMENTS AND OBSERVATIONS ON THE SINGING OF BIRDS, BY THE HON. DAINES BARRINGTON. IN A LETTER TO MATHEW MATY, M.D. SEC. R. S. 1773.

From the PHILOSOPHICAL TRANSACTIONS, Vol. LXIII.

DEAR SIR,

AS the experiments and obſervations I mean to lay before the Royal Society relate to the ſinging of birds, which is a ſubject that hath never before been ſcientifically treated of *, it may not be improper to prefix an explanation of ſome uncommon terms, which I ſhall be obliged to uſe, as well as others which I have been under a neceſſity of coining.

* *Kircher*, indeed, in his *Muſurgia*, hath given us ſome few paſſages in the ſong of the nightingale, as well as the call of a quail and cuckow, which he hath engraved in muſical characters. Theſe inſtances, however, only prove that ſome birds have in their ſong, notes which correſpond with the intervals of our common ſcale of the muſical octave.

To

To *chirp*, is the firſt ſound which a young bird utters, as a cry for food, and is different in all neſtlings, if accurately attended to; ſo that the hearer may diſtinguiſh of what ſpecies the birds are, though the neſt may hang out of his ſight and reach.

This cry is, as might be expected, very weak and querulous; it is dropped entirely as the bird grows ſtronger, nor is afterwards intermixed with its ſong, the *chirp* of a nightingale (for example) being hoarſe and diſagreeable.

To this definition of the *chirp*, I muſt add, that it conſiſts of a ſingle ſound, repeated at very ſhort intervals, and that it is common to neſtlings of both ſexes.

The *call* of a bird, is that ſound which it is able to make, when about a month old; it is, in moſt inſtances (which I happen to recollect) a repetition of one and the ſame note, is retained by the bird as long as it lives, and is common, generally, to both the cock and hen *.

The next ſtage in the notes of a bird is termed, by the bird-catchers, *recording*, which word is

* For want of terms to diſtinguiſh the notes of birds, *Bellon* applies the verb *chantent*, or ſing, to the gooſe and crane, as well as the nightingale. " Pluſieurs oiſeaux *chantent* la nuit, comme eſt l'oye, la grue, & le roſſignol. " *Bellon*'s Hiſt. of Birds, p. 50.

Xx2 probably

probably derived from a muſical inſtrument, formerly uſed in *England*, called a recorder *.

This attempt in the neſtling to ſing, may be compared to the imperfect endeavour in a child to babble. I have known inſtances of birds beginning to *record* when they were not a month old.

This firſt eſſay does not ſeem to have the leaſt rudiments of the future ſong; but as the bird grows older and ſtronger, one may begin to perceive what the neſtling is aiming at.

Whilſt the ſcholar is thus endeavouring to form his ſong, when he is once ſure of a paſſage, he commonly raiſes his tone, which he drops again when he is not equal to what he is attempting; juſt as a ſinger raiſes his voice, when he not only recollects certain parts of a tune with preciſion, but knows that he can execute them.

What the neſtling is not thus thoroughly maſter of, he hurries over, lowering his tone, as if he did not wiſh to be heard, and could not yet ſatisfy himſelf.

I have never happened to meet with a paſſage in any writer, which ſeems to relate to this ſtage of

* It ſeems to have been a ſpecies of flute, and was probably uſed to teach young birds to pipe tunes.

Lord *Bacon* deſcribes this inſtrument to have been ſtrait, to have had a leſſer and greater bore, both above and below, to have required very little breath from the blower, and to have had what he calls a *fipple*, or ſtopper. See his ſecond Century of Experiments.

ſinging

ſinging in a bird, except, perhaps, in the following lines of *Statius:*

------------ "Nunc volucrum novi
"Queſtus, inexpertumque carmen,
"Quod tacitâ ſtatuere brumâ."

Stat. Sylv. L. IV. Ecl. 5.

A young bird commonly continues to *record* for ten or eleven months, when he is able to execute every part of his ſong, which afterwards continues fixed, and is ſcarcely ever altered *.

When the bird is thus become perfect in his leſſon, he is ſaid to *ſing his ſong round*, or in all its varieties of paſſages, which he connects together, and executes without a pauſe.

I would therefore define a bird's *ſong* to be a ſucceſſion of three or more different notes, which are continued without interruption during the ſame interval with a muſical bar of four crotchets in an adagio movement, or whilſt a pendulum ſwings four ſeconds.

By the firſt requiſite in this definition, I mean to

* The bird called a *Twite* * by the bird-catchers commonly flies in company with linnets, yet theſe two ſpecies of birds never learn each other's notes, which always continue totally different.

* *Br. Zool.* Vol. II. p. 315. 8*vo.* preſent edition, I. p. 293.

exclude the call of a cuckow, or *clucking* of a hen *, as they consist of only two notes; whilst the short bursts of singing birds, contending with each other (called *jerks* by the bird-catchers) are equally distinguished from what I term *song*, by their not continuing for four seconds.

As the notes of a cuckow and hen, therefore, though they exceed what I have defined the *call* of a bird to be, do not amount to its *song*, I will, for this reason, take the liberty of terming such a succession of two notes as we hear in these birds, the *varied call*.

Having thus settled the meaning of certain words, which I shall be obliged to make use of, I shall now proceed to state some general principles with regard to the singing of birds, which seem to result from the experiments I have been making for several years, and under a great variety of circumstances.

Notes in birds are no more innate, than language is in man, and depend entirely upon the master under which they are bred, as far as their organs will enable them to imitate the sounds which they have frequent opportunities of hearing.

Most of the experiments I have made on this subject have been tried with cock linnets, which were fledged and nearly able to leave their nest, on

* The common hen, when she lays, repeats the same note very often, and concludes with the sixth above, which she holds for a longer time.

account

account not only of this bird's docility, and great powers of imitation, but becauſe the cock is eaſily diſtinguiſhed from the hen at that early period, by the ſuperior whiteneſs in the wing *.

In many other ſorts of ſinging birds the male is not at the age of three weeks ſo certainly known from the female; and if the pupil turns out to be a hen,

---------- "ibi omnis
"Effuſus labor."

The *Greek* poets made a ſongſter of the τεττιξ, whatever animal that may be, and it is remarkable that they obſerved the female was incapable of ſinging as well as hen birds:

Ειτ' εισιν οι τεττιγες ϗκ ευδαιμονες,
Ων ταις γυναιξιν ϗ δ'οτιϗν φωνης ενι;
Comicorum *Græcorum* Sententiæ, p. 452. Ed. Steph.

I have indeed known an inſtance or two of a hen's making out ſomething like the ſong of her ſpecies; but theſe are as rare as the common hen's being heard to crow.

I rather ſuſpect alſo, that thoſe parrots, magpies, &c. which either do not ſpeak at all, or very little, are hens of thoſe kinds.

* The white reaches almoſt to the ſhaft of the quill feathers, and in the hen does not exceed more than half of that ſpace: it is alſo of a brighter hue.

I have educated neſtling linnets under the three beſt ſinging larks, the *ſkylark*, *woodlark*, and *titlark*, every one of which, in ſtead of the linnet's ſong, adhered entirely to that of their reſpective inſtructors.

When the note of the *titlark-linnet* * was thoroughly *fixed*, I hung the bird in a room with two common linnets, for a quarter of a year, which were full in ſong; the *titlark-linnet*, however, did not borrow any paſſages from the linnet's ſong, but adhered ſtedfaſtly to that of the titlark.

I had ſome curioſity to find out whether an *European* neſtling would equally learn the note of an *African* bird: I therefore educated a young linnet under a *vengolina* †, which imitated its *African* maſter ſo exactly, without any mixture of the linnet ſong, that it was impoſſible to diſtinguiſh the one from the other.

* I thus call a bird which ſings notes he would not have learned in a wild ſtate; thus by a *ſkylark-linnet*, I mean a linnet with the ſkylark ſong; a *nightingale-robin*, a *robin* with the nightingale ſong, &c.

† This bird ſeems not to have been deſcribed by any of the ornithologiſts; it is of the *finch* tribe, and about the ſame ſize with our aberdavine (or ſiſkin). The colors are grey and white, and the cock hath a bright yellow ſpot upon the rump. It is a very familiar bird, and ſings better than any of thoſe which are not *European*, except the *American mocking bird*. An inſtance hath lately happened, in an aviary at *Hamſted*, of a *vengolina*'s breeding with a *Canary* bird.

This

This *vengolina-linnet* was abſolutely perfect, without ever uttering a ſingle note by which it could have been known to be a linnet. In ſome of my other experiments, however, the neſtling linnet retained the *call* of its own ſpecies, or what the bird-catchers term the linnet's *chuckle*, from ſome reſemblance to that word when pronounced.

I have before ſtated, that all my neſtling linnets were three weeks old, when taken from the neſt; and by that time they frequently learn their *own call* from the parent birds, which I have mentioned to conſiſt of only a ſingle note.

To be certain, therefore, that a neſtling will not have even the *call* of its ſpecies, it ſhould be taken from the neſt when only a day or two old; becauſe, though neſtlings cannot ſee till the ſeventh day, yet they can hear from the inſtant they are hatched, and probably, from that circumſtance, attend to ſounds, more than they do afterwards, eſpecially as the call of the parents announces the arrival of their food.

I muſt own, that I am not equal myſelf, nor can I procure any perſon to take the trouble of breeding up a bird of this age, as the odds againſt its being reared are almoſt infinite. The warmth indeed of incubation may be, in ſome meaſure, ſupplied by cotton and fires; but theſe delicate animals require, in this ſtate, being fed almoſt perpetually, whilſt the nouriſhment they receive ſhould not

not only be prepared with great attention, but given in very ſmall portions at a time.

Though I muſt admit, therefore, that I have never reared myſelf a bird of ſo tender an age, yet I have happened to ſee both a linnet and a goldfinch which were taken from their neſts when only two or three days old.

The firſt of theſe belonged to Mr. *Matthews*, an apothecary at *Kenſington*, which, from a want of other ſounds to imitate, almoſt articulated the words *pretty boy*, as well as ſome other ſhort ſentences: I heard the bird myſelf repeat the words *pretty boy*; and Mr. *Matthews* aſſured me, that he had neither the note or call of any bird whatſoever.

This talking linnet died laſt year, before which, many people went from *London* to hear him ſpeak.

The goldfinch I have before mentioned, was reared in the town of *Knighton* in *Radnorſhire*, which I happened to hear, as I was walking by the houſe where it was kept.

I thought indeed that a *wren* was ſinging; and I went into the houſe to inquire after it, as that little bird ſeldom lives long in a cage.

The people of the houſe, however, told me, that they had no bird but a goldfinch, which they conceived to ſing its own natural note, as they called it; upon which I ſtaid a conſiderable time in the room, whilſt its notes were merely thoſe of a *wren*, without the leaſt mixture of goldfinch.

On

On further inquiries, I found that the bird had been taken from the neſt when only a day or two old, that it was hung in a window which was oppoſite to a ſmall garden, whence the neſtling had undoubtedly acquired the notes of the wren, without having had any opportunity of learning even the *call* of the goldfinch.

Theſe facts, which I have ſtated, ſeem to prove very deciſively, that birds have not any innate ideas of the notes which are ſuppoſed to be peculiar to each ſpecies. But it will poſſibly be aſked, why, in a wild ſtate, they adhere ſo ſteadily to the ſame ſong, in ſo much, that it is well known, before the bird is heard, what notes you are to expect from him.

This, however, ariſes entirely from the neſtling's attending only to the inſtruction of the parent bird, whilſt it diſregards the notes of all others, which may perhaps be ſinging round him.

Young *Canary* birds are frequently reared in a room where there are many other ſorts; and yet I have been informed, that they only learn the ſong of the parent cock.

Every one knows, that the common houſe-ſparrow, when in a wild ſtate, never does any thing but chirp: this, however, does not ariſe from want of powers in this bird to imitate others; but becauſe he only attends to the parental note.

But, to prove this deciſively, I took a common ſparrow from the neſt when it was fledged,

and

and educated him under a linnet: the bird, however, by accident, heard a goldfinch alſo, and his ſong was, therefore, a mixture of the linnet and goldfinch.

I have tried ſeveral experiments, in order to obſerve, from what circumſtances birds fix upon any particular note when taken from the parents; but cannot ſettle this with any ſort of preciſion, any more than at what period of their *recording* they determine upon the ſong to which they will adhere.

I educated a young *robin* under a very fine nightingale; which, however, began already to be out of ſong, and was perfectly mute in leſs than a fortnight.

This *robin* afterwards ſung three parts in four *nightingale*; and the reſt of his ſong was what the bird-catchers call *rubbiſh*, or no particular note whatſoever.

I hung this robin nearer to the nightingale than to any other bird; from which firſt experiment I conceived, that the ſcholar would imitate the maſter which was at the leaſt diſtance from him.

From ſeveral other experiments, however, which I have ſince tried, I find it to be very uncertain what notes the neſtlings will moſt attend to, and often their ſong is a mixture; as in the inſtance which I before ſtated of the ſparrow.

I muſt own alſo, that I conceived, from the experiment of educating the *robin* under a nightingale, that the ſcholar would fix upon the note which

which it firſt heard when taken from the neſt; I imagined likewiſe, that, if the nightingale had been fully in ſong, the inſtruction for a fortnight would have been ſufficient.

I have, however, ſince tried the following experiment, which convinces me, ſo much depends upon circumſtances, and perhaps caprice in the ſcholar, that no general inference, or rule, can be laid down with regard to either of theſe ſuppoſitions.

I educated a neſtling robin under a woodlark-linnet, which was full in ſong, and hung very near to him for a month together: after which, the robin was removed to another houſe, where he could only hear a ſkylark-linnet. The conſequence was, that the neſtling did not ſing a note of woodlark (though I afterwards hung him again juſt above the woodlark-linnet) but adhered entirely to the ſong of the ſkylark-linnet.

Having thus ſtated the reſult of ſeveral experiments, which were chiefly intended to determine, whether birds had any innate ideas of the notes, or ſong, which is ſuppoſed to be peculiar to each ſpecies, I ſhall now make ſome general obſervations on their ſinging; though perhaps the ſubject may appear to many a very minute one.

Every poet, indeed, ſpeaks with raptures of the harmony of the groves; yet thoſe even, who have good muſical ears, ſeem to pay little attention to it, but as a pleaſing noiſe.

I am alſo convinced (though it may ſeem rather paradoxical

paradoxical), that the inhabitants of *London* distinguish more accurately, and know more on this head, than of all the other parts of the island taken together.

This seems to arise from two causes.

The first is, that we have not more musical ideas which are innate, than we have of language; and therefore those even, who have the happiness to have organs which are capable of receiving a gratification from this sixth sense (as it hath been called by some) require, however, the best instruction.

The *orchestra* of the opera, which is confined to the metropolis, hath diffused a good stile of playing over the other bands of the capital, which is, by degrees, communicated to the fidler and ballad-singer in the streets; the organs in every church, as well as those of the *Savoyards*, contribute likewise to this improvement of musical faculties in the *Londoners*.

If the singing of the ploughman in the country is therefore compared with that of the *London* blackguard, the superiority is infinitely on the side of the latter; and the same may be observed in comparing the voice of a country girl and *London* house-maid, as it is very uncommon to hear the former sing tolerably in tune.

I do not mean by this, to assert that the inhabitants of the country are not born with as good musical organs; but only, that they have not the same opportunities of learning from others, who play in tune themselves.

The

The other reaſon for the inhabitants of *London* judging better in relation to the ſong of birds, ariſes from their hearing each bird ſing diſtinctly, either in their own or their neighbours ſhops; as alſo from a bird continuing much longer in ſong whilſt in a cage, than when at liberty; the cauſe of which I ſhall endeavour hereafter to explain.

They who live in the country, on the other hand, do not hear birds ſing in their woods for above two months in the year, when the confuſion of notes prevents their attending to the ſong of any particular bird; nor does he continue long enough in a place, for the hearer to recollect his notes with accuracy.

Beſides this, birds in the ſpring ſing very loud indeed; but they only give ſhort jerks, and ſcarcely ever the whole compaſs of their ſong.

For theſe reaſons, I have never happened to meet with any perſon, who had not reſided in *London*, whoſe judgment or opinion on this ſubject I could the leaſt rely upon; and a ſtronger proof of this cannot be given, than that moſt people, who keep *Canary* birds do not know that they ſing chiefly either the titlark, or nightingale notes *.

Nothing,

* I once ſaw two of theſe birds which came from the *Canary Iſlands*; neither of which had any ſong at all; and I have been informed, that a ſhip brought a great many of them not long ſince, which ſung as little.

Moſt of thoſe *Canary* birds, which are imported from the *Tyrol*,

Nothing, however, can be more marked than the note of a nightingale called its *jug*, which moſt of the *Canary* birds brought from the *Tyrol* commonly have, as well as ſeveral nightingale *ſtrokes*, or particular paſſages in the ſong of that bird.

I mention this ſuperior knowledge in the inhabitants of the capital, becauſe I am convinced, that, if others are conſulted in relation to the ſinging of birds, they will only miſlead, inſtead of giving any material or uſeful information *.

Birds in a wild ſtate do not commonly ſing above ten weeks in the year; which is then alſo confined to the cocks of a few ſpecies; I conceive, that this laſt circumſtance ariſes from the ſuperior ſtrength of the muſcles of the larynx.

Tyrol, have been educated by parents, the progenitor of which was inſtructed by a nightingale; our *Engliſh Canary* birds have commonly more of the titlark note.

The traffick in theſe birds makes a ſmall article of commerce, as four *Tyroleze* generally bring over to *England* ſixteen hundred every year; and though they carry them on their backs one thouſand miles, as well as pay 20 l. duty for ſuch a number, yet, upon the whole, it anſwers to ſell theſe birds at 5 s. a piece.

The chief place for breeding *Canary* birds is *Inſpruck* and its environs, from whence they are ſent to *Conſtantinople*, as well as every part of *Europe*.

* As it will not anſwer to catch birds with clap-nets any where but in the neighbourhood of *London*, moſt of the birds which may be heard in a country town are neſtlings, and conſequently cannot ſing the ſuppoſed natural ſong in any perfection.

I pro-

I procured a cock nightingale, a cock and hen blackbird, a cock and hen rook, a cock linnet, as alſo a cock and hen chaffinch, which that very eminent anatomiſt, Mr. *Hunter*, F. R. S. was ſo obliging as to diſſect for me, and begged, that he would particularly attend to the ſtate of the organs in the different birds, which might be ſuppoſed to contribute to ſinging.

Mr. *Hunter* found the muſcles of the larynx to be ſtronger in the nightingale than in any other bird of the ſame ſize; and in all thoſe inſtances (where he diſſected both cock and hen) that the ſame muſcles were ſtronger in the cock.

I ſent the cock and hen rook, in order to ſee whether there would be the ſame difference in the cock and hen of a ſpecies which did not ſing at all. Mr. *Hunter*, however, told me, that he had not attended ſo much to their comparative organs of voice, as in the other kinds; but that, to the beſt of his recollection, there was no difference at all.

Strength, however, in theſe muſcles, ſeems not to be the only requiſite; the birds muſt have alſo great plenty of food, which ſeems to be proved ſufficiently by birds in a cage ſinging the greateſt part of the year *, when the wild ones do not (as

* Fiſh alſo which are ſupplied with a conſtant ſucceſſion of palatable food, continue in ſeaſon throughout the greateſt part of the year; trouts, therefore, when confined in a ſtew and

(as I obſerved before) continue in ſong above ten weeks.

The food of ſinging birds conſiſts of plants, inſects, or ſeeds, and of the two firſt of theſe there is infinitely the greateſt profuſion in the ſpring.

As for ſeeds, which are to be met with only in the autumn, I think they cannot well find any great quantities of them in a country ſo cultivated as *England* is; for the ſeeds in meadows are deſtroyed by mowing; in paſtures, by the bite of the cattle; and in arable, by the plough, when moſt of them are buried too deep for the bird to reach them *.

I know well that the ſinging of the cock-bird in the ſpring is attributed by many † to the motive only of pleaſing its mate during incubation.

They, however, who ſuppoſe this, ſhould recollect, that much the greater part of birds do not ſing at all: why ſhould their mate therefore be deprived of this ſolace and amuſement?

The bird in a cage, which, perhaps, ſings nine or ten months in a year, cannot do ſo from this inducement; and, on the contrary, it ariſes chiefly from contending with another bird, or indeed againſt almoſt any ſort of continued noiſe.

and fed with minnows, are almoſt at all ſeaſons of a good flavour, and are red when dreſſed.

* The plough indeed may turn up ſome few ſeeds, which may ſtill be in an eatable ſtate.

† See, amongſt others, *M. de Buffon*, in his lately-publiſhed Ornithology.

Superiority

Superiority in ſong gives to birds a moſt amazing aſcendency over each other; as is well known to the bird-catchers by the faſcinating power of their call-birds, which they contrive ſhould moult prematurely for this purpoſe.

But, to ſhew deciſively that the ſinging of a bird in the ſpring does not ariſe from any attention to its mate, a very experienced catcher of nightingales hath informed me, that ſome of theſe birds have *jerked* the inſtant they were caught. He hath alſo brought to me a nightingale, which had been but a few hours in a cage, and which burſt forth in a roar of ſong.

At the ſame time this bird is ſo ſulky on its firſt confinement, that he muſt be crammed for ſeven or eight days, as he will otherwiſe not feed himſelf; it is alſo neceſſary to tye his wings, to prevent his killing himſelf againſt the top or ſides of the cage.

I believe there is no inſtance of any bird's ſinging which exceeds our black bird in ſize; and poſſibly this may ariſe from the difficulty of its concealing itſelf, if it called the attention of its enemies, not only by bulk, but by the proportionable loudneſs of its notes *.

I ſhould rather conceive, it is for the ſame reaſon that no henbird ſings, becauſe this talent would be ſtill more dangerous during incubation; which

* For the ſame reaſon, moſt large birds are wilder than the ſmaller ones.

Y y 2 may

may poſſibly alſo account for the inferiority in point of plumage.

I ſhall now conſider how far the ſinging of birds reſembles our known muſical intervals, which are never marked more minutely than to half notes; becauſe, though we can form every gradation from half-note to half-note, by drawing the finger gently over the ſtring of a violin, or covering by degrees the hole of a flute; yet we cannot produce ſuch a minute interval at command, when a quarter-note for example might be required.

Ligon, indeed, in his hiſtory of *Barbadoes*, hath the following paſſage: "The next bird is of the "colour of the fieldfare; but the head is too large "for the body; and for that reaſon ſhe is called "a counſellor. She performs that with her voice, "which no inſtrument can play, or voice can ſing; "and that is quarter-notes, her ſong being com-"poſed of them, and every one a note higher than "another."

Ligon appears, from other parts of his work, to have been muſical; but I ſhould doubt much whether he was quite ſure of theſe quarter intervals, ſo as to ſpeak of them with preciſion.

Some paſſages of the ſong in a few kinds of birds correſpond with the intervals of our muſical ſcale (of which the cuckow is a ſtriking and known inſtance): much the greater part, however, of ſuch ſong is not capable of muſical notations.

This ariſes from three cauſes: the firſt is, that the

the rapidity is often ſo great, and it is alſo ſo uncertain when they may ſtop, that we cannot reduce the paſſages to form a muſical bar, in any time whatſoever.

The ſecond is, that the pitch of moſt birds is conſiderably higher * than the moſt ſhrill notes of thoſe inſtruments, which contain even the greateſt compaſs.

I have before ſaid, that our ideas of a voice, or inſtrument, being perfectly in tune or not, ariſe from comparing it with the muſical intervals to which we are moſt accuſtomed.

As the upper and lower parts of every inſtrument, however, are but ſeldom uſed, we are not ſo well acquainted with the intervals in the higheſt and loweſt octaves, as we are with thoſe which are more central; and for this reaſon the harpſichord-tuners find it more difficult to tune theſe extreme parts.

As a bird's pitch, therefore, is higher than that of an inſtrument, we are conſequently at a ſtill

* Dr. *Wallis* is miſtaken in part of what he ſuppoſes to be the cauſe of ſhrillneſs in the voice, "Nam ut tubus, ſic trachea longior, & ſtrictior, ſonum efficit magis acutum." Grammar, p. 3.

The narrower the pipe is, the more ſharp the pitch as he rightly obſerves; but the length of the tube hath juſt the contrary effect, becauſe players on the flute always inſert a longer middle-piece, when they want to make the inſtrument more flat.

greater loſs when we attempt to mark their notes in muſiçal characters, which we can ſo readily apply to ſuch as we can diſtinguiſh with preciſion.

The third, however, and unſurmountable difficulty is, that the intervals uſed by birds are commonly ſo minute, that we cannot judge at all of them from the more groſs intervals into which we divide our muſical octave.

It ſhould therefore be recollected, by thoſe who have contended that the *Greeks* and *Romans* were acquainted with ſuch more minute intervals of the octave, that they muſt inſiſt the ancients had organs of ſenſation, with which their degenerate poſterity are totally unprovided.

Though we cannot attain the more delicate and imperceptible intervals in the ſong of birds *, yet many of them are capable of whiſtling tunes with our more groſs intervals, as is well known by the common inſtances of piping bullfinches †, and *Canary* birds.

This, however, ariſes from mere imitation of what they hear when taken early from the neſt; for if the inſtrument from which they learn it is

* There have been inſtances indeed of perſons who could whiſtle the notes of birds, but theſe are two rare to be argued from.

† Theſe bullfinches alſo form a ſmall article of commerce, and are chiefly brought from the neighbourhood of *Cologne*.

out

out of tune, they as readily pipe the falſe, as the true notes of the compoſition.

The next point of compariſon to be made between our muſic and that of birds is, whether they always ſing in the ſame pitch.

This, however, I will not preſume to anſwer with any preciſion, for the reaſon I have before ſuggeſted; I ſhall, however, without reſerve, give the beſt conjectures I can form on this head.

If a dozen ſinging birds of different kinds are heard in the ſame room, there is not any diſagreeable diſſonance (which is not properly reſolved), either to my own ear, or to that of others, whoſe judgment on ſuch a point I can more rely.

At the ſame time, as each bird is ſinging a different ſong, it is extraordinary that what we call harmony ſhould not be perpetually violated, as we experience, in what is commonly called a *Dutch* concert, when ſeveral tunes are played together.

The firſt requiſite to make ſuch ſounds agreeable to the ear is, that all the birds ſhould ſing in the ſame key, which I am induced to believe that they do, from the following reaſons.

I have long attended to the ſinging of birds, but if I cannot have recourſe to an inſtrument very ſoon, I cannot carry the pitch of their notes in my memory, even for a very ſhort time.

I therefore deſired a very experienced harpſichord-tuner (who told me he could recollect any particular note which he happened to hear for ſeveral hours), to mark down when he returned home what he had obſerved on this head.

I had lately received an account from him of the following notes in different birds.

F. natural in woodlarks.

A. natural in common cocks.

C. natural in *Bantam* cocks.

B. flat in a very large cock.

C. falling to A. commonly in the cuckow.

A. in thruſhes.

D. in ſome owls.

B. flat in ſome others.

Theſe obſervations furniſh five notes, viz. A. B. flat, C. D. and F. to which I can add a ſixth, (viz. G.) from my own obſervations on a nightingale which lived three years in a cage. I can alſo confirm theſe remarks of the harpſichord-tuner by having frequently heard from the ſame bird C. and F.

As one ſhould ſpeak of the pitch of theſe notes with ſome preciſion, the B. flat of the ſpinnet I tried them by, was perfectly in tune with the great bell of *St. Paul's*.

The following notes, therefore, having been obſerved in different birds, viz. A. B. flat, C. D. F. and G. the E. is only wanting to complete the ſcale; the ſix other notes, however, afford ſufficient data for making ſome conjectures, at leaſt, with regard to the key in which birds may be ſuppoſed to ſing, as theſe intervals can only be found in in the key of F. with a ſharp third, or that of G. with a flat third.

I muſt own, I ſhould rather ſuppoſe it to be the latter, and for the following reaſons.

Lucretius ſays (and perhaps the conjecture is not only ingenious but well founded) that the firſt muſical notes were learned from birds:

"At liquidas avium voces imitarier ore
"Ante fuit multo, quam lævia carmina cantu
"Concelebrare homines poſſent, cantuque juvare."

Now, of all the muſical tones which can be diſtinguiſhed in birds, thoſe of the cuckow have been moſt attended to, which form a flat third, not only by the obſervations of the harpſichord tuner I have before mentioned, but likewiſe by thoſe of *Kircher*, in his *Muſurgia*.

I know well that there have been ſome late compoſitions, which introduce the cuckow notes in a ſharp third; theſe compoſers, however, did not trouble themſelves with accuracy in imitating theſe notes, and it anſwered their purpoſe ſufficiently, if there was a general reſemblance.

Another proof of our muſical intervals being originally borrowed from the ſong of birds, ariſes from moſt compoſitions being in a flat third, where muſic is ſimple, and conſiſts merely of melody.

The oldeſt tune I happen to have heard is a *Welſh*

Welſh one, called *Morva Rhydland* *, which is compoſed in a flat third; and if the muſic of the *Turks* and *Chineſe* is examined in *Du Halde* and Dr. *Shaw*, half of the airs are alſo in the minor third.

The muſic of two centuries ago is likewiſe often in a flat third, though ninety-nine compoſitions out of a hundred are now in the ſharp third.

The reaſon, however, of this alteration ſeems to be very clear: the flat third is plaintive, and conſequently adapted to ſimple movements, ſuch as may be expected in countries where muſic hath not been long cultivated.

There is on the other hand a moſt ſtriking brilliancy in the ſharp third, which is therefore proper for the amazing improvements in execution, which both ſingers and players have arrived at within the laſt fifty years.

When *Corelli*'s muſic was firſt publiſhed, our ableſt violiniſts conceived that it was too difficult to be performed; it is now, however, the firſt compoſition which is attempted by a ſcholar. Every year alſo now produces greater and greater prodigies upon other inſtruments, in point of execution.

I have before obſerved, that by attending to a nightingale, as well as a *robin* which was educated

* Or *Rhydland Marſh*, where the *Welſh* received a great defeat; *Rhydland* is in *Flintſhire*. We find alſo, by the *Orpheus Britannicus*, that even ſo late as the time of *Purcel*, two parts in three of his compoſitions are in the flat third.

under

under him, I always found that the notes reducible to our intervals of the octave were precisely the same; which is another proof that birds sing always in the same key.

In this circumstance, they differ much from the human singer; because they who are not able to sing from the notes, often begin a song either above or below the compass of their voice, which they are not therefore able to go through with. As birds, however, form the same passages with the same notes, at all times, this mistake of the pitch can never happen in them.

Few singers again can continue their own part, whilst the same passages are sung by another in a different key; or if other passages are played, though they may agree both in harmony and time.

As birds however adhere so stedfastly to the same precise notes in the same passages, though they never trouble themselves about what is called *time* or harmony in music; it follows that a composition may be formed for two piping bulfinches, in two parts, so as to constitute true harmony, though either of the birds may happen to begin, or stop, when they please.

I have therefore procured such an ingenious composition, by a very able musician*, which I send herewith; and it need scarcely be observed,

* Mr. *Zeidler*, who plays the violincello at *Covent Garden* theatre.

that

that there cannot poſſibly be much variety in the part of the ſecond bulfinch. See *Tab.* XI. in the *Philoſophical Tranſactions*, *Vol.* LXIII.

Though ſeveral birds have great muſical powers, yet they ſeem to have no delicacy of ſenſations, as the human ſinger hath; and therefore the very beſt of them cannot be taught to exceed the inſipidity of the upper part of the flute ſtop of an organ *, which hath not the modern improvement of a *ſwell.*

They are eaſily impoſed upon by that moſt imperfect of all inſtruments, a *bird-call*, which they often miſtake for the notes of their own ſpecies.

I have before obſerved, that perhaps no bird may be ſaid to ſing which is larger than a black bird, though many of them are taught to ſpeak: the ſmaller birds, however, have this power of imitation; though perhaps the larger ones have not organs which may enable them, on the other hand, to ſing.

We have the following inſtances of birds being taught to ſpeak, in the time of the *Greeks* and

* Lord *Bacon* mentions, that in the inſtrument called a *regall* (which was a ſpecies of portable organ) there was a *nightingale* ſtop, in which water was made uſe of to produce the ſtronger imitation of this bird's tone. See Cent. II. exper. 172. Though this inſtrument, as well as its *nightingale* ſtop, is now diſuſed, I have procured an organ pipe to be immerſed partly in water, which, when blown into, hath produced a tone very ſimilar to that of birds.

Romans

Romans, upon which we never try the ſame experiment. *Moſchus* addreſſes nightingales and ſwallows which were thus inſtructed:

> Αδονιδες, πασαι τε χελιδονες, ας ποκ' ετερπεν,
> Ας λαλειν εδιδασκε.
>
> *Moſchi Idyl.* III.

Pliny mentions both a cock, thruſh, and nightingales, which articulated *:

" Habebant & *Cæſares* juvenes *turdum* †, item
" *luſcinias* Græco atque Latino ſermone dociles,
" præterea meditantes in diem, & aſſidue nova lo-
" quentes longiore etiam contextu."

Statius alſo takes notice of ſome birds ſpeaking, which we never attempt to teach in this manner:

> " Huc doctæ ſtipentur aves, queis nobile fandi
> " Jus natura dedit, plangat Phœbeius ales,
> " Auditaſque memor penitus demittere voces
> " Sturnus, & *Aonio* verſæ certamine picæ;
> " Quique refert jungens iterata vocabula perdix,
> " Et quæ *Biſtonio* queritur ſoror orba cubili ‡."
>
> *Stat. Sylv.* lib. ii. ecl. 4.

As

* *Lib.* X. *c.* 21 & 42.

† Ibid. The other *turdus* belonged to the Empreſs *Agrippina*.

‡ Amongſt the five birds mentioned in theſe lines of *Statius*, there are four which are never taught to ſpeak at preſent, viz. the cock, the nightingale, the common, and the *red legged partridge*.

As

As we find, from theſe citations, that ſo many different ſorts of birds have learned to ſpeak, and as

As I ſuppoſe, however, that *perdix* ſignifies this laſt bird, and not the common partridge (as it is always tranſlated), it is proper I ſhould here give my reaſons why I diſſent from others, as alſo why I conceive that *ſturnus*, in this paſſage, is not a *ſtarling*, but the common partridge.

None of the ancients have deſcribed the plumage of the *perdix*; but *Ariſtotle*, *Ovid*, and *Pliny*, inform us of what materials the neſt of this bird is compoſed, as well as where it is placed.

Ariſtotle ſays, that the neſt is *fortified with wood* *; and in another chapter †, with *thorns and wood*; neither of which are uſed by the common partridge, which often builds in a country where they cannot be procured.

On the contrary, *M. de Buffon* informs us, that the red legged partridge, " ſe tiennent ſur les montagnes qui produiſent beaucoup de bruyeres, & de broſſailles ‡.

Ovid, therefore, ſpeaking of the *perdix*, ſays,

" ——— ponitque in ſepibus ova §,"

where the common partridge is ſeldom known to build.

Pliny again informs us, " perdices ſpinâ & frutice ſic mu-" niunt

* Επηλυγαζομεναι υλην. Lib. V. c. I. Which *Stephens* renders *making a covering of wood*.

† Lib. IX. c. 8. The common partridge, however, makes its neſt with hay and ſtraw.

‡ Orn. T. II. p. 433.

§ *Ovid. Met.* Lib. VIII. l. 258. I ſhall alſo refer to l. 237, of the ſame book:

" Garrula ramoſâ proſpexit ab ilice perdix :"

as it is well known that the common partridge never perches upon a tree, as I have shewn that a sparrow may be taught to sing the linnet's note, I scarcely know what species

"niunt receptaculum, ut contra feras abunde valentur *",
as also in the 52d chapter of his tenth book, that the *perdix* lays white eggs, which is not true of the common partridge.

But there are not wanting other proofs of the conjecture I have here made.

Aristotle speaking of this same bird, says, Των μεν περδικων, οι κακκαβιζουσιν, οι δε τριζουσι †.

Now, the word, κακκαβιζουσι is clearly formed from the *call* of the bird alluded to, which does not at all resemble that of the common partridge.

Thus also the author of the Elegy on the Nightingale, who is supposed by some to be *Ovid*, hath the following line:

"*Caccabat* hinc perdix, hinc gratitat improbus anser."

so that the call of the bird must have had something very particular, and have answered nearly, to the words κακκαβιζει and *caccabat*.

I find, indeed, that *M. de Buffon* contends ‡ that the περδιξ of *Aristotle* does not mean the common partridge, but the bartavel, with regard to which, I shall not enter into any discussion, but only observe, that most of his references are inaccurate, and that he entirely mistakes the materials of which the nest is composed, according to *Aristotle's* sixth book, and first chapter.

But the strongest proof that *perdix* signifies the red legged partridge is, that the *Italians* to this day call this bird *pernice*, and the common sort *starna* §.

This also now brings me to the proofs, of *sturnus* in this passage of *Statius* signifying the *common partridge, and not the*

* Lib. x. c. 23. † Lib. IV. c. 9. ‡ Orn. T. II. p. 422 § See *Olina*.

starling,

cies to fix upon, that may be confidered as incapable of fuch imitations; for it is very clear, from feveral experiments before ftated, that the utmoft endeavours will not be wanting in the bird, if he is endowed with the proper organs.

It can therefore only be fettled by educating a bird, under proper circumftances, whether he is thus qualified or not; for if one was only to determine this point by conjecture, one fhould fuppofe that a fparrow would not imitate the fong of the linnet, nor that a nightingale or partridge could be taught to fpeak.

And here it may not be improper to explain what I mean by birds learning to imitate the notes of others, or the human fpeech.

ftarling, which I muft admit are not fo ftrong as with regard to the import of the word *perdix*. If my arguments are not therefore fo convincing on this head, the number of birds taught to fpeak by the *Romans*, and not by us, muft be reduced to three, as the ftarling is frequently learned to talk in the prefent times.

As I cannot argue from the defcription of the habits of the *fturnus*, or the materials of its neft, as in the former inftance, I muft reft my conjecture (fuch as it is) on the two birds, almoft following each other in thefe lines of *Statius*; on the common partridge being called *ftarna* to this day by the *Italians*, and upon the *Romans* having had otherwife no name for our partridge (which is a very common bird in *Italy*), if *fturnus* is fuppofed to fignify only a *ftarling*.

If

If the birds differ little in ſhape or ſize (particularly of the beak *) the imitation is commonly ſo ſtrong, that

"Mirê

* It ſeems very obvious why the form and ſize of the beak may be material; but I have alſo obſerved, that the colour of a bird's bill changes, when in or out of ſong; and I am informed, that a cock ſeldom crows much, but when his comb is red.

When moſt of the finch tribe are coming into ſong, there is ſuch a gradual change in the colour of their bill; thus, thoſe of the chaffinch and linnet are then of a very deep blue, which fades away again, when the bird ceaſes to be in ſong.

This particular ſhould be attended to by the ornithologiſt, in his deſcription; becauſe, otherwiſe, he ſuppoſes the colour of the bill to be permanent, which is by no means ſo.

This alteration, however, rather ſeems to be the ſymptom than the cauſe of a bird's coming into ſong, or otherwiſe, and I have never attended to this circumſtance in the ſoft billed birds ſufficiently, to ſay whether it holds alſo with regard to them.

A very intelligent bird-catcher, however, was able to prognoſticate, for three winters together, when a nightingale, which I kept ſo long, was coming into ſong (though there was no change in the colour of the bill), by the dung's being intermixed with large bloody ſpots, which before was only of a dead white.

This ſame bird-catcher was alſo very ſucceſsful in his preſcriptions for ſick birds, with regard to the ingredients of which he was indeed very myſterious.

He ſaid, that as he could not feel their pulſe, the circumſtances which he chiefly attended to were their weight, as well as both the conſiſtence and colour of their dung.

He always frankly ſaid what he expected from his preſcriptions,

"Mirè ſagaces falleret hoſpites
"Diſcrimen obſcurum." Horat.

for, in ſuch inſtances, the paſſages are not only the ſame, but the tone.

Such was the event of the experiment I have before mentioned of the linnet educated under a vengolina.

In my experiment, however, of teaching the ſparrow the notes of the linnet, though the ſcholar imitated the paſſages of its maſter, yet the tone of the ſparrow had by no means the mellowneſs of the original.

The imitation might therefore be, in ſome meaſure, compared to the ſinging of an opera ſong by a black-guard, when, though the notes may be preciſely the ſame, yet the manner and tone would differ very much.

Thus alſo the linnet, which I heard repeat the words *pretty boy*, did not articulate like a *parrot*, though, at the ſame time, the words might be clearly diſtinguiſhed.

The education I have therefore been ſpeaking of will not give new organs of voice to a bird, and the inſtrument itſelf will not vary, though

tions, and that if ſuch and ſuch changes did not ſoon take place, the caſe was deſperate. He frequently alſo refuſed to preſcribe, if the bird felt too light in the hand, or he thought that there was not ſufficient time to bring about an alteration in the dung.

the

the notes or paſſages may be altered almoſt at pleaſure.

I tried once an experiment, which might indeed have poſſibly made ſome alteration in the tone of a bird, from what it might have been when the animal was at its full growth, by procuring an operator who caponiſed a young blackbird of about ſix weeks old; as it died, however, ſoon afterwards, and I have never repeated the experiment, I can only conjecture with regard to what might have been the conſequences of it.

Both * *Pliny* and the *London* poulterers agree that a capon does not crow, which I ſhould conceive to ariſe from the muſcles of the larynx never acquiring the proper degree of ſtrength, which ſeems to be requiſite to the ſinging of a bird, from Mr. *Hunter*'s diſſections.

But it will perhaps be aſked, why this operation ſhould not improve the notes of a neſtling, as much as it is ſuppoſed to contribute to the greater perfection of the human voice.

To this I anſwer, that caſtration by no means inſures any ſuch conſequence; for the voices of much the greater part of *Italian* eunuchs are ſo indifferent, that they have no means of procuring a livelihood but by copying muſic, and this is one of the reaſons why ſo few compoſitions are

* Lib. X. c. 21.

publiſhed in *Italy*, as it would ſtarve this refuſe of ſociety.

But it may be ſaid, that there hath been a *Farinelli* and a *Manzoli*, whoſe voices were ſo diſtinguiſhedly ſuperior.

To this I again anſwer, that the catalogue of ſuch names would be a very ſhort one; and that we attribute thoſe effects to caſtration, which ſhould rather be aſcribed to the education of theſe ſingers.

Caſtration commonly leaves the human voice at the ſame pitch as when the operation is performed; but the eunuch, from that time, is educated with a view only to his future appearance on the opera ſtage; he therefore manages his voice to greater advantage, than thoſe who have not ſo early and conſtant inſtruction.

Conſidering the ſize of many ſinging birds, it is rather amazing at what a diſtance their notes may be heard.

I think I may venture to ſay, that a nightingale may be very clearly diſtinguiſhed at more than half a mile *, if the evening is calm. I have alſo obſerved the breath of a *robin* (which exerted itſelf) ſo condenſed in a froſty morning, as to be very viſible.

* Monſ. *de Buffon* ſays, that the quadruped which he terms the *huarine*, may be heard at the diſtance of a league. *Ornith. Tom.* I.

To

To make the comparifon, however, with accuracy, between the loudnefs of a bird's and the human voice, a perfon fhould be fent to the fpot from whence the bird is heard; I fhould rather conceive that, upon fuch trial, the nightingale would be diftinguifhed further than the man.

It muft have ftruck every one, that, in paffing under a houfe where the windows are fhut, the finging of a bird is eafily heard, when, at the fame time, a converfation cannot be fo, though an animated one.

Moft people, who have not attended to the notes of birds, fuppofe that thofe of every fpecies fing exactly the fame notes and paffages, which is by no means true, though it is admitted that there is a general refemblance.

Thus the *London* bird-catchers prefer the fong of the *Kentifh* goldfinches, but *Effex* chaffinches; and when they fell the bird to thofe who can thus diftinguifh, inform the buyer that it hath fuch a note, which is very well underftood between them *.

* Thefe are the names which they give to fome of the nightingale's notes: *Sweet, Sweet jug, Jug fweet, Water bubble, Pipe rattle, Bell pipe, Scroty, Skeg, Skeg, Skeg, Swat fwat fwaty, Whitlow whitlow whitlow*, from fome diftant affinity to fuch words.

Some of the nightingale fanciers alſo prefer a *Surry* bird to thoſe of *Middleſex* *.

Theſe differences in the ſong of birds of the ſame ſpecies cannot perhaps be compared to any thing more appoſite, than the varieties of provincial dialects.

The nightingale ſeems to have been fixed upon, almoſt univerſally, as the moſt capital of ſinging birds, which ſuperiority it certainly may boldly challenge: one reaſon, however, of this bird's being more attended to than others is, that it ſings in the night †.

* Mr. *Henſhaw* informs us, that nightingales in *Denmark* are not heard till May, and that their notes are not ſo ſweet or various as with us. Dr. *Birch*'s Hiſtory of the Royal Society, Vol. III. p. 189. Whilſt Mr. *Fletcher* (who was miniſter from Q. *Elizabeth* to *Ruſſia*) ſays, that the nightingales in that part of the world have a finer note than ours. See *Fletcher*'s Life, in the *Biographia Britannica*.

I never could believe what is commonly aſſerted, that the *Czar Peter* was at a conſiderable expence to introduce ſinging birds near *Peterſburgh*; becauſe it appears, by the *Fauna Suecica*, that they have in thoſe latitudes moſt of the ſame birds with thoſe of *England*.

† The woodlark and reedſparrow ſing likewiſe in the night; and from hence, in the neighbourhood of *Shrewſbury*, the latter hath obtained the name of the willow-nightingale. Nightingales, however, and theſe two other birds, ſing alſo in the day, but are not then diſtinguiſhed in the general concert.

The

Hence *Shakeſpeare* ſays,

"The nightingale, if ſhe ſhould ſing by day,
"When every gooſe is cackling, would be thought
"No better a muſician than the wren."

The ſong of this bird hath been deſcribed, and expatiated upon, by ſeveral writers, particularly *Pliny* and *Strada*.

As I muſt own, however, that I cannot affix any preciſe ideas to either of theſe celebrated deſcriptions, and as I once kept a very fine bird of this ſort for three years, with very particular attention to its ſong; I ſhall endeavour to do it the beſt juſtice I am capable of.

In the firſt place, its tone is infinitely more mellow than that of any other bird, though, at the ſame time, by a proper exertion of its muſical powers, it can be exceſſively brilliant.

When this bird *ſang its ſong round*, in its whole compaſs, I have obſerved ſixteen different beginnings and cloſes, at the ſame time that the intermediate notes were commonly varied in their ſucceſſion with ſuch judgment, as to produce a moſt pleaſing variety.

The bird which approaches neareſt to the excellence of the nightingale, in this reſpect, is the ſky lark; but then the tone is infinitely inferior in point of mellowneſs: moſt other ſinging birds have not above four or five changes.

The next point of ſuperiority in a nightingale

is its continuance of ſong, without a pauſe, which I have obſerved ſometimes not to be leſs than twenty ſeconds. Whenever reſpiration, however, became neceſſary, it was taken with as much judgment as by an opera ſinger.

The ſkylark again, in this particular, is only ſecond to the nightingale *.

* I ſhall here inſert a table, by which the comparative merit of the *Britiſh* ſinging birds may be examined, the idea of which I have borrowed from *Monſ. de Piles*, in his *Cours de Peinture par Principes*. I ſhall not be ſurprized, however, if, as he ſuggeſts, many may diſagree with me about particular birds, as he ſuppoſes they will do with him, concerning the merits of painters

As I have five columns inſtead of the four which *M. de Piles* uſes, I make 20 the point of abſolute perfection, inſtead of 16, which is his ſtandard.

	Mellowneſs of tone.	Sprightly notes.	Plaintive notes.	Compaſs.	Execution.
Nightingale	19	14	19	19	19
Skylark	4	19	4	18	18
Woodlark	18	4	17	12	8
Titlark	12	12	12	12	12
Linnet	12	16	12	16	18
Goldfinch	4	19	4	12	12
Chaffinch	4	12	4	8	8
Greenfinch	4	4	4	4	6
Hedge-ſparrow	6	0	6	4	4
Aberdavine (or Siſkin)	2	4	0	4	4
Redpoll	0	4	0	4	4
Thruſh	4	4	4	4	4
Blackbird	4	4	0	2	2
Robin	6	16	12	12	12
Wren	0	12	0	4	4
Reed-ſparrow	0	4	0	2	2
Black-cap, or the Norfolk Mock nightingale *	14	12	12	14	14

* *Brit. Zool.* p. 262.

And

And here I muſt again repeat, that what I deſcribe is from a caged nightingale, becauſe thoſe which we hear in the ſpring are ſo rank, that they ſeldom ſing any thing but ſhort and loud jerks, which conſequently cannot be compared to the notes of a caged bird, as the inſtrument is overſtrained.

I muſt alſo here obſerve, that my nightingale was a very capital bird; for ſome of them are ſo vaſtly inferior, that the bird-fanciers will not keep them, branding them with the name of *Frenchmen* *.

I have made no mention of the bulfinch in this table, which is commonly conſidered as a ſinging bird; becauſe its wild note, without inſtructions, is a moſt jarring and diſagreeable noiſe.

I have likewiſe omitted * the redſtart (which is called by the French *Roſſignol de Muraille)*, as I am not ſufficiently acquainted with its ſong, though it is admired by many; I ſhould rather conceive, however, with *Zinanni*, that there is no very extraordinary merit in the notes.

The *London* bird-catchers alſo ſell ſometimes the yellow hammer, twite and brambling † as ſinging birds; but none of theſe will come within my definition of what may be deemed ſo.

* One ſhould ſuppoſe from this, that the nightingale-catcher had heard much of the French muſic; which is poſſibly the caſe, as ſome of them live in Spittal-fields.

* Il culo ranzo é un ucello, (per quanto dicono) molto canoro, ma io tale non lo ſtimo. Delle uova é del nidi, p. 53.

† They call this bird a *kate*.

But

But it is not only in tone and variety that the nightingale excells; the bird alſo ſings (if I may ſo expreſs myſelf) with ſuperior judgement and taſte.

I have therefore commonly obſerved, that my nightingale began ſoftly like the ancient orators; reſerving its breath to ſwell certain notes, which by this means had a moſt aſtoniſhing effect, and which eludes all verbal deſcription.

I have indeed taken down certain paſſages which may be reduced to our muſical intervals; but though by theſe means one may form an idea of ſome of the notes uſed, yet it is impoſſible to give their comparative durations in point of muſical time, upon which the whole effect muſt depend.

I once procured a very capital player on the flute to execute the notes which *Kircher* hath engraved in his *Muſurgia*, as being uſed by the nightingale; when, from want of not being able to ſettle their reſpective lengths, it was impoſſible to obſerve any traces almoſt of the nightingale's ſong.

It may not be improper here to conſider, whether the nightingale may not have a very formidable competitor in the *American* mocking-bird *; though almoſt all travellers agree, that the concert in the

* Turdus *Americanus* minor canorus. *Ray*'s *Syn*. It is called by the *Indians*, *Contlatolli*; which is ſaid to ſignify four hundred tongues. See alſo *Cateſby*.

European

European woods is ſuperior to that of the other parts of the globe*.

As birds are now annually imported in great numbers from *Aſia*, *Africa*, and *America*, I have frequently attended to their notes, both ſingly and in concert, which are certainly not to be compared to thoſe of *Europe*.

Thomſon, the poet, (whoſe obſervations in natural hiſtory are much to be depended upon) makes this ſuperiority in the *European* birds to be a ſort of compenſation for their great inferiority in point of gaudy plumage. Our goldfinch, however, joins to a very brilliant and pleaſing ſong, a moſt beautiful variety of colours in its feathers †, as well as a moſt elegant ſhape.

It muſt be admitted, that foreign birds, when brought to *Europe*, are often heard to a great diſadvantage; as many of them, from their great tameneſs, have certainly been brought up by hand, the conſequence of which I have already ſtated from ſeveral experiments. The ſoft-billed birds alſo cannot be well brought over, as the *ſuccedaneum* for

* See Rochefort's Hiſt. des Antilles, T. I. p. 366.—Ph. Tr. Abr. Vol. III. p. 563.—and Cateſby.

† I cannot but think, that there would be a demand for theſe birds in *China*, as the inhabitants are very ſedentary, and bird cages are commonly repreſented as hanging in their rooms. I have been informed, by a *Tyroleze*, that his beſt market for *Canary* birds was *Conſtantinople*.

inſects

insects (their common food) is fresh meat, and particularly the hearts of animals.

I have happened, however, to hear the *American* mocking-bird in great perfection at *Mess. Vogle*'s and *Scott*'s, in *Love-Lane*, *Eastcheap*.

This bird is believed to be still living, and hath been in *England* these six years. During the space of a minute, he imitated the woodlark, chaffinch, blackbird, thrush, and sparrow. I was told also, that he would bark like a dog; so that the bird seems to have no choice in his imitations, though his pipe comes nearest to our nightingale of any bird I have yet met with.

With regard to the original notes, however, of this bird, we are still at a loss; as this can only be known by those who are accurately acquainted with the song of the other *American* birds.

Kalm indeed informs us, that the natural song is excellent *; but this traveller seems not to have been long enough in *America* to have distinguished what were the genuine notes: with us, mimics do not often succeed but in imitations.

I have little doubt, however, but that this bird would be fully equal to the song of the nightingale in its whole compass; but then, from the attention which the *mocker* pays to any other sort of disagreeable noises, these capital notes would be always debased by a bad mixture.

* Vol. I. p. 219.

We

We have one mocking bird in *England*, which is the ſkylark; as, contrary to a general obſervation I have before made, this bird will catch the note of any other which hangs near it; even after the ſkylark note is *fixed*. For this reaſon, the bird-fanciers often place the ſkylark next one which hath not been long caught, in order, as they term it, to keep the caged ſkylark *honeſt*.

The queſtion, indeed, may be aſked, why the wild ſkylark, with theſe powers of imitation, ever adheres to the parental notes; but it muſt be recollected, that a bird when at liberty is for ever ſhifting its place, and conſequently does not hear the ſame notes eternally repeated, as when it hangs in a cage near another. In a wild ſtate therefore the ſkylark adheres to the parental notes; becauſe the parent cock attends the young ones, and is heard by them for ſo conſiderable a time, during which, they pay no regard to the ſong of any other bird.

I am aware alſo, that it may be aſked, how birds originally came by the notes which are peculiar to each ſpecies. My anſwer, however, to this is, that the origin of the notes of birds, together with its gradual progreſs, is as difficult to be traced, as that of the different languages in nations.

The loſs of the parent-cock at the critical time for inſtruction hath undoubtedly produced thoſe varieties, which I have before obſerved are in the ſong of each ſpecies; becauſe then the neſtling hath either attended to the ſong of ſome other birds; or or perhaps invented ſome new notes of its own, which are afterwards perpetuated from generation to generation, till ſimilar accidents produce other alterations. The organs of ſome birds alſo are probably ſo defective, that they cannot imitate properly the parental notes, as ſome men can never articulate as they ſhould do. Such defects in the parent bird muſt again occaſion varieties, becauſe theſe defects will be continued to their deſcendants, who (as I before have proved) will only attend to the parental ſong. Some of theſe deſcendants alſo may have imperfect organs; which will again multiply varieties in the ſong.

The truth is, as I have already obſerved, that ſcarcely any two birds of the ſame ſpecies have exactly the ſame notes, if any are accurately attended to, though there is a general reſemblance.

Thus moſt people ſee no difference between one ſheep and another, when a large flock is before them. The ſhepherd, however, knows each of them, and can ſwear to them, if they are loſt; as can the *Lincolnſhire* goſherd to each gooſe.

As I now draw towards a concluſion of both my experiments and obſervations on the ſinging of birds; it may be poſſibly aſked, what uſe reſults either from the trouble or expence which they have coſt me; both of which I admit to have been conſiderable.

I will readily own, that no very important advantages can be derived from them; and yet I ſhall not

not decline ſuggeſting what little profit they may poſſibly be of, though at beſt they ſhould rather be conſidered as what Lord *Bacon* terms, *experiments of light, than of fruit.*

In the firſt place, there is no better method of inveſtigating the human faculties, than by a comparifon with thoſe of animals; provided we make it without a moſt ungrateful wiſh of lowering ourſelves, in that diſtinguiſhed ſituation in which we are placed.

Thus we are referred to the ant for an example of induſtry and foreſight, becauſe it provides a magazine of food for the winter, when this animal is in a ſtate of torpidity during that ſeaſon; nor are we leſs willing to ſuppoſe the ſong of birds to be ſuperior to our own muſical powers.

The notes of many birds are certainly very pleaſing, but by no means ſtand in competition either with the human voice or our worſt muſical inſtruments; nor only from want of the ſtriking effects of harmony in many excellent compoſitions; but becauſe, even when compared to our ſimple melody, expreſſion is wanting *, without which muſic is ſo languid and inanimate.

But to return to the uſes (ſuch as they are) which may ariſe from attending to the ſong of birds, or from the experiments which I have given an account of.

* The nightingale, indeed, is perhaps an exception to this general obſervation.

The

The firſt of theſe is too much neglected by the naturaliſt; for, if the bird is not caught, the only means often by which either the ſex or the ſpecies can be determined is the ſong. For example, if *Monſ. Adanſon* had informed us whether the *European* ſwallows, which he conceived were to be ſeen during the winter at *Senegal*, had the ſame notes with thoſe of *Europe*, it would have been going one ſtep further in proof of the facts which he and others ſo much rely upon.

Theſe experiments, however, may be ſaid to be uſeful to all thoſe who happen to be pleaſed with ſinging birds; becauſe it is clear, that, by educating a bird under ſeveral ſorts, we may often make ſuch a mixture, as to improve the notes which they would have learned in a wild ſtate.

It reſults alſo from the experiment of the linnet being educated under the *Vengolina*, that we may introduce the notes of *Aſia*, *Africa*, and *America*, into our own woods; becauſe, if that linnet had been ſet at liberty *, the neſtlings of the next ſeaſon would have adhered to the *Vengolina* ſong, who would again tranſmit it to their deſcendants.

* I know well, that it is commonly ſuppoſed, if you ſet a caged bird at liberty, it will neither be able to feed itſelf, nor otherwiſe live long, on account of its being perſecuted by the wild ones. There is no foundation, however, for this notion; and I take it to ariſe from its affording an excuſe for continuing to keep theſe birds in confinement.

But

But we may not only improve the notes of birds by a happy mixture, or introduce thoſe which were never before heard in *Great Britain*; we may alſo improve the inſtrument with which the paſſages are executed.

If, for example, any one is particularly fond of what is called the ſong of the *Canary* bird, it would anſwer well to any ſuch perſon, if a neſtling linnet was brought up under a *Canary* bird, becauſe the notes would be the ſame, but the inſtrument which executes them would be improved.

We learn alſo, from theſe experiments, that nothing is to be expected from a neſtling brought up by hand, if he does not receive the proper inſtruction from the parent cock: much trouble and ſome coſt is therefore thrown away by many perſons in endeavouring to rear neſtling nightingales, which, when they are brought up and fed at a very conſiderable expence, have no ſong which is worth attending to.

If a woodlark, or ſkylark, was educated, however, under a nightingale, it follows that this charge (which amounts to a ſhilling per week *) might be in a great meaſure ſaved, as well as the trouble of chopping freſh meat every day.

* *Olina* ſpeaks of a paſte which is uſed in *Italy* for nightingales; but I cannot find that it ever anſwers with us; perhaps, they bring their nightingales up by hand, and ſo accuſtom them from their earlieſt infancy to ſuch food.

A a a A night-

A nightingale, again, when kept in a cage, does not live often more than a year or two; nor does he ſing more than three or four months; whereas the ſcholar pitched upon may not only be more vivacious, but will continue in ſong nine months out of the twelve.

I fear, however, that I have already dwelt too much upon theſe very minute and trifling advantages which may reſult from my experiments and obſervations; I ſhall therefore no longer defer ſubſcribing myſelf,

Dear Sir,

Your moſt faithful,

Humble Servant,

Daines Barrington.

No.

Compofitions for two piping Bullfinches.
Allegretto
1ſt B.
2d B.
1ſt B.
2d B.
Allegro
1ſt B.
2d B.
Allegro
1ſt B.
2d B.
Williams Sc.

No. VI.

Of the MIGRATION of BRITISH BIRDS.

Quam multæ glomerantur aves! ubi frigidus annus
Trans pontum fugat, et terris immittit apricis.

VIRGIL.

THE migration of birds, is a ſubject of ſo curious a nature, that every one who attempts to write the natural hiſtory of animals, ought to look upon it as an eſſential part of his inquiries, and at the ſame time ſhould endeavour to aſſign the cauſe why ſome birds prefer certain places for their ſummer, others for their winter reſidence.

To be qualified for this taſk, it is neceſſary that the inquirer ſhould confine himſelf to one certain tract the whole year; he ſhould be diligent in obſerving the arrival, and the diſappearance of birds; he ſhould commit every obſervation to paper, and compare them with the remarks of correſpondents, on the ſame ſubject, that lie on every ſide of him. He ſhould attend likewiſe to the weather; and to the plenty or failure of fruits and berries; as on thefe accidents many curious remarks may be founded. He fhould cultivate an acquaintance with the gentlemen of the navy, and other fea-faring people; he fhould confult their journals, to difcover what birds light on their fhips, at what feafons, in what latitudes, and in what weather, and from what points; and thus trace them in their very courfe.

A comparative view of the writings of thofe who fhould embrace this part of natural hiftory, would throw great light on the fubject. But it is to be lamented, that none, except two northern naturalifts, Mr. *Klein* and Mr. *Ekmarck*, have profeffedly treated on this point. The fouthern parts of *Europe*, which may be fuppofed to receive, during winter, many of our land birds, have as yet produced no *faunift* to affift the inquiries of the naturalifts, which muft account for the imperfect knowledge we have of the retreat of many of our birds.

We muft not omit, however, our acknowledgements to two eminent pens that have treated this fubject as far as it related to rural œconomy; and, in fuch a manner, as does honour to their refpective countries; we mean Mr. *Alex. Mal. Berger* and Mr. *Stillingfleet*: whom we fhould not mention a fecond time*, but to confefs the aid we here receive from their faithful attention to the fubject in queftion.

* *Vide* Preface.

We

We wiſh that any thing we could ſay, would induce others of our countrymen to follow their example: they need not fear that the matter is exhauſted, for every county will furniſh new obſervations; each of which, when compared, will ſerve to ſtrengthen and confirm the other. Such an amuſement is worthy of every one, beneath none; but would become no order of men better than our clergy, as they are (or ought to be) the beſt qualified, and the moſt ſtationary part of the community; and, as this is a mixed ſpecies of ſtudy (when conſidered as phyſico-theology) it is therefore particularly pertinent to their profeſſion. A moſt ingenious friend, whom modeſty prevents from putting his name to a work that renders obſervations of this kind of the utmoſt facility, has pointed out the way, and methodized every remark that can occur; the farmer, the ſportſman, and the philoſopher, will be led to the choice of materials proper to be inſerted in that uſeful companion, *the Naturaliſt's Journal* *.

From the obſervations of our friends, from thoſe made by ourſelves, and from the lights afforded us by preceding writers, we ſhall, in the brief relation we can pretend to give, proceed in a generical order, and as far as poſſible, trace each ſpecies of bird to its retreat.

* Printed for *W. Sandby*, *Fleet-Street*, *London*, 1767. Price One Shilling and Six-pence.

A few words will explain the cause of their disappearance in these northern regions; a defect of food at certain seasons, or the want of a secure asylum from the persecution of man during the time of courtship, incubation and nutrition.

HAWKS. Eagles, and all the ignoble species of this genus breed in *Great Britain*; of the *Falcons*, we only know that which is called the *Peregrine*, which builds its nest annually in the rocks of *Llandidno, Caernarvonshire*; and the *Gentil*, and the *Goshawk* which breed in *Scotland*.

OWLS. We are assured that every species breeds in *England*, except the *little Owl*, and *short eared Owl*. The last breeds in *Scotland*, and the *Orkney* isles, but migrates into *England* at the same season as the *Woodcocks* do. Hawks and owls are birds of prey, and having at all times in this island means of living, are not obliged to quit their quarters.

SHRIKES. The *Flusher*, or *red back Shrike*, and the great *Shrike*, breeds with us; we have not heard of the other, so suspect that it migrates.

CROWS. Of this genus, the *Hooded Crow* migrates regularly with the *Woodcock*. It inhabits *North Britain* the whole year: a few are said annually to breed on *Dartmoor*, in *Devonshire*. It breeds also in *Sweden* and *Austria*, in some of the *Swedish* provinces it only shifts its quarters, in others it re-

sides

ſides throughout the year. I am at a loſs for the ſummer retreat of thoſe which viſit us in ſuch numbers in winter, and quit our country in the ſpring. And for the reaſon why a bird, whoſe food is ſuch that it may be found at all ſeaſons in this country, ſhould leave us.

CUCKOO.

Diſappears early in autumn; the retreat of this and the following bird is quite unknown to us.

WRYNECK.

Is a bird that leaves us in the winter. If its diet be ants alone, as ſeveral aſſert, the cauſe of its migration is very evident. This bird diſappears before winter, and reviſits us in the ſpring a little earlier than the *Cuckoo*.

WOODPECKERS.

Continue with us the whole year; their food being the *larvæ* of inſects, which lodge themſelves at all times in the bark of trees.

KINGFISHER.

Continues here through all ſeaſons.

NUTHATCH.

Reſides in this country the whole year.

HOOPOE.

Comes to *England* but by accident: we once indeed heard of a pair that attempted to make their neſt in a meadow at *Selborne*, *Hampſhire*, but were frighted away by the curioſity of people. It breeds in *Germany*.

CREEPER.

Never leaves the country.

GROUS. The whole tribe, except the *Quail*, lives here all the year round: that bird either leaves us, or elſe retires towards the ſea-coaſts *.

BUSTARD. Inhabits our downs and their neighborhood all the year.

PIGEONS. Some few of the *Ring-doves* breed here; but the multitude that appears in the winter, is ſo diſproportioned to what continue here the whole year, as to make it certain that the greateſt part quit the country in the ſpring. It is moſt probable they go to *Sueden* to breed, and return from thence in autumn; as Mr. *Ekmark* informs us they entirely quit that country before winter †. Multitudes of the common *Wild Pigeons* alſo make the northern retreat, and viſit us in winter; not but numbers breed in the high cliffs in all parts of this iſland. We ſuſpect that the *Turtle* leaves us in the winter, at leſt changes its place, removing to the ſouthern counties.

STARE. Breeds here; poſſibly ſeveral remove to other countries for that purpoſe, ſince the produce of thoſe that continue here, ſeems unequal to the clouds of them that appear in winter. It is not unlikely that many migrate into *Sueden*, where Mr. *Berger* obſerves they return in ſpring.

* *Vide* p. 277. of this work.

† *Amæn. Acad.* IV. 592.

The

THRUSHES.

The *Fieldfare* and the *Redwing* breed and paſs their ſummers in *Norway*, and other cold countries; their food is berries, which abounding in our kingdoms, tempts them here in the winter. Theſe two and the *Royſton crow*, are the only land birds that regularly and conſtantly migrate into *England*, and do not breed here. The *Hawfinch* and *Croſs-bill* come here at ſuch uncertain times, as not to deſerve the name of birds of paſſage; and, on that account, rather merit a place in the appendix than in the body of the work.

CHATTERER.

The *Chatterer* appears annually about *Edinburgh* in flocks during winter; and feeds on the berries of the mountain aſh. In *South Britain* it is an accidental viſitant.

GROSBEAKS.

The *Groſbeak* and *Croſsbill* come here but ſeldom; they breed in *Auſtria*. I ſuſpect that the *Pine Groſbeak* breeds in the foreſts of the Highlands of *Scotland*.

BUNTINGS.

All the genus inhabits this kingdom throughout the year, except the greater *Brambling*, which is forced here from the north in very ſevere ſeaſons.

FINCHES.

All continue in ſome parts of theſe kingdoms, except the *Siſkin*, which is an irregular viſitant, ſaid to come from *Ruſſia*. The *Linnets* ſhift their quarters, breeding in one part of this iſland, and remove

remove with their young to others. All finches feed on the ſeeds of plants.

LARKS, FLY-CATCHERS, WAGTAILS, AND WARBLERS.

All of theſe feed on inſects and worms; yet only part of them quit theſe kingdoms; though the reaſon of migration is the ſame to all. The *Nightingale*, *Black-cap*, *Fly-catcher*, *Willow-wren*, *Wheatear*, and *White-throat*, leave us before winter, while the ſmall and delicate *Golden-creſted Wren* braves our ſevereſt froſts. We imagine that the migrants of this genus continue longeſt in *Great Britain* in the ſouthern counties, the winter in thoſe parts being later than in thoſe of the north; Mr. *Stillingfleet* having obſerved ſeveral *Wheat-ears* in the iſle of *Purbeck* the 18th of *November* laſt. As theſe birds are incapable of very diſtant flights, we ſuſpect that *Spain*, or the ſouth of *France*, is their winter aſylum.

TITMICE.

Never quit this country; they feed on inſects and their *larvæ*.

SWALLOWS, AND GOAT-SUCKER.

Every ſpecies diſappears at approach of winter.

WATER FOWL.

OF the vaſt variety of water fowl that frequent *Great Britain*, it is amazing to reflect how few are known

known to breed here: the caufe that principally urges them to leave this country, feems to be not merely the want of food, but the defire of a fecure retreat. Our country is too populous for birds fo fhy and timid as the bulk of thefe are: when great part of our ifland was a mere wafte, a tract of woods and fen; doubtlefs many fpecies of birds (which at this time migrate) remained in fecurity throughout the year. *Egrets*, a fpecies of *Heron*, now fcarce known in this ifland, were in former times in prodigious plenty; and the *Crane*, that has totally forfaken this country, bred familiarly in our marfhes: their place of incubation, as well as of all other *cloven footed water fowl* (the *Heron* excepted) being on the ground, and expofed to every one: as rural œconomy increafed in this country, thefe animals were more and more difturbed; at length, by a feries of alarms, they were neceffitated to feek, during the fummer, fome lonely fafe habitation.

On the contrary, thofe that build or lay in the almoft inacceffible rocks that impend over the *Britifh* feas, breed there ftill in vaft numbers, having little to fear from the approach of mankind: the only difturbance they meet with in general, being from the defperate attempts of fome few to get their eggs.

CLOVEN

CLOVEN FOOTED WATER FOWL.

HERONS.

THE *White Heron* is an uncommon bird, and visits us at uncertain seasons; the common kind and the *Bittern* never leave us.

CURLEWS.

The *Curlew* breeds sometimes on our mountains; but, considering the vast flights that appear in winter, we imagine the greater part retire to other countries: the *Whimbrel* breeds in the *Grampian Hills*, in the neighbourhood of *Invercauld.*

SNIPES.

The *Woodcock* breeds in the moist woods of *Sweden*, and other cold countries. Some *Snipes* breed here, but we believe the greatest part retire elsewhere; as do every other species of this genus.

SANDPIPERS.

The *Lapwing* continues here the whole year; the *Ruff* breeds here, but retires in winter; the *Redshank* and *Sandpiper* breed in this country, and reside here. All the others absent themselves during summer.

PLOVERS AND OYSTER-CATCHER.

The *long legged Plover* and *Sanderling* visit us only in winter; the *Dottrel* appears in spring and in autumn, yet what is very singular we do not find it breeds in *South Britain.* The *oyster-catcher* lives

lives with us the whole year. The *Norfolk Plover* and *Sea Lark* breed in *England.* The *Green Plover* breeds on the mountains of the North of *England*, and on the *Grampian Hills.*

We muſt here remark, that every ſpecies of the *genera* of *Curlews*, *Woodcocks*, *Sandpipers* and *Plovers* *, that forſake us in the ſpring, retire to *Sweden*, *Poland*, *Pruſſia*, *Norway*, and *Lapland* to breed; as ſoon as the young can fly, they return to us again; becauſe the froſts which ſet in early in thoſe countries totally deprive them of the means of ſubſiſting; as the dryneſs and hardneſs of the ground, in general, during our ſummer, prevent them from penetrating the earth with their bills, in ſearch of worms, which are the natural food of theſe birds.

RAILS AND GALLINULES.

Every ſpecies of theſe two *genera* continue with us the whole year; the *Land Rail* excepted, which

* Mr. *Ekmarck* ſpeaks thus of the retreat of the whole tribe of cloven footed water fowl out of his country (*Sweden*) at the approach of winter; and Mr. *Klein* gives much the ſame account of thoſe of *Poland* and *Pruſſia*.

Grallæ (tanquam conjuratæ) unanimiter in fugam ſe conjiciunt, ne earum unicam quidem inter nos habitantem invenire poſſumus. *Amæn. Acad.* IV. 588.

Scolopaces et *Glareolæ* incredibilibus multitudinibus verno tempore in *Polonia* et *Boruſſia* nidulantur; appropinquante autumno turmatim evolant. *Klein de av. errat.* 187.

is

is not ſeen here in winter. It likewiſe continues in *Ireland* only during the ſummer months, when they are very numerous, as Mr. *Smith* tells us in the hiſtory of *Waterford*, p. 336. Great numbers appear in *Angleſea* the latter end of *May*; it is ſuppoſed that they paſs over from *Ireland*, the paſ- age between the two iſlands being but ſmall. As we have inſtances of theſe birds lighting on ſhips in the *Channel* and the *Bay* of *Biſcay*, we conjecture their winter quarters to be in *Spain*.

FINNED FOOTED WATER BIRDS.

PHALAROPES. VISIT us but ſeldom; their breeding place is *Lapland**, and other arctic regions.

COOT. Inhabits *Great Britain* the whole year.

GREBES. The *great creſted Grebe*, the *black* and *white Grebe*, and *little Grebe* breed with us, and never migrate; the others viſit us accidentally, and breed in *Lapland*.

WEB-FOOTED BIRDS.

AVOSET. BREED near *Foſsdike* in *Lincolnſhire*; but quit their quarters, in winter. They are then ſhot in

* *Amæn. Acad.* IV. 590.

different

different parts of the kingdom, which they visit I believe not regularly but accidentally.

AUKS AND GUILLEMOTS.

The *great Auk* or *Pinguin* sometimes breeds in *St. Kilda*. The *Auk*, the *Guillemot* and *Puffin* inhabit most of the maritime cliffs of *Great Britain*, in amazing numbers, during summer. The *black Guillemot* breeds in the *Bass Isle*, and in *St. Kilda*, and sometimes in *Llandidno* rocks. We are at a loss for the breeding place of the other species; neither can we be very certain of the winter residence of any of them, excepting of the *lesser Guillemot* and *black-billed Auk*, which, during winter, visit in vast flocks the *Frith of Forth*.

DIVERS.

These chiefly breed in the lakes of *Sweden* and *Lapland*, and some in countries nearer the *Pole* *; but some of the *red throated Divers*, the *northern* and the *imber*, may breed in the north of *Scotland* and its isles.

GULLS.

I am uncertain where the *black toed Gull* breeds. The *Skua* is confined to the *Shetland Isles*, the *Rock Foula*, and perhaps *St. Kilda*. The *Arctic* breeds in the *Orknies* and in the *Hebrides*. The rest of the tribe breed dispersedly on all the cliffs of *Great Britain*. The *black headed* on our fens and lakes.

* *Faun. Suec.* No. 150. *Crantz. Greenl.* I. 82. 83.

Every

TERNS. Every ſpecies breeds here; but leaves us in the winter.

PETRELS. The *Fulmar* breeds in the iſle of *St. Kilda*, and continues there the whole year, except *September* and part of *October*; the *Shearwater* viſits the *Iſle of Man* in *April*, breeds there, and leaving it in *Auguſt* or the beginning of *September*, diſperſes over all parts of the *Atlantic Ocean*. The *Stormfinch* is ſeen at all diſtances from land on the ſame vaſt watery tract, nor is ever found near ſhore except by ſome very rare accident, unleſs in the breeding ſeaſon. We found it on ſome little rocky iſles, off the north of *Skie*. It alſo breeds in St. *Kilda*. We alſo ſuſpect that it neſtles on the *Blaſquet* iſles off *Kerry*, and that it is the *Gourder* of Mr. *Smith* *.

MERGANSERS. This whole genus is mentioned among the birds that fill the *Lapland* lakes during ſummer. I have ſeen the young of the *Red-breaſted* in the north of *Scotland*: a few of theſe, and perhaps of the *Gooſanders* may breed there.

DUCKS. Of the numerous ſpecies that form this genus, we know of few that breed here. The *Swan* and *Gooſe*, the *Shield Duck*, the *Eider Duck*, a few *Shovelers*, *Garganies*, and *Teals*, and a very ſmall portion of the *wild Ducks*.

* *Smith's hiſt. Kerry*, 186.

The

The rest contribute to form that amazing multitude of water fowl, that annually repair from most parts of *Europe* to the woods and lakes of *Lapland* and other *arctic* regions*, there to perform the functions of incubation and nutrition in full security. They and their young quit their retreat in *September*, and disperse themselves over *Europe*. With us they make their appearance the beginning of *October*; circulate first round our shores, and when compelled by severe frost, betake themselves to our lakes and rivers. Of the web-footed fowl there are some of hardier constitutions than others; these endure the ordinary winters of the more northern countries, but when the cold reigns there with more than common rigor, repair for shelter to these kingdoms: this regulates the appearance of some of the *Diver* kind, as also of the *wild Swans*,

* *Barentz* found the *Bernacles* with their nests in great numbers in *Nova Zembla*. *Collect. voy. Dutch East-India Company*, 8vo. 1703. p. 19. *Clusius* in his *Exot.* 368. also observes, that the *Dutch* discovered them on the rocks of that country and in *Waygate Straits*. They, as well as the other species of *wild Geese*, go very far north to breed, as appears from the histories of *Greenland* and *Spitzbergen*, by *Egede* and *Crantz*. These birds seem to make *Iceland* a resting place, as *Horrebow* observes, few continue there to breed, but only visit that island in the spring, and after a short stay, retire still further north.

The *Swallow tailed Shield Duck* breeds in the *Icy Sea*, and is forced southward only in the very hard winters. *Amæn. Acad.* IV. 585.

the *Swallow tailed Shield Duck*, and the different ſorts of *Gooſanders* which then viſit our coaſts.

CORVORANTS.

The *Corvorant* and *Shag* breed on moſt of our high rocks: the *Gannet* in ſome of the *Scotch* iſles, and on the coaſt of *Kerry:* the two firſt continue on our ſhores the whole year. The *Gannet* diſperſes itſelf all round the ſeas of *Great-Britain*, in purſuit of the *Herring* and *Pilchard*, and even as far as the *Tagus* to prey on the *Sardina*.

But of the numerous ſpecies of fowl here enumerated, it may be obſerved how very few entruſt themſelves to us in the breeding ſeaſon; and what a diſtant flight they make to perform the firſt great dictate of nature.

There ſeems to be ſcarcely any but what we have traced to *Lapland*, a country of lakes, rivers, ſwamps and alps *, covered with thick and gloomy foreſts, that afford ſhelter during ſummer to theſe fowls, which in winter diſperſe over the greateſt part of *Europe*. In thoſe *arctic* regions, by reaſon of the thickneſs of the woods, the ground remains moiſt and penetrable to the *Woodcocks*, and other ſlender billed fowl: and for the web-footed birds †, the waters afford *larvæ* innumerable of the torment-

* *Flora Lapponica* Lectori et Proleg.

† A diſciple of *Linnæus*, ſpeaks thus of their food, *Lapponia*, ubi victum ex *larvis* et *pupis* culicum, altrix paravit numinis

tormenting *Knat*. The days there are long; and the beautiful meteorous nights indulge them with every opportunity of collecting ſo minute a food: whilſt mankind is very ſparingly ſcattered over that vaſt northern waſte.

Why then ſhould *Linnæus*, the great explorer of theſe rude deſerts, be amazed at the myriads of water fowl that migrated with him out of *Lapland?* Which exceeded in multitudes the army of *Xerxes*; covering, for eight whole days and nights, the ſurface of the river *Calix* *. His partial obſervation as a botaniſt, would confine their food to the vegetable kingdom, almoſt denied to the *Lapland* waters; inattentive to a more plenteous table of inſect food, which the all bountiful Creator had ſpread for them in the wilderneſs †.

numinis munificentia. *Amæn. acad.* IV. 1. 5. M. *de Maupertuis* makes the ſame obſervation, Ce ruiſſeau nous conduiſit a un lac ſi rempli de petits grains jaunatres de la groſſeur du *Mil* que toute ſon eau en etoit teinte. Je pris ces grains pour la *Chryſalide* de quelque inſecte, &c. *Oeuvres de M. de Maupertuis*, III. 116.

* *Flora Lapponica*, 273. *Amæn. acad.* IV. 570.

† It may be remarked, that the lakes of mountanous rocky countri s in general are deſtitute of plants: few or none are ſeen on thoſe of *Switzerland*; and *Linnæus* makes the ſame obſervation in reſpect to thoſe of *Lapland*; having, during his whole tour, diſcovered only a ſingle ſpecimen of a *lemna triſulca*, or ivy leaved *duck's meat*. *Flora Lap.* No. 470. a few of the *ſcirpus lacuſtris*, No. 18. or bullruſh; the *alopecurus geniculatus*, No. 38. or flote foxtail graſs; and the *ranunculus aquatilis*, No. 234. which are all he enumerates in his *Prolegomena* to that excellent performance.

Bbb2 No.

No. VII.

EXTRACTS FROM OLD ENGLISH WRITERS RELATING TO OUR ANIMALS.

MENTION having been ſo frequently made, in this work, of the old *Engliſh* feaſts, and the ſpecies of animals that formed the good cheer; we tranſcribe from *Leland* an account of that given at the *intronazation* of *George Nevell*, archbiſhop of *York*, in the reign of *Edward* IV. and of the *goodly proviſion made for the ſame.*

In wheat, 300 quarters.
In ale, - 300 tunne.
Wyne, - 100 tunne.
Of ypocraſſe - 1 pype.
In oxen, - - 104.
Wylde Bulles, - - 6.
Muttons, - - 1000.
Veales, - - - 304.
Porkes, - - 304.
Swannes, - - 400.
Geeſe, - - 2000.
Capons, - - 1000.
Pygges, - - 2000.
Plovers, - - 400.
Quales, - 100 dozen.
Of the foules called rees, 200 dozen.
In peacockes, - 104.
Mallardes and teales, 4000.
In cranes, - - 204.
In kyddes, - 204.
In chyckens, - 2000.
Pigeons, - - 4000.
Conyes, - - 4000.
In bittors, - - 204.
Heronſhawes, - 400.
Feſſauntes, - - 200.
Partriges, - - 500.
Wodcockes, - 400.
Curlewes, - - 100.
Egrittes, - - 1000.
Stagges, buck and roes, 500 and mo.
Paſties of veniſon colde, 4000.
Parted dyſhes of gellies, 1000.
Playne dyſhes of gellies, 3000.
Colde tartes baked, 4000.
Colde cuſtardes baked, 3000.
Hot paſties of veniſon, 1500.
Hot cuſtardes, 2000.
Pykes and breames, 608.
Porpoſes and ſeals, 12.
Spices, ſugared delicates, and wafers plentie.

Beſides

Besides the birds in the above list, there are mentioned, in the particular of the courses *, *Redshanks*, *Styntes*, *Larks* and *Martynettes rost*; if the last were the same with the martin swallow, our ancestors were as general devourers of small birds as the *Italians* are at present, to whom none come amiss.

We must observe, that in the order of the courses it appears, that only the greatest delicacies were served up, as we may suppose, to the table where the nobility, gentlemen, and gentlewomen of *worship* were seated; and those seemed to have been dressed with almost as much art and disguise as at present. They had likewise their desert, or, as the term was, *sutteltie*; which was in form of dolphins or other animals; and sometimes recourse was had to the kalendar to embellish the table, and St. *Paul*, St. *Thomas*, St. *Dunstan*, and a whole multitude of *angels*, *prophetes* and *patriarkes* †, were introduced as *suttelties* to honor the day.

As no mention is made among the dishes that composed two of the courses, of the geese, the pygges, the veales, and other more substantial food, those must have been allotted to the *franklins* and *head yeomen* in the *lower hall*: and those most singular provisions, the porposes and seales, inde-

* *Leland's collectanea*, vi. 2.

† Idem, 23.

licate as they may ſeem at preſent, in old times were admitted to the beſt tables: the former, at leſt, as we learn from doctor *Caius* *, who mentions it not only as a common food, but even deſcribes its ſauce.

A tranſcript from that curious publication, *The Regulations of the Houſhold of the fifth Earl of* NORTHUMBERLAND, *begun in* 1512, will be eſteemed a very proper appendage to a work of this nature. It will ſhew not only the birds then in high vogue at the great tables of thoſe days, but alſo how capricious a thing is taſte, ſeveral then of high price being at preſent baniſhed from our tables; and others again of uncommon rankneſs much valued by our anceſtors.

Thus *Wegions* (I give the ſpelling of the time) *See-pyes*, *Sholardes*, *Kyrlewes*, *Ternes*, *Cranys*, *Hearon-ſewys*, *Bytters*, *See-gulles* and *Styntes*, were among the delicacies for principal feaſts, or his Lordſhip's own *mees*.

Thoſe excellent birds the *Teylles* were not to be bought except no other could be got.

Feſauntes, *Bytters*, *Hearon-ſewys* and *Kyrlewes* were valued at the ſame price, twelve pence each.

The other birds admitted to his Lordſhip's table were *Buſtardes*, *Mallardes*, *Woodcokes*, *Wypes*, *Quayles*, *Snypes*, *Pertryges*, *Redeſhankes*, *Reys*, *Pacokes*, *Knottes*, *Dottrells*, *Larkys* and *ſmall byrdes*.

* *Caii opuſc.* 113.

The

The great byrdes, for the Lord's *mees*, for the Chambreleyn and Stewardes *mees* may be, as the ingenious editor conjectures, Fieldfares, Thrufhes and the like *.

The eftimation each fpecies was held in may be known by the following table, to which I have added the modern name, and the reference to it in this work.

	Page.	Price.
Cranys, the Crane,	534,	16d.
Hearon-fewys, the Heron,	355,	12d.
Mallards,	500,	2d.
Teylles, Teal,	513,	1d.
Woodcock,	365,	1d. or 1d.½.
Wypes, Lapwings,	381,	1d.
Sea-gulls, Black-headed Gull,	456,	1d. or 1d.½.
Styntes, Purrs,	397,	6d. a dozen.
Quails,	234,	2d.
Snipes,	378,	3d. a dozen.
Partridges,	233,	2d.
Red-fhanks,	376,	1d.
Bytters, Bitterns,	358,	12d.
Pheafants,	238,	12d.
Reys, Land Rails †,	410,	2d.

* P. 104. 424.

† I imagine the *Reys* to be the Land Rail, not the *Reeve* the female of the *Ruff*, for that bird feems not to be in vogue in thofe days. Old *Drayton* does not even mention it in his long catalogue of birds, but fets a high value upon

The *Rayle* which feldom comes but upon rich men's fpits *.

* *Polyolbion.* Canto XXV.

	Page.	Price.
Sholardes, Shovelers,	504,	6d.
Kyrlewes, Curlews,	362,	12d.
Peacocks,	236,	12d.
Sea Pies,	405.	
Wigeons,	509,	1d.
Knots,	387,	1d.
Dotrels,	401,	1d.
Bustards,	241.	
Terns,	459,	4d. a dozen.
Great birds,		Ditto.
Small birds,		12d. a dozen.
Larks,	12d. for two dozens.	

No.

No. VIII.

A SYSTEMATIC ARRANGEMENT OF THE BIRDS OF GREAT BRITAIN, WITH THE NAMES IN THE ANTIENT BRITISH.

GENUS I.

FALCON.

1. GOLDEN Eagle,	Eryr melyn.
2. Black Eagle,	Eryr tinwyn.
3. Sea Eagle,	Mor-Eryr.
4. Cinereous,	Eryr cynffonwyn.
5. Ofprey,	Pyfg Eryr: Gwalch y weilgi.
6. Gyrfalcon,	Hebog chwyldro.
7. Peregrine Falcon,	Hebog tramor, Cammin.
8. Grey,	Hebog, Gwalch.
*9. Gentil,	Hebog mirain.
10. Lanner,	Hebog gwlanog.
11. Gofhawk,	Hebog Marthin.
12. Kite,	Barcud.
13. Buzzard,	Bod teircaill.
14. Spotted,	Bod mannog.

15. Honey

15.	Honey Buzzard,	Bod y mel.
16.	Moor Buzzard,	Bod y gwerni.
17.	Hen-Harrier,	Barcud glâs.
18.	Ringtail,	Bod tinwyn.
19.	Keſtrel,	Cudyll côch.
20.	Hobby,	Hebog yr Hedydd.
21.	Sparrow Hawk,	Gwepia.
22.	Merlin,	Corwalch, Llymyſten.

II.

OWL.

* 1.	Eagle,	Y Ddylluan fawr.
2.	Long eared,	Dylluan gorniog.
3.	Short eared,	Dylluan gluſtiog.
4.	White,	Dylluan wen.
5.	Tawny,	Dylluan frech.
6.	Brown,	Aderyn y Cyrph.
7.	Little,	Coeg Ddylluan.

III.

SHRIKE.

1.	Great,	Cigydd mawr.
2.	Red backed,	Cigydd cefn-goch.
3.	Wood chat,	Cigydd glâs.

IV.

IV.

C R O W.

1. Raven,	Cigfran.
2. Carrion,	Brân dyddyn.
3. Rook,	Ydfran.
4. Hooded,	Bran yr Jwerddon.
5. Magpie,	Piogen.
6. Jay,	Screch y Coed.
7. Red legged,	Brân big gôch.
8. Jackdaw,	Cogfran.

V.

C U C K O O.

1. Cuckoo,	Cog.

VI.

W R Y N E C K.

1. Wryneck,	Gwas y gôg, Gwdd-fdro.

VII.

W O O D P E C K E R.

1. Green,	Cnocell y coed, Delor y derw.

2. Great

2.	Great ſpotted,	Delor fraith.
* 3.	Middle.	
4.	Leſt ſpotted,	Delor fraith beiaf.

VIII.

KINGFISHER.

1. Kingfiſher, Glâs y dorlan.

IX.

NUTHATCH.

1. Nuthatch, Delor y cnau.

X.

HOOPOE.

1. Hoopoe, Y Goppog.

XI.

CREEPER.

1. Creeper, Y Grepianog.

XII.

XII.

GROUS.

1. Wood,	Ceiliog coed.
2. Black,	Ceiliog dû.
3. Red,	Ceiliog Mynydd, Jâr fynydd.
4. Ptarmigan,	Coriar yr Alban.
5. Partridge,	Coriar, Petrifen.
6. Quail,	Sofliar, Rhinc.

XIII.

BUSTARD.

1. Great,	Yr araf ehedydd.
* 2. Leffer,	Araf ehedydd Lleiaf.
3. Thick-kneed,	Y Glin-braff.

XIV.

PIGEON.

1. Common,	Colommen.
2. Ring,	Yfguthan.
3. Turtle,	Colommen fair, Turtur.

XV.

XV.

STARE.

1. Stare, Drydwen, Drydwy.

XVI.

THRUSH.

1. Miſſel, Treſglen, Pen y Llwyn.
2. Fieldfare, Caſeg y ddryccin.
3. Throſtle, Aderyn bronfraith.
4. Redwing, Soccen yr eira, Y dreſ-clen gôch.
5. Blackbird, Mwyalch, Aderyn dû.
6. Ring-ouzel, Mwyalchen y graig.
7. Water-ouzel, Mwyalchen y dwfr.

XVII.

CHATTERER.

1. Waxen, Sidan-gynffon.

XVIII.

GROSBEAK.

1. Haw, Gylfinbraff.

* 2. Pine

* 2. Pine.	
3. Crofs-billed,	Gylfingroes.
4. Bulfinch,	Y Chwybanydd, Rhawn goch.
5. Green,	Y Gegid, Llinos werdd.

XIX.

BUNTING.

1. Common,	Brâs y ddruttan, Brâs yr yd.
2. Yellow,	Llinos felen.
3. Reed,	Golfan y cyrs.
4. Tawny,	Golfan rhudd.
5. Snow,	Golfän yr eira.
6. Mountain,	Yr Olfan leiaf.

XX.

FINCH.

1. Gold,	Gwas y Sierri.
2. Chaff,	Afgell arian, Winc.
3. Brambling,	Bronrhuddyn y mynydd.
4. Sparrow,	Aderyn y to, Golfan.

5. Tree

5. Tree Sparrow,	Golfan y mynydd.
6. Siſkin,	Y Ddreiniog.
7. Linnet,	Llinos.
8. Red-headed Linnet,	Llinos bengoch.
9. Leſs red-headed Linnet,	Llinos bengoch leiaf.
10. Twite,	Llinos fynydd.

XXI.

FLY-CATCHER.

1. Spotted,	Y Gwybedog.
2. Pied,	Clochder y mynydd.

XXII.

LARK.

1. Sky,	Hedydd, Uchedydd.
2. Wood,	Hedydd y coed.
3. Tit,	Cor Hedydd.
4. Field,	Hedydd y cae.
5. Red,	Hedydd rhudd.
6. Creſted,	Hedydd coppog.

XXIII.

WAGTAIL.

1. White,	Brith y fyches, Tin-ſigl y gwys.

2. Yel-

2. Yellow, Brith y fyches felen.
3. Grey, Brith y fyches lwyd.

XXIV.

WARBLERS.

1. Nightingale, Eos.
2. Redstart, Rhonell goch.
3. Redbreast, Yr Hobi goch. Bron-goch.
4. Blackcap, Penddu'r brwyn.
5. Pettychaps, Y Ffigysog.
6. Hedge, Llwyd y gwrych.
7. Yellow, Dryw'r helyg. Sywidw.

* 8. Scotch.

9. Golden-crested, Ysfwigw, Sywigw.
10. Wren, Dryw.
11. Sedge, Hedydd yr helyg.
12. Grasshopper, Gwich hedydd.
13. Wheatear, Tinwyn y cerrig.
14. Whinchat, Clochder yr eithin.
15. Stonechatter, Clochder y cerrig.
16. Whitethroat, Y gwddfgwyn.

*17. Dartford.

XXV.

TITMOUSE.

1. Great,	Y Benloyn fwyaf.
2. Blue,	Y Lleian.
3. Cole,	Y Benloyn lygliw.
4. Marſh,	Penloyn y cyrs.
5. Longtailed,	Y Benloyn gynffonhir.
6. Bearded,	Y Barfog.

XXVI.

SWALLOW.

1. Chimney,	Gwennol, Gwenfol.
2. Martin,	Marthin Penbwl.
3. Sand,	Gennol y glennydd.
4. Swift,	Marthin dû.

XXVII.

GOATSUCKER.

1. Nocturnal,	Aderyn y droell, Rhodwr.

XXVIII.

XXVIII.

HERON.

1. Common,	Cryr glâs.
2. Bittern,	Aderyn y bwnn.
	Bwmp y Gors.
3. White,	Cryr gwyn.

XXIX.

CURLEW.

1. Curlew,	Gylfinhir.
2. Whimbrel,	Coeg ylfinhir.

XXX.

SNIPE.

1. Woodcock,	Cyffylog.
2. Godwit,	Rhoftog.
* 3. Cinereous,	Rhoftog llwyd.
4. Red,	Rhoftog rhûdd.
5. Leffer,	Cwttyn dû.
6. Greenfhank,	Coefwerdd.
7. Redfhank,	Coefgoch.
* 8. Cambridge,	
9. Spotted,	Coefgoch mannog.

10. Common,	Yſnittan, y Fyniar.
* 11. Great,	Yſnid.
12. Jack,	Giach.

XXXI.

SANDPIPER

1. Lapwing,	Cornchwigl.
2. Grey,	Cwttyn llwyd.
3. Ruff,	Yr Ymladdgar.
4. Knot,	Y Cnut.
5. Aſh colored,	Y Pibydd glâs.
6. Brown,	Y Pibydd rhudd.
7. Spotted,	Y Pibydd mannog.
8. Black,	Y Pibydd dû mannog.
* 9. Gambet,	
10. Turnſtone,	Huttan y môr.
* 11. Hebridal,	
12. Green,	Y Pibydd gwyrdd.
13. Red,	Y Pibydd coch.
* 14. Aberdeen,	
15. Common,	Pibydd y traeth.
16. Dunlin,	Pibydd rhuddgôch.
17. Purre,	Llygad yr ych.
* 18. Little,	Y Pibydd lleiaf.

XXXII.

PLOVER.

1. Golden	Cwttyn yr aur.

2. Long

2. Long legged,	Cwttyn hîrgoes.
3. Dottrel,	Huttan.
4. Ringed,	Môr Hedydd.
5. Sanderling,	Llwyd y tywod.

XXXIII.

OYSTER CATCHER.

1. Pied,	Piogen y môr.

XXXIV.

RAIL.

1. Water,	Cwtiar.

XXXV.

GALLINULE.

1. Spotted,	Dwfriar fannog.
2. Crake.	Rhegen yr yd.
3. Common,	Dwfriar.

XXXVI.

PHALAROPE.

1. Grey,	Pibydd llwyd llydan-droed.

2. Red,	Pibydd côch llydandroed.

XXXVII.

COOT.

1. Common,	Jâr ddwfr foel.
2. Great,	Jâr ddwfr foel fwyaf.

XXXVIII.

GREBE.

1. Tippet,	Gwyach. Tindroed.
2. Great crefted,	Gwyach gorniog.
3. Eared,	Gwyach gluftiog.
4. Dufky,	Gwyach leiaf.
5. Little,	Harri gwlych dy bîg.
6. Blackchin,	Gwyach gwddfrhûdd.

XXXIX.

AVOSET.

1. Scooping,	Pîg mynawd.

XL.

AUK.

1. Great,	Carfil mawr.

2. Razor

2. Razor-bill,	Carfil, Gwalch y pen-waig.
3. Black-billed,	Carfil gylfinddu.
4. Puffin,	Pwffingen.
5. Little,	Carfil bâch.

XLI.

GUILLEMOT.

1. Foolifh,	Gwilym.
2. Leffer,	Chwilog.
3. Black,	Gwilym dû.

XLII.

DIVER.

1. Northern,	Trochydd mawr.
* 2. Imber,	Trochydd.
3. Speckled,	Trochydd bâck.
4. Red-throated,	Trochydd gwddfgoch.
5. Black-throated,	Trochydd gwddfdu.

XLIII.

GULL.

1. Black-backed,	Gwylan gefn-ddu.
2. Skua,	Gwylan frech.

3. Black-toed,	Yr Wylan yſgafn.
4. Arctic,	Gwylan y Gogledd.
5. Herring,	Gwylan benwaig.
6. Wagel,	Gwylan rûdd a gwyn.
7. Winter,	Gwylan y gweunydd.
8. Common,	Gwylan lwyd, Huccan.
9. Kittiwake.	
10. Tarrock,	Gwylan gernyw.
11. Black-head,	Yr wylan benddu.
12. Brown,	Yr wylan fechan.

XLIV.

TERN.

1. Great,	Y fôr-wennol fwyaf. Yſcraean.
2. Leſſer,	Y fôr-wennol leiaf.
3. Black,	Yſcraean ddû.

XLV.

PETREL.

1. Fulmar,	Gwylan y graig.
2. Shear-water,	Pwffingen Fanaw.
3. Stormy,	Cas gan Longwr.

XLVI.

XLVI.

MERGANSER.

1. Goosander,	Hwyad ddanheddog.
2. Red-breasted,	Trochydd danheddog.
3. Smew,	Lleian wen.
4. Red-headed,	Lleian ben-goch.

XLVII.

DUCK.

1. Wild Swan,	Alarch gwyllt.
2. Tame Swan,	Alarch.
* 3. Grey Lag,	Gwydd.
4. Bean Goose,	Elcysen.
5. White fronted,	Gwydd wyllt.
6. Bernacle,	Gwyran.
7. Brent,	Gwyran fanyw.
8. Eider,	Hwyad fwythblu.
9. Velvet,	Hwyad felfedog.
10. Scoter,	Y fôr-Hwyad ddû.
11. Tufted,	Hwyad goppog.
12. Scaup,	Llygad arian.
13. Golden eye,	Llygad aur.
* 14. Morillon,	Hwyad benllwyd.
15. Shieldrake,	Hwyad yr eithin, Hwyad fruith.

16. Mallard,

16. Mallard,	Cors Hwyad, Garan Hwyad, Hydnwy.
17. Shoveler,	Hwyad lydanbig.
18. Red breasted Shoveler,	Hwyad fron-goch lydanbig,
19. Pintail,	Hwyad gynffonfain.
20. Long tailed,	Hwyad gynffon gwennol.
21. Pochard,	Hwyad bengoch.
22. Ferruginous,	Hwyad frech.
23. Wigeon,	Chwiw.
*24. Bimaculated,	
25. Gadwall,	Y gors Hwyad lwyd.
26. Garganey,	Hwyad addfain.
27. Teal,	Cor Hwyad, Crach Hwyad.

XLVIII.

CORVORANT.

1, Corvorant,	Mûlfran, Môrfran.
2. Shag,	Y Fulfran leiaf.
3. Gannet,	Gan, Gans.

APPENDIX.

* 1. Rough legged Falcon,	
2. Roller,	Y Rholydd,

3. Nutcracker,

3.	Nutcracker,	Aderyn y cnau.
* 4.	Oriole,	Y Fwyalchen felan.
5.	Rose colored Ouzel,	Y Fwyalchen gôch..
6.	Crane,	Garan.
7.	Egret,	Cryr coppog lleiaf.
8.	Little Bittern,	Aderyn y bwnn lleiaf.
* 9.	Spoon-bill,	Y Llydan-big.

*** The birds marked * are not in the octavo edition, 1768.

No.

No. IX.

CATALOGUE OF THE EUROPEAN QUADRUPEDS, BIRDS, AND REPTILES, *Extra-Britannic*.

SINCE the great uſe of Mr. RAY's *Sylloge ſtirpium* EUROPÆARUM *extra Britannias* * has been ſo fully approved by the travelling Botaniſt, it is thought a ſimilar enumeration of the ſpecies of certain claſſes of the animal kingdom would be equally agreeable and ſerviceable to the travelling Zoologiſt. It comprehends the *Extra-Britannic* quadrupeds, birds, and reptiles of *Europe*, formed from the works of the general naturaliſts, from the *Fauna* of different countries, and from my own obſervations. The arrangement of the ſubjects are according to the excellent method of our countryman Mr. RAY, a little altered, or reformed. As there are not at this inſtant *Engliſh* names for moſt of the articles, we have been obliged to ſubſtitute thoſe uſed by *Linnæus* and other foreign writers; but to gratify the *Engliſh* reader's curioſity, who may wiſh for fuller accounts of the quadrupeds in his own language, we refer him in the ſecond column to our own *ſynopſis* of *Quadrupeds*; and in reſpect to the birds, to the *Engliſh* edition of Mr. WILLUGHBY's *Ornithology*.

* Stirpium *Europæarum* extra *Britannias* naſcentium Sylloge, 1694.

CLASS

CLASS I. QUADRUPEDIA.

QUADRUPEDS.

		Lin.	Syn. noſt. No.	Place.
I *Bos*	Urus	99	4	Lithuania
	Bubalis	*ibid.*	5	Italy
II *Ovis*	Strepſiceros	98	8	B. Hungary
	Laticauda			Calmuck country
III *Capra*	Rupicapra	95	10	Alps, Pyrenees
	Ibex	*ibid.*	9	Alps
	Ammon	97	11	Corſica, Sardinia
	Tartarica	*ibid.*	30	Ukraine
IV *Cervus*	Alces	92	35	N. of the Baltic
	Tarandus	93	36	*ibid.*
V *Sus*	Aper Sylveſtris	102	54	Germany, France, &c.
	II.			
VI *Canis*	Lupus	58	111	Almoſt all the continent
	Lagopus	59	113	Lapland
VII *Felis*	Lynx	62	135	Many parts of *Europe*

VIII

APPENDIX.

			Lin.	Syn. noſt. No.	Place.
VIII	*Urſus*	Arctos	69	138	Many parts of Europe
		Maritimus	70	139	Nova Zembla
		Luſcus, et	71		
		Muſtela Gulo	67	140	N. of the Baltic
IX	*Viverra*	Genetta	65	171	Spain
		Zibellina	68	156	Lapland
		Perouaſca		p. 233 *Note*,	Poland
X	*Lutra*	Muſtela Lutreola	66	174	Sweden
XI	*Caſtor*	Fiber	78	190	N. of Europe
		Moſchatus	79	192	Ruſſia
XII	*Hyſtrix*	Criſtata	76	193	Italy
XIII	*Marmotta*	Mus Marmotta	81	197	Alps, Poland
		Cricetus	82	200	Germany
		Souſlik		201	S. of Ruſſia
		Lemmus	80	202	Lapland
		Citellus	80	203	S. of Europe
		Zemni		204	Poland
XIV	*Sciurus*	Volans	88	221	Poland
		Glis	87	217	S. of Europe
		Mus quercinus	84	218	*ibid.*
XV	*Jerboa*	Mus Jaculus	85	223	Calmucks country
XVI	*Mus*	Gregarius	84	234	Germany, Sweden

III.

			Lin.	Syn. noſt. No.	Place.
XVII	*Trichechus*	Roſmarus	49	263	Within the polar circle

IV.

			Lin.	Syn. noſt. No.	Place.
XVIII	*Veſpertilio*	Serotina		288	France
		Pipiſtrilla		289	*ibid.*
		Barbaſtella		290	*ibid.*
				286	*ibid.*

CLASS

CLASS II. AVES.

BIRDS.

I. ACCIPITRES.

RAPACIOUS.

				Wil. orn.	Place.
I *Vultur*	Vultur	*Briff.* I. 453		66	Alps, Italy
	Percnopterus	*Raii fyn.* 10	64	67	Spain, Minorca
II *Falco*	Leucocephalus	*Lin.* 124			North †
	Melanæetos	*ibid.*		61	
	Morphno congener	*Raii fyn.* 7		63	
	Rufticolus	*Lin.* 125			Sweden
	St. Martini	*Briff.* I. 443			France
	Iflandicus	*Brunnich No.*			Iceland
	Vefpertinus	*Lin.* 129			Ingria
	Minutus	131			Malta
	Subfurcatus	*Kramer* 326 *No.* 5			
	Caftaneus	*Kramer* 327 — 6			Auftria
	Ferrugineus	*Kramer* 328 — 7			
	Cinereus	*Kramer* 329 — 12			
III *Strix*	* Scandiaca	*Lin.* 132			North
	Subaurita	*Kr.* 323 *No.* 3			Auftria

† Countries the other fide the *Baltic.*

* * Nyctea

			Wil.	orn.	Place.
	** Nyctea	*Lin.*	132		North
	Sylvestris	*Scop. No.* 13			Carniola
	Funerea	*Lin.*	133		North
IV *Lanius*	Infaustus	*Lin.*	138	197?	North?
	Major *Gesneri* 581	*Briss.* II.	146	88	Germany

II. PICÆ.

PIES.

			Wil.	orn.	Place.
V *Corvus*	Caryocatactes	*Lin.*	157	132	Germany N.
	Pyrrhocorax	—	158		Alps
VI *Coracias*	Garrulus	—	159	131	Europe *passim* ‡.
VII *Oriolus*	Galbula	—	160	198	*ibid.*
VIII *Cuculus*	Glandarius	—	169		Spain
IX *Picus*	Martius	—	173	135	Europe *passim*
	Tridactylus	—	177		Norway
X *Merops*	Apiaster	—	182	147	Ita. S. of Eu.
	Icterocephala	*Briss.* IV.	537	148	*ibid.*
XI *Certhia*	Muraria	—	184		Italy

‡ Those with this word refer to all the continent, except the extreme north, *Lapland*, &c.

III. GAL-

III. GALLINÆ.

GALLINACEOUS.

				Wil. orn.	Place.
XII *Tetrao*	* Nemeſianus	*Sco. No.*	171		Carniola
	Betulinus	*No.*	172		*ibid.*
	Lagopus	*Bru. No.*	199		Norway
	Bonaſia	*Lin.*	257	175	Europe *paſſim*
	** Rufus	—	276	167	S. of Eu.
	Francolinus	—	275	174	*ibid.*
	Alchata	—	276	167 *No.* 5	Pyrenees
	Græca (Perdix)	*Briſſ.* I.	241	169	S. of Eu.
	Montana		224		
	Tridactyla	*Shaw's tra.*	253		Spain
XIII *Otis*	Tetrax	*Lin.*	264	179	France, Italy

IV. PASSERES.

SMALL BIRDS.

				Wil. orn.	Place.
XIV *Sturnus*	Collaris	*Sco. No.*	192		Carniola, Spain
XV *Turdus*	Arundinaceus	*Lin.*	296	143	Europe *paſſim*
	Roſeus		294	194	Italy
	Saxatilis		294	197	
	Cyanus		296	191	Italy, Spain
	Cœruleus	*Belon*		192	Alps

				Wil.	orn.	Place.
XVI	*Alauda*	Criſtata	*Lin.*	288	209	Europe *paſſim*
		Spinoletta		288	209	Italy
		Calandra		288		Italy, Spain
		Alpeſtris		299		Poland
		Luſitanica	*nova*			Portugal
		Craſſiroſtris	*nova*			*ibid.*
XVII	*Emberiza*	Hortulanus	*Lin.*	309	270	S. of Eu.
		Cia		310	271	*ibid.*
		Cirlus		311	269 zivolo	*ibid.*
		Barbata	*Sco. No.*	210		Carniola
		Brumalis	*No.*	213		*ibid.*
XVIII	*Fringilla*	Lapponica	*Lin.*	317		North
		Lulenſis		318		Sweden
		Citrinella		320	265	S. of Eu.
		Serinus		320	265	*ibid.*
		Petronia		322	267	*ibid.*
		Paſſer Campeſtris	*Briſſ.* III.	82	251 Friquet	*ibid.*
		Torquatus		85	250 *No.* 7	*ibid.*
		Stultus		87	249 — 2	*ibid.*
		Bononienſis		91	250 — 4	*ibid.*
		Albicilla		92	250 — 5	*ibid.*
		Paſſerculus		93	252 — 13	*ibid.*
		Sclavonicus		24	250 — 6	Dalmatia
		Argentoratenſis		146		Straſbourg
		Grau-fink	*Friſch.* I.	3		Germany
XIX	*Metacilla*	Schœnobanus	*Lin.*	329		Italy
		Curruca		329		Sweden
		Ficedula	*Lin.*	330		S. of Eu.

Stapazina

			Wil. orn.	Place.
	Stapazina	331	Strapazino 233	*ibid.*
	Dumetorum	334		Auſtria
	Erithacus	335	3tia Alder. 218	Sweden
	Suecica	336		North
	Curruca minor *Bri.* III.	374	Borin. *Wil.* 216	Italy
	cineraria	376		Italy, &c.
	rufa	387		Germany
	nævia	389	Boarina 217	Italy
	Ruticilla Gibraltar	407		S. of Eu.
	Phænicurus torqua	411		*ibid.*
	Rubecula Bononien.	422	Spipola 234	Bologna
	Curruca griſea næv.	*App.* VI. 112		
	Tithys *Sco. No.*	233		Carniola
	Zya —	234		*ibid.*
	Muſcipeta —	236		*ibid.*
	Luſitanica *nova*			Portugal
	Hiſpanica *nova*			Spain
XX *Parus*	Criſtatus *Lin.*	340	242	Germany
	Pendulinus	342		Auſtria
	Ignotus *Brun.* p.	73		North
XXI *Hirundo*	Melba *Lin.*	345		Spain
	Rupeſtris *Sco. No.*	167		Spain, & Carniola

V. AQUATICÆ FISSIPEDES.

CLOVEN FOOTED WATER FOWL.

				Wil.	*orn.*		*Place.*
XXII	*Platalea*	Leucorodia	*Lin.*	231	289		Europe *paſſim*
XXIII	*Ardea*	Grus	—	234	274		
		Ciconia		235	286		
		Nigra		*ibid.*	*ibid.*		
		Nyċticorax		*ibid.*	279		
		Purpurea		236			S.of Eu.
		Garzetta		237	280		*ibid.*
		Griſea		239			
		Minuta		240			
		Candida minor	*Bri.* V.	438	280		
		Torquata		440	282		
		Botaurus major		455	283		
		Botaurus minor		453			
		Botaurus ſtriatus		454			
		Botaurus rufus		458	283		
		Botaurus nævius		462			
		Cancrophagus		466	281	*No.* 9	Italy
		Cancroph. caſtaneus		468			*ibid.*
		Cancroph. rufus		469	281	— 7	
		Cancroph. nævius		471			
		Cancroph. luteus		472	281	— 8	Italy
		Viridis Belgica	*nova*				Holland
		Ardea alba	*Sco. No.*	127			Carniola
XXIV	*Tantalus*	Falcinellus	*Lin.*	241	295		Germany
XXV	*Numenius*	Danicus	*novus?*				Denmark
		Paſſerinus	*novus*				Holland

XXVI

			Wil. orn.	Place.
XXVI *Scolopax*	Fuſca	*Lin.*	243	
	Auſtralis	*Sco. No.*	94	Carniola
XXVII *Tringa*	Gambetta	*Lin.*	248	
	Striata?			
	Calidris			
	Helvetica			
	Varia			
	Totanus nævius	*Briſſ.* V.	200	
	Cinclus torquatus		216	
	Calidris griſea		233	
	Calidris nævia		229	
	Bononienſis major		110	
	Erythropus	*Sco. No.*	146	
	Undata	*Brun.* —	183	
XXVIII *Pratincola*	Krameria	*Kramer* 381 *L.* 345	*No.* 12	Auſtria
XXIX *Charadrius*	Alexandrinus	*Lin.*	253	
	Apricarius		254	
	Luteus	*novus*		France
XXX *Gallinula*	Grinetta	*Wil. orn.*	315	Italy
	Serica		*ibid.*	*ibid.*
	Major		313	
	Porphyrio	*Balearicus novus*		Minorca

VI. PEDIBUS PINNATIS.

WITH FINNED FEET.

XXXI *Phalaropus*	Platyrhynchus	*Brunnich No.*	172	North

VII. PEDIBUS PALMATIS.

WITH WEBBED FEET.

				Wil.	orn.	Place.
XXXII	*Phœnicopterus*	Ruber	*Lin.*	230	320	S. of Fr.
XXXIII	*Corrira*	Longipes	*Raii ſyn.*	118	231	Italy
XXXIV	*Mergus*	Caſtor				
		Æthiops	*Scop. No.*	90		Carniola
XXXV	*Larus*	Albus	*Scop. No.*	106		*ibid.*
		Merulinus		108		*ibid.*
		Bicolor		110		*ibid.*
XXXVI	*Sterna*	Cinerea	*Bri.* VI.	220		
		Nævia		216		
XXXVII	*Anas*	Niveus (anſer)	*Bri.* IV.	228		North
		Moſcoviticus ?		277	360	
		Spectabilis	*Lin.*	195		North
		Glaucion		201	367	
		Hiſtrionica		204		North
		Muſcaria	*Raii ſyn.*	146	375	
		Ferroënſis	*Bri.* VI.	466		Ferroe iſles
		Subterranea	*Scop. No.*	83		Carniola
		Cinerea	*K.* 341 *No.*	14		Auſtria
XXXVIII	*Pelecanus*	Onocrotalus	*Lin.*	215	327	Danube, Po.

CLASS.

CLASS III. REPTILIA.

REPTILES.

* PEDATA:

WITH FEET.

					Place.
Rana	Bombina	*Lin.*	355		Sweden
	Arborea		357	*Raii syn. qua.* 231	Germany
Lacerta	Chamæleon	*Lin.*	364	*Raii syn. qua.* 276	S. of Europe
	Salamandra		371	273	*ibid.*
	Orbicularis		365	264	Naples
	Marmorata		368		S. of Europe
	Aurata		*ibid.*		*ibid.*
	Umbra?		367		*ibid.*
	Seps		363		*ibid.*
	Chalcides		366	*Raii syn. qua.* 272	*ibid.*
Testudo	Corticata	*Rondel. pisc.*	445		*Mediterranean*
	Orbicularis	*Lin.*	351		S. of Europe
	Græca		352	*Raii syn. qua.* 253	*ibid.*
	Lutaria		*ibid.*	254	*ibid.*

** APODIA.

WITHOUT FEET.

Serpentes Anguis *Æsculapii*, *Plinii* lib. xxix. c. 4. *Raii syn. qua.* 291 Italy

Coluber Chersea	*Lin.* 377	*Wulff. Boruss.* 10	Sweden
Aspis	*Lin.* 378	*Strom. Sondm.* 193	Fr. Norway
Jaculus	*Wulff. Boruss.* 13		Prussia

ADDITIONS AND CORRECTIONS

TO

BRITISH ZOOLOGY.

[To be inſerted immediately before the Index of Birds.]

VOL. I. Page 280.

OF COCK-FIGHTING.

SOME account of the barbarous cuſtom of Cock-fighting, ſo frequent, till of late years, a favorite amuſement among ſome of all ranks in this kingdom, will be no improper appendage to the hiſtory of our domeſtic birds.

IF it can be any apology for ſo cruel a diverſion, we may plead that it was in uſe among the moſt polite people of antiquity: firſt invented, in all probability, by the *Athenians*, and borrowed from them by other nations, in particular by the *Romans*, who introduced it into our iſlands.

AT *Athens* was an annual feaſt, attended with Cock-fighting, inſtituted by *Themiſtocles* in honor of the birds from whoſe fighting he received an omen of his ſucceſs againſt the *Perſians*. He obſerved, that theſe birds fought for mere glory; neither for the gods of their country, nor tombs of of their anceſtors, nor yet for their children: * ſetting before his ſoldiers every motive to excite their valor, which they had ſuperior to theſe birds. This feſtival was ſtiled Αλεκτρυων αγων; and became anniverſary.

The Cock-pit, or Τηλία, was in the theatre where the public games were exhibited, and was in form of a ſquare ſtage, not round, like the modern pits. The game of Cock-fighting laſted but one day; for originally it was conſidered partly as a religious and partly as a political inſtitution.

But the cuſtom was ſoon abuſed, and Cock-matches grew frequent among private people. The barber *Meidias* and *Callias* fought a main: theſe gentlemen were, in all probability, alſo celebrated Cock-feeders, or at leſt Quail-feeders, being called Ορτυγοτροφοι; for it is certain that the antients prepared their birds for battle: great ſums were layed on the event; and the *Laniſtæ*, or Cockers, frequently totally ruined by their purſuits of the diverſion †.

The cuſtom ſpread ſoon, as is ſuſpected, from *Athens* to *Pergunius* and *Troas*. In the firſt were annual Cock-matches: and their neighbours, the *Dardanii Troas*, ſeem equally addicted to the diverſion, as is evident from their coins, which had on them two fighting cocks.

On two antient gems, in the collection of Mr.

* *Ælian. Var. Hiſt.* ii. c. 20.

† *Columella*, lib. viii. c. 2.

William Hamilton §, are ſtrong memorials of this cuſtom: on one is a Cock, with his head creſt, carrying in his bill a palm-branch, in token of victory over another, which ſtanding before with a drooping head. On the other, are two in the action of fighting, and a mouſe above, running away with an ear of corn, the cauſe of the battle: from both theſe repreſentations, it is evident that the antients neither trimmed their Cocks, nor cut off their combs and wattles.

THE race of birds moſt eſteemed by the antients, was that of *Tanagra*, a city of *Bœotia*, the Iſle of *Rhodes*, *Chalcis* in *Euboea*, and the country of *Media* *. They preferred the larger kind, or what we call *Shakebags*. The hens of *Alexandria* in *Egypt*, called Μονόσιροι, were highly valued for breeding ſpirited chickens †.

FROM *Greece* the diverſion was carried to *Rome*: but did not arrive at the heighth of folly as it did at *Athens*. The *Romans* delighting more in quail fightings, as the *Chineſe* do at this time. But we are told, that the fraternal hatred between *Baſſianus* and *Geta*, ſons of the emperor *Severus*, began when they were boys, from a quarrel they had about their Quails and Cocks ‡.

THE *Britons* had poultry before the arrival of *Cæſar*, but they owe the barbarous cuſtom of

§ *Archæologia*, vol. iii. tab. ix.
* *Plin. Nat. Hiſt.* lib. x. c. 21.
† *Geoponic.* lib. xiv. c. 7.
‡ *Herodian.* iii. § 33.

Cocking

Cocking to the *Romans*. Yet it does not occur among our writers, till the time of *Henry* II. when *Fitz-Stephens* § mentions it as the ſchool-boys diverſion on *Carnelevaria*, or *Shrove-Tueſday*. *Edward* III. diſapproved and prohibited Cock-fighting ‖. But that barbarous prince *Henry* VIII. gave it ſo much encouragement as to build a theatre, near *Whitehall*, for that purpoſe, to this day known by the name of the *Cockpit*. At length, *Oliver Cromwell*, in 1654, by a humane edict, ſuppreſſed theſe diſgraceful meetings; which, after his time, revived with full fury: yet it is ſome conſolation, in this profligate age, that whatſoever other follies flouriſh, this loſes credit, and drops (excepting among the dregs of the people) into the utmoſt diſrepute *.

§ p. 45. ‖ *Maitland London*, i. 131.

* It will be injuſtice not to ſay, that almoſt the whole of this is borrowed from the memoir on this ſubject, by that able antiquary the Rev. Mr. *Pegge*. See *Archæologia*, vol. iii. 132.

VOL.

STOCK DOVE, or WOOD PIGEON.

VOL. I. Page 290.

Of the STOCK DOVE, or WOOD PIGEON.

Œnas ſive vinago. *Wil. Orn.* 185.

THIS bird has been confounded with the Wild Pigeon, and the Rock Pigeon, and made the origin of the domeſtic kind. I firſt had an opportunity of correcting my error, * by ſeeing the true Stock Dove in the LEVERIAN *Muſeum*, which ſatisfied me that Mr. *Willughby*, with great juſtice, deſcribed it as a diſtinct ſpecies.

SIZE.

IT is equal in ſize to the common kind, perhaps larger. The weight of a male is fourteen ounces: its extent of wing, twenty-ſix inches: its length, fourteen.

COLOR.

The bill is of a light red: the head, neck, and upper part of the back, of a dark grey; the lower part, and rump, changes into an elegant light grey: the primary feathers of the wings, are duſky: coverts and ſecondaries, deep grey, marked with two black ſpots on the exterior webs: the lower half of the exterior webs of the two outermoſt feathers of the tail, are white: the reſt, cinereous, with their ends black.

THE ſides of the neck, of a variable gloſſy green: the breaſt, of pale purpliſh or vinaceous color: the belly cinereous: legs red.

PLACE.

BREEDS in hollow ſtocks of trees, and ſometimes on the tops: from which it derives its name of

* By LINNÆUS alſo, who makes it ſynonymous with the Tame Pigeon.

Stock

STOCK DOVE, or WOOD PIGEON.

Stock Dove, or *Wood Pigeon*: in oppoſition to the other, which breeds in holes of rocks, ſteeples, and towers. Thoſe are reſident in this kingdom: the former generally migratory: a few breed in the woods in *Suſſex*, * and perhaps other ſouthern parts of *Great Britain*. Their eggs have been hatched under tame Pigeons; but the young, as ſoon as they could fly, have betook themſelves to their ſavage ſtate. Theſe perch on trees: the true Wild or Rock Pigeon rarely or never. It has alſo marks different: in particular the lower part of the back, and the rump, are never of any other color than white. Yet, as Pigeons are frequently ſeen among our tame flocks, with grey back and rumps, it is highly probable, that notwithſtanding the above experiment may ſometimes fail, yet both kinds may have contributed to ſtock our pigeon-houſes.

GENTLEMEN who have pigeon-houſes near ſome of the lofty cliffs which impend over the ſea, ſeldom preſerve the pigeons in them the whole year: tempted by food, they will viſit and continue ſometime in the houſe, but uſually fly to the rocks to breed.

MIGRATION

THE STOCK DOVES migrate into the ſouth of *England*, in great multitudes, in *November*; and while the beech woods were ſuffered to cover large tracts of ground, came in *myriads*, reaching in ſtrings of a mile in length, when they went in

* Mr. *Latham Rivett*.

the

the morning to feed. They retire in the ſpring: I ſuppoſe into *Sweden*; for Mr. *Ekmark* makes their retreat from that kingdom coincide with the time of their appearance here.

VOL. I. Page 385.

GREY WHEAT-EAR.

Cul blanc gris, *Briſſon* iii. 552. tab. xxi.

A Bird of this ſpecies was ſhot near *Uxbridge*.

THE crown and back were of a tawny brown; the under ſide of the neck, of a dull browniſh yellow: from bill to eye paſſed an obſcure duſky line. Quil feathers and ſecondaries black, edged with tawny and white: tail, like the common Wheat-Ear; but the edges were marked with pale tawny.

IN the LINNÆAN *Syſtem*, p. 332. it is made a variety of the common Wheat-Ear.

VOL. II. Page 574.

TRANSFER to the GREY-LAG GOOSE, p. 570, all the ſynonyms prefixed, by miſtake, to the BEAN GOOSE.

VOL.

SULA. GREEN COD.

VOL. II. Page 620.

SULA.

THIS variety of the GANNET was ſent to me in *Auguſt* 1779, by *Hugh Stodart*, eſq; of *Treganwy*, in *Caernarvonſhire*. I do not recollect that it has been obſerved in *Europe* ſince the days of Dr. *Hoier*, a phyſician at *Bergen*, who procured it from the *Ferroe Iſles*, and ſent it to his friend *Cluſius*. It has ſince been ſeen frequently in *Falkland Iſland*, and in the South Seas, eſpecially on the coaſts of *New Holland* and *New Zealand*. Seamen call it the *Port Egmont Hen*.

THIS bird differs from the common GANNET only in thoſe particulars: in having ſome of the ſecondaries feathers black; and the middle feathers of the tail of the ſame color: whereas both, in the common ſort, are entirely white.

VOL. III. Page 179.

GENUS COD.

Gadus virens, *Lin. Syſt.* 438. *Faun. Suec.* N° 309.

GREEN.

THE Green Cod-Fiſh is beardleſs; ſmooth, of duſky green on the back; and ſilvery in every other part: jaws, of equal lengths: ſide line, ſtrait: tail forked.

I WAS favored by Sir *John Cullum*, bart. with the notice of this ſpecies being *Britiſh*; he obſerved numbers of them which had been taken in the *German* ocean; none exceeded ſeven inches in length. LINNÆUS does not attribute to them a greater ſize than that of a Perch.

I N D E X.

A.

B.

B.

Buck,

C.

Crake,

D.

Elk,

Fox,

G.

Goofander,

Himantopus,

J.

K.

L.

Latax

M.

N.

O.

Ofprey,

P.

Q.

Quail,

Shag,

T.

Taurosthenes

U.

URCHIN,

Winter

Y.

Z.

INDEX.

Z.

THE END.